Fungal Endophytes Volume I

Kamel A. Abd-Elsalam • Synan F. AbuQamar
Editors

Fungal Endophytes Volume I

Biodiversity and Bioactive Materials

Springer

Editors
Kamel A. Abd-Elsalam
Plant Pathology Research Institute
Agricultural Research Center
Cairo, Egypt

Synan F. AbuQamar
Department of Biology
College of Science, United Arab Emirates University
Al Ain, Abu Dhabi, United Arab Emirates

ISBN 978-981-97-7311-4 ISBN 978-981-97-7312-1 (eBook)
https://doi.org/10.1007/978-981-97-7312-1

This Springer imprint is published by the registered company Springer Nature Singapore Pte Ltd.
The registered company address is: 152 Beach Road, #21-01/04 Gateway East, Singapore 189721, Singapore

Preface

Fungal endophytes have long been known for their contribution to plant health and defense against abiotic and biotic stresses. They inhabit within plant tissues without causing harmful effects and frequently give great advantages to their host plants. As researchers delve further into the realm of fungal endophytes, they discover a plethora of fresh knowledge on the diverse species, their interactions with plants, and the bioactive substances they create. This book, *Fungal Endophytes: Biodiversity and Bioactive Materials*, takes us on a journey to discover the astonishing variety and bioactive potential of these secret partners. Over the last few decades, the importance of fungal endophytes has grown, exposing their complex roles in plant ecosystems. These fungi have developed complicated interactions with their host plants, resulting in symbiotic partnerships that benefit both. Fungal endophytes can provide their host with multiple benefits, including enhanced growth, improved tolerance to environmental challenges, better nutrient uptake, and disease protection. Furthermore, these fungal residents have been discovered to create a remarkable variety of bioactive substances, including antibacterial agents, anticancer compounds, immunomodulators, and enzymes with industrial uses. This book is a comprehensive resource that explores the various elements of fungal endophytes, including taxonomy, ecology, and functional relevance. We have assembled a team of eminent specialists from across the world, each providing the knowledge and views to shed light on the various aspects of fungal endophytes. Through a series of chapters, we hope to give readers a comprehensive overview on the topic, including both basic and applied research.

The first part of this book will look at the diversity of fungal endophytes, including the species that have been discovered previously and their significance in plant ecosystems. Readers will learn about the complex associations between plants and fungi, as well as how these interactions affect plant health and vitality. *The second part* will look at the bioactive compounds generated by fungal endophytes, emphasizing their potential use in agriculture, medicine, and biotechnology. Fungal endophytes may create a wide range of bioactive molecules, indicating their role in tackling some of the world's most urgent concerns. Book chapters include the following topics: Biodiversity of fungal endophytes, diversity of endophytes from medicinal plants and their effect in mitigating agricultural problems, endophytic fungi interactions with plant, endophytic insect-pathogenic fungi, root endophytic fungi, foliar endophytic fungi, marine algal-derived endophytic fungi, fungal endophytes: a potent bioactive metabolites reservoir, bioprospecting endophytic fungi from Fusarium genus as sources of bioactive metabolites, volatile compounds from endophytic fungi and their role in sustainable agriculture, exploring the potential of endophytic fungal terpenoids: from biosynthesis to bioactivity, endophytic fungal alkaloids, endophytic fungi as a promising source for sulfur-containing compounds, endophytic fungi as a source for plant hormones, recent trends in phytotoxic compounds from endophytic fungi, endophytic fungi: a biofactory for the synthesis of nanomaterials.

This book's chapters cover several topics, including the discovery and identification of fungal endophytes from various plant species and ecosystems, their ecological roles and interactions with host plants, biosynthesis of bioactive compounds, and potential applications in agriculture, medicine, and industry. The writers have methodically collated their study findings, offering cutting-edge discoveries, methodological advances, and thought-provoking viewpoints that challenge our knowledge. We believe that this book will not only be a useful resource for academics, students, and professionals in mycology, plant biology, and biotechnology, but also spark fresh research and application ideas. Unraveling the secrets of fungal endophytes has the potential to reveal a wealth of bioactive materials that might affect the future of sustainable development and food security. We are grateful to the contributing writers for their devotion and competence in sharing their knowledge. We also want to thank the editorial and publishing staff for their diligent efforts to make this book a reality. We warmly appreciate the Springer personnel, particularly Dr. Naren Aggarwal, Editorial Director-Books, Springer, Asia, Associate Editor Sanchi Bhimrajka, Ph.D., and Production Editor Mrs. Nandhini Viswanathan, for their kind help and hard effort in making this book possible. Finally, I want to express my deepest gratitude to the readers for their interest in this work. It is my honest goal that the materials offered within these pages will stimulate further investigation and ignite new ideas that lead to the creation of sustainable and eco-friendly solutions.

We would like to express our gratitude to the contributing authors for their dedication and expertise in sharing their knowledge. We also extend our thanks to the editorial and publication teams who have worked tirelessly to bring this book to fruition. We hope that you find this book as fascinating and inspiring as we do, and that it sparks new ideas and conversations about the incredible potential of fungal endophytes.

Cairo, Egypt — Kamel A. Abd-Elsalam
Al Ain, Abu Dhabi, UAE — Synan F. AbuQamar

Contents

Editors and Contributors

About the Editors

Kamel A. Abd-Elsalam, Ph.D. is currently a Research Professor and Deputy Director for Research at the Plant Pathology Research Institute, Agricultural Research Center, Giza, Egypt. Dr. Kamel's research interests include developing and improving plant biosecurity diagnostic tools, understanding fungal pathogen genomes, and creating eco-friendly hybrid nanomaterials for controlling toxigenic fungi and plant diseases as well as applications in agroecosystems. Kamel earned his Ph.D. in Molecular Plant Pathology from Christian Albrechts University of Kiel (Germany) and Suez Canal University (Egypt). In 2008, he was awarded a postdoctoral fellowship from the same institution. Dr. Kamel served as a visiting associate professor at Mae Fah Luang University in Thailand, the Institute of Microbiology at TUM in Germany, the Laboratory of Phytopathology at Wageningen University in the Netherlands, and the Plant Protection Department at Sassari University in Italy.

He has published 35 books related to nanobiotechnology applications in agriculture and plant protection with major publishing houses such as Springer, Taylor & Francis, and Elsevier. Since 2019, he has served as the Editor-in-Chief of the Elsevier book series *Nanobiotechnology for Plant Protection* and as the Series Editor for *Applications of Genome Modified Plants and Microbes in Food and Agriculture*. Additionally, he has been an active member of the Elsevier Advisory Panel since 2020, providing feedback and suggestions to improve Elsevier's products and services. Since 2023, he has served as the Editor-in-Chief of the Springer book series *Agroecosystem Dynamics and Sustainable Practices*. He has published more than 302 scientific research articles in international and regional specialized journals with high impact factors and has an H-index of 49, an i-10 index of 156, and over 9326 citations. He served as a Guest Editor for the *Journal of Fungi, Plants, and Microorganisms* and as a Review Editor for *Frontiers in Genomic Assay Technology, Agrochemicals, and Plant Nano Biology*. He was ranked in the top 2% of the most influential scientists in the world in nanobiotechnology for 2021, 2022, 2023, and 2024 by Elsevier and Stanford University. In 2014, he was awarded the Federation of Arab Scientific Research Councils Prize for excellent scientific research in biotechnology (fungal genomics), ranking first. Additionally, he received the Egyptian State Award for Agricultural Scientific Excellence from ASRT, Egypt in 2022 with a silver medal.

Synan F. AbuQamar, Ph.D. is currently a Professor in the Department of Biology, College of Science, United Arab Emirates University (UAEU), Al Ain, UAE. Previously, Prof. AbuQamar completed his Ph.D. from the Department of Botany and Plant Pathology at Purdue University in 2007, and postdoc from the same university in 2008. As a Plant Molecular Geneticist, his research focuses on understanding the molecular and cellular factors that control plant defense mechanisms against necrotrophic pathogens using host plants (*Arabidopsis*, tomato, and date palm).

Synan has published more than 90 peer-reviewed papers related to plant biotechnology, biological control agents, and nanobiotechnology applications in agriculture. Due to his expertise in the area of molecular plant–microbe interaction, Prof. AbuQamar also serves as an Associate Editor in BMC Biology, a Review Editor in *Frontiers in Plant Science* and *Frontiers in Microbiology*, and an Editor in *Scientific Reports*. Synan has an h-index of 41, an i-10 index of 65, with over 6650 citations. In 2020, Synan was awarded Khalifa Award in Education. He was awarded Khalifa International Award for Date Palm and Agricultural Innovation and King Hamad Prize in Agricultural Development, both in 2023. According to Elsevier and Stanford University, Prof. AbuQamar has been ranked in the top 2% of the most influential scientists in the world in Agriculture, and Plant Science in 2024. Prof. Synan AbuQamar has tremendous contribution in several committees related to Plant Biotechnology Centers.

Contributors

Amer M. Abdelaziz Faculty of Science, Botany and Microbiology Department, Al-Azhar University, Cairo, Egypt

Toka E. Abdelkhalek Biotechnology English Program, Faculty of Agriculture, Cairo University, Giza, Egypt

Kamel A. Abd-Elsalam Plant Pathology Research Institute, Agricultural Research Center (ARC), Giza, Egypt

Synan F. AbuQamar Department of Biology, College of Science, United Arab Emirates University, Al Ain, UAE

Faijuddin Ahammad Department of Plant Pathology, Sam Higgingbottom University of Agriculture, Technology and Science, Prayagraj, Uttar Pradesh, India

Sk Naim Aktar Department of Agronomy, Palli Siksha Bhavana, Visva Bharati University, Sriniketan, Bolpur, West Bengal, India

Fazilet Özlem Albayrak Biology Department, Science Faculty, Mersin University, Mersin, Turkey

Mahatab Ali Department of Genetics and Plant Breeding, Bidhan Chandra Krishi Viswavidyalaya (BCKV), Mohanpur, Nadia, West Bengal, India

Hassan Almoammar Wellness and Preventive Medicine Institute, Health Sector, King Abdulaziz City for Science and Technology (KACST), Riyadh, Saudi Arabia

Mohammad Ashfaq Department of Biotechnology, University Centre for Research & Development, Chandigarh University, Gharaun, Mohali, Punjab, India

Mohamed S. Attia Faculty of Science, Botany and Microbiology Department, Al-Azhar University, Cairo, Egypt

Arghya Banerjee Department of Plant Pathology, School of Agriculture and Allied Science, The Neotia, University, Jhinga, South 24, Parganas, West Bengal, India

Debdulal Banerjee Centre for Life Sciences, Vidyasagar University, Midnapore, West Bengal, India

Microbiology and Microbial Biotechnology Laboratory, Department of Botany and Forestry, Vidyasagar University, Midnapore, West Bengal, India

Narasimhan Banu Department of Botany, Bharathi Women's College, George Town, Chennai, Tamil Nadu, India

Divya Chuhan Department of Drinking Water and Sanitation, Ministry of Jal Shakti, New Delhi, India

Surekha R. Deshpande Basaveshwara Science College, Bagalkote, Karnataka, India

Ahmed S. Doghish Department of Biochemistry, Faculty of Pharmacy, Badr University in Cairo (BUC), Badr, Cairo, Egypt

Biochemistry and Molecular Biology Department, Faculty of Pharmacy (Boys), Al-Azhar University, Nasr City, Cairo, Egypt

Ali A. A. El-Sayed Plant Protection Research Institute, Agricultural Research Center, Dokki, Giza, Egypt

Francisco J. Fernández Biotechnology Department, Universidad Autónoma Metropolitana (Unidad Iztapalapa), Avenida Ferrocarril San Rafael Atlixco, Mexico City, Mexico

Juan Moisés Galindo-Solís Biotechnology Department, Universidad Autónoma Metropolitana (Unidad Iztapalapa), Avenida Ferrocarril San Rafael Atlixco, Mexico City, Mexico

Nihal Doğruöz Güngör Department of Biology, Fundamental and Industrial Microbiology, Faculty of Science, Istanbul University, Istanbul, Turkey

Amr H. Hashem Faculty of Science, Botany and Microbiology Department, Al-Azhar University, Cairo, Egypt

Rawan K. Hassan Biotechnology English Program, Faculty of Agriculture, Cairo University, Giza, Egypt

Saidul Islam Nadia Krishi Vigyan Kendra, Bidhan Chandra Krishi Viswavidyalaya (BCKV), Gayeshpur, Nadia, West Bengal, India

Nandita Jana Centre for Life Sciences, Vidyasagar University, Midnapore, West Bengal, India

Sibashish Kityania Department of Life Science and Bioinformatics, Assam University, Silchar, Assam, India

Jayeeta Mitra Department of Botany, Arunachal University of Studies, Namsai, Arunachal Pradesh, India

Bholanath Mondal Department of Plant Pathology, Palli Siksha Bhavana, Visva Bharati University, Sriniketan, Bolpur, West Bengal, India

Sourav Mondal Rathindra Krishi Vigyan Kendra, Palli Siksha Bhavana, Visva Bharati University, Sriniketan, Bolpur, West Bengal, India

B. Shankar Naik Department of P.G. Studies and Research in Applied Botany, Bio Science Complex, Kuvempu University, Jnana Sahyadri, Shimoga, Karnataka, India

Government PU College for Girls, Basavanahalli, Chikkamagaluru, Karnataka, India

Deepa Nath Department of Botany, Gurucharan College, Silchar, Assam, India

Priyakshi Nath Department of Life Science and Bioinformatics, Assam University, Silchar, Assam, India

Rajat Nath Department of Life Science and Bioinformatics, Assam University, Silchar, Assam, India

Rupshikha Nath Department of Life Science and Bioinformatics, Assam University, Silchar, Assam, India

Rishabh Anand Omar Centre for Environmental Science and Engineering, Indian Institute of Technology Kanpur, Kanpur, India

İbrahim Ortaş Department of Soil Science and Plant Nutrition, Faculty of Agriculture, University of Cukurova, Adana, Turkey

Kaathambari Purushothaman School of Biosciences and Technology, VIT, Vellore, India

Nahdhoit Ahamada Rachid Institute of Graduate Studies in Sciences, Istanbul University, Istanbul, Turkey

Kalyanaraman Rajagopal Department of Botany, Ramakrishna Mission Vivekananda College, Chennai, Tamil Nadu, India

D. Rekha Department of P.G. Studies and Research in Applied Botany, Bio Science Complex, Kuvempu University, Jnana Sahyadri, Shimoga, Karnataka, India

Nihal Gören Sağlam Biology Department, Science Faculty, Istanbul, Turkey

Hiran Kanti Santra Microbiology and Microbial Biotechnology Laboratory, Department of Botany and Forestry, Vidyasagar University, Midnapore, West Bengal, India

Mythili Sathiavelu School of Biosciences and Technology, VIT, Vellore, India

Maksud Hasan Shah Department of Agronomy, Bidhan Chandra Krishi Viswavidyalaya (BCKV), Mohanpur, Nadia, West Bengal, India

Esaimozhi Sivasankar School of Biosciences and Technology, VIT, Vellore, India

V. . G. Sowmeya School of Biosciences and Technology, VIT, Vellore, India

Meenambiga Setti Sudharsan Department of Bioengineering, School of Engineering, Vels Institute of Science, Technology and Advanced Studies (VISTAS), Chennai, Tamil Nadu, India

Neetu Talreja Department of Science, Faculty of Science and Technology, Alliance University, Anekal, Bengaluru, Karnataka, India

Anupam Das Talukdar Department of Life Science and Bioinformatics, Assam University, Silchar, Assam, India

Nivetha Venkatesan Department of Bioengineering, School of Engineering, Vels Institute of Science, Technology and Advanced Studies (VISTAS), Chennai, Tamil Nadu, India

Fatima Younas School College of Pharmacy, University of the Punjab, Lahore, Pakistan

Hajira Younas School of Integrative Plant Sciences, Cornell University, Ithaca, NY, USA

Endophytic Fungi: Exploring Biodiversity and Bioactive Potential

1

Kamel A. Abd-Elsalam, Hassan Almoammar, Amr H. Hashem, and Synan F. AbuQamar

Abstract

Fungal endophytes are microorganisms that live in plants' tissues without causing any major harm. They were found to be very diverse and highly beneficial for the health and growth of their host plants. This has driven attention to potential of fungal endophytes that are generating a lot of interest from various sectors such as agriculture, industrial, and pharmaceutical, where it is believed that fungi can produce multiple bioactive molecules. Some screening and isolation methods have demonstrated the diversity of secondary metabolites synthesized by them. From this study, the results showed that bioactive compounds secreted by fungal endophytes have antibacterial, antioxidant, anticancer, insecticidal properties, etc., while also indicating their usefulness in biotechnology applications such as agriculture or medicine manufacture. Finally, the production of biologically active materials by these fungi can be useful for biotechnological processes including agriculture or drugs preparation. Consequently, it is stated that fungal endophytes are capable of accumulating a vast array of novel bioactive compounds. This review paper focuses on two topics: diversification of fungal endophytes and biosynthetic pathways for bioactive molecules.

K. A. Abd-Elsalam (✉)
Plant Pathology Research Institute, Agricultural Research Center (ARC), Giza, Egypt
e-mail: kamel.abdelsalam@arc.sci.eg

H. Almoammar
Wellness and Preventive Medicine Institute, Health Sector, King Abdulaziz City for Science and Technology (KACST), Riyadh, Saudi Arabia

A. H. Hashem
Faculty of Science, Botany and Microbiology Department, Al-Azhar University, Cairo, Egypt

S. F. AbuQamar
Department of Biology, College of Science, United Arab Emirates University, Al Ain, UAE

K. A. Abd-Elsalam, S. F. AbuQamar (eds.), *Fungal Endophytes Volume I*,
https://doi.org/10.1007/978-981-97-7312-1_1

Keywords

Fungal endophytes · Biodiversity · Metabolites · Bioactive materials

1.1 Introduction

An interestingly diverse group of microorganisms known as fungal endophytes inhabit plant roots that do not appear to harm their hosts. It is recognized that this symbiotic association plays an important role in the ecology of terrestrial ecosystems, where endophytes are essential for plant development, stress tolerance, and pathogen defense (Sudheep et al. 2017) on materials. To thoroughly review their potential as sources, they wrote this book "Fungal Endosomes: Biodiversity and Bioactive Materials." Chapters in this book will cover a variety of topics, such as the identification of endophytic fungi and isolation, endophytic host relationship functions, and bioactive substances produced by endosomes can be broad range, such as antioxidants, immunomodulators, anticancer, and microbiological substances, so the study of fungal endosomes has received much attention in recent years. Endosomes are a potential source of bioactive materials because these bioactives provide sufficient material for new drug development and biotechnological applications. Observation of endophytes, their role in plant ecology, and their biological importance for various applications are discussed in this introductory chapter. We also discuss methods for isolation and characterization of endophytic fungi, and factors a possibilities and challenges associated with investigating these microorganisms of interest and highlighting the role of fungal endophytes in this regard.

1.1.1 What Are Fungal Endophytes?

Fungal endophytes are a remarkable collection of microorganisms that inhabit plant roots without causing any visible damage. These fungi have a symbiotic relationship with their host plants, colonizing various parts including leaves, stems, and roots. Unlike pathogenic bacteria, fungal endophytes have evolved to coexist with their host plants, resulting in benefits for both parties (Rodriguez et al. 2009; Aly et al. 2011). Fungal endophytes are highly diverse, encompassing several taxonomic groups including Ascomycota, Basidiomycota, and Zygomycota (Higginbotham et al. 2013) and have been found on a variety of plants including trees, shrubs, herbs, and algae. The discovery of fungal endophytes has changed our knowledge of plant–microbial interactions and ecosystem formation.

1.1.1.1 Endophyte Colonization

Fungal endophytes form in plant tissues when their spores or mycelium penetrate through natural tissues or wounds. Once inside, they form dense hyphae that penetrate between the plant cells. This colonization process is highly selective, with different fungal strains having a bias towards certain plant tissues or organs (Andrade-Linares and Franken 2013). The colonization potential of most of fungal endophytes may be influenced by several factors, such as host plant physiology, environmental factors, and the presence of plant-microbiome. This basic knowledge suggests an evolutionary relationship between fungal endophytes and their host plants.

1.1.1.2 Benefits to Host Plants

Fungal endophytes provide many benefits to host plants, improving their overall health and fitness. One of the best-known benefits is the creation of phytochemicals, which can help protect the plant against disease, herbivory, and environmental challenges (Johnson et al. 2013). These active compounds, which have antimicrobial, anticancer, anti-inflammatory, and immunomodulatory properties, are high in pharmaceutical and agricultural application (Caruso et al. 2022). Not only do fungal endophytes produce bioactive compounds but can also increase nutrient uptake and plant growth. They can synthesize enzymes that degrade naturally occurring complexes, making them more susceptible to plants. Some endophytes can fix atmospheric nitrogen, which is an essential nutrient for the host plant. In addition, fungal endophytes can improve plants. In addition, fungal endophytes can enhance plant tolerance to abiotic stresses such as drought, salinity, and heavy metal toxicity (Fig. 1.1).

1.1.1.3 Coevolution with Host Plants

Fungal endophytes and host plants have resulted from million-year coevolution and have demonstrated an impact on the course of evolution in due course, though the coevolutionary process has been believed to alter its incidence and specialization of fungal endophyte associations with host plants. Evidence shows there has been a complex balance of reciprocal interactions and plant defense responses as a result of coevolutionary processes among fungal endophytes and host plants. On the other hand, promising endophytic fungi have been characterized by Lane et al. (2000) who opened up locales for dealing with beneficial endophytic fungi while further safeguarding possible host plants concurrently. These defense mechanisms can be induced through antiviral synthesis, cell wall armoring, and/or the induction of defense-linked genes. Meanwhile, the fungal endophytes counter-evolved mechanisms to prevent or break through host plants' defense mechanisms, consequently leading to sustained long-term symbiotic relationships.

1.1.1.4 Ecological Significance

Fungal endophytes play a crucial role in the regulation of ecosystem function and plant community dynamics within several systems. By increasing plant species' richness and functional variability, ecological diversity is also increased. The structure and content of a plant community may also be changed by the influence of fungal endophytes, thus facilitating the necessary support for the establishment and survival of some important plant species. Further tropic cascade effects may occur due to the fungal endophytes within higher food-chain trophic levels. For instance, some of the endophytes produce anti-herbivore secondary metabolites, hence benefiting the host plant itself. These molecules may as well change interactions of the

Fig. 1.1 A schematic diagram depicting the diversity of fungal endophytes found in plant tissues. Each plant has a varied range of fungal endophytes that play critical roles in plant development, defense, and survival. These endophytes create a wide range of bioactive compounds that have potential applications in medicine, agriculture, and environment. (Designed by ideogram AI image generator)

plants with pollinators or seed dispersers due to some modifications on both the reproduction output of the host plant and the animal partner.

1.1.2 Importance of Fungal Endophytes in Biodiversity

Fungal endophytes, therefore, represent a subgroup of endophytic fungi that are nonpathogenic to the host and have been reported from many plant species across a wide range of habitats. Their syndrome of endophytic lifestyle, relative to hosts and plants, strongly affects ecosystem health and function. Therefore, it is expected that their success in biodiversity is related to this fact.

1.1.2.1 Enhancing Plant Adaptations

Suryanarayanan and Shaanker (2021), also state that fungal endophytes confer improved host plant adaptation as well as survival in harsh environments. They make plants resistant to different abiotic stresses, among which the important ones are drought, salt, and extreme temperatures. These interactions make plant communities more resilient so as to be able to thrive in diverse ecological settings (Rodriguez et al. 2010).

1.1.2.2 Facilitating Nutrient Cycling

Fungal endophytes play an important role in the nitrogen cycle and contribute to ecosystem functioning through the decomposition of organic litter, thereby releasing key nutrients back into the soil (Saikkonen et al. 2015). It has been considered particularly important in nutrient-poor sites because it would increase nutrient availability not only to plants but also to other biota.

Furthermore, some endophytes can fix atmospheric nitrogen, which may be a significant source of this nutrient to the plants. The fungal endophytes enhance nutrient recycling processes that are associated with ecosystem productivity and stability.

1.1.2.3 Supporting Plant Diversity and Community Structure

The presence of fungal endophytic associations has been positively linked with a high diversity of plant species and community structures. On the one hand, endophytes may change the course of competition in plants due to their influence on the growth and development of host plants. On the other hand, they can affect interactions of plants with other organisms, such as pollinators and herbivores, creating shifts in plant community structure and functioning. In addition, endophytes drive or stimulate the growth of rare and endangered plant species for the conservation of biodiversity (Oono et al. 2020).

1.1.2.4 Bioprospecting and Biotechnological Applications

Over the years, fungal endophytes have gained huge attention for their bioprospection and biotechnological applications (Yan et al. 2018). The majority of them can produce bioactive compounds with a broad spectrum of pharmacologic activity, ranging from antibacterial, anticancer, antioxidative, and immunomodulatory activities, among others. These bioactive products lead toward the development of novel drugs, agricultural products, and industrial applications. There is a huge potential in the study of bioactive chemicals derived from endophytes with respect to the discovery of new molecules and developing sustainable solutions for medicine, agriculture, and other fields.

1.1.2.5 Conservation and Sustainable Use

The habitats of endophytes are very sensitive to any form of degradation in environmental conditions, pollution, and climatic changes; this may also lead to the extinction of rare and valuable varieties of microorganisms. Conservation efforts should be targeted toward protecting habitats and host plants sustaining the different groups of endophytes (Ballesteros et al. 2023).

Moreover, sustainable means of agriculture and forestry should be developed to ensure that endophyte– host plant cohabitation is sustainable and guaranteed for ecosystem stability and production throughout time.

1.1.3 Bioactive Materials Produced by Fungal Endophytes

Some of the current great promises given by bioactive compounds produced by fungal endophytes include medicinal, agricultural, and industrial applications. These bioactive compounds are secondary metabolites that are made by the fungi in relation to their symbiotic connection with host plants (Nagarajan et al. 2021). Many researchers have discovered a wide diversity of bioactive substances, in general, over the years, which have varied chemical structures and biological activity derived from fungal endophytes.

1.1.3.1 Antimicrobial Compounds

Fungal endophytes have been shown to produce some of the most noted and researched bioactive chemicals that include antimicrobial compounds. These possess the ability to arrest the growth of a great number of microorganisms including viruses, fungus, and bacteria. Important application of these substances is to protect the host plant from the destructive pathogens. Fungal endophytes can produce different types of antimicrobial chemicals including peptides, polyketides, terpenoids, and alkaloids. These substances have been highly promising in demonstrating antibacterial activity against drug-resistant strains, which could be potential compounds for the development of new-generation antibiotics.

1.1.3.2 Anticancer Compounds

On the other hand, fungal endophytes harbor an enormous number of bioactive metabolites, to which the anticancer property is attributed. They may inhibit or suppress the growth of cancerous cells or induce a programmed death of such cells called apoptosis. They could be used as an adjuvant in the treatment of cancer in the future or in the creation of new anticancer medications. Among the bioactive substances produced by fungi endophytes are those with anticancer effects such as vinblastine, taxol, camptothecin, and podophyllotoxin (Chen et al. 2016). The cytotoxic effects of these chemicals have been thoroughly studied on various cancer cell lines.

1.1.3.3 Antioxidant Compounds

The other closely related activity by metabolites derived from antioxidant fungal endophytes is being linked to health benefits and/or health-promoting effects (Selim et al. 2018). These metabolites elicit free radical scavenging activities and thereby protect the cells from oxidative damage. Available literature has well established that many diseases, such as cancer, heart disease, and some of the neurodegenerative diseases, are linked to oxidative stress. All these productions of phenolic metabolites, flavonoids, and carotenoids that are known to show potent antioxidant activity have possible uses as natural antioxidants in the food and pharmaceutical industry.

1.1.3.4 Immunomodulatory Compounds

It has also been found that fungal endophytes produce bioactive compounds with immunomodulatory characteristics. Such compounds have the capability of altering the immunological responses; they can cause stimulation of them or inhibition. Immunomodulatory compounds produced by fungal endophytes have been proven to bring improvements in allergy, immunodeficiency disorders, and autoimmune diseases. Some of the common examples of immunomodulatory compounds from the fungal endophytes are polysaccharides, glycoproteins, and peptides (Kumar et al. 2005). These substances have been shown to induce phagocytosis, enhance cytokine synthesis, and regulate the activity of immune cells.

1.1.3.5 Other Bioactive Activity

In addition to the discussed bioactive products, fungal endophytes synthesize a wide number of other bioactive molecules with different biological activities. Among these are major ones that display activities such as insecticidal, antiviral, antifungal, and anti-inflammatory. The bioactive molecules include peptides, polyketides, terpenoids, and alkaloids. Such diverse bioactive products have found application in the development of natural insecticides, antiviral drugs, and anti-inflammatory therapeutics. (Rai et al. 2021). The discovery of bioactive compounds produced by fungal endophytes has set out new avenues in business and the development of agriculture and medicine.

1.1.4 Ways to Explore Fungal Endophytes

The area of research relating to endophytic fungi is an intricate and enthralling area where the convergence of both conventional and contemporary methodologies is evident. When one uses those methodologies, it becomes apparent that endophytes are rich in the production of bioactive compounds from a secret realm. Some of the hugely important techniques for the study related to fungal endophytes will be presented herein (Dos Reis et al. 2022).

1.1.4.1 Isolation Techniques

The primary prerequisite for investigating the diversity and bioactivity potential of fungal endophytes is their isolation from plant tissues. Scientists usually apply a range of isolation methodologies based on the nature of plant material and the research objectives at hand. One such most common used method by researchers in isolating endophytes from plant tissues is the surface sterilization method. It entails cleansing the plant tissue by use of disinfectants to kill the outside impurities (Sun and Guo 2012). After sterilization, the tissue is then sectioned into small portions and put on substrate, which enhances the growth of fungi while killing the rest of the microbes. The plant tissue is ground and diluted; these are plated out onto agar plates by another process referred to as dilution plating. Through this, individual colonies of fungi can be isolated and subjected to further characterization.

1.1.4.2 Culture Media and Conditions

Fungal endophytes have to be cultured in the laboratory environment after their isolation for studying growth characteristics and potential bioactivity. Crucial to proper growth of this group of fungi is that culture medium is chosen correctly and that a set of environmental factors be chosen which will support optimal growth. Composite media is often used by researchers so as to cultivate fungi from their subjects, since different species of fungi have various nutritional requirements. Fungal endophytes are most commonly grown on media such as malt extract agar (MEA), potato dextrose agar (PDA), and synthetic media like Czapek-Dox agar. In addition to the medium content, induction in the growth of fungal endophytes is also based on various environmental parameters like temperature, pH, and light intensity. Individual species have to be cultured in their ideal condition for successful culture.

1.1.4.3 Molecular Identification

The classical methods of identification of endophytic fungi were based on physical characters, for example, spore color and spore morphology. While these are of great use in identifying several species, in less-known or closely related taxa, they might frequently bear disadvantages in the exact identification of species. Molecular techniques of identification have really revolutionized the scenario by making possible the identification of fungal endophytes purely on the bases of their DNA sequences. For stronger analysis, the polymerase chain reaction will be done, since there are several sections of DNA in fungi that amplify it. The sequences could then be translated into the existing databases, enabling the identification of species through DNA sequencing on the internal transcribed spacer (ITS) region.

Besides, other available molecular approaches, such as metagenomics and DNA fingerprinting, explore the diversity and fungal endophyte compositions applied in other studies (Manias et al. 2020).

1.1.4.4 Microscopy and Imaging

Fungal endophytes are identified based on morphological studies and their interaction with the host plant. The most commonly used method of imaging is microscopy. It is possible to use light microscopy to observe structures or organs and reproductive organs for the identification of the various fungal species. Higher magnification photographs can be achieved in electron microscopy that allows easy observation of the cellular structure. The two forms of electron microscopy are transmission electron microscopy shortened as TEM and scanning electron microscopy known as SEM. The application of these techniques has allowed the study of how fungal endophytes colonize plant tissues and observed which alterations of cellular interaction occur between the fungus and their hosts.

1.1.4.5 Metabolomics and Bioassays

The powerful techniques are able to study the synthesis of bioactive compounds by fungal endophytes in post-metabolomics. This field of metabolomics focuses on small compounds produced by fungi, better known as metabolites, which harbor potential biological activities. The chemicals are characterized and quantified using metabolic techniques such as nuclear magnetic resonance (NMR) spectroscopy and liquid chromatography-mass spectrometry (LC-MS). These secondary metabolites can be tested for bioactivity upon infections or cancerous cells, among other targets, using bioassays. In this manner, these methods allow one to understand the potential applications of fungal endophytes in agriculture, medicine, and other fields.

1.1.4.6 Genomics and Transcriptomics

Advances in genomics and transcriptomics now provide critical information on the genetic content and expression of fungal endophytes. Whole-genome sequencing allows for obtaining the whole DNA sequence of a species of fungus, which encompasses information concerning its genes and probable functions. Meanwhile, transcriptomics studies gene expression patterns under a diversity of conditions. These methodologies enable researchers to elucidate the genetic foundation of the bioactivity and ecological functions of fungal endophytes (Wang et al. 2015a, b).

1.2 Diversity of Fungal Endophytes

1.2.1 Fungal Endophyte Taxonomy

Endophytes are considered to be a large taxonomic unit of microorganisms living inside plant tissues without apparently acting negatively toward their host. Higher plant species, namely trees, shrubs, herbs, and even algae, harbor fungi. Fungal endophyte taxonomy is complex and under continuous evolution because it is still common to have new species continually discovered and described (Naranjo-Ortiz and Gabaldón 2019).

1.2.1.1 Classification of Fungal Endophytes

These are seen in many taxonomic categories of fungi. These are found in Ascomycota, Basidiomycota, as well as in Zygomycota. The taxon that is most heavily laden with fungal endophytes is Ascomycota—a phylum. A couple of the more familiar species in this phylum would be Cladosporium, Penicillium, and Fusarium. What is unique about these fungus is that they produce spores in a structure that looks like a bag, and this is known as asci. Many fungal endophytes are within Ascomycota, in the class Dothideomyc. It harbors vast genera diversity, including Alternaria, Colletotrichum, and Phomopsis, and these fungi often occur resident in many diverse plant hosts in which bioactive traits are well studied. A vast majority of the endophytes belonging to the major subclass of fungi are from the phylum Basidiomycota.

The famous genera of this phylum include Trichoderma and Ganoderma. One identifying feature of these fungi is that the basidomycetes produced basidia on which spores were also produced and these are called basidiospores. These have been associated with woody plants and proved to be important for both the defense and the health of the plant. Though not as well studied in the Ascomycota and Basidiomycota, fungal endophytes of immature zygomycota do have representatives. One of the important characteristics of zygomycetes is their unique ability to produce thick-walled zygospores during the sexual phase of their life cycle. Diversity and ecological roles of zygomycota endophytes are less researched, but they have been isolated from very diverse plant hosts and sometimes show unique bioactive properties (Aamir et al. 2020).

1.2.1.2 Molecular Approaches to Fungal Endophyte Taxonomy

In the earlier times, identification and classification of fungal endophytes were based on morphological features such as spore size, shape, and reproductive structures. But it was when science stepped into the molecular era that the real taxonomy of these organisms came to light. DNA sequencing is one of the most popular molecular techniques that helped scientists delve deeper into the world of fungal endophytes. For example, through comparing genetic sequences of ITS region among different isolates—researchers can establish taxonomical affiliations which sometimes unveil recognized species as new ones altogether due to their diversity. The wealth of diversity in fungal endophytes has been practically revealed through other molecular methods like DNA fingerprinting and PCR apart from sequencing. DNA fingerprinting and PCR come in handy by amplifying specific regions of the fungal genome that can be studied further; this helps bring out more information about these organisms' identities and classifications, providing critical insights into genetic diversity within a population or community of a fungal endophyte (Vasundhara et al. 2016).

1.2.1.3 Challenges in Fungal Endophyte Taxonomy

There are still a number of difficulties in the taxonomy of fungal endophytes despite the advances in molecular methods. The absence of thorough reference databases for fungal endophytes is one of the primary obstacles. The taxonomic connections of a large number of fungal species remain poorly defined and their characteristics are still lacking. Because of this, it is challenging to correctly identify and categorize fungal endophytes using only their genetic sequences. The existence of cryptic species in fungal endophyte communities presents another difficulty. Despite having similar physical characteristics, cryptic species are genetically diverse and difficult to recognize from one another. These obscure species can impede precise taxonomic categorization and cause the diversity of fungal endophytes to be underestimated. Moreover, taxonomy is hampered by the cultivation bias in the study of fungal endophytes. Since many fungal endophytes are challenging to cultivate in the lab, several taxa are underrepresented in culture collections. This makes it more difficult to accurately classify fungal endophytes taxonomically and restricts our knowledge of their real diversity (Singh 2019).

1.2.1.4 Future Directions in Fungal Endophyte Taxonomy

Moreover, the application of high-throughput sequencing methods, including next-generation sequencing, might significantly improve our comprehension of the variety of fungal endophytes. A more complete image of the fungal endophyte communities connected to various plant hosts and ecosystems may be obtained thanks to these technologies, which enable the quick sequencing of many fungal isolates (Sagita et al. 2021).

1.2.2 Habitats and Host Plants of Fungal Endophytes

Fungal endophytes are a broad class of microorganisms that live inside the plant tissues and don't seem to harm their hosts. According to Giauque et al. (2019), these amazing creatures can be found in a variety of habitats, including terrestrial, aquatic, and even harsh settings like deserts and Polar Regions. Since various plant species offer distinct ecological niches for these microorganisms to flourish, the variety of fungal endophytes is intimately correlated with the diversity of their host plants.

1.2.2.1 Terrestrial Habitats

The terrestrial habitats are the most abundant as well as well-researched environments under the existence of endophytic fungi. They include; cultivated fields, marshes, forests as well as grasslands. In the given habitats, endophytic fungi can be found in the differential plant organs, including leaves, stems, roots, and seeds. Due to the lack of monophyletic groups, the distribution of a plant is dependent on the species. Fungal endophytes provide excellent examples. Some prefer the roots, while others colonize the leaves heavily. Fungal endophytes are common in deciduous and evergreen forest trees. They contribute importantly to the overall health and survival of them through the improvement of resistance to infections, acquisition of nutrients, and on occasions by growth stimulation. Mostly, the Fungi are endophytes found inside hosts that belong to the grass family; for example, they are found mostly in grasslands. Endophytic fungi are found with most crop plants under agricultural situations. They help a plant resist various biotic and abiotic stresses, and hence can increase productivity besides reducing chemical inputs that a plant may require.

1.2.2.2 Aquatic Habitats

Endophytic fungi occur naturally within a wide range of aquatic environments; however, they have been much less studied in comparison to terrestrial ecosystems (Rudgers and Clay 2005). These habitats range from sea and estuarine habitats to freshwater habitats, such as lakes, rivers, and ponds. The aquatic endophytic fungi are harbored by the aquatic plants, which comprise floating, emergent, and submerged aquatic plants. In an environment, a study conducted by Sandberg et al. (2014). Fungal endophytes have been isolated from various freshwater aquatic plants, such as the water lilies, water hyacinths, and other forms of algae. They aid these plants with protection from infections, enhanced nutrition, and triggering development—all indispensable for their overall health and prosperity. Marine fungal endophytes are also isolated from the sea grass beds and seaweed growing in maritime habitats, as well as mangroves. These endophytes play very vital roles in the adaptation and survival of these plants under harsh marine conditions characterized by high salinity and wave action (Bärlocher and Boddy 2016).

1.2.2.3 Extreme Environments

There are also exotic places with extreme conditions where fungal endophytes can be found. They help their host plants by providing support for survival in these environments, and they do so through special adaptations developed in response to plant habitat challenges that inhibit plant growth. Fungal endophytes have been discovered in cacti and succulents typical of desert regions—supporting their hosts under harsh weather conditions due to low availability of water plus strong exposure to sunlight. Fungal endophytes have also been found on alpine flora growing at high altitudes—aiding plants against strong winds coupled with freezing temperatures, as well as high UV levels. Arctic and Antarctic vegetation have also been found to harbor fungal endophytes; these support the plant's life under nutrient-poor soils compounded by cold climatic conditions favoring survival strategies via freezing (Magan 2007).

1.2.2.4 Host Specificity

Fungal endophytes are frequently linked to certain plant species or groups of closely related plants, demonstrating variable degrees of host specificity. While some endophytes are limited to a single plant species and have a small host range, others are more widely distributed and can be found in several plant species either within the same family or even in distinct plant families (Apigo and Oono 2018). A number of variables, such as the host plant's physiological and biochemical traits and the environmental circumstances in which it grows, affect the host specificity of fungal endophytes. The tissues of different plant species have different microenvironments, with differences in secondary metabolite profiles, pH levels, and nutrient availability. These microenvironments can select for specific groups of fungal endophytes that are adapted to thrive in those conditions (Arnold and Lutzoni 2007). Moreover, the variety and makeup of the plant community can be impacted by the existence of fungal endophytes. Certain endophytes generate bioactive substances that might hinder the expansion of rival plant species, providing a competitive edge to their host plant. Endophyte-induced resistance is a phenomena that can have significant effects on the dynamics of plant communities and the efficiency of ecosystems.

1.2.3 Factors Influencing Fungal Endophyte Diversity

One major group of microorganisms, known as endophytic fungi, lives inside the plant tissues without apparently damaging them. They have been discovered in high numbers within diverse plants, that is, from marshes, grasslands, and woodlands—even in extreme settings like arctic and desert regions. Here, we outline some of the most important determinants of variability in fungal endophytes: host plant traits, environmental circumstances, and the interactions of these fungi with other microbes associated with them (Lau et al. 2013).

1.2.3.1 Host Plant Characteristics

Plant host determinants play a pivotal role in shaping the diversity of fungal endophytes. Specific plant species provide unique tiny ecosystems in their tissues that favor the growth of certain types of fungi. The selection of endophyte colonizers is contingent upon plant physiochemical properties such as nutrient availability and secondary metabolite production: it varies greatly based on these factors. Some plants have been found to harbor more diverse endophyte populations than others, indicating differential host support capabilities for fungal colonization (Aly et al. 2011). Apart from plant age, health status also dictates endophyte composition diversity because younger plants are less established and therefore have fewer endophytes assemblage than mature plants that are well developed over time with different fungus species. Additionally, because physiological changes in stressed or sick plants can decrease resource availability and change the microenvironment within their tissues, they may also show changes in their endophyte populations.

1.2.3.2 Environmental Conditions

For development as well as survival numerous fungal varieties have various requirements for dampness web content together with temperature level. Subsequently, modifications in these ecological variables might have an influence on endophyte neighborhoods' structure. As an example contrasted to completely dry or cool setups the range of fungal endophytes is regularly greater in exotic rain forests with high moisture plus temperature level (Ahlholm et al. 2002; Giauque and Hawkes 2013). The pH coupled with nutrient accessibility of the dirt can have a result on the variety of fungal endophytes. According to Bokati et al. (2016) some fungal types might have certain nutritional demands or pH choices which might affect their capability to occupy plants in different dirt kinds. Additionally, the colonization patterns of fungal endophytes can be affected by communications with various other microorganisms existing in the dirt consisting of germs along with various other fungi.

1.2.3.3 Plant–Endophyte Interactions

The variety of endophyte communities can be greatly influenced by interactions between fungal endophytes and plants. While certain plants may have defense mechanisms that prevent some fungal species from colonizing, others may have evolved mechanisms to recruit endophytes. Many factors, including chemical signaling, the presence of other microbes, and plant genetics, can affect these interactions (Wani et al. 2015). Plant colonization by endophytes is largely conditioned by undefined plant genetic components that favor endophyte accommodation in the rhizosphere. Fungal species will be attracted or repelled depending on whether the plant produces a specific class of chemical compounds or if their cell wall compositions are distinct. Moreover, different genotypes of plants within a species could also vary in their ability to interact with endophytes affecting the structure and diversity of each plant individual's endophyte community. Endophyte–endophyte and endophyte–plant chemical signals are involved in the formation of endophyte communities. Other plants give off chemical cues—volatile organic molecules, or VOCs—that tell some kinds of fungi to come interact with that host plant, and other compounds to make the fungi steer clear.

1.2.3.4 Interactions with Other Microorganisms

Rather, endophytic fungi are surrounded by fellow microorganisms, including bacteria, viruses, and other fungus, that share the plant tissues with them. The interactions might have good and bad impact on the diversity of fungal endophytes. Under other conditions, competition for nutrients with microbes may inhibit the colonization and diversification of endophytes. On the other hand, certain microbes may facilitate the growth and establishment of endophytes leading to an increase in diversity (van Overbeek and Saikkonen 2016). Studies, as an example, show that bacterial endophytes also benefit through association with fungal endophytes.

These interactions can be antagonistic, in which one creature prevents the growth of the other, or mutualistic, in which both species gain. Certain bacterial species can affect the diversity and colonization patterns of fungal endophytes in plant tissues. Additionally, the variety of fungal endophytes can be impacted by the existence of s or microorganisms linked with plants. According to Zabalgogeazcoa (2008), pathogens can create poisons that prevent endophytes from growing or they can compete with them for resources. On the other hand, it has been shown that some endophytes generate antimicrobial substances that can inhibit the growth of plant s, improving plant health and endophyte variety.

1.2.4 Ecological Roles of Fungal Endophytes

In order for plant communities to function ecologically, fungal endophytes are essential. These unseen plant residents have developed intricate connections with their host plants, impacting a range of ecological functions. The ecological functions of fungal endophytes and their contributions to the general well-being and efficiency of ecosystems will be discussed in this section.

1.2.4.1 Plant Growth Promotion

Fungal endophytes play an important ecological role in promoting plant growth. In their host plants, some fungal endophytes can boost nutrient intake, increase tolerance to water stress, and boost resistance to hosts. Fungal endophytes aid in the establishment and growth of their host plants by producing chemicals that promote growth, such as enzymes and phytohormones. Because endophytes can give their host plants a competitive edge in nutrient-poor situations, these interactions are especially crucial (Omomowo and Babalola 2019).

1.2.4.2 Nutrient Cycling

In ecosystems, fungal endophytes also aid in the cycling of nutrients. They have the ability to break down organic debris, releasing nutrients that are vital to the growth of plants. Endophytes are essential to the recycling of carbon, nitrogen, and other elements because they break down complex organic compounds. In addition to helping the host plant, this process affects the microbial populations and the availability and cycling of nutrients in the nearby soil (Saikkonen et al. 2015)

1.2.4.3 Suppression

Plants have been reported to be significantly suppressed by fungal endophytes. Plant pathogens are inhibited in their growth and activity by a variety of bioactive molecules that they can create, including antimicrobial agents. These bioactive substances have the direct ability to stop infections from growing or to strengthen the defenses of the host plant. Fungal endophytes enhance the general health and robustness of plant communities by defending their host plants against illnesses (O'Hanlon et al. 2012)

1.2.4.4 Stress Tolerance

Many environmental stressors, including salt, drought, and high temperatures, are frequently experienced by plants. By generating chemicals that are resistant to stress, fungal endophytes can assist their host plants in overcoming these challenges. A summary of endophyte-mediated systems in agricultural plants for managing salt and drought stress is given in (Fig. 1.2). These substances can improve a plant's resistance to harsh environments by controlling the water balance, scavenging reactive oxygen species, and altering the expression of genes linked to stress. Under adverse environmental circumstances, the existence of endophytes can greatly enhance a plant's chances of survival and growth (Verma et al. 2021).

1.2.4.5 Allelopathy

Through allelopathy, fungal endophytes can also affect how plants interact with one another. The term "allelopathy" describes a plant's or microbe's capacity to emit substances that have an impact on the growth and development of nearby plants. Certain endophytes generate allelochemicals that give their host plants a competitive edge by preventing the germination and growth of rival plant species. According to Aschehoug et al. (2014), these allelochemicals have the power to shape the dynamics of plant communities and aid in the emergence of dominant plant species within a particular ecosystem.

1.2.4.6 Mutualistic Interactions

While some fungal endophytes may provide ecological services like enhancing plant growth or protection, they can also form mutualistic relationships with the plants in which they reside. Mutualistic endophytes are beneficial to the host enhancing herbivory resistance, stress tolerance, and nutrient acquisition. In return, the endophytes get both some protection and a plethora of nutrients from the host plants which allow them to grow and reproduce. This fulfillment of the endophytes

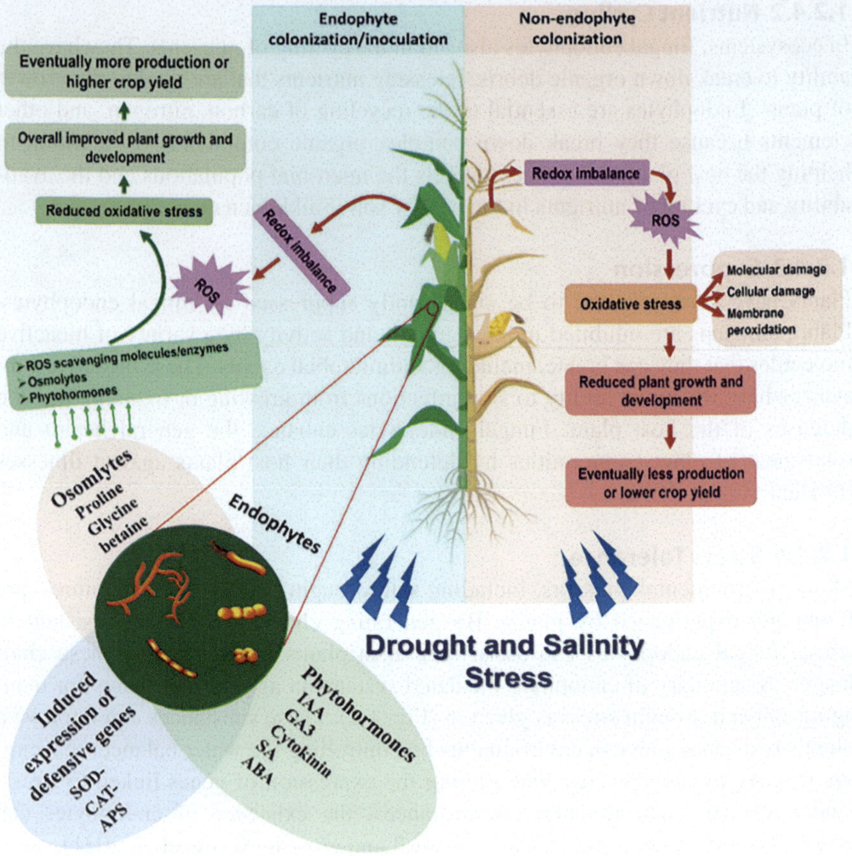

Fig. 1.2 Summary of endophyte-mediated processes in crop plants that regulate salinity and drought stress. (Reprinted from Verma et al. 2021)

can take place in the absence of this relationship, and their host plants' ability to thrive is supported by Schulz (2006).

1.2.4.7 Ecosystem Functioning

Fungal endophytes have ecological roles that are not tied to individual hosts, but to how ecosystems work. Endophytes play a significant role in the structure and function of ecosystems by helping to enhance productivity, stability, and resilience through their influence on plant growth, nutrient cycling, suppression, and stress tolerance. They can shape the unique plant communities that make up broad and diverse cereals, as well as how plants interface with other living beings. Structure and function of entire ecosystems may be affected by the presence or absence of specific endophyte species, as these other studies show (Hardoim et al. 2015).

1.3 Bioactive Compounds from Fungal Endophytes

1.3.1 Antimicrobial Compounds

The ability of fungal endophytes to produce an enormous diversity of bioactive compounds including antibacterial compounds has attracted much recent attention (Deshmukh et al. 2022). Synthetic antimicrobial agents are no longer reliable because of the rising antibiotic resistance. Thus, these substances can be used as alternative forms. The antimicrobial chemicals released by the fungal endophytes have been effective against many different types of microorganisms including bacteria, fungi, and some viruses. Due to their broad spectrum, they are preferred for developing new antibacterial drugs (Hussain et al. 2014). Fungal endophytes can generate many antibacterial polyketides. Polyketides make up a big group of naturally occurring substances that have strong bacteriostatic properties. Penicillin, a well-known drug produced by endophytic fungus named Penicillium is among such polyketides that are made by fungal endophytes.

Besides polyketides, fungal endophytes can also make other types of antibiotics including peptides, terpenoids, and alkaloids. It has been reported that these compounds have antibacterial activity against different organisms, such as drug-resistant pathogens. This antibacterial effect of the substances is associated with plant defense against harmful microorganisms. Antimicrobial chemicals produced by fungal endophytes help prevent dangerous pathogens from colonizing plant tissues thereby increasing their resistance to diseases. Furthermore, there might be a role for human medicines in the bactericides which these American tree root fungi produce. Some of these compounds have been shown to be effective in humans against methicillin-resistant *Staphylococcus aureus* (MRSA) and vancomycin-resistant Enterococcus (VRE), among other antibiotic resistant bacteria. Identification of antibacterial substances from fungal endophytes has opened up new avenues for development of novel antimicrobials, which can help overcome barriers created by antibiotic resistance and provide alternative therapeutic options for infectious diseases (Adeleke and Babalola 2021).

It can be difficult to separate and identify antimicrobial chemicals from fungal endophytes, though. It can be difficult to get enough quantities of bioactive substances for additional research because many endophytic fungi are slow-growing and difficult to culture in the lab (Sun and Guo 2012). To address these issues, scientists have created a number of methods for isolating and growing fungal endophytes. These methods include the use of culture conditions that are optimized and selective media to encourage the growth of particular endophytic fungi. Accurate fungal species identification and assessment of their potential for bioactive chemical production have also been achieved by the use of molecular identification techniques, such as DNA sequencing.

1.3.2 Anticancer Compounds

The quest for efficacious and secure anticancer medications persists, notwithstanding progress in medical research and treatment alternatives. Investigating the possibility of fungal endophytes as a source of new anticancer chemicals has gained traction in recent years. It has been shown that these harmless microbes that live inside plant tissues create a variety of bioactive substances with intriguing anticancer qualities (Kharwar et al. 2011).

1.3.2.1 Discovery of Anticancer Compounds from Fungal Endophytes

New directions in the creation of drugs have been made possible by the identification of anticancer substances from fungal endophytes. From these bacteria, various bioactive chemicals with strong anticancer properties have been extracted and discovered by researchers (Pandey et al. 2018). These substances are excellent prospects for more research since they are frequently distinct and structurally varied. Finding anticancer chemicals from fungal endophytes is a multi-step procedure. First, a variety of isolation procedures are employed to separate the endophytes from their host plants. After being isolated, the endophytes are cultivated in carefully monitored lab settings in order to speed up the synthesis of bioactive substances. Optimizing the yield and diversity of chemicals generated is contingent upon the circumstances and culture media used (Bedi et al. 2018).

Subsequently, the bioactive substances are isolated from the fungal cultures and put through a battery of screening tests to gauge their potential anticancer effects. To ascertain the drugs' cytotoxicity and selectivity, these assays usually include comparing them to a panel of cancer cell lines. Chemical structures of compounds with notable anticancer action are clarified by employing sophisticated analytical methods including mass spectrometry and nuclear magnetic resonance (NMR) spectroscopy (Chandra 2012).

1.3.2.2 Mechanisms of Action of Anticancer Compounds

Fungi living inside plants produce chemicals that can fight cancer in ways. Research has shown that some of these substances trigger apoptosis, a type of programmed cell death, in cancer cells. Apoptosis is a process in the body to eliminate irregular cells. These substances have been found to halt tumor growth and spread by inducing apoptosis in cancer cells (Kousar et al. 2022). Moreover, other anticancer compounds from fungal endophytes have been found to hinder cancer cell growth by disrupting functions. Signaling pathways vital for the survival and growth of cancer cells. By blocking these targets these compounds effectively inhibit the proliferation of cancer cells. Prevent ones, from forming. Furthermore, certain substances derived from fungal endophytes exhibit angiogenic properties. The process of angiogenesis, which is the formation of new blood vessels, is essential to the growth and metastasis of tumors. Angiogenesis-inhibiting substances can efficiently deprive tumors of blood flow, halting their growth and dissemination.

1.3.2.3 Examples of Anticancer Compounds from Fungal Endophytes

Microbial endophytes have actually generated a wide range of anticancer chemicals much of which have actually shown motivating end results in preclinical as well as professional examinations. One such is the popular anticancer drug taxol, which was very first removed from the endophytic fungus *Taxomyces andreanae* (Wang et al. 2015a, b). One more image is camptothecin a solid anticancer compound initially recognized from the endophytic fungus *Camptotheca acuminata*. Substantial anticancer activity has actually been shown by camptothecin as well as its by-products versus a range of cancer cells kinds, such as lung, pancreatic as well as colon cancer cells (Amna et al. 2012). Many even more anticancer chemicals have actually been discovered from fungal endophytes along with taxol as well as camptothecin. Podophyllotoxin, vinblastine plus epothilones are a few of these. Every one of these materials has a unique setting of activity together with the possibility of being utilized in medication.

1.3.3 Antioxidant Compounds

Compounds called antioxidants are essential in defending cells from the harm that free radicals can inflict. Because they can induce oxidative stress, which can result in a number of free radicals and other substances are extremely reactive chemicals. According to Huang et al. (2007), fungal endophytes have been discovered to generate a diverse array of antioxidant chemicals that have garnered substantial interest because of their possible medical advantages.

1.3.3.1 Types of Antioxidant Compounds

Fungal endophytes can produce a bunch of different antioxidant chemicals that help prevent cell damage by getting rid of free radicals. These include types of phenolic compounds, flavonoids, alkaloids, terpenoids, and polyketides. Stilbenes and other phenolics are pretty well-known for being good antioxidants, since they can donate electrons or hydrogen to stop free radicals. Flavonoids are also antioxidants but they work a bit differently by scavenging the free radicals. Alkaloids terpenoids and polyketides produced by the fungal endophytes also give them antioxidant abilities, research has found. So basically, fungal endophytes make a variety of helpful antioxidant chemicals using different ways to prevent cell damage from free radicals floating around (Goutam et al. 2018).

1.3.3.2 Sources of Antioxidant Compounds

Numerous plant species, including herbs, trees, and medicinal plants, have had fungal endophytes isolated from them (Adeleke and Babalola 2021). It has been shown that these endophytes generate antioxidant chemicals that are advantageous for the host plant as well as perhaps having uses in human health. Examples of plants containing fungal endophytes that produce antioxidants are *Ginkgo biloba, Camellia sinensis,* and *Taxus brevifolia.*

1.3.3.3 Mechanisms of Antioxidant Activity

The capacity of fungal endophytes to scavenge free radicals and prevent oxidative stress is thought to be responsible for their antioxidant function. By giving away hydrogen atoms or electrons, these endophytes create antioxidant chemicals that can directly neutralize free radicals. Moreover, they have the ability to increase the activity of endogenous antioxidant enzymes that support the antioxidant defense system, including glutathione peroxidase (GPx), catalase (CAT), and superoxide dismutase (SOD) (Hamilton et al. 2012).

1.3.3.4 Health Benefits of Antioxidant Compounds

There is potential health benefit from the antioxidant chemicals that fungal endophytes produce. They have been shown to have neuroprotective, cardioprotective, anti-inflammatory, and anti-cancer effects. Certain chemicals have the potential activity to prevent chronic s like cancer, cardiovascular, and neurological disorders by decreasing oxidative stress and neutralizing free radicals. According to Abo Nahas (2019), they can also strengthen the immune system and advance general health.

1.3.3.5 Potential Applications

Fungal endophytes produce antioxidant chemicals that have a wide range of possible applications. To increase product shelf life and guard against oxidative damage, they can be utilized as natural antioxidants in the food and cosmetics industries (Darwish 2019). The pharmaceutical industry may investigate these chemicals for their possible use as medicinal agents (Table 1.1). In order to enhance health and prevent, they can also be used in the production of functional foods and dietary supplements.

1.3.4 Other Bioactive Compounds

Many other bioactive compounds have been found and investigated in addition to the well-known bioactive chemicals produced by fungal endophytes, such as antibacterial, anticancer, antioxidant, and immunomodulatory compounds. These substances contain a diverse array of biological activity and could be used in business, agriculture, and medicine. We will examine a few of these lesser-known bioactive substances made by fungal endophytes in this section (Bhardwaj and Agrawal 2014).

1.3.4.1 Enzymes

Enzymes having a wide range of functions are known to be produced by fungal endophytes. Proteins called enzymes are important players in a variety of biological processes because they catalyze chemical reactions. Enzymes including cellulases, hemicellulases, ligninases, lipases, proteases, and amylases have been reported to be produced by fungal endophytes. Waste management, food processing, textile processing, and biofuel manufacturing are just a few of the industries that can use these enzymes. For instance, plant cell walls can be broken down by cellulases and

Table 1.1 Antioxidant compounds produced by endophytic fungi and their potential benefits

Compound class	Examples	Potential benefits
Phenolics	Flavonoids, phenolic acids, stilbenes	– Free radical scavenging – Metal chelation – Modulation of cellular signaling pathways
Flavonoids	Catechins, quercetin, resveratrol	– Anti-inflammatory activity – Neuroprotective effects – Improved cardiovascular health
Phenolic acids	Gallic acid, chlorogenic acid, ellagic acid	– Anticancer properties – Antibacterial activity – Reduced risk of chronics
Stilbenes	Resveratrol, piceatannol	– Anti-aging effects – Improved glucose metabolism – Protection against neurodegeneratives
Terpenoids	Monoterpenes, diterpenes, triterpenes	– Anti-inflammatory activity – Immunomodulatory effects – Anticancer properties
Monoterpenes	Limonene, α-pinene, β-pinene	– Antimicrobial activity – Wound healing properties – Respiratory health improvement
Diterpenes	Taxol, gibberellins, phorbol esters	– Some with potent anticancer activity (Taxol) – Plant growth regulation (gibberellins) – Can have toxic effects (phorbol esters)
Triterpenes	Betulinic acid, ursolic acid, ganoderic acid	– Anti-inflammatory activity – Anticancer properties – Liver protection
Alkaloids	Ergoline alkaloids, indole alkaloids, pyrrole alkaloids	– Broad spectrum of biological activities
Ergoline alkaloids	Ergotamine, Claviceps alkaloids	– Can be toxic at high doses (ergotamine) – Some with potential in neurological disorders (limited research)
Indole alkaloids	Vinblastine, vincristine, ajmaline	– Potent anticancer drugs (vinblastine, vincristine) – Antiarrhythmic properties (ajmaline)
Pyrrole alkaloids	Festuclavine, lolitrem B	– Can be toxic to grazing animals – Lolitrem B has insecticidal properties
Other antioxidants	Quinones, glutathione, superoxide dismutase (SOD)	– Play a role in cellular antioxidant defense mechanisms

(continued)

Table 1.1 (continued)

Compound class	Examples	Potential benefits
Quinones	Ubiquinone (coenzyme Q10), juglone	– Involved in mitochondrial electron transport – May have benefits for skin health and energy production
Glutathione	– Major cellular antioxidant – Detoxification processes	
Superoxide dismutase (SOD)	– Enzyme that breaks down superoxide radicals – Protects cells from oxidative damage	

hemicellulases, which makes them valuable in the lignocellulosic biomass generation of biofuels (Suryanarayanan et al. 2012).

1.3.4.2 Plant Growth-Promoting Compounds

Additionally, chemicals that support plant growth and development can be produced by fungal endophytes. These substances include gibberellins, auxins, and cytokinins-phytohormones that control the processes involved in plant growth. Siderophores are another type of molecules that fungal endophytes can make. These compounds help plants absorb iron by chelating it. Furthermore, a few types of fungal endophytes generate volatile organic compounds (VOCs) that have the ability to promote plant development and create systemic resistance to infections. These substances that stimulate plant development may be applied to agriculture to increase crop yield while lowering the demand for artificial fertilizers and pesticides (Hassan 2017).

1.3.4.3 Insecticidal Compounds

According to Han et al. (2013), several fungal endophytes generate bioactive substances with insecticidal qualities. As an alternative to manufactured chemical insecticides that may be hazardous to the environment and nontarget creatures, these substances can function as natural insecticides. It has been discovered that fungal endophytes generate substances that are poisonous to a variety of insect pests, including disease-carrying insects and agricultural pests. To reduce crop damage and control insect populations, integrated pest management systems might make use of these insecticidal chemicals.

1.3.4.4 Antiviral Compounds

In recent years, there has been a renewed interest in the hunt for antiviral chemicals from natural sources, such as fungal endophytes. Fungal endophytes have been discovered to create bioactive chemicals with antiviral properties against a wide range of viruses, including humans. These chemicals can block viral replication and entry into host cells. The finding of antiviral chemicals in fungal endophytes offers hope for the development of new antiviral medications and treatments (Lacerda et al. 2022).

1.3.4.5 Anti-inflammatory Compounds

Inflammation is a natural immunological reaction that helps the body defend itself against infection and harm. Chronic inflammation has been linked to the development of a number of s, including cancer and autoimmune disorders. Fungal endophytes are known to produce bioactive chemicals with anti-inflammatory effects. These chemicals can reduce the synthesis of pro-inflammatory molecules and influence the immune response. The identification and characterization of anti-inflammatory chemicals found in fungal endophytes could lead to the development of new treatments for inflammatory illnesses (Pal et al. 2020).

1.3.4.6 Neuroprotective Compounds

Neurodegenerative diseases, like Alzheimer's and Parkinson's, are distinguished by the progressive loss of neurons in the brain. Fungal endophytes have been studied for their ability to create bioactive chemicals with neuroprotective effects. These chemicals can protect neurons from oxidative stress, inflammation, and other harmful events. The finding of neuroprotective chemicals in fungal endophytes could lead to the development of medicines for neurodegenerative illnesses (Takahashi et al. 2019).

1.3.4.7 Antidiabetic Compounds

Diabetes is a chronic metabolic condition marked by elevated blood sugar levels. Fungal endophytes have been studied for their ability to create bioactive chemicals that have antidiabetic effects. These chemicals can help to manage blood sugar, improve insulin sensitivity, and protect pancreatic beta cells. The discovery of antidiabetic chemicals in fungal endophytes may aid in the development of new diabetes medicines (Agrawal et al. 2022).

Fungal endophytes produce a variety of bioactive compounds in addition to the well-known antimicrobial, anticancer, antioxidant, and immunomodulatory compounds, such as enzymes, plant growth-promoting compounds, insecticidal compounds, antiviral compounds, anti-inflammatory compounds, neuroprotective compounds, and antidiabetic compounds (Table 1.2).

1.4 Biotechnological Applications of Fungal Endophytes

1.4.1 Bioprospecting for Novel Compounds

Bioprospecting is the act of investigating and discovering novel molecules or substances from natural sources, such as plants, animals, and microbes, for a variety of purposes. Fungal endophytes, which live in plant tissues and cause no harm, have received a lot of attention in recent years as a possible source of novel bioactive chemicals. These compounds have demonstrated interesting biological activities such as antibacterial, anticancer, antioxidant, and immunomodulatory characteristics (Gakuubi et al. 2021)

Table 1.2 Various bioactive compounds produced by endophytic fungi and their potential applications

Compound class	Examples	Potential applications
Enzymes	Cellulases, xylanases, lacases, lipases	– Biomass degradation and biofuel production (cellulases, xylanases) – Food and textile industry applications (lacases) – Bioremediation and biocatalysis (lipases)
Cellulases and xylanases	Break down cellulose and xylan (plant cell wall components)	– Can be used to convert plant biomass into fermentable sugars for bioethanol production
Lacases	Oxidize various phenolic compounds	– Used in food processing to improve wine clarity and shelf life, textile dye decolorization, and degradation of environmental pollutants
Lipases	Break down fats and oils	– Used in bioremediation of oil spills, detergent production, and synthesis of biofuels and other bioproducts
Plant growth-promoting compounds	Auxins, gibberellins, cytokinins, phosphate solubilizing enzymes	– Enhance plant growth and development
Auxins	Promote cell division, root elongation, and stem development	
Gibberellins	Stimulate stem elongation, leaf growth, and flowering	
Cytokinins	Promote cell division, shoot growth, and nutrient uptake	
Phosphate solubilizing enzymes	Release bound phosphate in soil, making it more available for plants	
Anti-inflammatory compounds	Terpenoids, phenolics, alkaloids	– Treatment of chronic inflammatory s (arthritis, inflammatory bowel)
Terpenoids	Triterpenes (e.g., betulinic acid)	– May reduce inflammation by inhibiting pro-inflammatory pathways
Phenolics	Flavonoids (e.g., quercetin)	– Possess anti-inflammatory and antioxidant properties
Alkaloids	Specific alkaloids with anti-inflammatory effects (limited research on endophytic fungi)	– Some alkaloids may target specific inflammatory pathways
Neuroprotective compounds	Terpenoids, phenolics	– Potential for treatment of neurodegeneratives (Alzheimer's, Parkinson's)
Terpenoids	Some terpenoids exhibit antioxidant and neuroprotective properties	

(continued)

Table 1.2 (continued)

Compound class	Examples	Potential applications
Phenolics	Flavonoids may protect neurons from oxidative damage	
Antidiabetic compounds	Alkaloids, polysaccharides	– Potential for development of new diabetes medications
Alkaloids	Specific alkaloids with blood sugar-lowering effects (limited research on endophytic fungi)	– May improve insulin sensitivity or mimic insulin action
Polysaccharides	Some polysaccharides may help regulate blood sugar levels	
Insecticides	Alkaloids, terpenoids	– Development of eco-friendly pest control alternatives
Alkaloids	Lolitrem B (produced by some grass endophytes)	– Toxic to certain insects that graze on infected plants
Terpenoids	Beauvericin, spinosad (produced by some soil-borne fungi)	– Act as insecticides by disrupting insect molting or nerve function
Antiviral agents	Polysaccharides, alkaloids	– Potential for development of novel antiviral drugs
Polysaccharides	Beta-glucans	– May stimulate the immune system and potentially have antiviral effects
Alkaloids	Specific alkaloids with antiviral activity (limited research on endophytic fungi)	– Some alkaloids may target viral replication or entry into host cells

1.4.1.1 Isolation and Cultivation of Fungal Endophytes

The initial stage in bioprospecting for new chemicals from fungal endophytes is to isolate and cultivate these microorganisms. Isolation approaches entail surface cleaning of plant tissues to remove any external contamination, followed by endophytic fungus extraction from internal tissues. Various procedures, such as tissue maceration, grinding, and dilution plating, can be used to extract fungal endophytes from plant tissues. Once isolated, fungal endophytes can be cultivated on media that closely resembles their native environment. Culture media and circumstances have a significant impact on fungal endophyte development and bioactive chemical synthesis. Optimizing pH, temperature, carbon and nitrogen sources, and incubation duration can improve the production of desirable chemicals (Götz et al. 2006).

1.4.1.2 Screening for Bioactive Compounds

Following successful isolation and cultivation, the fungal endophytes are screened for the presence of medicinal substances. Secondary metabolites are extracted from fungal cultures and tested for biological activity against a variety of target organisms or cell lines. Various bioassays can be used to assess the antibacterial, anticancer, antioxidant, and immunomodulatory properties of the isolated chemicals. These assays aid in determining the potential of fungal endophytes to produce bioactive chemicals for therapeutic or industrial purposes (Vasundhara et al. 2016; Rana et al. 2019).

1.4.1.3 Dereplication and Compound Characterization

Once bioactive compounds have been discovered, dereplication is performed to determine whether they are novel or previously documented in the literature. Dereplication involves comparing the chemical fingerprints of isolated molecules to known compounds in databases or utilizing spectroscopic techniques such as nuclear magnetic resonance (NMR) and mass spectrometry (MS). Compound characterization is a critical stage in bioprospecting since it provides detailed information on the chemical structure and properties of bioactive compounds. Chemical structure and chemical identity can be identified using techniques including NMR, MS, infrared spectroscopy (IR), and X-ray crystallography (Gupta et al. 2023).

1.4.1.4 Scale-up and Production

Following the identification and characterization of bioactive chemicals, the next issue is to increase their production for future research or commercial applications. This entails optimizing fermentation parameters such as fermentation vessel selection, aeration, agitation, and nutrient supplementation in order to maximize the output of desired chemicals. Fungal endophytes can produce bioactive chemicals more efficiently using bioprocess engineering approaches like fed-batch and continuous fermentation. These strategies seek to increase the productivity, efficiency, and cost-effectiveness of the fermentation process (Ganeshan et al. 2021).

1.4.2 Biological Control of Plant Pathogens

Fungal endophytes have received a lot of attention in recent years because of their potential as biological control agents for plant infections. Pathogenic fungus in plants can cause severe crop losses, reduced agricultural production, and an increasing dependency on chemical pesticides. However, the use of chemical pesticides has prompted concerns about environmental contamination, human health risks, and the emergence of pesticide-resistant organisms. As a result, there is an increasing interest in developing sustainable and ecologically friendly plant management methods, and fungal endophytes appear to be a possible answer (Jaber and Ownley 2018).

1.4.2.1 Mechanisms of Biological Control

Fungal endophytes can affect biological processes through a variety of methods. One of the most prevalent processes is the synthesis of antifungal chemicals, which slow the growth and development of plant infections. These chemicals can either be directly harmful to pathogens or disrupt their vital biological processes. Some fungal endophytes produce volatile organic compounds (VOCs), which have antibacterial qualities and can prevent harmful fungus from growing in their surroundings.

Another way that fungal endophytes can suppress plant diseases is through competition for nutrients. Endophytes can outcompete infections for resources and space, lowering their population number and suppressing their pathogenicity. Additionally, fungal endophytes can induce systemic resistance in host plants, priming them to mount a more effective defense response against pathogen attacks (Latz et al. 2018).

1.4.2.2 Examples of Fungal Endophytes as Biological Control Agents

Several fungal endophytes have been found as potent biological control agents for a variety of plant diseases. One such example is the endophytic fungus *Trichoderma* spp., which has been intensively researched for its biocontrol potential. *Trichoderma* spp. produce a number of antifungal chemicals, including chitinases, glucanases, and proteases, which can break down pathogenic fungi's cell walls. These enzymes not only directly suppress pathogen growth but also strengthen the plant's defense response. Another example is the endophytic fungus *Beauveria bassiana*, which is well-known for controlling insect infestations. However, recent research has shown that it can also act as a biocontrol agent against plant infections. *B. bassiana* produces a variety of secondary metabolites, including destruxins and beauvericins, which have antifungal properties (Fig. 1.3). These compounds can disrupt the cell membranes of pathogenic fungi, leading to their death (Mantzoukas and Eliopoulos 2020). Four putative fungal endophytes (strains WF4–7) isolated from finger millet and showed anti-fungal activity against *F. graminearum* (Mousa et al. 2016).

1.4.2.3 Application of Fungal Endophytes in Agriculture

In agriculture, the utilization of fungal endophytes as biocontrol agents has trivial advantages over chemical pesticides. For starters, these are environmentally safe and they do not pose a serious threat to noneconomic species or the health of humans. Fungal endophytes act as a biodegradable pesticide which leaves no toxic residues in the environment, unlike chemical pesticides that can persist and accumulate in food webs. The second, would be if it will be possible to manage fungal endophytes over the long-term and sustainably. Unlike chemical pesticides that may create pesticide-resistant diseases, fungal endophytes work via multiple modes of action limiting the development of pathogen resistance. Furthermore, fungal endophytes have the ability to form enduring relationships with host plants, offering persistent defense against infections. Additionally, employing fungal endophytes can support sustainable farming practices by lowering the usage of chemical pesticides. Farmers can lessen their influence on the environment, including water pollution and the extinction of pollinators and beneficial insects, by using less chemical pesticides (Sharma and Bhattarai 2019).

1.4.3 Plant Growth Promotion

It is known that fungal endophytes can promote plant development and health. Through symbiotic relationships with their host plants, these beneficial microorganisms provide a range of growth-promoting strategies that can significantly influence

ID	Tissue	Taxonomic classification
WF5	Shoot	*Fusarium fujikuroi*
WF6	Root	*Penicillium chrysogenum*
WF7	Root	*Penicillium expansum*

Fig. 1.3 Evaluation of the potential fungal endophytes that were previously identified from finger millet and the efficaciousness of their extracts against fungi. The finger millet plant is shown in (**a**); the fungal endophytes WF4, WF5, and WF7 that were isolated from the plant are shown in (**b–d**); the predicted fungal endophyte nomenclatures are shown in (**e**) along with the corresponding plant tissue from which the fungal endophytes were isolated and the best BLAST match taxonomic identification based on 18S rDNA sequencing; (**f–h**) Illustrations of each endophyte fermenting on the designated rice media; (**i**) Measuring the impact of each endophyte extract (250 mg/mL) on *F. graminearum* growth in relation to the fungicide nystatin (10 μg/mL) (inhibition zone diameter in millimeters, $n = 3$); (**j–l**) Sample photos showing the outcomes of each endophytic fungus's in vitro co-incubation with *F. graminearum*. (Reproduced from Mousa et al. 2016)

plant growth and development. This section addresses the diverse methods in which fungal endophytes facilitate plant growth promotion (Omomowo and Babalola 2019). Plant growth promotion mechanisms are summarized in Table 1.3.

1.4.3.1 Enhanced Nutrient Uptake

Increased nutrient uptake is one of the main ways that fungal endophytes support plant growth. By solubilizing and mobilizing nutrients, these endophytes increase their availability for plant absorption. For instance, some fungal endophytes can generate phosphatases, which are enzymes that break down organic phosphorus molecules and increase phosphorus availability for plants. Similar to this, certain endophytes have the ability to fix atmospheric nitrogen and change it into a form that is useful to plants. Fungal endophytes aid plants in overcoming nutrient shortages and fostering healthy growth by making vital nutrients more readily available (García-Latorre et al. 2021)

1.4.3.2 Hormone Production and Regulation

Through the synthesis and control of plant hormones, fungal endophytes are also essential for the encouragement of plant growth. These microbes are capable of producing a variety of hormones that stimulate plant growth, including gibberellins, cytokinins, and auxins. Cell division, elongation, and differentiation are among the

Table 1.3 Plant growth promotion mechanism via endophytic fungi

Mechanism	Description	Benefits for plants
Nutrient acquisition		
Phosphate Solubilization	Endophytes produce enzymes that convert insoluble phosphate in the soil into a form readily available for plants	– Increased phosphorus uptake, a crucial nutrient for plant growth and development
Mycorrhiza-like associations	Some endophytes form symbiotic relationships with plant roots similar to mycorrhizal fungi, enhancing nutrient uptake (particularly phosphorus and nitrogen)	– Improved access to essential nutrients from the soil
Stress tolerance		
Drought tolerance	Endophytes can produce compatible solutes that help plants retain water during dry periods	– Improved drought resistance and survival under water stress conditions
Resistance	Endophytes may produce antibiotics or other antifungal compounds that suppress plant pathogens	– Enhanced protection against various fungal pathogens in the host plant
Production of phytohormones	Endophytes can synthesize plant growth hormones like auxins, gibberellins, and cytokinins	– Stimulation of root development, shoot growth, and overall plant vigor

(continued)

Table 1.3 (continued)

Mechanism	Description	Benefits for plants
Improved photosynthesis	Some endophytes may enhance the efficiency of photosynthesis in the host plant, leading to increased growth	– Increased plant productivity and biomass production
Defense against herbivores	Endophytes can produce secondary metabolites that deter herbivores from feeding on the host plant	– Reduced insect damage and improved protection from herbivory

processes in plant growth and development that are regulated by these hormones. Fungal endophytes have the ability to increase overall plant vigor, boost nutrient absorption, and stimulate root and shoot growth by generating these hormones (Fonseca et al. 2018).

1.4.3.3 Induced Systemic Resistance

Inducing systemic resistance is a crucial strategy by which fungal endophytes facilitate plant growth. Certain endophytes have the ability to colonize plants and incite a defense mechanism in the host plant, increasing the plant's resistance to pests and diseases. The synthesis of several defense-related substances, including phytoalexins and pathogenesis-related proteins, is a component of this induced systemic resistance (ISR) mechanism. Fungal endophytes can shield their host plant from harm and encourage healthy growth by triggering the plant's defensive mechanisms (Shoresh et al. 2010).

1.4.3.4 Stress Tolerance

In adverse environments, fungal endophytes have also been shown to promote plant growth. These microbes aid in the tolerance of a number of abiotic conditions by plants, including salt, drought, and heavy metal toxicity. Reactive oxygen species (ROS) can be scavenged by endophytes' enzymes and metabolites, which lessens oxidative stress in plants (Fig. 1.4). Additionally, they have the ability to increase the synthesis of osmoprotectants like proline and glycine betaine, which support plants in preserving the water balance inside their cells and guard against dehydration stress. Fungal endophytes enhance a plant's ability to withstand stress, which helps the plant develop and survive under harsh conditions (Dastogeer and Wylie 2017).

1.4.3.5 Biocontrol of Plant Pathogens

Fungal endophytes not only directly stimulate plant growth but also indirectly do so by shielding plants from harmful organisms. The growth and activity of plant pathogens can be inhibited by the antibacterial chemicals produced by these endophytes. They can hinder pathogen establishment and development by outcompeting them for nutrition and space by invading the plant tissues. In addition to shielding plants from pests, the biocontrol function of fungal endophytes enables plants to devote more resources to expansion and development.

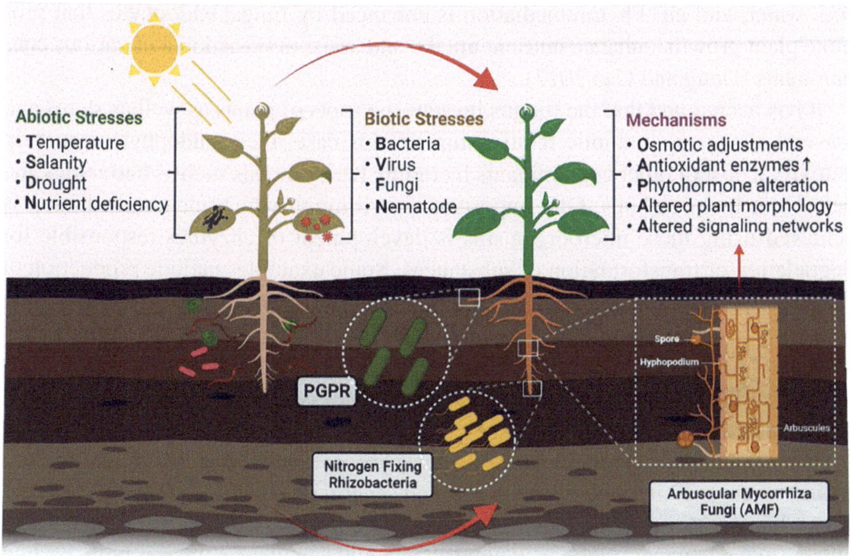

Fig. 1.4 An overview diagram of main environmental stressors and the processes of plant–microbe interactions that impact plant fitness under these conditions. Tolerance to these stimuli is primarily achieved by the increase of antioxidant defense enzymes, osmotic adjustment, and phytohormone synthesis. (Reprinted from Kamran et al. 2022)

1.4.3.6 Improved Soil Health

By enhancing the health of the soil, fungal endophytes can also aid in the promotion of plant growth. By producing organic acids and enzymes that break down organic matter and increasing the availability of nutrients for plants, these microbes can improve soil fertility. By generating polysaccharides and other substances that encourage soil aggregation and water infiltration, they can also enhance soil structure. Fungal endophytes promote the establishment of healthy plant communities and foster a suitable environment for plant growth by improving the fertility and structure of the soil (Harman and Uphoff 2019).

1.4.4 Phytoremediation and Biodegradation

Phytoremediation and biodegradation are two essential processes that employ fungal endophytes to clean and destroy contaminants in the environment. These techniques have received a lot of interest in recent years because of their promise for long-term and environmentally benign solutions to pollution problems.

1.4.4.1 Phytoremediation

Phytoremediation is a process by which plants, and their associated microbes called fungal endophytes, are utilized to get rid of, degrade, or immobilize pollutants from soil, water, and air. Phytoremediation is enhanced by fungal endophytes that promote plant growth, enhance nutrient uptake and assist in breakdown of various contaminants (Deng and Cao 2017).

It has been noted that the fungus invades the roots of plants as well as stems and leaves leading to symbiotic relationship. In this case, these endophytes can help detoxify a host of other contaminants including heavy metals such as herbicides and petroleum hydrocarbons. One important way through which phytoremediation is achieved using these microorganisms is development of enzymes responsible for degradation or transformation of substances. Some examples include production of lignin peroxidase, manganese peroxidase laccase, etc., by certain fungi within their cells which act on complex organic pollutants such as polycyclic aromatic hydrocarbons (PAHs) or chlorinated compounds. Apart from producing enzymes that can degrade different toxic materials, fungal endophytes can also raise the plant's capacity to resist toxins.

They can benefit plants by boosting their absorption of nutrients, releasing hormones that encourage development, and energizing their defensive systems. According to Marín et al. (2018), the symbiotic interaction between fungal endophytes and plants produces a mutually favorable environment for both parties, improving the results of phytoremediation. In addition, fungal endophytes can help to phytostabilize polluted areas. By immobilizing contaminants in the soil, restricting their mobility, and lowering their bioavailability, phytostabilization entails the utilization of plants and the microbes that are linked with them. By creating a barrier of defense around the roots, limiting the absorption of pollutants into the plant and their transport to the aboveground portions, fungal endophytes can aid in phytostabilization.

1.4.4.2 Biodegradation

The breaking down and changing of organic pollutants into simpler and safer chemicals is called biodegradation; it happens when microorganisms like fungal endophytes perform this work. Fungal endophytes have shown great potential in biodegradation, because they produce a range of enzymes capable of breaking down complex chemical compounds. For instance, fungal endophytes manufacture cellulases, hemicellulases as well as ligninolytic enzymes which play a role in degrading organic contaminants and plant biomass. These enzymes make it possible for the microbes to use these smaller molecules easily by facilitating their breakdown into simple organic compounds that can be used for growth and energy by the microbes. In this case, they break down complex organic substances.

Several studies indicate that fungal endophytes are capable of degrading organic pollutants. Certain types of fungal endophytes, for example, have been found to degrade polycyclic aromatic hydrocarbons (PAHs) that are persistent organic pollutants (POPs) in contaminated soils and sediments. According to Russell et al. (2011), such fungi produce enzymes involved in the degradation and detoxification of PAHs through disruption of their aromatic rings. Furthermore, fungal endophytes have also been reported to degrade different types of organic pollutants, including PAHs, pesticides, herbicides, and petroleum hydrocarbons. Compared to conventional techniques, the employment of fungal endophytes in biodegradation has a number of benefits. To begin with, fungal endophytes are organic, eco-friendly agents that can break down contaminants naturally without the use of harsh chemicals or energy-intensive procedures. Furthermore, fungal endophytes are easily isolated from plants and may be cultivated in a lab, which makes them easily accessible for applications involving biodegradation. Last but not least, fungal endophytes are adaptable agents for bioremediation since they can break down a variety of contaminants. To sum up, fungal endophytes are important for the processes of phytoremediation and biodegradation. They are useful instruments for the cleanup of contaminated areas because of their capacity to promote plant growth, generate enzymes, and break down contaminants. Fungal endophytes are used in phytoremediation and biodegradation to provide environmentally acceptable and long-lasting solutions for environmental pollution, paving the way for a cleaner and healthier future.

In order to protect plants from various abiotic stresses, such as temperature fluctuations, salinity, low moisture levels, and resistance to fungicides in soil treatments, Trichoderma endophytes are employed as biological controllers. These traits make Trichoderma formidable rivals because, as a defense mechanism for plants, they can create siderophores that release chelated iron (Fe^{3+}) from the surrounding environment to stop the growth of other fungi (Fig. 1.5).

Siderophores, their secondary metabolites, have been demonstrated to be essential to phytoremediation and pollution removal. Organisms such as fungi, bacteria, and plants manufacture siderophores, which are small molecular weight chelators, to scavenge iron from their surroundings. Fungal siderophores are

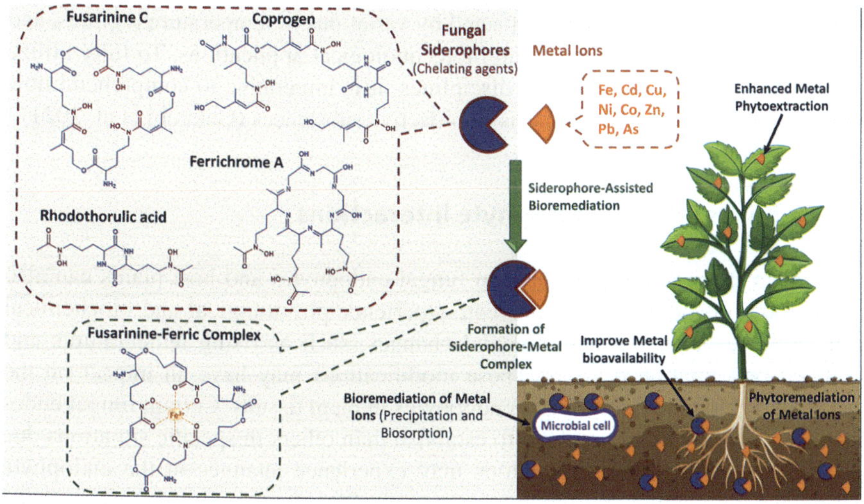

Fig. 1.5 Conceptual representation of the phytoremediation of metal ions by fungal siderophores. The chemical structures of selected fungal siderophores (fusarinine C, ferrichrome A, coprogen, and rhodothorulic acid) and an example of a siderophore–metal complex (Fusarinine C-ferric complex). (Reprinted from Elhamouly et al. 2022)

hydroxamate-derived from ornithine, an amino acid that is not proteinogenic. Fusarinines, ferrichromes, coprogens, and rhodothorulic acid are the four categories of hydroxamate-type siderophores that have been classified according to their chemical structure (Rabani et al. 2022)

1.5 Climate Change and Fungal Endophytes

One of the most important worldwide issues of our day is climate change, which has profound effects on ecosystems and biodiversity. It is critical to comprehend the effects of the Earth's warming climate on various creatures, especially fungal endophytes. It is crucial to look into how these unseen allies are adjusting to climate change because they are so important to the health and functioning of plants (Suryanarayanan and Shaanker 2021).

1.5.1 Effects of Temperature

One important factor affecting the development and activity of fungal endophytes is temperature. It is anticipated that the distribution and number of these bacteria would change as global temperatures rise. Research has indicated that elevated temperatures have the potential to modify the structure and various fungal endophyte communities found in plants. While some species would do well in warmer climates, others might find it difficult to survive. Moreover, fungal endophytes' ability to produce bioactive chemicals can potentially be impacted by elevated temperatures. These chemicals may be impacted by variations in temperature regimes, and they have significant ecological and biotechnological applications. To fully utilize fungal endophytes in a variety of disciplines, it is imperative to comprehend how climate change affects the synthesis of bioactive substances (Gakuubi et al. 2021).

1.5.2 Shifts in Host–Endophyte Interactions

Changes in the relationships between fungal endophytes and host plants can also result from climate change. Plants can experience physiological and biochemical changes in response to environmental changes, such as rising temperatures and modified precipitation patterns. These modifications may have an impact on the variety and colonization of fungal endophytes in plant tissues. Certain fungal endophyte species may be more likely to establish than others in specific situations due to climate change. Plant populations may experience changes in the endophyte communities' structure and function as a result. Plant health, nutrient cycling, and ecosystem dynamics may all be impacted in a cascade manner by these modifications (Pathak et al. 2022)

1.5.3 Resilience and Adaptation

Fungal endophytes have developed a variety of adaptations to adapt to shifting environmental circumstances. The synthesis of osmolytes, heat-shock proteins, and other stress-related compounds is one of these pathways. Due to these modifications, fungal endophytes may flourish in a variety of environmental conditions. It's still unclear, though, if fungal endophytes can quickly adapt to changing climate conditions. According to certain research, some species may be able to withstand environmental stresses better than others, while others might find it difficult. Predicting how fungal endophytes will react to potential future climate scenarios requires an understanding of their adaptive capabilities (Giauque and Hawkes 2013).

1.5.4 Implications for Plant Health and Ecosystem Functioning

Fungal endophytes' reaction to climate change has significant effects on the health of plants and the efficiency of ecosystems. These microbes are essential for increasing plants' resistance to environmental stresses like salinity and drought. These beneficial connections may be disrupted by changes in the activity and composition of fungal endophytes, which could have detrimental effects on plant health. Fungal endophytes also support the cycling of nutrients and the health of ecosystems. They participate in the intake of nutrients, the breakdown of organic matter, and the synthesis of bioactive substances. Variations in the variety and abundance of fungal endophytes can impact ecosystem processes and nutrient dynamics in a cascade manner (Grover et al. 2011).

1.5.5 Conservation and Management Strategies

Since fungal endophytes are essential to the health of plants and the functioning of ecosystems, it is imperative that conservation and management strategies take the effects of climate change into account. Plant populations can be made more resilient to shifting environmental conditions by preserving and rehabilitating environments that support a variety of fungal endophyte communities. Fungal endophytes can also benefit from the promotion of sustainable land management techniques like organic farming and agroforestry. According to Wemheuer et al. (2020), these methods can increase the variety and quantity of fungal endophytes in agricultural systems, which will benefit plant health and productivity.

1.6 Challenges and Opportunities

We learn more about the realm of fungal endophytes and the chances and difficulties that await us. There are many obstacles in the way of studying these amazing species, but there is also a lot of promise for many other professions. This section discusses the difficulties faced by scientists and the fascinating opportunities that come with researching fungal endophytes.

1.6.1 Taxonomic Challenges

The taxonomic identification of fungal endophytes is a major challenge in their study. The study of fungi is a complicated and dynamic field, and new species are found on a regular basis. Because fungal endophytes lack distinguishable physical traits and are minute in nature, precisely identifying and describing them can be a challenging process. Although molecular methods—like DNA sequencing—have made identification much easier, naming recently found species is still a difficult task. Furthermore, the taxonomy of fungal endophytes is made more difficult by the large number of species that have not yet been reported due to their great diversity.

1.6.2 Cultivation Difficulties

The cultivation of fungal endophytes poses a substantial obstacle to the research of these organisms. Numerous endophytic fungi have particular needs for growth and feeding, and they are frequently picky eaters. Because of this, it is difficult to separate and grow them in lab environments. Certain endophytic species may not be able to flourish in traditional culture media, which would result in an underrepresentation of their diversity. In order to overcome these obstacles and raise the success rate of fungal endophyte production, researchers are continuously investigating novel approaches and culture conditions.

1.6.3 Functional Characterization

Comprehending the purposes of fungal endophytes is an essential component of their research. Even while we have come a long way in identifying the bioactive substances these fungi produce, there is still much to discover about the precise roles and modes of action of these molecules. It is still difficult to ascertain the ecological functions of fungal endophytes, including how they interact with host plants and other microbes. While sophisticated methods like transcriptomics and metagenomics are being used to decipher the intricate relationships and roles of fungal endophytes, more investigation is required to completely understand their ecological contributions.

1.6.4 Conservation and Sustainable Use

Conservation of fungal and endophytic diversity will promote health and preserve the potential benefits of the ecosystem inhabited by these organisms. However, there are unique challenges in preserving them. Unlike large, beautiful animals, fungal endophytes are often overlooked and ignored when considering conservation strategies. The safety of these organisms is complicated by their associations with plants and the specific requirements of their habitats. Strategies should be developed to conserve fungal endophytes and their host plants and to raise public awareness of the benefits of these organisms Continuous use of fungal endophytes should also be carefully considered to avoid damage or biodegradation of natural habitats.

1.6.5 Expanding Applications

The study of fungal endophytes offers a wealth of prospects for several professions, despite certain obstacles. These fungi create bioactive chemicals that have enormous potential use in biotechnology, medicine, and agriculture. New paths in drug research have been opened by the identification of unique antibacterial, anticancer, antioxidant, and immunomodulatory chemicals from fungal endophytes. These substances could offer substitutes for traditional antibiotics, anti-cancer medications, and other medicinal substances. Furthermore, myco-endophytes have several agricultural biotechnologies uses in phytoremediation, biofertilizers, and develop biocontrol agents using fungal endophytes to their full capacity can result in eco-friendly and sustainable agricultural and environmental remediation practices.

1.6.6 Climate Change and Fungal Endophytes

Fungal endophytes face both possibilities and challenges as a result of climate change. Variations in temperature, precipitation patterns, and other environmental conditions can have a substantial effect on fungal endophyte distribution and variety. Under changed circumstances, some species might flourish while others might become extinct. It is essential to comprehend how fungal endophytes react to climate change in order to forecast their future distribution and evaluate possible contributions to ecosystem functioning. Furthermore, by interacting with host plants, fungal endophytes may hold the secret to creating crops that are climate resilient. Mitigate and adapt to climate change tactics by investigating these relationships and utilizing fungal endophytes' adaptive capabilities.

1.7 Conclusion

A group of microorganisms called fungal endophytes inhabit plant roots, providing host plants with a variety of benefits such as nutrient uptake, biochemical production, and stress resistance. Their interactions with host plants support biological function and ecology. Understanding their role in plant–microbial interactions is essential to successfully exploit fungal endophytes in various industries such as agriculture, medicine, and conservation. Interdisciplinary design using both traditional and innovative approaches include important to study fungal endophytes. Researchers can discover the hidden world of fungal endo-microorganisms and explore their biology and bioactive potential using tools such as isolation methods, culture media, molecular detection methods, microscopic imaging, metabolomics, genomics, and transcriptomics with the help of fungi. Endobionts can be found in different environments and exhibit different host specificity. They have significant ecological effects on plant communities by modifying plant–plant interactions, forming s, enhancing nutrient cycling, and promoting plant growth. More studies are needed to understand the complexity of these relationships and how it effectively affects ecosystem resilience and sustainability. Further research is needed to address the problems of isolation, culture, and clinical analysis. In summary, there are opportunities and obstacles in comprehending the potential of fungal endophytes in biotechnology, medicine, and agriculture. Maintaining the diversity of fungal endophytes and optimizing their advantages requires conservation initiatives and sustainable use of these organisms. We can unleash the potential of fungal endophytes and clear the path for a better future by tackling these obstacles and seizing the chance.

References

Aamir M, Rai KK, Zehra A, Kumar S, Yadav M, Shukla V, Upadhyay RS (2020) Fungal endophytes: classification, diversity, ecological role, and their relevance in sustainable agriculture. In: Microbial endophytes. Woodhead Publishing, pp 291–323

Abo Nahas HH (2019) Endophytic fungi: a gold mine of antioxidants. Microb Biosyst 4(1):58–79

Adeleke BS, Babalola OO (2021) Pharmacological potential of fungal endophytes associated with medicinal plants: A review. J Fungi 7(2):147

Agrawal S, Samanta S, Deshmukh SK (2022) The antidiabetic potential of endophytic fungi: future prospects as therapeutic agents. Biotechnol Appl Biochem 69(3):1159–1165

Ahlholm JU, Helander M, Henriksson J, Metzler M, Saikkonen K (2002) Environmental conditions and host genotype direct genetic diversity of Venturia ditricha, a fungal endophyte of birch trees. Evolution 56(8):1566–1573

Aly AH, Debbab A, Proksch P (2011) Fungal endophytes: unique plant inhabitants with great promises. Appl Microbiol Biotechnol 90:1829–1845

Amna T, Amina M, Sharma PR, Puri SC, Al-Youssef HM, Al-Taweel AM, Qazi GN (2012) Effect of precursors feeding and media manipulation on production of novel anticancer pro-drug camptothecin from endophytic fungus. Braz J Microbiol 43:1476–1489

Andrade-Linares DR, Franken P (2013) Fungal endophytes in plant roots: taxonomy, colonization patterns, and functions. In: Symbiotic endophytes. Berlin, Heidelberg, Springer, pp 311–334

Apigo A, Oono R (2018) Dimensions of host specificity in foliar fungal endophytes. Endophytes of forest trees: biology and applications, pp 15–42

Arnold AE, Lutzoni F (2007) Diversity and host range of foliar fungal endophytes: are tropical leaves biodiversity hotspots? Ecology 88(3):541–549

Aschehoug ET, Callaway RM, Newcombe G, Tharayil N, Chen S (2014) Fungal endophyte increases the allelopathic effects of an invasive forb. Oecologia 175:285–291

Ballesteros GI, Newsham KK, Acuña-Rodríguez IS, Atala C, Torres-Díaz C, Molina-Montenegro MA (2023) Extreme environments as sources of fungal endophytes mitigating climate change impacts on crops in Mediterranean-type ecosystems. Plants People Planet 6:148–161

Bärlocher F, Boddy L (2016) Aquatic fungal ecology–how does it differ from terrestrial? Fungal Ecol 19:5–13

Bedi A, Adholeya A, Deshmukh SK (2018) Novel anticancer compounds from endophytic fungi. Curr Biotechnol 7(3):168–184

Bhardwaj A, Agrawal P (2014) A review fungal endophytes: as a store house of bioactive compound. World J Pharm Pharm Sci 3(9)

Bokati D, Herrera J, Poudel R (2016) Soil influences colonization of root-associated fungal endophyte communities of maize, wheat, and their progenitors. J Mycol 2016:8062073

Caruso DJ, Palombo EA, Moulton SE, Zaferanloo B (2022) Exploring the promise of endophytic fungi: a review of novel antimicrobial compounds. Microorganisms 10(10):1990

Chandra S (2012) Endophytic fungi: novel sources of anticancer lead molecules. Appl Microbiol Biotechnol 95:47–59

Chen L, Zhang QY, Jia M, Ming QL, Yue W, Rahman K et al (2016) Endophytic fungi with antitumor activities: their occurrence and anticancer compounds. Crit Rev Microbiol 42(3):454–473

Darwish AMG (2019) Fungal mycotoxins and natural antioxidants: two sides of the same coin and significance in food safety. Microb Biosyst 4(1):1–16

Dastogeer KM, Wylie SJ (2017) Plant–fungi association: role of fungal endophytes in improving plant tolerance to water stress. Plant-microbe interactions in agro-ecological perspectives: volume 1: fundamental mechanisms, methods and functions, pp 143–159

Deng Z, Cao L (2017) Fungal endophytes and their interactions with plants in phytoremediation: a review. Chemosphere 168:1100–1106

Deshmukh SK, Dufossé L, Chhipa H, Saxena S, Mahajan GB, Gupta MK (2022) Fungal endophytes: a potential source of antibacterial compounds. Journal of Fungi 8(2):164

Dos Reis JBA, Lorenzi AS, do Vale HMM (2022) Methods used for the study of endophytic fungi: a review on methodologies and challenges, and associated tips. Arch Microbiol 204(11):675

Elhamouly NA, Hewedy OA, Zaitoon A, Miraples A, Elshorbagy OT, Hussien S et al (2022) The hidden power of secondary metabolites in plant-fungi interactions and sustainable phytoremediation. Front Plant Sci 13:1044896

Fonseca S, Radhakrishnan D, Prasad K, Chini A (2018) Fungal production and manipulation of plant hormones. Curr Med Chem 25(2):253–267

Gakuubi MM, Munusamy M, Liang ZX, Ng SB (2021) Fungal endophytes: A promising frontier for discovery of novel bioactive compounds. J Fungi 7(10):786

Ganeshan S, Kim SH, Vujanovic V (2021) Scaling-up production of plant endophytes in bioreactors: concepts, challenges and perspectives. Bioresour Bioprocess 8:1–16

García-Latorre C, Rodrigo S, Santamaría O (2021) Endophytes as plant nutrient uptake-promoter in plants. Endophytes 3:247–265

Giauque H, Hawkes CV (2013) Climate affects symbiotic fungal endophyte diversity and performance. Am J Bot 100(7):1435–1444

Giauque H, Connor EW, Hawkes CV (2019) Endophyte traits relevant to stress tolerance, resource use and habitat of origin predict effects on host plants. New Phytol 221(4):2239–2249

Götz M, Nirenberg H, Krause S, Wolters H, Draeger S, Buchner A et al (2006) Fungal endophytes in potato roots studied by traditional isolation and cultivation-independent DNA-based methods. FEMS Microbiol Ecol 58(3):404–413

Goutam J, Singh R, Vijayaraman RS, Meena M (2018) Endophytic fungi: carrier of potential antioxidants. Fungi and their role in sustainable development: current perspectives, pp 539–551

Grover M, Ali SZ, Sandhya V, Rasul A, Venkateswarlu B (2011) Role of microorganisms in adaptation of agriculture crops to abiotic stresses. World J Microbiol Biotechnol 27:1231–1240

Gupta A, Meshram V, Gupta M, Goyal S, Qureshi KA, Jaremko M, Shukla KK (2023) Fungal endophytes: microfactories of novel bioactive compounds with therapeutic interventions; a comprehensive review on the biotechnological developments in the field of fungal endophytic biology over the last decade. Biomol Ther 13(7):1038

Hamilton CE, Gundel PE, Helander M, Saikkonen K (2012) Endophytic mediation of reactive oxygen species and antioxidant activity in plants: a review. Fungal Divers 54:1–10

Han LR, Wang ZH, Zhang HJ, Xue LS, Feng JT, Zhang X (2013) Isolation of endophytic fungi from Tripterygium wilfordii and their insecticidal activities. Afr J Microbiol Res 7:771–776

Hardoim PR, Van Overbeek LS, Berg G, Pirttilä AM, Compant S, Campisano A et al (2015) The hidden world within plants: ecological and evolutionary considerations for defining functioning of microbial endophytes. Microbiol Mol Biol Rev 79(3):293–320

Harman GE, Uphoff N (2019) Symbiotic root-endophytic soil microbes improve crop productivity and provide environmental benefits. Scientifica 2019:1

Hassan SED (2017) Plant growth-promoting activities for bacterial and fungal endophytes isolated from medicinal plant of Teucrium polium L. J Adv Res 8(6):687–695

Higginbotham SJ, Arnold AE, Ibañez A, Spadafora C, Coley PD, Kursar TA (2013) Bioactivity of fungal endophytes as a function of endophyte taxonomy and the taxonomy and distribution of their host plants. PLoS One 8(9):e73192

Huang WY, Cai YZ, Xing J, Corke H, Sun M (2007) A potential antioxidant resource: endophytic fungi from medicinal plants. Econ Bot 61(1):14–30

Hussain H, Kliche-Spory C, Al-Harrasi A, Al-Rawahi A, Abbas G, Green IR et al (2014) Antimicrobial constituents from three endophytic fungi. Asian Pac J Trop Med 7:S224–S227

Jaber LR, Ownley BH (2018) Can we use entomopathogenic fungi as endophytes for dual biological control of insect pests and plant pathogens? Biol Control 116:36–45

Johnson LJ, de Bonth AC, Briggs LR, Caradus JR, Finch SC, Fleetwood DJ et al (2013) The exploitation of epichloae endophytes for agricultural benefit. Fungal Divers 60:171–188

Kamran M, Imran QM, Ahmed MB, Falak N, Khatoon A, Yun BW (2022) Endophyte-mediated stress tolerance in plants: a sustainable strategy to enhance resilience and assist crop improvement. Cells 11(20):3292

Kharwar RN, Mishra A, Gond SK, Stierle A, Stierle D (2011) Anticancer compounds derived from fungal endophytes: their importance and future challenges. Nat Prod Rep 28(7):1208–1228

Kousar R, Naeem M, Jamaludin MI, Arshad A, Shamsuri AN, Ansari N et al (2022) Exploring the anticancer activities of novel bioactive compounds derived from endophytic fungi: mechanisms of action, current challenges and future perspectives. Am J Cancer Res 12(7):2897

Kumar DSS, Lau CS, Wan JM, Yang D, Hyde KD (2005) Immunomodulatory compounds from Pestalotiopsis leucothes, an endophytic fungus from Tripterygium wilfordii. Life Sci 78(2):147–156

Lacerda ÍCDS, Polonio JC, Golias HC (2022) Endophytic fungi as a source of antiviral compounds–a review. Chem Biodivers 19(6):e202100971

Lane GA, Christensen MJ, Miles CO (2000) Coevolution of fungal endophytes with grasses: the significance of secondary metabolites. Microbial endophytes, pp 341–388

Latz MA, Jensen B, Collinge DB, Jørgensen HJ (2018) Endophytic fungi as biocontrol agents: elucidating mechanisms in suppression. Plant Ecol Divers 11(5–6):555–567

Lau MK, Arnold AE, Johnson NC (2013) Factors influencing communities of foliar fungal endophytes in riparian woody plants. Fungal Ecol 6(5):365–378

Magan N (2007) Fungi in extreme environments. Mycota 4:85–103

Manias D, Verma A, Soni DK (2020) Isolation and characterization of endophytes: biochemical and molecular approach. In: Microbial endophytes. Woodhead Publishing, pp 1–14

Mantzoukas S, Eliopoulos PA (2020) Endophytic entomopathogenic fungi: A valuable biological control tool against plant pests. Appl Sci 10(1):360

Marín F, Navarrete H, Narvaez-Trujillo A (2018) Total petroleum hydrocarbon degradation by endophytic fungi from the Ecuadorian Amazon. Adv Microbiol 8(12):1029–1053

Mousa WK, Schwan AL, Raizada MN (2016) Characterization of antifungal natural products isolated from endophytic fungi of finger millet (Eleusine coracana). Molecules 21(9):1171

Nagarajan K, Ibrahim B, Ahmad Bawadikji A, Lim JW, Tong WY, Leong CR, Khaw KY, Tan WN (2021) Recent developments in metabolomics studies of endophytic fungi. J Fungi 8(1):28

Naranjo-Ortiz MA, Gabaldón T (2019) Fungal evolution: diversity, taxonomy and phylogeny of the fungi. Biol Rev 94(6):2101–2137

O'Hanlon KA, Knorr K, Jørgensen LN, Nicolaisen M, Boelt B (2012) Exploring the potential of symbiotic fungal endophytes in cereal suppression. Biol Control 63(2):69–78

Omomowo OI, Babalola OO (2019) Bacterial and fungal endophytes: tiny giants with immense beneficial potential for plant growth and sustainable agricultural productivity. Microorganisms 7(11):481

Oono R, Black D, Slessarev E, Sickler B, Strom A, Apigo A (2020) Species diversity of fungal endophytes across a stress gradient for plants. New Phytol 228(1):210–225

Pal PP, Shaik AB, Begum AS (2020) Prospective leads from endophytic fungi for anti-inflammatory drug discovery. Planta Med 86(13/14):941–959

Pandey PK, Singh S, Singh MC, Singh AK, Yadav SK, Pandey AK, Heisnam P (2018) Diversity, ecology, and conservation of fungal and bacterial endophytes. Microbial resource conservation: conventional to modern approaches, pp 393–430

Pathak P, Rai VK, Can H, Singh SK, Kumar D, Bhardwaj N et al (2022) Plant-endophyte interaction during biotic stress management. Plan Theory 11(17):2203

Rabani MS, Hameed I, Mir TA, Wani BA, Gupta MK, Habib A et al (2022) Chapter 5—microbial-assisted phytoremediation. In: Bhat RA, Tonelli FMP, Dar GH, Hakeem K (eds) Phytoremediation. Academic Press, pp 91–114. https://doi.org/10.1016/B978-0-323-89874-4.00006-6

Rai N, Kumari Keshri P, Verma A, Kamble SC, Mishra P, Barik S et al (2021) Plant associated fungal endophytes as a source of natural bioactive compounds. Mycology 12(3):139–159

Rana KL, Kour D, Sheikh I, Dhiman A, Yadav N, Yadav AN, et al (2019) Endophytic fungi: biodiversity, ecological significance, and potential industrial applications. Recent advancement in white biotechnology through fungi: volume 1: diversity and enzymes perspectives, pp 1–62

Rodriguez RJ, White JF Jr, Arnold AE, Redman ARA (2009) Fungal endophytes: diversity and functional roles. New Phytol 182(2):314–330

Rodriguez RJ, Woodward C, Redman RS (2010) Adaptation and survival of plants in high stress habitats via fungal endophyte conferred stress tolerance. Symbioses and stress: joint ventures in biology, pp 461–476

Rudgers JA, Clay K (2005) Fungal endophytes in terrestrial communities and ecosystems. Mycol Ser 23:423

Russell JR, Huang J, Anand P, Kucera K, Sandoval AG, Dantzler KW et al (2011) Biodegradation of polyester polyurethane by endophytic fungi. Appl Environ Microbiol 77(17):6076–6084

Sagita R, Quax WJ, Haslinger K (2021) Current state and future directions of genetics and genomics of endophytic fungi for bioprospecting efforts. Front Bioeng Biotechnol 9:649906

Saikkonen K, Mikola J, Helander M (2015) Endophytic phyllosphere fungi and nutrient cycling in terrestrial ecosystems. Curr Sci:121–126

Sandberg DC, Battista LJ, Arnold AE (2014) Fungal endophytes of aquatic macrophytes: diverse host-generalists characterized by tissue preferences and geographic structure. Microb Ecol 67:735–747

Schulz B (2006) Mutualistic interactions with fungal root endophytes. Microbial root endophytes, pp 261–279

Selim KA, Elkhateeb WA, Tawila AM, El-Beih AA, Abdel-Rahman TM, El-Diwany AI, Ahmed EF (2018) Antiviral and antioxidant potential of fungal endophytes of Egyptian medicinal plants. Fermentation 4(3):49

Sharma P, Bhattarai C (2019) Applications of fungal endophytes in agriculture and medicine field. Think India J 22(34):946–977

Shoresh M, Harman GE, Mastouri F (2010) Induced systemic resistance and plant responses to fungal biocontrol agents. Annu Rev Phytopathol 48:21–43

Singh BP (2019) Advances in endophytic fungal research: present status and future challenges. Springer

Sudheep NÁ, Marwal A, Lakra N, Anwar K, Mahmood S (2017) Fascinating fungal endophytes role and possible beneficial applications: an overview. Plant-microbe interactions in agro-ecological perspectives: volume 1: fundamental mechanisms, methods and functions, pp 255–273.

Sun X, Guo LD (2012) Endophytic fungal diversity: review of traditional and molecular techniques. Mycology 3(1):65–76

Suryanarayanan TS, Shaanker RU (2021) Can fungal endophytes fast-track plant adaptations to climate change? Fungal Ecol 50:101039

Suryanarayanan TS, Thirunavukkarasu N, Govindarajulu MB, Gopalan V (2012) Fungal endophytes: an untapped source of biocatalysts. Fungal Divers 54(1):19–30

Takahashi JA, Sande D, da Silva Lima G, e Moura MAF, Lima MTNS (2019) Fungal metabolites as promising new drug leads for the treatment of Alzheimer's. Stud Nat Prod Chem 62:1–39

van Overbeek LS, Saikkonen K (2016) Impact of bacterial–fungal interactions on the colonization of the endosphere. Trends Plant Sci 21(3):230–242

Vasundhara M, Kumar A, Reddy MS (2016) Molecular approaches to screen bioactive compounds from endophytic fungi. Front Microbiol 7:1774

Verma H, Kumar D, Kumar V, Kumari M, Singh SK, Sharma VK et al (2021) The potential application of endophytes in management of stress from drought and salinity in crop plants. Microorganisms 9(8):1729

Wang X, Wang C, Sun YT, Sun CZ, Zhang Y, Wang XH, Zhao K (2015a) Taxol produced from endophytic fungi induces apoptosis in human breast, cervical and ovarian cancer cells. Asian Pac J Cancer Prev 16(1):125–131

Wang X, Zhang X, Liu L, Xiang M, Wang W, Sun X et al (2015b) Genomic and transcriptomic analysis of the endophytic fungus Pestalotiopsis fici reveals its lifestyle and high potential for synthesis of natural products. BMC Genomics 16:1–13

Wani ZA, Ashraf N, Mohiuddin T, Riyaz-Ul-Hassan S (2015) Plant-endophyte symbiosis, an ecological perspective. Appl Microbiol Biotechnol 99:2955–2965

Wemheuer F, Berkelmann D, Wemheuer B, Daniel R, Vidal S, Bisseleua Daghela HB (2020) Agroforestry management systems drive the composition, diversity, and function of fungal and bacterial endophyte communities in Theobroma cacao leaves. Microorganisms 8(3):405

Yan L, Zhao H, Zhao X, Xu X, Di Y, Jiang C et al (2018) Production of bioproducts by endophytic fungi: chemical ecology, biotechnological applications, bottlenecks, and solutions. Appl Microbiol Biotechnol 102:6279–6298

Zabalgogeazcoa I (2008) Fungal endophytes and their interaction with plant pathogens: a review. Span J Agric Res 6:138–146

Biodiversity of Fungal Endophytes

2

Amer M. Abdelaziz, Amr H. Hashem, Kamel A. Abd-Elsalam, and Mohamed S. Attia

Abstract

Endophytic fungi are a diverse group of microorganisms that reside asymptomatically within the tissues of various plant species. This chapter aims to provide a comprehensive overview of the diversity of endophytic fungi and their host plants. The diversity of fungal endophytes is influenced by a multitude of factors, including environmental variables, host tissue, plant phylogeny, geographic location, time of year, farming practices (organic or conventional), and the composition of surrounding vegetation. Agricultural management practices can alter soil properties, pesticide use, and nutrient availability, thereby affecting the diversity and composition of endophytic fungi. The presence of neighboring plants can influence resource availability, competition, and facilitation among these fungi. In conclusion, this chapter offers a detailed exploration of the diversity, ecological relationships, and influencing factors of endophytic fungi. It provides valuable insights into fungal biodiversity conservation and the sustainable management of plant ecosystems.

Keywords

Biodiversity · Endophytic fungi · Environment

A. M. Abdelaziz · A. H. Hashem (✉) · M. S. Attia
Faculty of Science, Botany and Microbiology Department, Al-Azhar University, Cairo, Egypt
e-mail: amr.hosny86@azhar.edu.eg

K. A. Abd-Elsalam
Plant Pathology Research Institute, Agricultural Research Center (ARC), Giza, Egypt

K. A. Abd-Elsalam, S. F. AbuQamar (eds.), *Fungal Endophytes Volume I*,
https://doi.org/10.1007/978-981-97-7312-1_2

2.1 Introduction

Fungal endophyte biodiversity refers to the vast variety of fungal species that reside inside plant tissues including leaves, stems, fruit, seeds, and roots without evidently harming the host plant. These endophytic fungi contribute significantly to plant health and ecological interactions, forming an interesting and intricate ecology. Fungal endophytes live inside plant tissues and are essential to the plant micro-ecosystem because they have a favorable effect on the physiological functions of the plants they inhabit. This includes functions such as hormone synthesis, nutrient acquisition, and defense against invasive pathogens and stress (Attia et al. 2022a, b, c, d, e). Endophytic fungi are thought to be present in practically all plants, including those in a variety of environments such as tropical rainforests, oceans, and deserts, grasses and trees to herbs and shrubs (Niskanen et al. 2023). Due to the potential applications of endophytic fungi in various industries such as bioprospecting, human and plant pathogens biological control, and bioremediation, the relationship between endophytic fungi and their host plants is quite interesting (Hashem et al. 2023b).

The variety of fungal endophytes greatly depends on the host plant's genotype and physiology, the growth season, and environmental conditions. The type and location of the host plant also impact this diversity; endophytic fungal diversity is highest in tropical and subtropical climates. Endophytic fungi, due to their great diversity, produce a large number of bioactive compounds, making them a valuable source of bioactive chemicals for various purposes. Fungal endophytes have a wide range of potential applications, including the production of bioactive substances with medicinal properties and the synthesis of fungal inoculants to promote plant growth, promoting overall ecosystem resilience (Attia et al. 2022a, b, c, d, e). Studying fungal endophyte biodiversity is essential for understanding their ecological roles and potential applications thus the major focus of this chapter is the biodiversity of fungal endophytes and the diverse ecological and mutualistic relationships between endophytic fungi and their host plants.

2.2 Distribution of Fungal Endophytes

Fungal endophytes are a broad category of fungi that live asymptomatically inside plant tissues. These fungi have been identified in a wide range of plant species across diverse global habitats. There may be millions of distinct species of fungal endophytes, according to estimates of their enormous diversity (Rungjindamai and Jones 2024). They belong to several taxonomic groups, including Zygomycota, Ascomycota, and Basidiomycota. Fungal endophytes are found in various ecosystems, from aquatic settings to meadows and from tropical rainforests to deserts. As a result, their distribution is not restricted to any one area or type of habitat (Sonam et al. 2023). The health of plants and the efficiency of ecosystems, which encompass plant development, production, and overall ecosystem dynamics, can be significantly impacted by the distribution of fungal endophytes. Numerous fungal endophytes form mutualistic relationships with their host plants, offering advantages (see Fig. 2.1) such as improved nutrient uptake, increased stress tolerance, and defense against herbivores and diseases (Portman et al. 2024).

Studies show that fungal endophytes exhibit a high degree of host specificity. A particular endophyte species may exclusively inhabit a certain plant species or a group of closely related plants. This specificity suggests that the diversity and distribution of fungal endophytes have been significantly influenced by co-evolutionary processes with their host plants (Hong et al. 2024). The understanding of the diversity and distribution of endophytic fungi has been revolutionized by advances in molecular techniques, such as metagenomics and DNA sequencing. These resources have allowed scientists to delve into the little-known world of endophytic fungi and reveal their enormous diversity in terms of taxonomy and function. However, little is currently known about the ecology of endophytic fungi and their participation in plant communities. Therefore, the complex relationships between fungal endophytes, their host plants, and the environment require further research and studies (Rana et al. 2024).

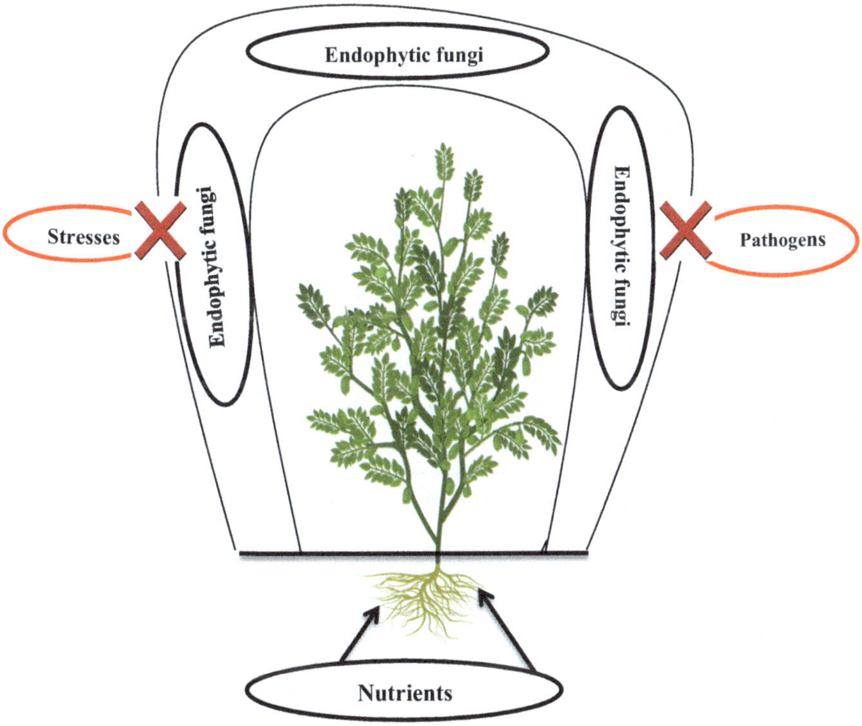

Fig. 2.1 Advantages of mutualistic effect of fungal endophytes on host plants

2.3 Factors Affecting Fungal Endophyte Diversity

The distribution of fungal endophytes is influenced by several factors, including host plant species, habitat, environmental conditions, and the interactions between the fungi and their hosts (see Fig. 2.2).

The presence of specific endophyte species within a plant is influenced by factors such as plant genotype, tissue type, and physiological status. For example, certain fungal endophytes prefer specific tissues, such as leaves, stems, or roots. Environmental conditions, including temperature, humidity, and nutrient availability, also play a crucial role in the distribution of fungal endophytes within plant communities (Guo et al. 2024). The diversity and distribution of fungal endophytes are influenced by various factors, which can be summarized as follows:

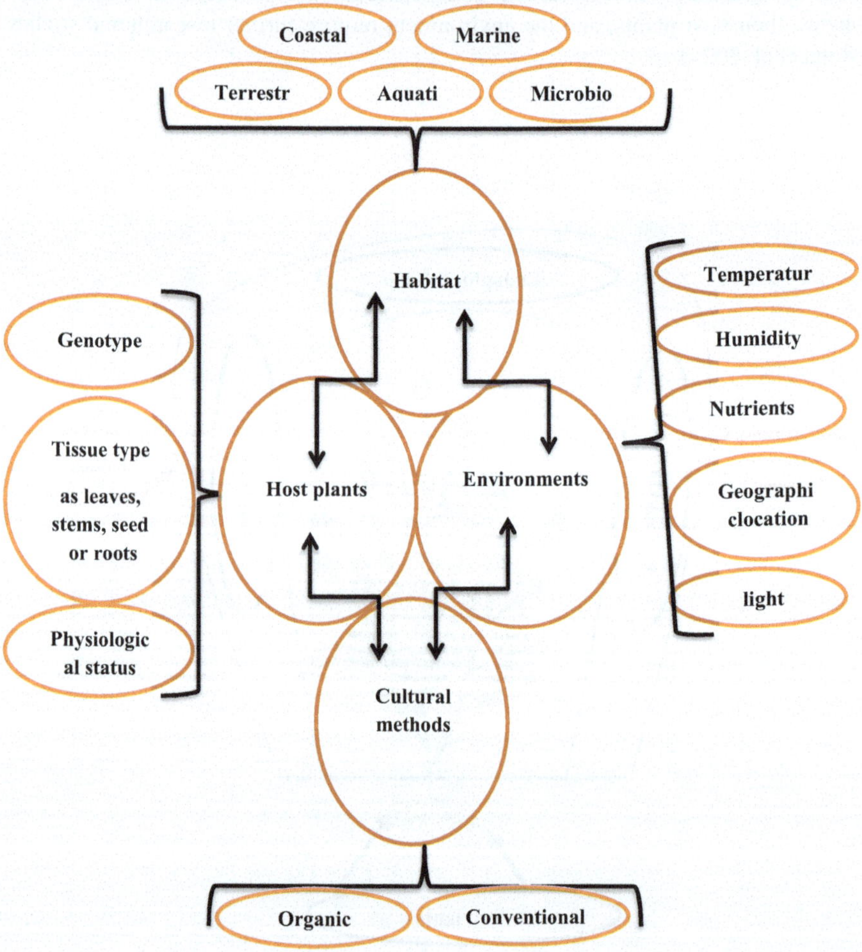

Fig. 2.2 Factors affecting endophytic fungal biodiversity

(a) Host plant species: Different plant species can harbor distinct communities of fungal endophytes. Each plant species provides a unique habitat and nutrient environment that favors the growth and colonization of specific fungal species (Hashem et al. 2023b). The genetic traits and physiological characteristics of the host plant can also influence the types of fungal endophytes that establish symbiotic relationships (see Table 2.1). A broad variety of fungal endophytes are hosted by the diverse group of plants known as tropical orchids. Tropical orchid fungal endophyte populations are impacted by environmental variables, host tissues, and plant phylogeny. There are differences in the relative richness and diversity of fungal endophyte assemblages across plant species in different tissue types (Harrison and Griffin 2020). Endophytic fungi, abundant in eucalyptus trees, are significant contributors to microbial biodiversity. In nature, fungi frequently colonize terrestrial plants, and endophytic fungi can potentially originate from any plant species (Zheng et al. 2016).

(b) Environmental conditions and Geographic location: A variety of environmental parameters, such as temperature, humidity, light availability, soil composition, irrigation rate, and nutrient levels, can significantly impact the diversity of fungal endophytes as well as fungal pathogens (Abdelaziz et al. 2021; Khalil et al. 2015; Muhammad et al. 2023). Changes in these conditions can affect the distribution of certain fungus species, as they may thrive in specific environmental settings. Certain fungal endophytes are more prevalent in tropical rainforests than in other areas, and some have evolved to withstand harsh desert conditions (Li et al. 2024).

(c) Microbiota: The diversity of fungal endophytes may be influenced by interactions among themselves, host plants, and other species. The development and longevity of certain fungal endophytes within plant tissues can be affected by competition with other microbiota, such as bacteria or fungi (Lailheugue et al. 2024). Additionally, herbivores and diseases can influence the specific composition and distribution of fungal endophytes.

(d) Dispersal and evolutionary history: The capacity for dispersal and the evolutionary history of fungal endophytes can impact their diversity. Some fungal endophytes may have coevolved with specific plant lineages, leading to host-specific relationships. The ability of fungal endophytes to colonize new host plants or geographic regions may be determined by dispersal processes such as wind, water, or animal vectors, which can in turn affect the distribution patterns of endophytic fungi (Cheng et al. 2024).

2.4 Methods for Investigating the Biodiversity of Fungal Endophytes

There are many methods used to study the diversity of endophytic fungi, each with its own advantages and applications (see Fig. 2.3). Isolation and culturing is the primary approach for researching fungal endophytes. To isolate these organisms, plant samples must be collected and surface-sterilized to eliminate external

Table 2.1 Effect of host plant on biodiversity of endophytic fungi

Host plant	Endophytes	References
Moringa oleifera	*Aspergillus alabamensis, Aspergillus oryzae, and A. tubingensis*	Attia et al. (2022d)
Moringa oleifera	*Aspergillus fumigatus*	Abdelaziz et al. (2022a)
Ocimum basilicum	*Aspergillus terreus*	Attia et al. (2022c)
Ocimum basilicum	*Aspergillus flavus, Aspergillus fumigatus,* and *Aspergillus nidulans*	Abdelaziz et al. (2022b)
Phaseolus vulgaris	*Aspergillus terreus*	Abdelaziz et al. (2023a)
Avicennia marina	*Alternaria tenuissima, Chaetomium, Aspergillus ochraceus, Aspergillus hiratsukae, Curvularia lunata*	Khalil et al. (2021)
Azadirachta indica A. Juss	*Alternaria alternata*	Chatterjee et al. (2019)
Rosa damascena	*Aspergillus aculeatus*	Farooq et al. (2020)
Acacia nilotica	*Aspergillus awamori*	Adeleke and Babalola (2021)
Artemisia vulgaris	*Chalara* sp.	Lösgen et al. (2008)
Huperzia serrata	*Cladosporium cladosporioides*	Zhang et al. (2011)
Salsola oppostifolia	*Coniothyrium* sp.	Sun et al. (2013)
Artemisia mongolica	*Colletotrichum gloeosporioides*	Zou et al. (2000)
Salsola oppostifolia	*Coniothyrium* sp.	Sun et al. (2013)
Tvipterigeum wilfordii	*Cryptosporiopsis quercina*	Strobel et al. (1999)
Picea mariana	*Diaporthe* sp.	Kouipou Toghueo and Boyom (2019)
Theobroma cacao	*Epicoccum* sp.	Talontsi et al. (2013)
Anvillea garcinii	*Fusarium chlamydosporium*	Ibrahim et al. (2016)
Eleusine coracana	*Fusarium fujikuroi*	Mousa et al. (2016)
Ginkgo biloba	*Fusarium oxysporum*	Jia et al. (2016)
Populus deltoides Marsh	*Hyalodendriella* sp.	Mao et al. (2017)
Sinopodophyllum hexandrum	*Mucor fragilis*	Ejaz et al. (2020)
Cinnamomum zeylanicum	*Muscodor albus*	Ezra et al. (2004)
Smilax china	*Mycosphaerella nawae*	Wang et al. (2017)
Erica arborea	*Nodulisporium* sp.	Dai et al. (2009)
Ginkgo biloba	*Nodulisporium* sp.	Cao et al. (2016)
Terminalia catappa	*Oidium* sp.	Strobel et al. (2008)
Senna spectabilis	*Phaeoacremonium* sp.	Silva et al. (2017)
Solanum surattense	*Penicillium* sp.	Ikram et al. (2019)
Taxus cuspidata	*Periconia* sp.	Kim et al. (2004)

(continued)

Table 2.1 (continued)

Host plant	Endophytes	References
Cistanche deserticola	*Pestalotiopsis theae*	Lin et al. (2008)
Forsythia viridissima	*Pezicula* sp.	Wang et al. (2014)
Crocus sativus	*Phialophora mustea*	Nalli et al. (2015)
Artimisia maritime	*Plectophomella* sp.	Krohn et al. (2007)
Platycladus orientalis	*Phyllosticta spinarum*	Wijeratne et al. (2008)
Chinese mangrove	*Rhizophora mucronata*	Xu et al. (2009)
Rosa varieties	*Seimatosporium* sp.	Zhao et al. (2020)

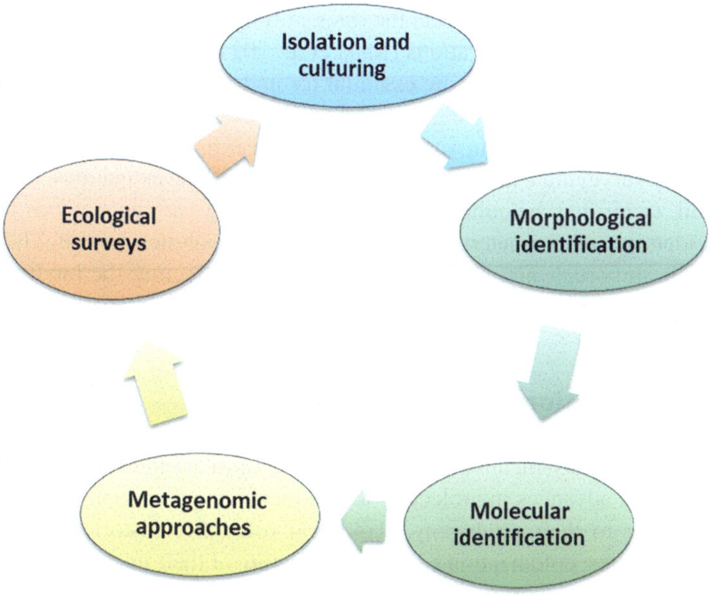

Fig. 2.3 Methods for investigating the biodiversity of fungal endophytes

contamination. Then, tissue segments must be cultured or extracted using various techniques. Molecular methods and other techniques can then be employed to study and characterize the isolated fungal endophytes (Rungjindamai and Jones 2024). Additional useful instruments for researching the diversification of fungal endophytes are microscopy methods. Morphological identification: Researchers can observe the shape of fungal endophytes using light microscopy, which can aid in their identification and categorization (Attia et al. 2023). More sophisticated methods, such as electron and confocal microscopy, enable extensive analysis of fungal endophytes at the cellular and subcellular levels with higher-resolution imaging. These microscopy methods provide visual information that complements other ecological and molecular techniques (Shoayb et al. 2023). Molecular identification: To fully understand the biodiversity of fungal endophytes, molecular approaches are essential. These methods illuminate the evolutionary links and diversity of fungal endophytes. By using DNA-based techniques such as DNA sequencing and polymerase chain reaction (PCR), scientists can identify and categorize fungal endophytes based on their genetic composition. The internal transcribed spacer (ITS) region is one specific DNA region that researchers target to amplify and sequence fungal DNA, enabling the identification of the endophytes' taxonomic classification (Abdelaziz et al. 2023b).

Metagenomic approaches are used to understand the secondary metabolites produced by fungal endophytes and their genetic potential, metabolomics and genomics techniques. Genetics focuses on studying the genetic makeup and potential of fungi, while metabolomics profiles the chemical compounds produced by the endophytes. These approaches illuminate the roles of fungal endophytes in the environment and their potential applications (Rana et al. 2024). Ecological surveys: Field research and ecological surveys are essential for understanding the patterns of fungal endophyte species distribution. Researchers can evaluate the diversity and abundance of fungal endophytes in various plant species, habitats, and environmental conditions by surveying a range of environments. In field investigations, samples of plants, soil, or air may be collected and analyzed using various methods to determine the identity and characteristics of the fungal communities present. These findings provide important new insights into the factors affecting the biodiversity of fungal endophytes and their potential ecological roles (Nguyen et al. 2024).

2.5 Classification of Fungal Endophytes

Fungal endophyte classification is a dynamic and intricate field that has benefited immensely from advances in molecular techniques. Fungal endophytes are categorized according to their phylogeny, ecological roles, life histories, host ranges, reproductive tactics, colonization sites, and the nature of their interactions with their hosts (see Fig. 2.4). The original taxonomy of fungal endophytes divided them into two major groups: clavicipitaceous and non-clavicipitaceous endophytes, based on phylogeny, ecological roles, and life cycle characteristics (Bhunjun et al. 2024). The primary basis for the traditional taxonomic classification of fungal endophytes is their morphological and physiological traits. Endophytic fungi have been identified and categorized into distinct taxonomic groups using morphological characteristics such as spore size, shape, color, and hyphal structures. Additionally, cultural traits have been used for classification, including growth patterns, colony morphology, and biochemical tests (Abdelaziz et al. 2022b; Daigham et al. 2023). However, depending solely on shape and physiology is limited because fungal endophytes can have cryptic or sterile stages that make accurate identification and classification difficult. They can also exhibit phenotypic plasticity. Molecular techniques have supplemented traditional taxonomic approaches based on morphology and physiology, allowing for more precise identification and classification.

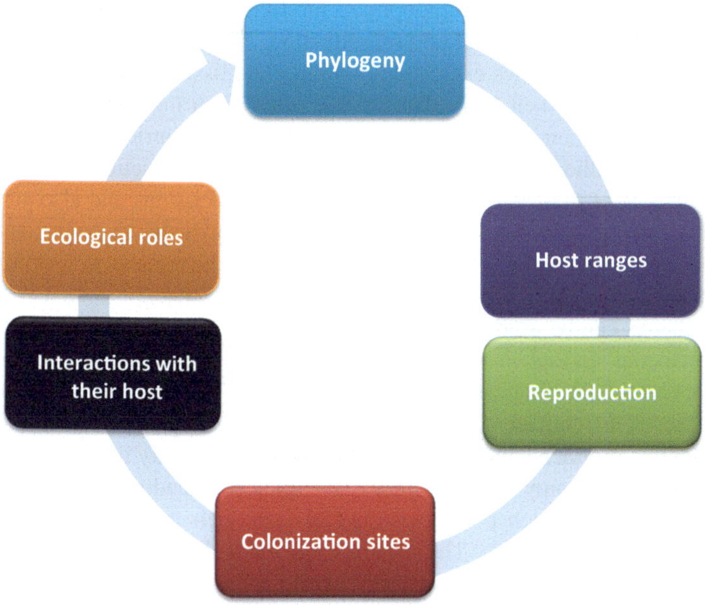

Fig. 2.4 Criteria for categorizing fungi

Fungal endophytes are known to produce various bioactive substances, such as enzymes, antibacterial agents, pesticides, and anticancer drugs. Classification data can aid in understanding biosynthetic pathways, identifying novel bioactive compounds, and streamlining fermentation processes for large-scale production. Furthermore, a robust classification system facilitates focused bioprospecting, allowing scientists to concentrate on specific taxonomic groups likely to contain significant metabolites. Some studies have demonstrated that endophytic Aspergillus fungi can produce secondary metabolites which stimulate plant growth and induce resistance in plants against biotic and abiotic stresses (Attia et al. 2022d; Badawy et al. 2021).

Fungal endophytes are categorized based on the spectrum of hosts they inhabit, which provides important information about their ecological interactions and patterns of distribution. Fungal species known as generalist endophytes can colonize a diverse array of host plants. These endophytes are frequently found in a variety of plant taxa, ranging across families, genera, and even orders. Specialist endophytes exhibit a restricted host range and are typically associated with specific plant species or closely related plant taxa. These endophytes have evolved specific adaptations to colonize and interact with their preferred host plants. Specialist endophytes often play important roles in plant health, growth promotion, and defense mechanisms (Rodriguez et al. 2009). Fungal species that can colonize a variety of host plants for a brief time, but do not form persistent or long-term relationships, are known as transient or opportunistic endophytes. These endophytes can colonize plant species opportunistically and may not be highly dependent on their host plant

(Moricca et al. 2012). Obligate endophytes are fungal species that depend entirely on their host plants for survival and reproduction. These endophytes have evolved to complete their life cycles exclusively within plant tissues and cannot survive or reproduce outside of their host plants. They often form mutualistic or symbiotic relationships with their hosts, providing benefits such as nutrient acquisition, protection against pathogens, and enhanced stress tolerance. Facultative endophytes can colonize plant tissues to establish endophytic associations and also survive as free-living fungi in soil or other environments. They may exhibit varying degrees of host specificity, from generalists that colonize multiple host plants to specialists that prefer specific host species (Schouten 2019).

Four classes can be used to categorize endophytic fungi (see Fig. 2.5): Class 1: Clavicipitaceous endophytes are a class of fungal endophytes belonging to the Ascomycota phylum and the Clavicipitaceous family. Their symbiotic relationships with grasses. Clavicipitaceous endophytes, potentially important for both ecology and agronomy, produce a broad variety of bioactive secondary metabolites, including alkaloids. Taxonomy indicates that clavicipitaceous endophytes belong to the order Hypocreales, which comprises several taxa such as *Embellisia, Epichloë, Neotyphodium*, and *Balansia*. The genera *Epichloe* and *Neotyphodium* include the most researched and economically significant clavicipitaceous endophytes (Leistner and Steiner 2018). Class 2: fungal endophytes, which are not members of the clavicipitaceous family, are known as non-clavicipitaceous endophytes. They are linked

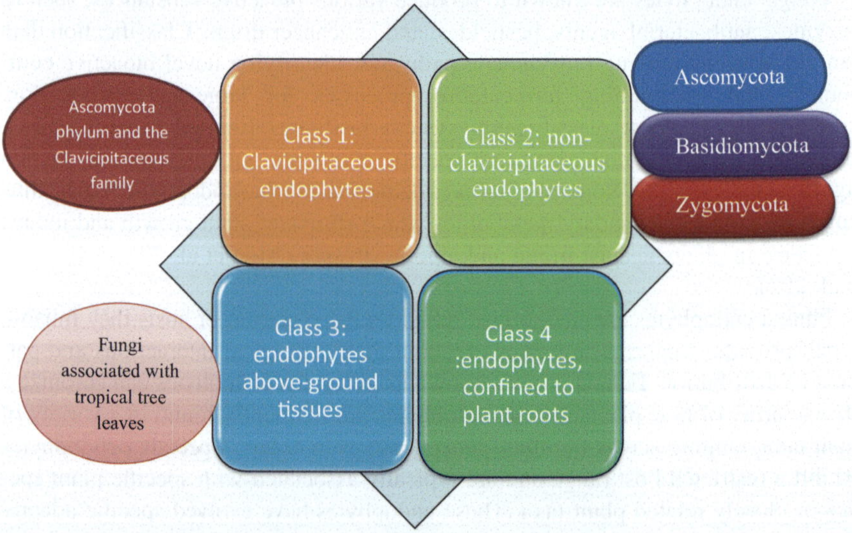

Fig. 2.5 Classes of endophytic fungi including; Clavicipitaceae, Xylariaceae, Hypocreaceae, Chaetomiaceae, and Pleosporaceae are endophytic fungi found in various plant families. They are symbiotic with grasses and plants, producing antimicrobial and anticancer metabolites. They are also associated with bioactive compounds with potential pharmaceutical applications. Chaetomiaceae, found in decaying plant material, plays a role in organic matter decomposition and ecosystem nutrient cycling

to a wide range of plant species, including both grasses and non-grass plants, and encompass a broad spectrum of fungal taxa. Non-clavicipitaceous endophytes exhibit a variety of biological characteristics and produce a large array of metabolites. Within the fungal kingdom, non-clavicipitaceous endophytes are classified into distinct taxonomic groups. These may include several fungal phyla such as Ascomycota, Basidiomycota, and Zygomycota. Non-clavicipitaceous endophytes are further categorized into various orders, families, and genera within these phyla based on their genetic traits. Frequently investigated non-clavicipitaceous endophytes include those belonging to the genera *Phoma, Fusarium, Trichoderma,* and *Colletotrichum* (Christian et al. 2017). Class 3 endophytes are distinguished by their presence primarily or exclusively in above-ground tissues, horizontal transmission, the ability to confer benefits on hosts that are not habitat-specific, and a high level of plant biodiversity. Endophytic fungi associated with tropical tree leaves fall into this category (Rodriguez et al. 2009). Class 4 endophytes, confined to plant roots and characterized by deeply melanized septa, are recognized as a distinct functional category. These ascomycetous fungi, which are either conidial or sterile, produce melanized structures such as intracellular and intercellular hyphae and microsclerotia within the roots. Class 4: endophytes exhibit little host or ecological specialization and have been associated with over 600 plant species (Rodriguez et al. 2009).

Ascomycota, one of the largest and most diverse phyla of fungi, is home to the majority of endophytic fungi. Due to its high degree of environmental adaptability and ability to grow on a variety of plant hosts, Ascomycota is abundant and diverse across terrestrial, marine, and coastal plant habitats. The ascomycetes fungi *Aspergillus, Cladosporium, Penicillium, Trichoderma, Filobasidium,* and *Alternaria* are most commonly found growing inside plants. These fungal genera have been isolated from several plant species and genera at various times of the year and from different parts of the host plants (Hashem et al. 2023a; Ju et al. 2022). *Phaseolus vulgaris* seeds, *Ocimum basilicum, Moringa oleifera, and Avicennia marina* leaves were used for the isolation of the endophytic fungus *Aspergillus*. The by-products of these endophytic fungal isolates have been used in medicine to combat human-pathogenic bacteria and fungi as well as anti-cancer agents. In agriculture, the by-products of these fungal isolates have been used as alternative fertilizers to promote plant growth and increase plant tolerance to both biotic and abiotic stressors (Hashem et al. 2023b). *Viola odorata* roots exhibited the greatest diversity of endophytes, with a dominance of fungi, followed by leaves and leaf nodes. Most of the endophytic fungi from *Viola odorata* belong to the phylum Ascomycota, class Dothideomycetes, including *Colletotrichum and Aspergillus* (Katoch et al. 2017). Class Dothideomycetes is within the phylum Ascomycota. This class includes many plant-associated fungi, such as endophytes, pathogens, and saprophytes (Schoch et al. 2009). A complex and diversified fungal population, including endophytic fungi, is present in Brachiaria and it plays a variety of roles in the host plant fitness and growth. *Acremonium* is the third most common genus after *Microsphaeropsis* and *Cladosporium* (Ghimire et al. 2015).

Moreover, endophytic fungi can be classified according to culture to cultivable and those that are not. In vitro cultivation of cultured endophytic cells enables comprehensive morphological and physiological characterization. Nevertheless, a large number of endophytes are non-cultivable and can only be identified and recognized by molecular methods (Sun and Guo 2012).

2.6 Variations in Seasonal Fungal Endophyte Communities

The richness and composition of fungal endophyte communities are greatly influenced by seasonal dynamics. The improved understanding of climate and its intra-annual fluctuations is reflected in the link between endophyte richness and seasonality. Seasonality also affects the makeup of fungal endophyte communities, indicating that endophytes in the tropics and elsewhere are subject to similar factors mediated by seasonality. Severe dry seasons can act as a physiological filter for these horizontally spreading endophytes, influencing their populations in various ways (Oita et al. 2021). A study published in Frontiers in Microbiology demonstrated that plant identity, plant tissue, and environmental conditions dynamically shape the fungal endophyte communities of crucifer crops, varying with the seasons. The investigation revealed that autumn had the highest colonization rate and abundance, while winter had the lowest. Significant seasonal fluctuations in the assemblage of fungal endophytes were observed in the spring, with higher abundance and species richness compared to summer. The genetic makeup of the host plant, location, time of year, and composition of surrounding vegetation all impact the structure and dispersion patterns of the endophyte community (Chen et al. 2020). It demonstrated how these communities change in response to environmental radiation stress and seasonal variation. The previous study illustrated the effect of seasonal variation on endophyte populations in a desert habitat by presenting the frequencies of co-occurrence connections between endophyte bacteria and fungi across four different seasons (Zhu et al. 2021). Another significant source of endophytic fungi is grasses. Plant identity, plant tissue, and environmental conditions shape the dynamic and seasonal populations of fungal endophytes found in cruciferous crops. Autumn yielded the highest colonization rate and quantity, while winter produced the lowest (Aamir et al. 2020).

2.7 Diversity of Foliar Fungal Endophytes in Plant Communities

The characteristics of the host plants and their geographical location influence foliar endophyte fungal populations. Foliar endophytes differ significantly from other plant-associated, nonpathogenic fungi. Compared to endo-, ecto-, and mycorrhizal fungi, leaf endophytes can be dominated by distinct primary lineages of Ascomycota in certain situations. In other instances, they represent unique genotypes or species whose close relatives colonize plant tissues such as roots (Arnold 2007).

Although a great deal of research has been conducted, we still do not fully understand the variety of fungi that can develop endophytic relationships with leaves.

Endophytic fungi, due to their wide range of associations with plant species, are significant contributors to microbial biodiversity. Fungi frequently colonize terrestrial plants in nature, and endophytic fungi could theoretically originate from any plant (Arnold and Lutzoni 2007).

2.8 Diversity of Root Fungal Endophytes in Plant Communities

Compared to the stem and leaf, the diversity and composition of endophytic fungi in the root can vary. Studies indicate that the growth habit of the host affects the relative richness of fungal endophytes in different tissues, with some plant parts harboring richer fungal endophyte assemblages than others (Rodriguez et al. 2009). The diversity and composition of endophytic fungi in roots can differ from those in stems and leaves, leading to the emergence of new properties in the fungi growing inside the roots. Endophytic fungi are more likely to colonize plant tissues in the roots. The ecological significance of these colonizations lies in their capacity to impact multiple facets of plant health and output. These properties have environmental significance and potential pharmaceutical and agricultural importance (Zuo et al. 2021). A specific siderophore receptor protein enables root endophytic fungi to recognize Fe^{3+}. The siderophore chelates Fe3+ from the soil and transports it into microbial cells. Plant roots are shielded from harmful organisms by siderophore-producing endophytic fungi, which chelate available iron, making it inaccessible to pathogens. In a process known as siderophore-mediated suppression of plant infections, endophytic fungi can be selected for biological control by competing for iron with pathogens that produce fewer siderophores with lower affinity for iron (Abd Alhakim et al. 2022). Endophytic fungi often promote plant growth by colonizing the root system and reducing or preventing harmful Rhizosphere infections. To outcompete pathogens, the endophytic fungus must colonize the plants' Rhizosphere effectively. A well-known requirement for biocontrol is colonization, and the root surface should be colonized by a biocontrol agent. Also, due to their capacity to colonize the surfaces of plant roots, endophytic fungi affect plant metabolism (Harman et al. 2004).

2.9 Symbiotic Associations of Endophytic Fungi with Host Plants

Fungal endophytes form mutualistic associations with many types of organisms, including cyanobacteria, algae, plants, and animals. In mutualism, two species engage in a long-term interaction that is beneficial to both. Fungus–plant mutualism is one of the most well-known examples. Some orchids remain mycorrhizal throughout their life cycle. The absence of symbiotic fungi from the soil can significantly impact plant growth (Abdelaziz et al. 2022a).

In nature's complex web of relationships, organisms often engage in multiple symbiotic interactions simultaneously. There are two basic types of symbiosis: obligate and facultative. In obligate symbiosis, one or both organisms completely depend on the relationship for survival and cannot live without their partner. Conversely, in facultative symbiosis, organisms can exist independently of each other. In natural ecosystems, fungal endophytes and plants appear to work in harmony. The effects of this incredibly varied collection of fungi on plant communities and ecosystems can be substantial. Fungal endophytes can establish mutualistic, symbiotic, or neutral interactions with their host plants. More research is required to close the information gaps regarding the distribution, abundance, and ecological roles of endophytic fungi in a range of plant species and habitats. Endophytes and plants engage in mutualistic partnerships that are beneficial to both parties. Endophytes protect plants from diseases and pests and can accelerate plant growth (Abd Alhakim et al. 2022). These symbiotic relationships shield endophytes from biotic and abiotic challenges while providing protection and, in some cases, nutrition for the plant. This reciprocal interaction is often necessary for the survival of both partners. Certain endophytes can trigger plant defense mechanisms, while others inhibit the growth of diseases by producing and secreting secondary compounds (Attia et al. 2022b). Fungal endophytes and plants can engage in mutualistic or neutral symbiotic interactions, where both parties benefit from the contact, but neither appears to be adversely affected. Endophytic fungi can protect plants against harmful microorganisms through strategies such as competition, antibiosis, mycoparasitism, and induced resistance. Numerous variables, including the identity of the host plant, geographic location, and environmental conditions, affect the symbiotic interactions between endophytes and their hosts. Geographic location, time of year, type of farming (organic or conventional), and the composition of surrounding vegetation all impact the community structure and distribution patterns of endophytic fungi in host plants. Comprehending the ecological role of fungal endophytes is vital to developing this symbiotic relationship, which is considered an effective tool for managing habitats and formulating sustainable farming approaches (Khalil et al. 2015).

2.10 Conclusions

Endophytic fungi play a crucial role in plant ecosystems, residing asymptomatically within the tissues of various plant species. This chapter has provided a comprehensive overview of the diversity, ecological relationships, and influencing factors of endophytic fungi. Additionally, understanding the intricate associations between endophytic fungi and their host plants, as well as the environmental and ecological variables that shape their communities, is essential for endophytic fungal biodiversity conservation. Future studies on endophytic fungi should focus on understanding their physiological functions, examining the effects of climate change, investigating biotechnological applications, and utilizing their potential for ecological restoration.

References

Aamir M, Rai KK, Zehra A, Kumar S, Yadav M, Shukla V, Upadhyay RS (2020) Chapter 12—Fungal endophytes: classification, diversity, ecological role, and their relevance in sustainable agriculture. In: Kumar A, Singh VK (eds) Microbial endophytes. Woodhead Publishing, pp 291–323

Abd Alhakim A, Hashem A, Abdelaziz AM, Attia MS (2022) Impact of plant growth promoting fungi on biochemical defense performance of tomato under fusarial infection. Egypt J Chem 65(13):291–301

Abdelaziz AM, Dacrory S, Hashem AH, Attia MS, Hasanin M, Fouda HM et al (2021) Protective role of zinc oxide nanoparticles based hydrogel against wilt disease of pepper plant. Biocatal Agric Biotechnol 35:102083

Abdelaziz AM, El-Wakil DA, Attia MS, Ali OM, AbdElgawad H, Hashem AH (2022a) Inhibition of Aspergillus flavus growth and aflatoxin production in Zea mays L. using endophytic Aspergillus fumigatus. J Fungi 8(5):482

Abdelaziz AM, Kalaba MH, Hashem AH, Sharaf MH, Attia MS (2022b) Biostimulation of tomato growth and biocontrol of Fusarium wilt disease using certain endophytic fungi. Bot Stud 63(1):1–14

Abdelaziz AM, El-Wakil DA, Hashem AH, Al-Askar AA, AbdElgawad H, Attia MS (2023a) Efficient role of endophytic Aspergillus terreus in biocontrol of Rhizoctonia solani causing damping-off disease of Phaseolus vulgaris and Vicia faba. Microorganisms 11(6):1487

Abdelaziz AM, Sharaf MH, Hashem AH, Al-Askar AA, Marey SA, Mohamed FA et al (2023b) Biocontrol of Fusarium wilt disease in pepper plant by plant growth promoting Penicillium expansum and Trichoderma harzianum. Notulae Botanicae Horti Agrobotanici Cluj-Napoca 51(3):1–23

Adeleke BS, Babalola OO (2021) Pharmacological potential of fungal endophytes associated with medicinal plants: a review. J Fungi 7(2):147

Arnold AE (2007) Understanding the diversity of foliar endophytic fungi: progress, challenges, and frontiers. Fungal Biol Rev 21(2–3):51–66

Arnold AE, Lutzoni F (2007) Diversity and host range of foliar fungal endophytes: are tropical leaves biodiversity hotspots? Ecology 88(3):541–549

Attia MS, Abdelaziz AM, Al-Askar AA, Arishi AA, Abdelhakim AM, Hashem AH (2022a) Plant growth-promoting fungi as biocontrol tool against fusarium wilt disease of tomato plant. J Fungi 8(8):775

Attia MS, El-Wakil DA, Hashem AH, Abdelaziz AM (2022b) Antagonistic effect of plant growth-promoting fungi against Fusarium wilt disease in tomato: in vitro and in vivo study. Appl Biochem Biotechnol 194(11):5100–5118

Attia MS, Hashem AH, Badawy AA, Abdelaziz AM (2022c) Biocontrol of early blight disease of eggplant using endophytic Aspergillus terreus: improving plant immunological, physiological and antifungal activities. Bot Stud 63(1):26

Attia MS, Salem MS, Abdelaziz AM (2022d) Endophytic fungi Aspergillus spp. reduce fusarial wilt disease severity, enhance growth, metabolism and stimulate the plant defense system in pepper plants. Biomass Convers Biorefinery 14:1–11

Attia MS, Salem MS, Abdelaziz AM (2022e) Endophytic fungi Aspergillus spp. reduce fusarial wilt disease severity, enhance growth, metabolism and stimulate the plant defense system in pepper plants. Biomass Convers Biorefinery 14:16603–16613. https://doi.org/10.1007/s13399-022-03607-6

Attia MS, Sharaf MH, Hashem AH, Mahfouz AY, Daigham GE, Al-Askar AA et al (2023) Application of Rhizopus microsporus and Aspergillus oryzae to enhance the defense capacity of eggplant seedlings against Meloidogyne incognita. Notulae Botanicae Horti Agrobotanici Cluj-Napoca 51(3):1–23

Badawy AA, Alotaibi MO, Abdelaziz AM, Osman MS, Khalil AM, Saleh AM et al (2021) Enhancement of seawater stress tolerance in barley by the endophytic fungus Aspergillus ochraceus. Meta 11(7):428

Bhunjun CS, Phukhamsakda C, Hyde KD, McKenzie EH, Saxena RK, Li Q (2024) Do all fungi have ancestors with endophytic lifestyles? Fungal Divers 125:73–73

Cao L-L, Zhang Y-Y, Liu Y-J, Yang T-T, Zhang J-L, Zhang Z-G et al (2016) Anti-phytopathogenic activity of sporothriolide, a metabolite from endophyte Nodulisporium sp. A21 in Ginkgo biloba. Pestic Biochem Physiol 129:7–13

Chatterjee S, Ghosh R, Mandal NC (2019) Production of bioactive compounds with bactericidal and antioxidant potential by endophytic fungus Alternaria alternata AE1 isolated from Azadirachta indica A. Juss. PLoS One 14(4):e0214744

Chen J, Akutse KS, Saqib HSA, Wu X, Yang F, Xia X et al (2020) Fungal endophyte communities of crucifer crops are seasonally dynamic and structured by plant identity, plant tissue and environmental factors. Front Microbiol 11:1519

Cheng Y, Yang J, Li T, Li J, Ye M, Wang J et al (2024) Endosymbiotic fungal diversity and dynamics of the brown planthopper across developmental stages, tissues, and sexes revealed using circular consensus sequencing. Insects 15(2):87

Christian N, Whitaker BK, Clay K, Dighton J, White J (2017) Chapter—A novel framework for decoding fungal endophyte diversity. In: The fungal community: its organization and role in the ecosystem, 4th edn. CRC Press Taylor & Francis Group, Boca Raton, FL, pp 65–78

Dai J, Krohn K, Draeger S, Schulz B (2009) New Naphthalene-Chroman Coupling Products from the Endophytic Fungus, Nodulisporium sp. from Erica arborea. Wiley Online Library

Daigham GE, Mahfouz AY, Abdelaziz AM, Nofel MM, Attia MS (2023) Protective role of plant growth-promoting fungi Aspergillus chevalieri OP593083 and Aspergillus egyptiacus OP593080 as biocontrol approach against Alternaria leaf spot disease of Vicia faba plant. Biomass Convers Biorefinery:1–17

Ejaz M, Javed S, Hamza M, Tabassum S, Abubakar M, Ullah I (2020) Fungal endophytes are effective alternatives and novel sources of anticancer drugs. Punjab Univ J Zool 35:13–24

Ezra D, Hess W, Strobel GA (2004) New endophytic isolates of Muscodor albus, a volatile-antibiotic-producing fungus. Microbiology 150(12):4023–4031

Farooq S, Qayum A, Nalli Y, Lauro G, Chini MG, Bifulco G et al (2020) Discovery of a secalonic acid derivative from Aspergillus aculeatus, an endophyte of Rosa damascena Mill., triggers apoptosis in MDA-MB-231 triple negative breast cancer cells. ACS Omega 5(38):24296–24310

Ghimire SR, Njuguna J, Kago L, Ahonsi M, Njarui DM (2015) Fungal endophytes from the aerial tissues of important tropical forage grasses Brachiaria spp. in Kenya

Guo B, Zhang H, Liu Y, Chen J, Li J (2024) Assembly of cereal crop fungal communities under water stress determined by host niche. Environ Exp Bot 219:105650

Harman GE, Howell CR, Viterbo A, Chet I, Lorito M (2004) Trichoderma species—opportunistic, avirulent plant symbionts. Nat Rev Microbiol 2(1):43–56

Harrison JG, Griffin EA (2020) The diversity and distribution of endophytes across biomes, plant phylogeny and host tissues: how far have we come and where do we go from here? Environ Microbiol 22(6):2107–2123

Hashem AH, Abdelaziz AM, Attia MS (2023a) Trichoderma a promising biofungicide biofungicides. In: Eco-safety and future trends. CRC Press, pp 166–189

Hashem AH, Attia MS, Kandil EK, Fawzi MM, Abdelrahman AS, Khader MS et al (2023b) Bioactive compounds and biomedical applications of endophytic fungi: a recent review. Microb Cell Factories 22(1):1–23

Hong P, Dash CK, Ghafar MA, Haq IU, Lu L, Zhou C et al (2024) Demography and population projection of Tetranychus urticae (Tetranychidae) on Phaseolus vulgaris (Fabaceae) colonized by Entomopathogenic fungal endophytes. Insects 15(1):73

Ibrahim SRM, Elkhayat ES, Mohamed GAA, Fat'hi SM, Ross SA (2016) Fusarithioamide A, a new antimicrobial and cytotoxic benzamide derivative from the endophytic fungus Fusarium chlamydosporium. Biochem Biophys Res Commun 479(2):211–216. https://doi.org/10.1016/j.bbrc.2016.09.041

Ikram M, Ali N, Jan G, Hamayun M, Jan FG, Iqbal A (2019) Novel antimicrobial and antioxidative activity by endophytic Penicillium roqueforti and Trichoderma reesei isolated from Solanum surattense. Acta Physiol Plant 41(9):1–11
Jia M, Chen L, Xin H-L, Zheng C-J, Rahman K, Han T, Qin L-P (2016) A friendly relationship between endophytic fungi and medicinal plants: a systematic review. Front Microbiol 7:906
Ju M, Zhang Q, Wang R, Yan S, Li Z, Li P, Gu P (2022) Correlation in endophytic fungi community diversity and bioactive compounds of Sophora alopecuroides. Front Microbiol 13:955647
Katoch M, Singh A, Singh G, Wazir P, Kumar R (2017) Phylogeny, antimicrobial, antioxidant and enzyme-producing potential of fungal endophytes found in Viola odorata. Ann Microbiol 67(8):529–540
Khalil A, Ahmed AF, Mahmoud EE, Abdelaziz AM (2015) Influence of organic farming system on microbial biomass and fungal communities of agricultural soil. Afr J Mycol Biotechnol 20:23–40
Khalil A, Abdelaziz A, Khaleil M, Hashem A (2021) Fungal endophytes from leaves of Avicennia marina growing in semi-arid environment as a promising source for bioactive compounds. Lett Appl Microbiol 72(3):263–274
Kim S, Shin D-S, Lee T, Oh K-B (2004) Periconicins, two new fusicoccane diterpenes produced by an endophytic fungus Periconia sp. with antibacterial activity. J Nat Prod 67(3):448–450
Kouipou Toghueo RM, Boyom FF (2019) Endophytic fungi from Terminalia species: a comprehensive review. J Fungi 5(2):43
Krohn K, Hussain H, Flörke U, Schulz B, Draeger S, Pescitelli G et al (2007) Massarilactones E-G, new metabolites from the endophytic fungus Coniothyrium sp., associated with the plant Artimisia maritima. Chirality 19(6):464–470
Lailheugue V, Darriaut R, Tran J, Morel M, Marguerit E, Lauvergeat V (2024) The genotypes of the two components of grafted grapevines influence the rhizosphere and root endophyte microbiomes, with rootstocks having a greater impact
Leistner E, Steiner U (2018) The genus Periglandula and its symbiotum with morning glory plants (Convolvulaceae). Physiology and genetics: Selected basic and applied aspects, pp 131–147
Li M, He C, Wei M, Long J, Wang J, Yang X et al (2024) Temporal and spatial dynamics and functional metabolism of dark septate endophytes of Gymnocarpos przewalskii Maxim. in Northwest Desert, China. Appl Soil Ecol 194:105194
Lin Z, Wen J, Zhu T, Fang Y, Gu Q, Zhu W (2008) Chrysogenamide A from an endophytic fungus associated with Cistanche deserticola and its neuroprotective effect on SH-SY5Y cells. J Antibiot 61(2):81–85
Lösgen S, Magull J, Schulz B, Draeger S, Zeeck A (2008) Isofusidienols: novel Chromone-3-oxepines produced by the endophytic fungus Chalara sp. Wiley Online Library
Mao Z, Lai D, Liu X, Fu X, Meng J, Wang A et al (2017) Dibenzo-α-pyrones: a new class of larvicidal metabolites against Aedes aegypti from the endophytic fungus Hyalodendriella sp. Ponipodef12. Pest Manage Sci 73(7):1478–1485
Moricca S, Ginetti B, Ragazzi A (2012) Species-and organ-specificity in endophytes colonizing healthy and declining Mediterranean oaks. Phytopathol Mediterr:587–598
Mousa WK, Schwan AL, Raizada MN (2016) Characterization of antifungal natural products isolated from endophytic fungi of finger millet (Eleusine coracana). Molecules 21(9):1171
Muhammad M, Waheed A, Wahab A, Majeed M, Nazim M, Liu Y-H et al (2023) Soil salinity and drought tolerance: an evaluation of plant growth, productivity, microbial diversity, and amelioration strategies. Plant Stress 100319
Nalli Y, Mirza DN, Wani ZA, Wadhwa B, Mallik FA, Raina C et al (2015) Phialomustin A–D, new antimicrobial and cytotoxic metabolites from an endophytic fungus, Phialophora mustea. Rsc Adv 5(115):95307–95312
Nguyen M-P, Lehosmaa K, Toth K, Koskimäki JJ, Häggman H, Pirttilä AM (2024) Weather in two climatic regions shapes the diversity and drives the structure of fungal endophytic community of bilberry (Vaccinium myrtillus L.) fruit. Environ Microbiome 19(1):7

Niskanen T, Lücking R, Dahlberg A, Gaya E, Suz LM, Mikryukov V et al (2023) Pushing the frontiers of biodiversity research: unveiling the global diversity, distribution, and conservation of fungi. Annu Rev Environ Resour 48:149–176

Oita S, Ibáñez A, Lutzoni F, Miadlikowska J, Geml J, Lewis LA et al (2021) Climate and seasonality drive the richness and composition of tropical fungal endophytes at a landscape scale. Communic Biol 4(1):313

Portman TA, Arnold AE, Bradley RG, Fehmi JS, Rasmussen C, Tfaily MM (2024) Fungal endophytes of the invasive grass Eragrostis lehmanniana shift metabolic expression in response to native and invasive grasses. Fungal Ecol 68:101327

Rana S, Kumar M, Maurya DK, Singh SK (2024) Fungal metagenomics: an emerging approach to determine the imperceptible facets of fungal diversity. In: Applied mycology for agriculture and foods. Apple Academic Press, pp 19–36

Rodriguez R, White J Jr, Arnold A, Redman RS (2009) Fungal endophytes: diversity and functional roles. New Phytol 182(2):314–330

Rungjindamai N, Jones EG (2024) Why are there so few basidiomycota and basal fungi as endophytes? A review. J Fungi 10(1):67

Schoch C, Crous PW, Groenewald JZ, Boehm E, Burgess TI, De Gruyter J et al (2009) A class-wide phylogenetic assessment of Dothideomycetes. Stud Mycol 64(1):1–15

Schouten A (2019) Endophytic fungi: definitions, diversity, distribution and their significance in plant life. In: Endophyte biotechnology: potential for agriculture and pharmacology. CABI, Wallingford, pp 6–31

Shoayb M, Soliman HG, Abdelghany TM, Abdelaziz AM (2023) Occurrence of heavy metals in Qarun Lake and its influence on microbial biodiversity. Al-Azhar J Agric Res 48(2):1–5

Silva GH, Zeraik ML, De Oliveira CM, Teles HL, Trevisan HC, Pfenning LH et al (2017) Lactone derivatives produced by a Phaeoacremonium sp., an endophytic fungus from Senna spectabilis. J Nat Prod 80(5):1674–1678

Sonam W, Liu Y, Guo L (2023) Endophytic bacteria in the periglacial plant Potentilla fruticosa var. albicans are influenced by habitat type. Ecol Process 12(1):57

Strobel GA, Miller RV, Martinez-Miller C, Condron MM, Teplow DB, Hess W (1999) Cryptocandin, a potent antimycotic from the endophytic fungus Cryptosporiopsis cf. quercina. Microbiology 145(8):1919–1926

Strobel GA, Spang S, Kluck K, Hess W, Sears J, Livinghouse T (2008) Synergism among volatile organic compounds resulting in increased antibiosis in Oidium sp. FEMS Microbiol Lett 283(2):140–145

Sun X, Guo L-D (2012) Endophytic fungal diversity: review of traditional and molecular techniques. Mycology 3(1):65–76

Sun P, Huo J, Kurtán T, Mándi A, Antus S, Tang H et al (2013) Structural and stereochemical studies of hydroxyanthraquinone derivatives from the endophytic fungus Coniothyrium sp. Chirality 25(2):141–148

Talontsi FM, Dittrich B, Schüffler A, Sun H, Laatsch H (2013) Epicoccolides: antimicrobial and antifungal polyketides from an endophytic fungus Epicoccum sp. associated with Theobroma cacao. Eur J Org Chem 2013(15):3174–3180

Wang J, Wang G, Zhang Y, Zheng B, Zhang C, Wang L (2014) Isolation and identification of an endophytic fungus Pezicula sp. in Forsythia viridissima and its secondary metabolites. World J Microbiol Biotechnol 30(10):2639–2644

Wang L-W, Wang J-L, Chen J, Chen J-J, Shen J-W, Feng X-X et al (2017) A novel derivative of (−) mycousnine produced by the endophytic fungus Mycosphaerella nawae, exhibits high and selective immunosuppressive activity on T cells. Front Microbiol 8:1251

Wijeratne EK, Paranagama PA, Marron MT, Gunatilaka MK, Arnold AE, Gunatilaka AL (2008) Sesquiterpene quinones and related metabolites from Phyllosticta spinarum, a fungal strain endophytic in Platycladus orientalis of the Sonoran Desert. J Nat Prod 71(2):218–222

Xu J, Kjer J, Sendker J, Wray V, Guan H, Edrada R et al (2009) Chromones from the endophytic fungus Pestalotiopsis sp. isolated from the Chinese mangrove plant Rhizophora mucronata. J Nat Prod 72(4):662–665

Zhang ZB, Zeng QG, Yan RM, Wang Y, Zou ZR, Zhu D (2011) Endophytic fungus Cladosporium cladosporioides LF70 from Huperzia serrata produces Huperzine A. World J Microbiol Biotechnol 27:479–486
Zhao Y, Ji X-L, Shen T, Tang W-T, Li S-S, Zhu Z-Q et al (2020) Fungal endophytic communities of two wild Rosa varieties and the role of an endophytic Seimatosporium sp. in enhancing host plant powdery mildew resistance. Plant Soil 447:553–564
Zheng Y-K, Qiao X-G, Miao C-P, Liu K, Chen Y-W, Xu L-H, Zhao L-X (2016) Diversity, distribution and biotechnological potential of endophytic fungi. Ann Microbiol 66(2):529–542. https://doi.org/10.1007/s13213-015-1153-7
Zhu J, Sun X, Tang Q-Y, Zhang Z-D (2021) Seasonal dynamics and persistency of endophyte communities in Kalidium schrenkianum shifts under radiation stress. Front Microbiol 12:778327
Zou W, Meng J, Lu H, Chen G, Shi G, Zhang T, Tan R (2000) Metabolites of Colletotrichum gloeosporioides, an endophytic fungus in Artemisia mongolica. J Nat Prod 63(11):1529–1530
Zuo Y, Li X, Yang J, Liu J, Zhao L, He X (2021) Fungal endophytic community and diversity associated with desert shrubs driven by plant identity and organ differentiation in extremely arid desert ecosystem. J Fungi 7(7):578

Endophytic Fungi Interactions with Plants

3

Priyakshi Nath, Sibashish Kityania, Rajat Nath, Deepa Nath, and Anupam Das Talukdar

Abstract

Endophytic fungi exhibit beneficiary symbiotic relationship with various parts of its host plant and positively affect substantial number of activities in the host plant like hormone production, nutrient procurement and its biosynthesis for proper growth and development in plants. The interaction of endophytic fungi with plants drives microbes to secrete metabolites that help plants to adapt to abiotic-biotic stress factors and protects plants from herbivore and pathogenic invasion, and in turn, plants supply nutrients and habitat to the endophytic fungi. Fungal endophytes have been reported to secrete various kinds of extracellular enzymes like hydrolytic enzymes which provide resistance mechanism to defend against host plant's response and are capable of degrading many organic compounds, enzymes capable of degrading organic contaminants and reduce level of both phytotoxicity and evapotranspiration of volatile contaminants, as a way to contribute in the process of phytoremediation. Endophytic fungi possess the capability to synthesize a range of bioactive compounds that exhibit different biological activities such as antioxidant, antimicrobial, anti-tumour, anti-diabetic, anti-hypercholesterolemic and anti-proliferative properties, thus showing satisfactory demands in the pharmaceutical, medical, biotechnology, agriculture and other industrial fields. Another significant role played by endophytic fungi is in biotransformation process: Fungi isolated from roots and shoots of *Aphelandra tetragona* plant have been found capable of transforming benzoxazinones 2-benzoxazolinone to 2-hydroxy-1,4-benzoxazin-3-one. Studies revealed that

P. Nath · S. Kityania · R. Nath · A. D. Talukdar (✉)
Department of Life Science and Bioinformatics, Assam University, Silchar, Assam, India
e-mail: anupam@bioinfoaus.ac.in

D. Nath
Department of Botany, Gurucharan College, Silchar, Assam, India

K. A. Abd-Elsalam, S. F. AbuQamar (eds.), *Fungal Endophytes Volume I*,
https://doi.org/10.1007/978-981-97-7312-1_3

inoculation of endophytic fungal strains in plants enhanced water stress tolerance ability, in addition to increase in soluble sugar quantity, proline quantity, enzymatic activity and reduction in reactive oxygen species content and electrical conductivity. Fungal endophytes not only conttributes in various environmental factors but can also provide opportunities in development of novel compounds that would help in meeting increasing requirements globally.

Keywords

Endophytic fungi · Plants · Extracellular enzymes · Secondary metabolites

3.1 Introduction

Endophytes refer to endosymbiotic group of microorganisms which colonize either intercellular or intracellular locations of plants and help to increase host plant's growth and in nutrient accumulation (Gouda et al. 2016). The term endophyte "endo-within; phyto-plant" was coined by Anton De Barry (Father of Modern Plant Pathology) in 1866 (Collinge et al. 2022). Plants, mostly perennial, allow colonization of different types of endophytic microorganisms, which reside in tissues either for a certain time-period of life cycle or for their entire lives, without causing any morphological or visible damage to the host plants (Alam et al. 2021). Such endophytic microorganisms include both genres of bacteria and fungi, called as endophytic bacteria and endophytic fungi respectively. Endophytic fungi are studied for over a century now, and the first endophyte was identified and collected from *Lolium temulentum L.*(ryegrass) in 1898 (Vogl 1898). Endophytic fungi are fungal microorganisms which spend either a part or their entire life cycle colonizing the tissues of plants intracellular/intercellular, without causing any apparent disease symptoms (Radić and Štrukelj 2012). These organisms have the potential to positively influence the plant's physical activities in many ways, and in turn, plants supply nutrients and habitat to them (Wen et al. 2022). A few genera of endophytic fungi include *Aspergillus, Bipolaris, Cladosporium, Diaporthe, Fusarium,Mucor, Nigrospora, Penicillium, Piriformospora, Trichoderma, and Ulocladium* (del Carmen Orozco-Mosqueda and Santoyo 2021). Endophytic fungi exhibit high biodiversity with ubiquitous appearance in the environment. Studies have reported that majority of plants allow colonization of endophytes in them whether they are colonized plants in tropical rain forest, in desert, in Arctic and Antarctic region or plants of marine ecosystem (Ding et al. 2015). Fungal microorganisms can enter plant tissues through stomata, tissue wounds, root cracks, germinating radicles etc. Endophytic fungi, in order to colonize plant tissues, pave their entry through plant seed, roots, leaves, stomata, hydathodes, wounds etc., with the enzymes secreted using different biotic and abiotic mechanisms (Ji et al. 2022) and enter into the plant cell wall. The plant cell wall then undergoes modification and flexibility, thus allowing infiltration of plasma membrane and entry of fungal microbes inside the plant. However, endophytic fungi remain enclosed within the membrane compartment and, therefore, they do not come in direct contact with the cytoplasm of plant (Geurts et al. 2016). Molecular exchange between endophytic fungi and plant cytoplasm occurs through a functional compartment known as symbiotic interface (Lu et al. 2021). Transmission and colonization of endophytic fungi into plants through hyphae fragmentation or spores or by biotic and abiotic dispersive agents called as horizontal transmission (Wiewióra et al. 2015) and vertical transmission shows propagation of fungal microorganisms through seeds to enter into the next generation of plants whilst staying active in parent plants themselves (Hodgson et al. 2014). Endophytic fungi causes degradation of plant cell wall by secreting degradative enzymes namely cellulases, laccases, pectinase etc., thus, allowing infiltration, colonization and proliferation of fungal tissues within plant cell (Chow and Ting 2019). Systematic and expanded colonization of fungal endophytes in their specific host plants occurs in root region mostly, then in aerial leaves or stem. Endophytic fungi uses three different ways of interactions with host plants –mutualistic association, commensalistic association and neutralism association. Figure 3.1 discusses the way of different types of interaction whether beneficiary or no effect between the endophytic fungi and plant species. All these modes of interactions show increased concentration of fungal strains, with no palpable damage or effect in host plant. Fungal endophyte and plant interaction is based on stable and peculiar way of adaptation which allows them to develop harmony along with significant growth. Once, fungal endophytes enter inside the tissues of host plants, they establish quiescent or latent period, regardless for entire lifetime of the host plant or for certain period of time, until environmental conditions become favourable for endophytic fungi to proceed their activities (Kaur 2020). Mutualistic interaction between endophytic fungi and host plants allow beneficiary perpetuity with progress in evolutionary and ecological areas (Jia et al. 2016). Endophytic fungi promote growth and nutrient acquisition, improve drought and metal tolerance and modify metabolic mechanism of host

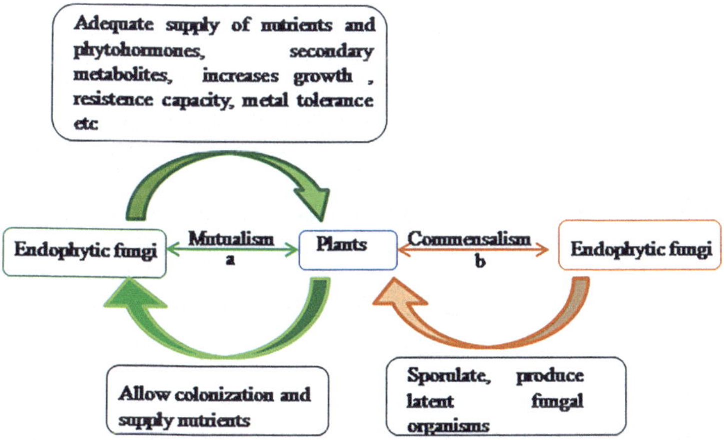

Fig. 3.1 Types of association—(*a*) Mutualistic association and (*b*) commensalistic association—between endophytic fungi and plants

plants (Poveda et al. 2021). They also increase the resistance capacity of the specific host plants against pathogens, pests and herbivores. In turn for mutualistic relationship, endophytic fungi provide the plants symbiotic shelter, adequate supply of nutrients and protection to complete their life cycle during colonization period (Alam et al. 2021). Endophytic fungi also benefit host plants by supplying adequate phytohormones such as auxin, cytokines and gibberellins, which are essential for proper plant growth (Kumar and Kaushik 2013). The resultant advantage of such symbiotic interaction allowed better growth and rate of survivality than non-symbiotic interaction between endophytic fungi and host plants. In plant–endophytic microbes association, endophytes escape pattern recognition receptors by the progression of microbial associated molecular patterns and the modifications of cell wall components thus causing suppression of plant immune system. Also, the endophytes secrete effectors such as extracellular ATP and apoplastic protein effectors to interfere with the plants immune signaling mechanism (Lu et al. 2021). All these strategies are pursued by the endophytes to establish beneficiary symbiotic signaling system with its host plants. Upon entering the immune system of plant body, plants recognize individual receptors (viz., chitin which is fungal cell wall) and initiate immune response (Mengistu 2020). Plants contain genetically engrained innate immune arrangement which usually responds to pathogen effectors and to MAMPs (microbe associated molecular patterns) or PAMPs (pathogen associated molecular patterns or DAMPs (damage associated molecular patterns) (Hacquard et al. 2017). Identification of MAMPs or PAMPs or DAMPs induces the first line of defense called as PTI (pattern-triggered immunity), which plays an integral role in controlling infection dispersed by adapted pathogens and prevent non-adapted pathogens to infect the susceptible host (Saijo et al. 2018; Lu et al. 2021). Many receptor-like protein and receptor-like kinases are deployed by plants as PRRs (pattern recognition receptors) (Tang et al. 2017). Receptor kinases domain structure is similar to the metazoan RTKs with extracellular domain which binds ligands, cytosolic protein kinase domain and transmembrane helix with single span (Jamieson et al. 2018) and receptor proteins which form bipartite receptors along with adapter RTKs, with cytomplasmic kinase domain lacking in it (Liebrand et al. 2013). Because of such domain structures, plasma membrane localized pattern recognition receptor complexes permit extracellular molecules to bind and interact across membranes into cytoplasmic matrix signaling events (Köster et al. 2022). Plant cell wall degraded components and compounds that arise from damaged cells or stressed cells, sometimes, behave as plants DAMPs signaling molecules (Mhlongo et al. 2018). BIK1 (botrytis-induced kinase1) effector kinase of PRR complex gets activated upon pathogen interaction, causing a surge in cyclic nucleotide-gated channel (CNGC) cytosolic Ca^{2+} level signaling to trigger PAMPs immunity at the time of PTI in plants (Tian et al. 2019). Pathogens engage wide range of virulence effectors in order to control pattern-triggered immunity and successfully constitute effector-triggered susceptibility (ETS) (Saijo et al. 2018; Naveed et al. 2020). In such cases, plant immune system initiates effector-triggered immunity (ETI) inside the plant cell to exhibit an amplified and powerful defense program against the pathogens

(Cui et al. 2015) by operating various defensive mechanisms (Lu et al. 2021). It is essential to realize that calcium (Ca^{2+})formation occurs as plant defense response at the time of PTI and ETI responses (Bhar et al. 2023). A momentary elevated peak of cytosolic Ca^{2+} ions for fraction of seconds is observed at the time of PTI while such transient elevated peak of Ca^{2+} ions is followed by persistent accumulation of cytosolic Ca^{2+} ions at the time of ETI (Marcec et al. 2019). Thus, pattern-triggered and effector-triggered immunity can influence plant defense mechanism by altering ion influx/efflux across plasma membrane, enhancing cytosolic Ca^{2+} ions, nitric oxide accumulation, stomatal closure, phytohormones formation, regulating ROS (reactive oxygen species) levels, active signaling of mitogen-activated protein kinase etc. (Yu et al. 2017; Lu et al. 2021). On recognition of MAMPs/DAMPs, PRR complex activates intracellular Ca^{2+} signaling with Ca^{2+}-dependent protein kinase as intermediate and defensive transcriptional factor (Yuan et al. 2017). Also, microbial extracellular ATP has the potential to induce Ca^{2+} signaling, and all of these transcriptional events affect the microbial (fungal/bacterial) colonization in plant body, rendering to be beneficiary for both survival of plant and endophytes (Yu et al. 2017). Pathogens tend to release certain types of toxins that are identified and removed by several classes of receptors and resistance genes (R) at the time of effector-triggered immunity response (Bhar et al. 2023). Effector binding proteins represent NLRs (nucleotide binding leucine rich repeat receptors) in plants which may help to develop immunogenic memories utilizing different mechanism of action (Bhar et al. 2021). LRRs (leucine rich repeats)–RLKs activate immune signaling to assist in plant resistance against microbial pathogens. LRRs-RLKs form complex with BAK1 which is a co-receptor of BRI1-associated kinase 1 necessary for activation of phytohormones signaling along with plant immunity against foreign pathogens (Yang et al. 2011), while LRR-RLP associate with SOBIR1 (suppressor of BIR1) in order to initiate downstream immunogenic signaling activities (Van Der Burgh et al. 2019). The primary immune system engages a series of signaling mechanisms and instigates intracellular cross-talks in order to create sustainable resistances against pathogens in plant system. In commensalistic association, endophytic fungi undergo sporulation and interact with host plants with or without significant beneficiary effects (Vionnie et al. 2016). Such endophytes prevail as latent pathogens in host plants under normal conditions while a few genera of endophytic fungi behave as active disease-causing agents under unfavourable physiological stress conditions (Photita et al. 2004). Such latent endophytic fungi reside inside the tissues of host plants long before the breakout of symptoms of the diseases. The latent period is important as it signifies time period during which endophytic fungi acts either harmless or as virulent pathogen (Alam et al. 2021). Factors such as malnutrition, abiotic and biotic stress, drastic climate change and senescence highly influence the balance between endophytic fungi and their host plants, causing endophytic fungi to transmute from latent mode prior to virulent mode but there hadn't been observed any obvious disease symptoms before the transitioning phase (Poveda et al. 2020). Several studies are carried out to isolate and distinguish endophytic fungi from various parts of plant structure such as leaf segments, roots, bark of stem, petiole, leaf buds and scale primordia (Manganyi 2018). A comparative study on ripe teak plant leaves and giant thibet plant showed endophytic fungi with diversified classes and high colonization incidence in comparison to the younger leaves (Chareprasert et al. 2006). Moreover, frequency of endophytic fungus increases during rainy season (Manganyi 2018). Genetic makeup of the host plant simultaneously affects the distribution pattern of endophytic fungi, while differences in geographical location and environmental conditions in which the host plant resides have resulted in variation of endophytic fungi communities. Therefore, all such factors significantly affect the growth of endophytic fungi within its host plant (Hashem et al. 2023). A thorough understanding of several crucial factors that affect host plants' growing conditions is essential as it impacts the colonizing properties of fungal endophytes and synthesis of bioactive compounds by them (Rai et al. 2021). Screening of endophytic fungi for essential secondary metabolites has led to discovery of useful compounds with beneficial biological activities. The ability of fungi to mimic various structures and functions of host species provided the scope to initiate large-scale screening of endophytic fungi (Strobel 2003). Fungal endophytes have high diversity and show relatively fast growth when employed on artificial media (Strobel 2006; Nair and Padmavathy 2014). Synthesis of bioactive compounds in plants by endophytic fungi ensures quality growth and sustainability and also provides protection in harsh environmental conditions (Rashmi et al. 2019). The discovery of novel bioactive compounds from endophytic microorganisms is an essential alternative way to overcome increased levels of drug resistance to several pathogens. Around 20,000 bioactive compounds isolated from microbes were studied till the year 2002 (Selvakumar and Panneerselvam 2018). Discovery of paclitaxel produced from endophytic fungus *Taxomyces andreanae* aroused attention of mycologists, pharmaceutical chemists and researchers on endophytic fungi to consider it as a reliable source of plant secondary metabolites and as new source of bioactive compounds (Wen et al. 2022). Another study that ignited focus on this area occurred with the discovery of the anticancer drug Taxol from the bark of *Taxus brevifolia* (Stierle et al. 1993). Endophytic fungi gained immense attention because of their capacity to secrete novel bioactive agents which exhibit various biological properties, and so, they are highly recommended in different sectors such as agriculture, medicine and pharmaceuticals (Manganyi and Ateba 2020). Endophytic fungi have gained immense interest not only as novel source of cytotoxic compounds but also as biostimulants for the biosynthesis of essential oil. Several studies with bioactive compounds are conducted throughout the world, mainly to enhance the need to search new efficient compounds with cost-effectivity, in order to contribute in treating human diseases and ultimately benefit mankind. In this review, we discuss about endophytic fungi's ability to serve as source of secondary metabolites and their role as biocatalysts in regulating phytoremediation, degrading and detoxifying organic compounds etc., all in association with herbaceous or medicinal plants as a way to remediate environment and provide sound health to for humans.

3.2 Endophytic Fungi: An Alternative Way for Valuable Plant Bioactive Compounds

Endophytic fungi comprises of wide range of bioactive compounds which possess the potential to mimic plant based secondary metabolites. Host plants significantly affect the changes in fungal colonization and extraction of fungal derived bioactive molecules (Rai et al. 2021). Such bioactive molecules proved to be useful for drug discovery to treat various ailments in humans. Although plants allow endophytic fungi to harbour in them, only a fraction of endophytes have the ability to synthesize essential metabolites. Secondary metabolites produced from endophytic fungi are classified into different classes such as polyketides (Zhao et al. 2023), terpenes, non-ribosomal peptides (Singh et al. 2017), ribosomal peptides and hybrid metabolites (Cruz et al. 2020). These classes synthesize particular set of enzymes required for the synthesis of secondary metabolites encoded by the cluster of genes called as biosynthetic gene clusters (BGC). Such biosynthetic gene clusters can exhibit either of two ways. First way, various types of gene clusters when work collectively, complex secondary metabolites are formed. Second way is the production of secondary metabolites occur when gene clusters remain situated adjacent to one another in fungal genome (Jha et al. 2023; Nützmann et al. 2018). Clustered genes contain spatial arrangement which are considered to be a fundamental requirement for synthesis of bioactive molecules linked to signaling pathway. The reason for such necessity is because the genes are regulated in close proximity through genomic arrangements in the signaling events (Osbourn 2010). The resultant gene structure may permit beneficial amalgations of alleles to co-inherit at different gene loci. Essential genes necessary for production of primary metabolites remain dispersed in the fungal genome while genes that encode enzymes for production of secondary metabolites which are required for regulating metabolic pathways in a continuous fashion (Cruz et al. 2020). Few notable and beneficial plant-derived compounds that are synthesized by endophytic fungi in synergistic interaction with plant species include paclitaxel, vinca alkaloids, saadamycin, cryptocandin, vincristine, emodin and azadirachtin (Rai et al. 2021). There are different mechanisms to study fungal biosynthetic gene clusters. One such strategy involves management of transcription and translationary mechanisms which are essential to regulate mutational changes in RNA POL(polymerase) and protein functions (Pfannenstiel and Keller 2019) while another strategy incorporates manipulation of global regulators to initiate activation of more than one pathway in order to ascertain multiple secondary metabolites (Baral et al. 2018). There have been discussed specific approaches to regulate BGCs expression. First, this approach needs adequate knowledge of genes necessary to encode assumed pathway which are specific for TFs, and regulating of such factors is of primary importance in order to increase BCGs' expression. Second, target biosynthetic gene clusters require a reporter-guided mutant selection pathway which permits prominent visualization of mutant. Third, refactoring by employing constitutive or inducible promoters instead of natural promoters. Fourth, once the interested BCG has been identified, genes get activated and express themselves in diverse hosts (Rutledge and Challis 2015; Pillay et al. 2022). However, approach of heterologous expression facilitates easier metabolite identification workflows and it would prove to be a better alternative approach for fungal biosynthetic gene expression (Pillay et al. 2022). Heterologous expression allows cryptic BGCs access from any organism and such approach particularly, focusses on set of bioinformatic tools for identification and classifying of BGCs in genomic data; assemble and expressing of synthetic DNA using biological tools in diverse host and restructuring BGCs expression using genetic tools (Harvey et al. 2018). Although using such genetic approaches has brought success in studying fungal species, their applications in fungi and beneficial interactions with other species are limited mainly due to lacking of genetic tools necessary for application in filamentous fungal organisms (Jiang et al. 2021). Different classes of functionalized metabolites that are studied and aggregated from various endophytic fungi with their biological activities are discussed as follows.

1. Polyketides: Polyketide compounds represent a group of secondary metabolites that exhibit significant diversity in their structure and function (Risdian et al. 2019). These compounds are formed as condensed activated primary metabolites that result into β-ketoacyl polymers which are joined to enzymes by thioester bonds (Santiago et al. 2014). They exhibit notable biological activities like immunosuppresion and antifungal, antiparasitic and anticancer activities (Debbab et al. 2011). Secondary metabolites reported as derivatives of polyketides include macrolides, α-pyrones, α-pyridones, chromones, phomaketides and pseurotins, detected from different genuses of endophytic fungi (Wang et al. 2018). Chromones and their derivatives isolated from specific host plants and endophytic fungi exhibited antagonistic antimicrobial activity towards certain pathogens and proved useful in reducing plant resource depletion in the ecosystem (Wen et al. 2022). 15 polyketides compounds are isolated from the culture of ascomyete *Chaetomium seminudum*, of which chaetosemins A (1) configuration was confirmed employing X-ray crystallography technique, compound 1 and compound 2 showed presence of L-cysteine and D-cysteine units, chaetosemin B showed antifungal activity against fungal species *Magnaporthe oryzae* and *Gibberella saubinettii* with minimum inhibitory concentrations of 6.25 μM and 12.5 μM respectively, while the compound chaetosemin C exhibited 50.7% DPPH free radical scavenging antioxidant activity at 50 μM and compound chaetoquadrin J showed weak sEH (soluble epoxide hydrolase) inhibition (Li et al. 2015). *Diaporthe* fungi and its anamorph *Phomopsis* have been valuable sources of secondary metabolites. A study reported 331 bioactive compounds isolated from the genus *Diaporthe* and *Phomopsis*, out of which, 143 compounds obtained from *Diaporthe* comprised 19.6% polyketides while 188 compounds isolated from *Phomopsis* consisted of 22.3% polyketide natural compounds (Jiang et al. 2023). Another study showed that two nigerapyrones compounds, nigerapyrone A and nigerapyrones B, collected from *Aspergillus niger* MA-132 which was collected from the mangrove host plant *Avicennia marina*, exhibited antagonistic antifungal activity towards tumour cell lines

(viz., HL60, A549) in preliminary biopsies with IC50 values ranging from 0.3 μM to 5.41 μM (Liu et al. 2013).

2. Alkaloids: Till today, a number of alkaloids are studied and discovered from endophytic fungi in their specific host plants which exhibit diverse biological properties such as antifungal, antitumor and antiviral properties. Camptothecin (CPT) is an alkaloid—cytotoxic compound produced by number of endophytic fungi viz., *Fusarium solani, Entrophospora infrequens, Neurospora* sp., *Nodulisiporium* sp. in roots of *Nethapodytes* and *Apodyte* grass species (Zhang et al. 2012). Alkaloid 7-*O*-methylvariecolortide A along with variecolortides A, variecolortides B, variecolortides C were isolated and identified from culture *Eurotium rubrum*, collected and isolated from the inner tissues of *Hibiscus tiliaceus* stems (Li et al. 2010). Around nine alkaloid compounds viz., diaporphasine A, diaporphasine B, diaporphasine C, diaporphasine D, meyeroguilline A, meyeroguilline C, meyeroguilline D, 5-deoxybostrycoidin and fusaristatin A, were isolated from endophytic fungus *Diaporthe phaseolorum* SKS019 in a study where all the compounds exhibited cytotoxic activity using in vitro human cell lines, while 5-deoxybostrycoidin and fusaristatin A exhibited cytotoxic activity and growth-inhibitory activity against human cancer cell lines (Cui et al. 2017). Chaetomine is another alkaloidal compound characterized from fermentation culture of an endophytic fungus *Chaetomium* sp., apparently on the leaves of *Adenophora axilliflora* plant, which exhibited cytotoxic activity against human leukemia K562 and colon cancer SW1116 cell lines (Jiao et al. 2006). Certain alkaloids isolated from endophytic fungi show broad spectrum insecticidal activity, having significant roles in the agricultural, medicinal and food industrial sectors. Many fungal endophytes synthesize alkaloids which either increases hosts' resistance to insects or found to be toxic for the grazing mammals.

3. Terpenoids: Terpenoids represent a group of secondary metabolites which contain terpene molecules, synthesized by plants and fungal endophytes (Azizan et al. 2016). Gibberellic acid (GA3) is a diterpenoid, isolated from the fungus *Gibberella fujikuroi*, which is an essential plant growth hormone and plays an effective role in seed germination and cell proliferation (Datta et al. 2023). Biotechnological application has resulted into huge industrialization of gibberellic acid in recent times, as the fungus *G. Fujikuroi* synthesizes higher amount of GA3 in comparison to plants.

   Sesquiterpene compound trichoderic acid, cyclonerodiol, 2b-hydroxytrichoacorenol, cyclonerodiol oxide and sorbicillin were collected from *Trichoderma* sp. PR-35 culture, the fungal endophyte being isolated from *Paeonia delavayi* stems. These compounds exhibited moderate to weak antimicrobial activities when tested using agar-diffusion method (Wu et al. 2011). A study reported Fusidic acid being collected from the endophytic fungus *Acremonium pilosum* F47 culture, endophyte being isolated from *Mahonia fortunei* stem. Fusidic acid showed essential antimicrobial activity when tested and evaluated against four selected bacterial pathogens (*Staphylococcus aureus* ATCC 6538, *Bacillus subtilis* ATCC 9372, *Escherichia coli* ATCC 25922), of which Fusidic acid exhibited strong inhibitory effect on *Staphylococcus aureus* and *B. Subtilis* microorganisms (Tian et al. 2021).

4. Others: *Phomopsis* sp. is a rich source of bioactive metabolites. (14b,22E)-9,14-dihydroxyergosta-4,7,22-triene-3,6-dione,(5a,6b,15b,22E)-6-ethoxy-5,15-dihydroxyergosta-7,22-dien-3-one, with calvasterol A, calvasterol B as well as ganodermaside D compounds, all these were identified and isolated from the culture of endophyte Phomopsis sp., isolated from the chinese plant Aconitum carmichaeli, upon experimentation, have exhibited moderate to weak antifungal activity (Wu et al. 2013). Ergosterol compound Fusaristerol A 200 collected from rice culture medium *Fusarium* sp. endophytic fungus, isolated from *Mentha longifolia* L. plant roots. Fusaristerol A compound showed notable antifungal property against *Candida albicans* with 8.3 μg/disc minimum inhibitory concentration (MIC) value. Moreover this compound showed cytotoxic activity against HCT-116 cell line with 0.21 μM IC50 value, thus behaving as a promising compound for potent cytotoxic and antimicrobial activity (Mohamed Ibrahim et al. 2018). Butenolide compounds namely Terrusnolide CA, Terrusnolide B, Terrusnolide C, Terrusnolide D from endophyte Aspergillus sp. (isolated from roots of chinese herb Tripterygium wilfordii) which showed potent inhibition on the synthesis of nitric oxide, tumor necrosis factor , IL-1 in polysaccharide induced RAW264.7 macrophage cells with IC50 values arraying (Qi et al. 2018).

Nowadays, many drugs are formed successfully from endophytic fungi origin namely, lovastatin, griseofulvin, cyclosporine A and β-lactam antibiotics, which are used in treating ailments (Selvakumar and Panneerselvam 2018). Isolation of bioactive compounds from the extracts of endophytic fungi for purification and characterization are done using sophisticated processes viz., thin-layer chromatography (TLC), high-performance liquid chromatography (HPLC), fast-atom bombardment, nuclear magnetic resonance (NMR), electron spray ionisation (ESI) and matrix-assisted laser desorption/ionisation–time of flight (Rai et al. 2021). Compounds isolated from endophytic fungi exhibited similar biological properties to that of plant species, validating the ability of endophytic fungi as an alternative source of bioactive molecules other than plants. Although isolation and identification of endophytic fungi is relatively an integral method, screening of quality fungi which possess the ability to synthesize essential metabolites is, however, a clumsy process (Vasundhara et al. 2016). So, there are chances of missing essential strains of endophytic fungi which have the potential to produce important secondary metabolites. Various methods are utilized for the screening of fungal endophytes by bioassays that helps in purification, characterization of secondary metabolites and determine biological activities of the metabolites (Garyali et al. 2013). Table 3.1 represents the summary of compounds isolated from various endophytic fungi that exhibit similar activity as plants.

Endophytic fungi act as a reservoir of bioactive compounds. Bioactive compounds isolated and derived from such fungal endophytes are mostly used for drug discovery to help in administering and treating many diseases. Synthesis of novel compounds from endophytic fungi raise questions about innate ability of these

Table 3.1 Bioactive compounds produced by endophytic fungi in association with its host plants

Endophytic fungi	Host plant	Bioactive compound	Bioactivity	Reference
Lophiostoma Sp Eef-7	*Eucalyptus exserta*	Scorpinone, 5-deoxybostrycoidin, 4-methyl-5,6-dihydro-2 *H*-pyran-2-one (compounds 1–3)	Antimicrobial	Mao et al. (2021)
Cryptosporiopsis quercina	Tripterigeum wilfordii	Cryptocandin (compound 4)	Antifungal	Strobel et al. (1999)
Phomopsis sp	Odontoglossum sp	Sabinene (compound 5)	Antifungal	Singh et al. (2011)
Serratia marcescens	*Rhyncholacis penicillata*	Oocydin A (compound 6)	Antioomycetous	Strobel (2003)
Aspergillus allahabadii	*Cinnamomum subavenium*	Allahabadolactone A, Allahabadolactone B, cytochalasin D (compounds 7–9)	Antibacterial, cytotoxicity	El-Hawary et al. (2020)
Chaetomium sp.	*Salvia officinalis*	Cochliodinol, Isocochliodinol (compounds 10, 11)	Cytoxicity	Debbab et al. (2009)
Talaromyces radicus	*Catharanthus roseus*	Vincristine, vinblastine (compounds 12, 13)	Cytotoxicity	Palem et al. (2015)
Phaeosphaeria Fuckelii	*Phlomis Umbrosa*	Phaeosphaones A, Chetoseminudin B, Polanrazine B, Leptosin D (compounds 14–17)	Tyrosinase inhibitory activity	Zhai et al. (2020), Momtaz et al. (2008)
Alternaria sp	*P. incarnata*	Chrysin (compound 18)	Anticancer, antioxidant	Seetharaman et al. (2017)

fungal organisms to produce secondary metabolites (Vasundhara et al. 2016). Production of secondary products causes physiological alterations in host plants, which not only strengthens the immune system of plants (Alam et al. 2021) but also assist the host plants in their chemical defence actions (Ogbe et al. 2020). Studies reported that bioactive compounds yielded by endophytic fungi synthesize gene encoding enzymes and biosynthetic pathways parallel to that of host plants (Vasundhara et al. 2016). In order to surpass the hurdles in study of secondary metabolites, it is vitally important to adopt interdisciplinary methods. Omics represents scientific areas which measure biological compounds in an elevated throughput process such as phenomics, metatranscriptomics, metabolomics, epigenomics and metagenomics. All these fields when studied in a collective manner help immensely in the discovery of bioactive compounds from various biological fields, which is called the multi-omics approach (Jha et al. 2023). With multiple-omics approach, scientific experiments can be done distinctly to regulate the structure and function of the secondary metabolites (Hasin et al. 2017). One such approach is using genomics in the study and synthesis of secondary metabolites. Whole genome sequencing helps in considering genes which instigate either directly or indirectly many bioactivity features in endophytes or specific host plants that are responsible for synthesis of secondary metabolites. High-throughput methods are offered as powerful tools for the identification of essential genes that are accountable for activating signaling pathways, which are responsible for the synthesis of bioactive metabolites in organisms (Alam et al. 2021). Whole genome analysis aids in determination of antibiotic resistance gene, plant growth gene, endophytic secretory gene, and other interrelated mechanisms (Kaul et al. 2016). Endophytic fungi belonging to order sebacinales depicts model for the study of symbiotic interaction between plants and endophytes. For instance, *Pirifomospora indica* genomic sequence analysis has discovered its ability as a probiotic agent (Qiang et al. 2012); while fungus *Harpophora* oryzae acts as valuable endosymbiont of *Zizania* sp. (wild rice). Genome sequence analysis of *Harpophora oryzae* showed comparatively high amount of transposable elements (TEs) and shared ortholog groups with symbiotic fungi, gain or loss of genes, expansion of genes involved in intracellular signaling and transport mechanisms etc. All such factors have been essential in the evolutionary pathway of this endosymbiont and made it distinct from its pathogenic ancestor (Xu et al. 2014). Metagenomics has been quite valuable in separating uncultured microbiome of endophytes and helpful in finding out information other than the genetic information of individual taxa. In this process, extraction of genetic material (DNA) is made from the whole population and gene content is analyzed. Using metagenomics approach, there are reports that have revealed functional features from diversity of rice endophytic microbial community. Unexpectedly, a few taxa have been reported *using metagenomic study* such as *Verrucomicrobia* and *Fusobacteria, which were found inside the roots of rice plant (Sessitsch et al.* 2012*). So, several metabolic alterations of endophytic community either from genomic analyses to knowing about their microhabitat, about biotic-abiotic stress resistance or intracellular signaling mechanism or transport and many other integral activities are reported using different omic techniques.*

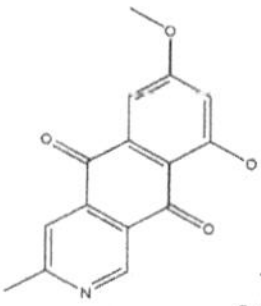

Scorpinone (1)

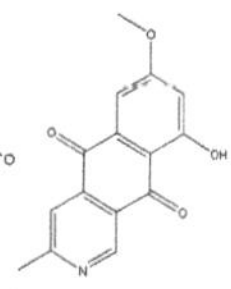

5-deoxybostrycoidin (2)

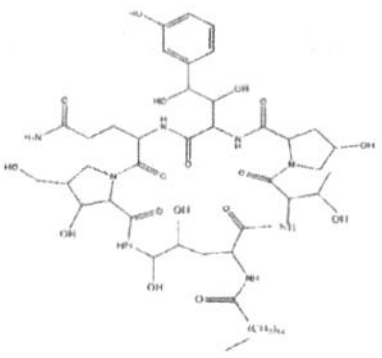

Cryptocandin (4)

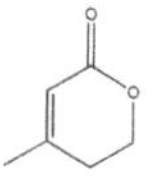

4-methyl-5,6-dihydro-2H-pyran-2-one (3)

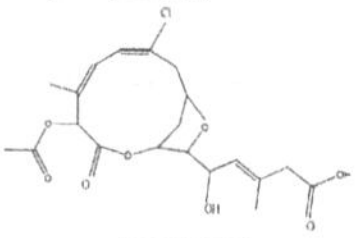

Oocydin A (6)

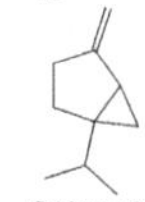

Sabinene (5)

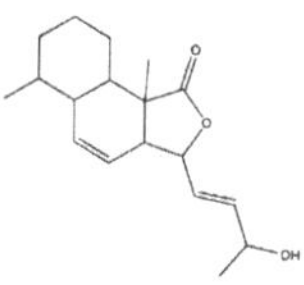

Allahabadolactone A (7)

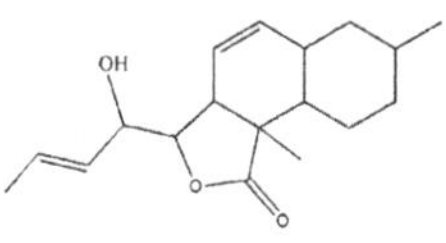

Allahabadolactone B (8)

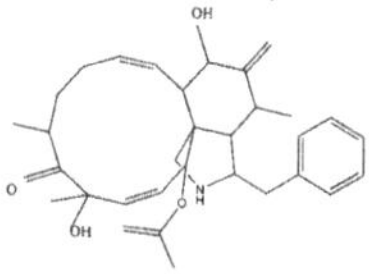

Cytochalasin D (9)

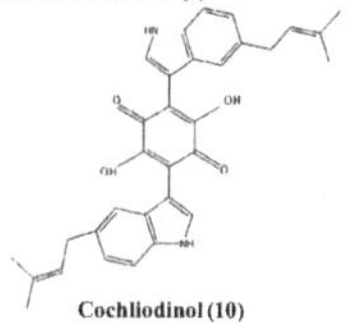

Cochliodinol (10)

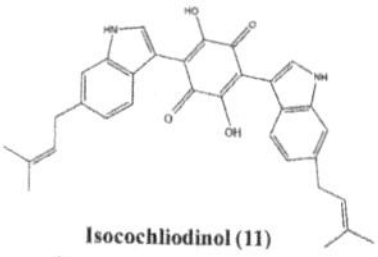

Isocochliodinol (11)

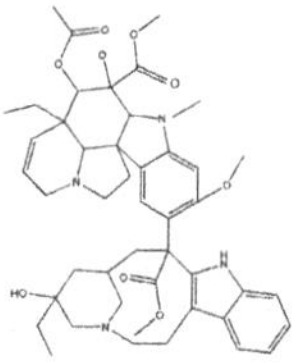

Vinblastine (13)

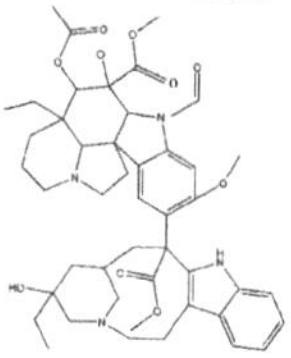

Vincristine (12)

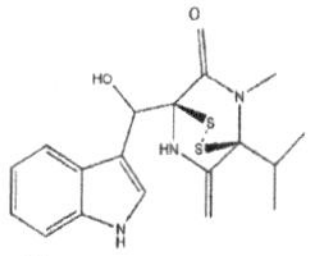

Phaeosphaones A (14)

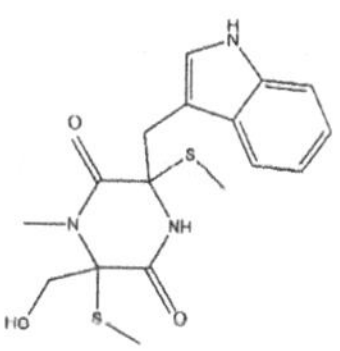

Chetoseminudin B (15)

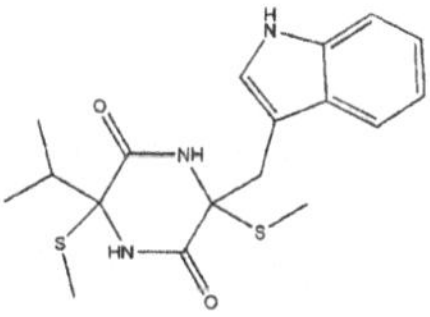

Polanrazine B (16)

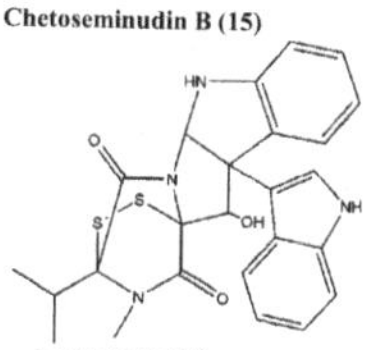

Leptosin D (17)

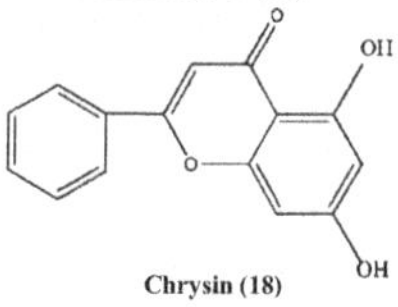

Chrysin (18)

Chemical structures of compounds (1–18) of Table 3.1

3.3 Role of Endophytic Fungi in Phytoremediation

3.3.1 Metal Resistance

Increased human habitation, volant industrial development and modernized agronomic practices have globally caused huge bioaccumulation and toxicity in the environment (Qin et al. 2021). Excessive use of pesticides, fertilizers, deteriorated water irrigation condition, soil pollution etc., have contributed to the degradation of crop safety and accumulation of heavy metals in large amount in the agricultural lands. As a result, deposition of toxic heavy metals in agricultural lands elevated several harmful impacts on crop plants. Use of hyper-accumulating plants during phytoremediation is a feasible and renewable way of treating deteriorated agricultural lands (Su et al. 2021). Combination of such plants with their associated endophytic fungus have readily helped in reducing contaminants and provided cost-effective, in situ technology (Tiodar et al. 2021). During symbiotic association, fungal endophytes either instigate host plant resistance to heavy alloys or diffusely enhances its participation through sufficient water uptake and nourishment in plants which allows increase in shoot biomass and certain transformations in root structure (Su et al. 2021). Table 3.2 provides details of how a few endophytic fungi exhibit metal resistance. Thus, endophytic fungi in association with host plants exhibit

Table 3.2 List of endophytic fungi with metal resistance capacity

Endophytic fungi	Host plant	Heavy metal tolerance	Mode of activity	Reference
Piriformospora indica	*Nicotiana tabacum* (Common name: Tobacco)	Cadmium (Cd)	Improves GSH (glutathione) content and POD (peroxidase) activity in roots of tobacco plant	Su et al. (2021)
Piriformospora indica *Pseudomonas putida*	*Triticum aestivum* L.	Zinc (Zn)	Useful in improving fitness of L-Zn genotype enhanced peroxidase (POD) activity by 40% and 85.7% resp.	Abadi et al. (2021)
Mucor sp	*Arabidopsis sp.*	Zn, Cd, Pb	Suppresses toxic-metal-induced oxidative stress, induces plant-defence action and stress-related response, reduces fungal abundance in shoots etc.	Domka et al. (2019)
Aspergillus fumigates, *Rhizopus* sp.	*Lactuca sativa L.*	Cr	Useful as biofertilizer for healthy and good crop production in chromium-contaminated soil	Hussain et al. (2018)
Sordariomycetes sp.	*Oryza sativa*	Pb	Lead tolerance of rice increases by the regulation of photosystem II using Chl a fluorescence technique	Li et al. (2015)

resistance to several metals such as cadmium, manganese, cobalt, zinc and chromium indicating that such microorganisms have the potential to clear toxic metals from soil and are helpful in improving deteriorated agricultural fields, thus ensuring crop safety and human health (Pietro-Souza et al. 2020).

3.3.2 Organic Contaminants Degradation Capability

Urbanization and industrialization are two main factors which have contributed to increased anthropogenic activities with release of hazardous organic contaminants into environment, thus imposing an imminence impact on human health. Collective effect of plant growth promoting factors and pollutant degradation properties of endophytic fungi significantly assist in phytoremediation of biotic pollutants in soils (Glick 2015). Research studies have provided humankind with methods for decontamination of organic pollutants from soil, which are economically viable and ecologically beneficial as well (Feng et al. 2017). Plants possess the ability to tolerate organic pollutants, resulting in their impaired growth, but they could hardly participate in remediation of contaminated soils because of the phytotoxicity and hydrophobicity of organic pollutants, which inhibit plants' capability to gain nutrients and water (Khan et al. 2013). Table 3.3 presents the effects of various endophytic fungi on their host plants in detoxifying and degrading organic contaminants. Endophytic fungi play beneficiary roles in detoxifying and degrading organic contaminants in plants by improving stress tolerance, increasing procurement of nutrients, increasing efficiency in quality of organic contaminants degradation etc., and also help plants to fight oxidative stress by regulating anti-oxidative stress mechanism in plants (Feng et al. 2017).

Table 3.3 Bio-transformation of organic compounds by endophytic fungi

Endophytic fungi	Plant	Conversion of organic compounds	Transformed products	References
Phomopsis liquidambari	*Bischofia polycarpa*	Sinapic acid (compound 19)	H_2O, CO_2	Xie et al. (2016)
Neocosmospora solani	*Asarum heterotropoides*	Aristolochic acid I (AAI) (compound 20)	O-demethylate of AAI	Wang et al. (2022)
Penicillium sp. F5	*Polygonum cuspidatum*	Resveratrol (compound 21)	Pterostilbene (compound 22)	Xu et al. (2020)
Epicoccum nigrum	*Salix* sp	Flavonoid	Kaempferol-*O*-diglycide	Harwoko et al. (2019)
Phomopsis sp.	*Pinus taeda*	Limonene (compound 23)	Limonene-1,2-diol (compound 24)	Bier et al. (2017)
Umbelopsis isabellina	*Huperzia serrata*	Ursolic acid (compound 25)	3b,7b-dihydroxy-urs-11-en-28,13-lactone, 1b,3b-dihydroxy-urs-11-en-28,13-lactone, 3b-hydroxy-urs-11-en--28,13-lactone (compounds 26–28)	Fu et al. (2011)
Fusarium oxysporum	*Catharanthus roseus*	Vinblastine	Vincristine	Ashutosh and Absar (2013)

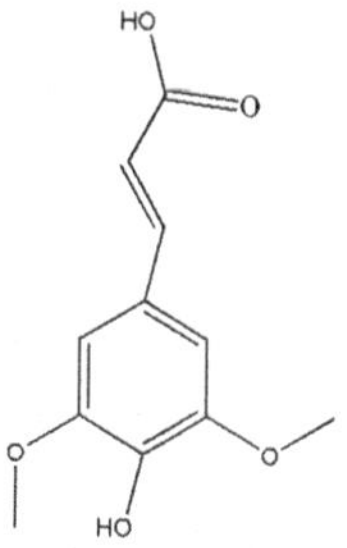

Sinapic acid (19)

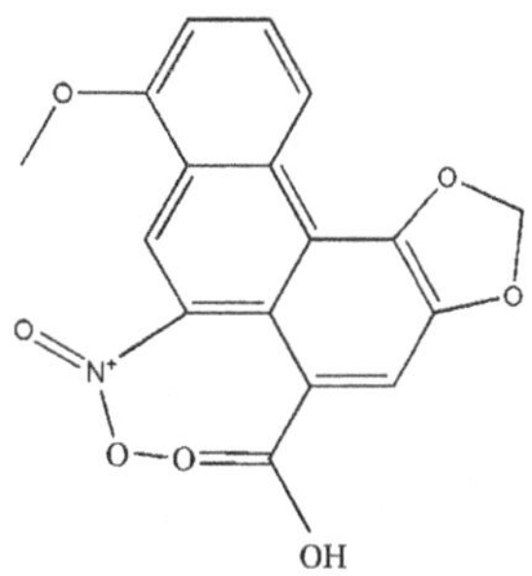

Aristolochic acid I(20)

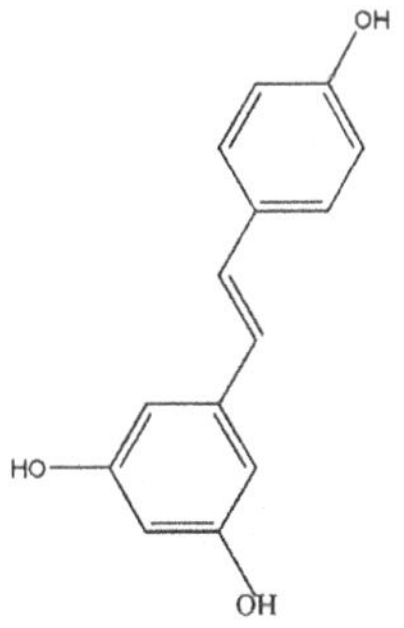

Resveratrol (21)

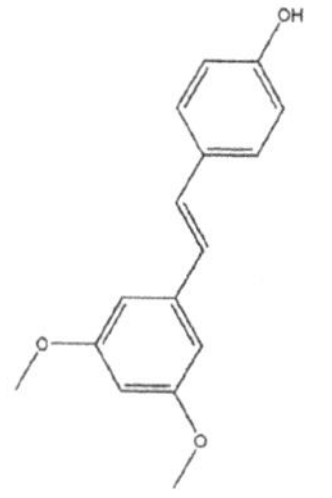

Pterostilbene (22)

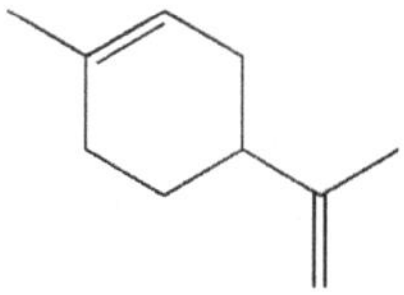

Limonene (23)

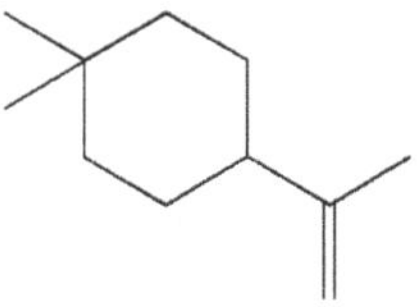

Limonene-1-2-diol (24)

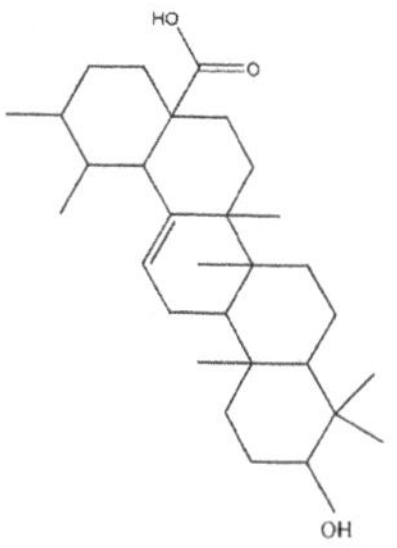

Ursolic acid (25)

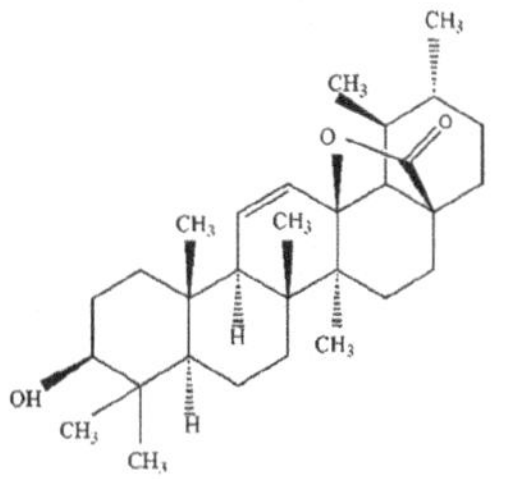

3b,7b-dihydroxy-urs-11-en-28,13-lactone (26)

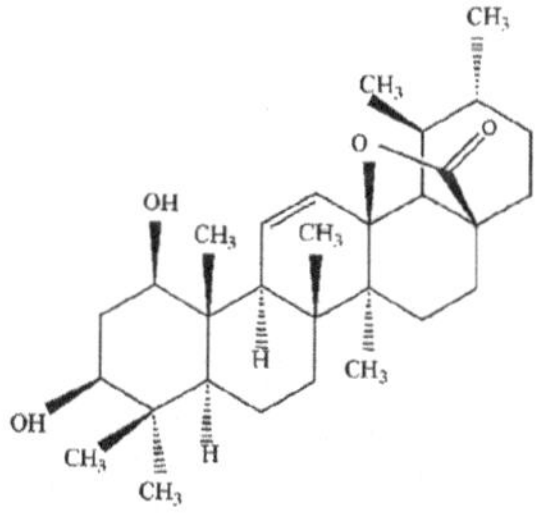

1b,3b-dihydroxy-urs-11-en-28,13-lactone (27)

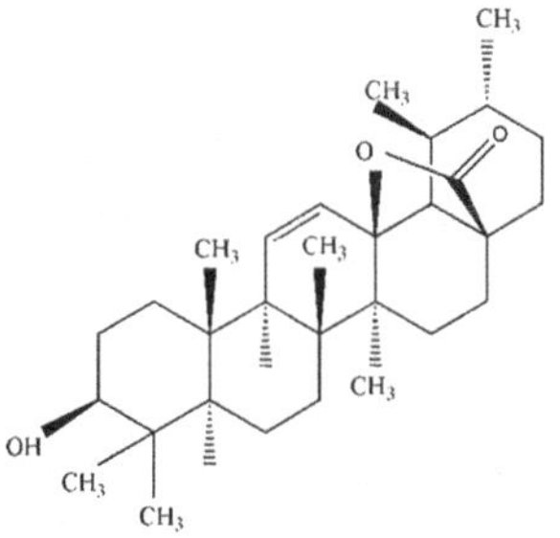

3b-hydroxy-urs-11-en-28,13-lactone (28)

Chemical structures of compounds 19–28 of Table 3.3

3.3.3 Role of endophytic fungi and plant interaction assisted in synthesis of exozymes

Endophytic fungi in synergistic communication along with their host plants have undergone co-evolution and emerged as influential tool in the field of genetic diversity. They can act as biocatalysts i.e., implementing microbial biotransformation capability in the area of catalysis (Choudhary et al. 2021). Of all microbial organisms, fungal endophytes, with their enzymatic capabilities, have become key constituents for assisting in synthetic process for molecule production as per industrial demand. Such endophytic organisms in association with the herbaceous plants have co-evolved throughout their life history, and now their distribution can be seen in different segments of plant body similar to rhizosphere (as in roots), phylloplane (leaves), caulosphere (stems), carposphere (fruits), spermosphere (seeds), anthosphere (flowers) etc. (Khan et al. 2017). Studies reported that plant stress can be successfully reduced by fungal endophytes the production of phytohormones and extracellular enzymes/exozymes as integral biochemical resources. However, extracellular enzymes play important role in food production industry, fermentation process and other biotechnological processes (Traving et al. 2015). Exozymes are produced on an industrial scale because of their capabilities to obtain nutrition from host plants, provide defense against pathogens in plants, be useful in detoxification or removal of wastes from environment, treat several diseases etc. Several endophytic fungi isolated from different plant sources are studied and described for their competence to produce various exozymes. Table 3.4 shows the list of a few of exozymes produced by fungal endophytes. In Industrial sectors, more than hundreds of fungal endophytic enzyme based products have been produced commercially. For instance, laccase is widely utilized in textile industry, protease is used in dairy and fermentation industry, lipase and amylase are extensively utilized in waste management sectors, while exozymes like cellulase and xylanases are used in paper industry widely (Ghosh et al. 2023). Enzymatic properties such as thermostability of enzymes, pH stability and inertness of the enzymes are essential features to focus upon for the production of exozymes on an industrial level. The term 'enzyme' was first named in the year 1877 by Wilhelm Friedrich Kuhne (Gurung et al. 2013), and since ancient times this term has widely been in use. In the year 1947, James B. Sumner identified, isolated and crystallized Urease enzyme from legume plant *Canavalia ensiformis* in its pure form (Simoni et al. 2002). Nearly, majority of the industrially categorized enzymes possess hydrolysis property in them. Alpha-amylase being the most indispensable and flexible enzyme during the process (Farooq et al. 2021) of conversion of starch to the synthesis of cyclodextrins (Biwer et al. 2002). Protease enzyme is the most extensively utilized enzyme in biotechnological sector because of its features like maximum yield, minimum space requirement, imperious genetic manipulation and cost-effectiveness (Nisha and Divakaran 2014). In baking industrial sector, for faster dough preparation, heat-labile fungal protease is used for partial hydrolysis of gluten and early inactivation of the enzyme in later baking stages (Miguel et al. 2013). Microbial alkaline proteases are used in preparation of protein hydrolysates with good nutritional value and plays a

Table 3.4 List of extracellular enzymes isolated and derived from endophytic fungi

Endophytic fungi	Plants	Extracellular enzyme	Industrial utilization	References
Aspergillus flavus, *Aspergillus fumigatus*, *Aspergillus niger*, *Cladosporium* sp., *Rhizopus* sp., and *Mucor* sp.	*Ziziphus spina*	Amylase, protease, pectinase, cellulase, lipase and laccase	Helps in food preservation and increases shelf life of food	Hawar (2022)
Cylindrocephalum sp.	*Alpinia calcarata*	Amylase	Hydrolysis of starch, in formulation of detergents, highly utilized in brewing, baking industries etc.	Sunitha et al. (2012)
Fusarium sp., *Cercospora* sp., *Diaporthe* sp., *Lasiodiplodia* sp., *Colletotrichum* sp., *Cryptococcus* sp., *Hypoxylon* sp., *Ophioceras* sp., *Ampelomyces* sp., *Pseudofusicoccum* sp., *Paraconiothyrium* sp., *Corynespora* sp., *Nigrospora* sp., *Phoma*sp, *Trichoderma* sp., *Penicillium* sp., *Guignardia* sp., *Cladosporium* sp., *Chaetomium* sp.	*Cananga odorata*, *Terminalia catappa*, *Terminalia mantaly*	Amylase, lipase, laccase, cellulase	Starch degradation, lignin degradation, biological catalyst etc.	Toghueo et al. (2017)
Talaromyces pinophilus	*Curcuma amada*	L-asparaginase	As chemotherapeutic agent and food processing agent	Krishnapura and Belur (2016) Vimal and Kumar (2017)
Penicillium oxalicum R4	*Taxus cuspidata*	Cellulase	Cellulose hydrolysis, renewable fuel raw material etc.	Li et al. (2021)

(continued)

Table 3.4 (continued)

Endophytic fungi	Plants	Extracellular enzyme	Industrial utilization	References
Aspergillus terreus KP900973	*Corchorus olitorius*	Xylanase	Lignocellulosic biomass degradation, biofuel production, utilized as value added products in food products, soft drinks, several pharmaceutical products, bio-processing of textiles, bio-bleaching of wood pulp etc.	Ahmed et al. (2016) Basit et al. (2020)

significant role in fortification of juices, carbonated drinks, and in infant food formula preparation along with other proteases industrially (Hamosh 1997). The antioxidant activity of protein hydrolysates increases because of the their high solubility (Chi et al. 2015; Razzaq et al. 2019). Protein hydrolysates are often used in food industry as they act as excellent flavour enhancers with potent bioactive properties, mainly as immunosuppressant, antithrombotic, antidiabetic etc. (Mahomoodally and Irfaan 2022). Pentosanases, β-glucanases, amyloglucosidases, amylases, pullanases, ALDC (α-acetolactate-decarboxylases) etc., are the enzymes utilized in brewing industry, and utilization of these enzymes causes better yield and production of good quality beer and beverages. β-glucanase enzyme causes hydrolyzation of glucans into oligomers (soluble in nature) and thus, offering lower viscosity with refined filtration ability (Caseiro et al. 2022); ALDC (α-acetolactate-decarboxylases) prevents diacetyl formation and thus reducing the period of fermentation process to attain the appropriate taste of beer and beverages (Li et al. 2012; Choi et al. 2015). The commercial and large-scale production of microbial enzymes from fungal, bacterial and archael microorganisms has been most efficiently exploited in the fields of life-science, microbiology, biotechnology, medicine and pharmacology. Microbial lipases are predominantly utilized in the production of inhibitors or antagonists of biological system (Carvalho et al. 2015). They act as flexible biocatalysts and are utilized in various bioconversion processes. The most alluring characteristic of lipase enzyme is its capability to undergo processing of different types of glycerides, fatty acids in transesterification process, its ability to tolerate fluctuations in temperature and pH change, excellent working potential in non-aqueous medium and lipase enzyme stays stable and active in organic solvents without theuse of any cofactors (Kumar et al. 2012; Verma et al. 2012; Ali et al. 2023). Such versatile characteristics of microbial enzymes accelerate enzyme-assisted applications at large scale in industries with better yield, quality and economically effective techniques.

3.3.4 Role of Eendophytic Fungi as Biocatalyst

Endophytic fungi are found to be a reliable source for creating high-purity compounds with high yield in an environment-friendly manner. This has attracted much interest in research and industrial sectors. Fungal endophyte catalysis and transformation are done to overcome boundaries in chemical production, reduce toxicity of Pb^{2+}drugs and assist in developing structure–activity relationship study of drugs (Liu et al. 2021). Endophytic fungi mediates biotransformation process as extracellular enzyme or in a multi-step process as whole cell catalysis which have significantly helped in development of pharmaceutically bioactive molecules (Choudhary et al. 2021). Studies have reported the remarkable biochemical capability of endophytic fungi in different productive areas. For instance, study on *Alternaria eureka* endophytic fungus showed its capability to modify compounds and biotransformation of the major bioactive compound sapogenin into neoruscogenin, along with the generation of 14 new biotransformation products that were isolated and their structures interpreted by NMR and HRESIMS data analyses (Özçınar et al. 2018). Applying endophytic fungi as whole-cell bioprocess has provided an alternative source for generating limited bioactive natural compounds. In a study, researchers showed novel bioconversion route for calebin-A production from curcumin compound isolated from *Curcuma caesia* plant in association with *Ovatospora brasiliensis* strain EPE-10 MTCC 25236 (Majeed et al. 2019). B-glucuronidase (cg-GUS) screened from fungalAQ22 endophytic strain Chaetomium globosum DX-THS3 have undergone biotransformation of glycyrrhizin (GL) to glycyrrhetinic acid monoglucuronide (GAMG) showing maximum enzyme activity of enzyme at temp. 45 °C and pH 6.0 providing a good source in food and medicinal industries (Zhang et al. 2020).

3.4 Future Prospects

Endophytic fungi represent versatile microorganisms that are diversely present in association with almost all plants throughout nature. They can synthesize various secondary metabolites on interaction with host plants which are efficient and less toxic in nature. Most of the compounds possess the potentiality of parent molecules or could precisely be developed into drugs for curing various diseases. Although, due to limitations in due to invitro studies, compounds are needed to inquire using invivo approaches and study proper mode of action and signaling pathways to understand the mechanism of the new compound in treating a particular disease or a disorder. Thorough studies are needed to focus on symbiotic controlling mechanisms at the gene, transcriptional and post-transcriptional levels. There is a necessity to focus on the studying of specific signaling pathways that are needed for maintaining endophytic fungi genes and their influence in activating virulence mechanisms. Because of the difficulty and intricacy that appears in plant-endophytes interaction, it would be beneficial to use meta-omic techniques or tools to explore the various mechanisms, physiology to understand genetic expression and secondary metabolite synthesis and defensive actions at both ends.

3.5 Conclusion

Since recent decades, plants are studied greatly to isolate biologically active compounds and are researched thoroughly for the development of novel drugs. Endophytic fungi-assisted plants have been found to be the probable origin of numerous compounds which exhibit biological properties similar to compounds that are isolated from plants. The identification, isolation and assortment of compounds from such endophytic fungi have paved ways for development of drug candidates with pharmacological importance, as they are effortless to control, govern and engage, in comparison to plants. The advanced and improved techniques AQ26for molecular-level characterization, modulation of enzymes using omic tools and algorithmic platforms in accordance to applied biology are beneficial. The necessity to develop time-efficient, safe, cost-effective, low in toxicity drugs from natural resources has pushed researchers and industries to shift their focus to the area of endophytes. Hence, more efficient studies are necessitated to illustrate the right mechanism of action for harnessing drug-like properties in these compounds using biochemical studies, bioinformatics toolkits, pharmacokinetics and pre-clinical approaches.

References

Abadi VAJM, Sepehri M, Khatabi B, Rezaei M (2021) Alleviation of zinc deficiency in wheat inoculated with root endophytic fungus Piriformospora indica and rhizobacterium Pseudomonas putida. Rhizosphere 17:100311

Ahmed SA, Saleh SA, Mostafa FA, Abd El Aty AA, Ammar HA (2016) Characterization and valuable applications of xylanase from endophytic fungus Aspergillus terreus KP900973 isolated from Corchorus olitorius. Biocatal Agric Biotechnol 7:134–144

Alam B, Khan MA, Gōng J, Yuán Y, Gǒng W (2021) Endophytic fungi: from symbiosis to secondary metabolite communications or vice versa? Front Plant Sci 12:791033

Ali S, Khan SA, Hamayun M, Lee IJ (2023) The recent advances in the utility of microbial lipases: a review. Microorganisms 11(2):510

Ashutosh K, Absar A (2013) Biotransformation of vinblastine to vincristine by the endophytic fungus Fusarium oxysporum isolated from Catharanthus roseus. Biocatal Biotransformation 31(2):89–93

Azizan MS, Zamani AI, Stahmann KP, Ng CL (2016) Fungal metabolites and their industrial importance: a brief review. Malays J Biochem Mol Biol 19:15–23

Baral B, Akhgari A, Metsä-Ketelä M (2018) Activation of microbial secondary metabolic pathways: avenues and challenges. Synth Syst Biotechnol 3(3):163–178

Basit A, Jiang W, Rahim K (2020) Xylanase and its industrial applications. Biotechnological applications of biomass

Bhar A, Chakraborty A, Roy A (2021) Plant responses to biotic stress: old memories matter. Plan Theory 11(1):84

Bhar A, Chakraborty A, Roy A (2023) The captivating role of calcium in plant-microbe interaction. Front Plant Sci 14:1138252

Bier MCJ, Medeiros ABP, Soccol CR (2017) Biotransformation of limonene by an endophytic fungus using synthetic and orange residue-based media. Fungal Biol 121(2):137–144

Biwer A, Antranikian G, Heinzle E (2002) Enzymatic production of cyclodextrins. Appl Microbiol Biotechnol 59:609–617

Carvalho ACLDM, Fonseca TDS, Mattos MCD, Oliveira MDCFD, Lemos TLGD, Molinari F, Romano D, Serra I (2015) Recent advances in lipase-mediated preparation of pharmaceuticals and their intermediates. Int J Mol Sci 16(12):29682–29716

Caseiro C, Dias JNR, de Andrade Fontes CMG, Bule P (2022) From cancer therapy to winemaking: the molecular structure and applications of β-glucans and β-1, 3-glucanases. Int J Mol Sci 23(6):3156

Chareprasert S, Piapukiew J, Thienhirun S, Whalley AJS, Sihanonth P (2006) Endophytic fungi of teak leaves Tectona grandis L. and rain tree leaves Samanea saman Merr. World J Microbiol Biotechnol 22:481–486

Chi CF, Wang B, Wang YM, Zhang B, Deng SG (2015) Isolation and characterization of three antioxidant peptides from protein hydrolysate of bluefin leatherjacket (Navodonseptentrionalis) heads. J Funct Foods 12:1–10

Choi EJ, Ahn HW, Kim WJ (2015) Effect of α-acetolactate decarboxylase on diacetyl content of beer. Food Sci Biotechnol 24:1373–1380

Choudhary M, Gupta S, Dhar MK, Kaul S (2021) Endophytic fungi-mediated biocatalysis and biotransformations paving the way toward green chemistry. Front Bioeng Biotechnol 9:664705

Chow YY, Ting ASY (2019) Influence of fungal infection on plant tissues: FTIR detects compositional changes to plant cell walls. Fungal Ecol 37:38–47

Collinge DB, Jensen B, Jørgensen HJ (2022) Fungal endophytes in plants and their relationship to plant disease. Curr Opin Microbiol 69:102177

Cruz JS, da Silva CA, Hamerski L (2020) Natural products from endophytic fungi associated with Rubiaceae species. J Fungi 6(3):128

Cui H, Tsuda K, Parker JE (2015) Effector-triggered immunity: from pathogen perception to robust defense. Annu Rev Plant Biol 66:487–511

Cui H, Yu J, Chen S, Ding M, Huang X, Yuan J, She Z (2017) Alkaloids from the mangrove endophytic fungus Diaporthe phaseolorum SKS019. Bioorg Med Chem Lett 27(4):803–807

Datta R, Roy A, Paul S (2023) Phytohormone signaling in osmotic stress response. In: Plant hormones in crop improvement. Academic Press, pp 89–108

Debbab A, Aly AH, Edrada-Ebel R, Müller WE, Mosaddak M, Hakikj A, Ebel R, Proksch P (2009) Bioactive secondary metabolites from the endophytic fungus Chaetomium sp. isolated from Salvia officinalis growing in Morocco. Biotechnol Agron Soc Environ 13(2):229–234

Debbab A, Aly AH, Proksch P (2011) Bioactive secondary metabolites from endophytes and associated marine derived fungi. Fungal Divers 49(1):1–12

del Carmen Orozco-Mosqueda M, Santoyo G (2021) Plant-microbial endophytes interactions: scrutinizing their beneficial mechanisms from genomic explorations. Curr Plant Biol 25:100189

Ding WJ, Wang SS, Ren JQ, Li G, Zhan JP (2015) Progress on plant endophyte. Curr Biotechnol 5:425–428

Domka A, Rozpądek P, Ważny R, Turnau K (2019) Mucor sp.—An endophyte of Brassicaceae capable of surviving in toxic metal-rich sites. J Basic Microbiol 59(1):24–37

El-Hawary SS, Moawad AS, Bahr HS, Abdelmohsen UR, Mohammed R (2020) Natural product diversity from the endophytic fungi of the genus Aspergillus. RSC Adv 10(37):22058–22079

Farooq MA, Ali S, Hassan A, Tahir HM, Mumtaz S, Mumtaz S (2021) Biosynthesis and industrial applications of α-amylase: a review. Arch Microbiol 203:1281–1292

Feng NX, Yu J, Zhao HM, Cheng YT, Mo CH, Cai QY, Li YW, Li H, Wong MH (2017) Efficient phytoremediation of organic contaminants in soils using plant–endophyte partnerships. Sci Total Environ 583:352–368

Fu SB, Yang JS, Cui JL, Feng X, Sun DA (2011) Biotransformation of ursolic acid by an endophytic fungus from medicinal plant Huperzia serrata. Chem Pharm Bull 59(9):1180–1182

Garyali S, Kumar A, Reddy MS (2013) Taxol production by an endophytic fungus, Fusarium redolens, isolated from Himalayan yew. J Microbiol Biotechnol 23(10):1372–1380

Geurts R, Xiao TT, Reinhold-Hurek B (2016) What does it take to evolve a nitrogen-fixing endosymbiosis? Trends Plant Sci 21(3):199–208

Ghosh U, Mathur P, Chaturvedi P, Sharma C, Bhatnagar P (2023) Bioprospection of endophytic fungi for extracellular enzymes. In: Fungal resources for sustainable economy: current status and future perspectives. Springer Nature Singapore, Singapore, pp 127–146

Glick BR (2015) Beneficial plant-bacterial interactions, vol 243. Springer, Heidelberg

Gouda S, Das G, Sen SK, Shin HS, Patra JK (2016) Endophytes: a treasure house of bioactive compounds of medicinal importance. Front Microbiol 7:219261

Gurung N, Ray S, Bose S, Rai V (2013) A broader view: microbial enzymes and their relevance in industries, medicine, and beyond. BioMed Res Int 2013:329121

Hacquard S, Spaepen S, Garrido-Oter R, Schulze-Lefert P (2017) Interplay between innate immunity and the plant microbiota. Annu Rev Phytopathol 55:565–589

Hamosh M (1997) Introduction: should infant formulas be supplemented with bioactive components and conditionally essential nutrients present in human milk? J Nutr 127:971S–974S

Harvey CJ, Tang M, Schlecht U, Horecka J, Fischer CR, Lin HC, Li J, Naughton B, Cherry J, Miranda M, Li YF (2018) HEx: a heterologous expression platform for the discovery of fungal natural products. Sci Adv 4(4):eaar5459

Harwoko H, Hartmann R, Daletos G, Ancheeva E, Frank M, Liu Z, Proksch P (2019) Biotransformation of host plant flavonoids by the fungal endophyte Epicoccum nigrum. ChemistrySelect 4(45):13054–13057

Hashem AH, Attia MS, Kandil EK, Fawzi MM, Abdelrahman AS, Khader MS, Khodaira MA, Emam AE, Goma MA, Abdelaziz AM (2023) Bioactive compounds and biomedical applications of endophytic fungi: a recent review. Microb Cell Factories 22(1):107

Hasin Y, Seldin M, Lusis A (2017) Multi-omics approaches to disease. Genome Biol 18:1–15

Hawar SN (2022) Extracellular enzyme of endophytic fungi isolated from Ziziphus spina leaves as medicinal plant. Int J Biomater 2022:2135927

Hodgson S, de Cates C, Hodgson J, Morley NJ, Sutton BC, Gange AC (2014) Vertical transmission of fungal endophytes is widespread in forbs. Ecol Evol 4(8):1199–1208

Hussain A, Hamayun M, Rahman H, Iqbal A, Shah M, Irshad M, Qasim M, Islam B (2018) Bioremediation of hexavalent chromium by endophytic fungi; safe and improved production of Lactuca sativa L. Chemosphere 211:653–663

Jamieson PA, Shan L, He P (2018) Plant cell surface molecular cypher: receptor-like proteins and their roles in immunity and development. Plant Sci 274:242–251

Jha P, Kaur T, Panja A, Paul S, Kumar V, Malik T (2023) Endophytic fungi: hidden treasure chest of antimicrobial metabolites interrelationship of endophytes and metabolites. Front Microbiol 14:1227830

Ji X, Xia Y, Zhang H, Cui JL (2022) The microscopic mechanism between endophytic fungi and host plants: from recognition to building stable mutually beneficial relationships. Microbiol Res 261:127056

Jia M, Chen L, Han T (2016) A friendly relationship between endophytic fungi and medicinal plants: a systematic review. Front Microbiol 7:179220

Jiang C, Lv G, He B (2021) Applications of CRISPR/Cas9 in the synthesis of secondary metabolites in filamentous fungi. Front Microbiol 12:638096

Jiang L, Ma Q, Li A, Sun R, Tang G, Huang X, Pu H (2023) Bioactive secondary metabolites produced by fungi of the genus diaporthe (phomopsis): structures, biological activities, and biosynthesis. Arab J Chem 16:105062

Jiao RH, Xu S, Liu JY, Ge HM, Ding H, Xu C, Zhu HL, Tan RX (2006) Chaetominine, a cytotoxic alkaloid produced by endophytic Chaetomium sp. IFB-E015. Org Lett 8(25):5709–5712

Kaul S, Sharma T, K Dhar M (2016) "Omics" tools for better understanding the plant–endophyte interactions. Front Plant Sci 7:183603

Kaur T (2020) Fungal endophyte-host plant interactions: role in sustainable agriculture. Sustainable crop production, p 211

Khan S, Afzal M, Iqbal S, Khan QM (2013) Plant–bacteria partnerships for the remediation of hydrocarbon contaminated soils. Chemosphere 90(4):1317–1332

Khan AL, Shahzad R, Al-Harrasi A, Lee IJ (2017) Endophytic microbes: a resource for producing extracellular enzymes. Endophytes: crop productivity and protection: Volume 2, pp 95–110

Köster P, DeFalco TA, Zipfel C (2022) Ca2+ signals in plant immunity. EMBO J 41(12):e110741
Krishnapura PR, Belur PD (2016) Partial purification and characterization of L-asparaginase from an endophytic Talaromycespinophilus isolated from the rhizomes of Curcuma amada. J Mol Catal B Enzym 124:83–91
Kumar S, Kaushik N (2013) Endophytic fungi isolated from oil-seed crop Jatropha curcas produces oil and exhibit antifungal activity. PLoS One 8(2):e56202
Kumar A, Sharma P, Kanwar SS (2012) Lipase catalyzed esters syntheses in organic media: a review. Int J Inst Pharm Life Sci 2(2):91–119
Li DL, Li XM, Proksch P, Wang BG (2010) 7-O-Methylvariecolortide A, a new spirocyclic diketopiperazine alkaloid from a marine mangrove derived endophytic fungus, Eurotium rubrum. Nat Prod Commun 5(10):1934578X1000501014
Li S, Yang X, Yang S, Zhu M, Wang X (2012) Technology prospecting on enzymes: application, marketing and engineering. Comput Struct Biotechnol J 2(3):e201209017
Li H, Tian JM, Tang HY, Pan SY, Zhang AL, Gao JM (2015) Chaetosemins A–E, new chromones isolated from an Ascomycete Chaetomium seminudum and their biological activities. RSC Adv 5(37):29185–29192
Li H, Dou M, Wang X, Guo N, Kou P, Jiao J, Fu Y (2021) Optimization of cellulase production by a novel endophytic fungus Penicillium oxalicum R4 isolated from Taxus cuspidata. Sustain For 13(11):6006
Liebrand TW, van den Berg GC, Zhang Z, Smit P, Cordewener JH, America AH, Sklenar J, Jones AM, Tameling WI, Robatzek S, Thomma BP (2013) Receptor-like kinase SOBIR1/EVR interacts with receptor-like proteins in plant immunity against fungal infection. Proc Natl Acad Sci 110(24):10010–10015
Liu D, Li XM, Li CS, Wang BG (2013) Nigerasterols A and B, antiproliferative sterols from the mangrove-derived endophytic fungus Aspergillus niger MA-132. Helv Chim Acta 96(6):1055–1061
Liu X, Zhou ZY, Cui JL, Wang ML, Wang JH (2021) Biotransformation ability of endophytic fungi: from species evolution to industrial applications. Appl Microbiol Biotechnol 105:1–19
Lu H, Wei T, Lou H, Shu X, Chen Q (2021) A critical review on communication mechanism within plant-endophytic fungi interactions to cope with biotic and abiotic stresses. J Fungi 7(9):719
Mahomoodally MF, Irfaan MM (2022) Protein hydrolysates. In: Antioxidants effects in health. Elsevier, pp 451–460
Majeed A, Majeed M, Thajuddin N, Arumugam S, Ali F, Beede K, Adams SJ, Gnanamani M (2019) Bioconversion of curcumin into calebin-A by the endophytic fungus Ovatospora brasiliensis EPE-10 MTCC 25236 associated with Curcuma caesia. AMB Express 9:1–13
Manganyi MC (2018) Diversity of endophytic fungi possessing bioactive compounds isolated from selected medicinal plants (Doctoral dissertation, North-West University (South Africa))
Manganyi MC, Ateba CN (2020) Untapped potentials of endophytic fungi: a review of novel bioactive compounds with biological applications. Microorganisms 8(12):1934
Mao Z, Zhang W, Wu C, Feng H, Peng Y, Shahid H, Cui Z, Ding P, Shan T (2021) Diversity and antibacterial activity of fungal endophytes from Eucalyptus exserta. BMC Microbiol 21(1):155
Marcec MJ, Gilroy S, Poovaiah BW, Tanaka K (2019) Mutual interplay of Ca2+ and ROS signaling in plant immune response. Plant Sci 283:343–354
Mengistu AA (2020) Endophytes: colonization, behaviour, and their role in defense mechanism. Int J Microbiol 2020:6927219
Mhlongo MI, Piater LA, Madala NE, Labuschagne N, Dubery IA (2018) The chemistry of plant–microbe interactions in the rhizosphere and the potential for metabolomics to reveal signaling related to defense priming and induced systemic resistance. Front Plant Sci 9:314703
Miguel AM, Martins-Meyer TS, Figueiredo EVDC, Lobo BWP, Dellamora-Ortiz GM (2013) Enzymes in bakery: current and future trends. Food industry, pp 287–321
Mohamed Ibrahim S, Mohamed G, Al Haidari R, Soliman El-Kholy A, Asfour H, Zayed M (2018) Fusaristerol a: a new cytotoxic and antifungal ergosterol fatty acid ester from the endophytic fungus Fusarium sp. associated with Mentha longifolia roots. Pharmacogn Mag 14(56)

Momtaz S, Lall N, Basson A (2008) Inhibitory activities of mushroom tyrosine and DOPA oxidation by plant extracts. S Afr J Bot 74(4):577–582
Nair DN, Padmavathy S (2014) Impact of endophytic microorganisms on plants, environment and humans. ScientificWorldJournal 2014:250693
Naveed ZA, Wei X, Chen J, Mubeen H, Ali GS (2020) The PTI to ETI continuum in Phytophthora-plant interactions. Front Plant Sci 11:593905
Nisha NS, Divakaran J (2014) Optimization of alkaline protease production from Bacillus subtilis NS isolated from sea water. Afr J Biotechnol 13(16)
Nützmann HW, Scazzocchio C, Osbourn A (2018) Metabolic gene clusters in eukaryotes. Annu Rev Genet 52:159–183
Ogbe AA, Finnie JF, Van Staden J (2020) The role of endophytes in secondary metabolites accumulation in medicinal plants under abiotic stress. S Afr J Bot 134:126–134
Osbourn A (2010) Gene clusters for secondary metabolic pathways: an emerging theme in plant biology. Plant Physiol 154(2):531–535
Özçınar O, Tağ O, Yusufoglu H, Kivçak B, Bedir E (2018) Biotransformation of neoruscogenin by the endophytic fungus Alternaria eureka. J Nat Prod 81(6):1357–1367
Palem PP, Kuriakose GC, Jayabaskaran C (2015) An endophytic fungus, Talaromycesradicus, isolated from Catharanthus roseus, produces vincristine and vinblastine, which induce apoptotic cell death. PLoS One 10(12):e0144476
Pfannenstiel BT, Keller NP (2019) On top of biosynthetic gene clusters: how epigenetic machinery influences secondary metabolism in fungi. Biotechnol Adv 37(6):107345
Photita W, Lumyong S, Lumyong P, McKenzie EHC, Hyde KD (2004) Are some endophytes of Musa acuminata latent pathogens. Fungal Divers 16:131–140
Pietro-Souza W, de Campos Pereira F, Mello IS, Stachack FFF, Terezo AJ, da Cunha CN, White JF, Li H, Soares MA (2020) Mercury resistance and bioremediation mediated by endophytic fungi. Chemosphere 240:124874
Pillay LC, Nekati L, Makhwitine PJ, Ndlovu SI (2022) Epigenetic activation of silent biosynthetic gene clusters in endophytic fungi using small molecular modifiers. Front Microbiol 13:815008
Poveda J, Zabalgogeazcoa I, Soengas P, Rodríguez VM, Cartea ME, Abilleira R, Velasco P (2020) Brassica oleracea var. acephala (kale) improvement by biological activity of root endophytic fungi. Sci Rep 10(1):20224
Poveda J, Eugui D, Abril-Urías P, Velasco P (2021) Endophytic fungi as direct plant growth promoters for sustainable agricultural production. Symbiosis 85(1):1–19
Qi C, Gao W, Wang J, Liu M, Zhang J, Chen C, Hu Z, Xue Y, Li D, Zhang Q, Lai Y (2018) Terrusnolides AD, new butenolides with anti-inflammatory activities from an endophytic Aspergillus from Tripterygium wilfordii. Fitoterapia 130:134–139
Qiang X, Weiss M, Kogel KH, Schäfer P (2012) Piriformospora indica—a mutualistic basidiomycete with an exceptionally large plant host range. Mol Plant Pathol 13(5):508–518
Qin G, Niu Z, Yu J, Li Z, Ma J, Xiang P (2021) Soil heavy metal pollution and food safety in China: effects, sources and removing technology. Chemosphere 267:129205
Radić N, Štrukelj B (2012) Endophytic fungi—the treasure chest of antibacterial substances. Phytomedicine 19(14):1270–1284
Rai N, Kumari Keshri P, Verma A, Kamble SC, Mishra P, Barik S, Kumar Singh S, Gautam V (2021) Plant associated fungal endophytes as a source of natural bioactive compounds. Mycology 12(3):139–159
Rashmi M, Kushveer JS, Sarma VV (2019) A worldwide list of endophytic fungi with notes on ecology and diversity. Mycosphere 10(1):798–1079
Razzaq A, Shamsi S, Ali A, Ali Q, Sajjad M, Malik A, Ashraf M (2019) Microbial proteases applications. Front Bioeng Biotechnol 7:110
Risdian C, Mozef T, Wink J (2019) Biosynthesis of polyketides in Streptomyces. Microorganisms 7(5):124
Rutledge PJ, Challis GL (2015) Discovery of microbial natural products by activation of silent biosynthetic gene clusters. Nat Rev Microbiol 13(8):509–523

Saijo Y, Loo EPI, Yasuda S (2018) Pattern recognition receptors and signaling in plant–microbe interactions. Plant J 93(4):592–613

Santiago C, Sun L, Munro MHG, Santhanam J (2014) Polyketide and benzopyran compounds of an endophytic fungus isolated from C innamomummollissimum: biological activity and structure. Asian Pac J Trop Biomed 4(8):627–632

Seetharaman P, Gnanasekar S, Chandrasekaran R, Chandrakasan G, Kadarkarai M, Sivaperumal S (2017) Isolation and characterization of anticancer flavone chrysin (5, 7-dihydroxy flavone)-producing endophytic fungi from Passiflora incarnata L. leaves. Ann Microbiol 67:321–331

Selvakumar V, Panneerselvam A (2018) Bioactive compounds from endophytic fungi. Fungi and their role in sustainable development: current perspectives, pp 699–717

Sessitsch ANGELA, Hardoim PABLO, Döring J, Weilharter A, Krause ANDREAS, Woyke TANJA, Mitter B, Hauberg-Lotte L, Friedrich F, Rahalkar M, Hurek T (2012) Functional characteristics of an endophyte community colonizing rice roots as revealed by metagenomic analysis. Mol Plant-Microbe Interact 25(1):28–36

Simoni RD, Hill RL, Vaughan M (2002) Urease, the first crystalline enzyme and the proof that enzymes are proteins: the work of James B. Sumner. J Biol Chem 277(35):e1–e2

Singh SK, Strobel GA, Knighton B, Geary B, Sears J, Ezra D (2011) An endophytic Phomopsis sp. possessing bioactivity and fuel potential with its volatile organic compounds. Microb Ecol 61:729–739

Singh R, Mittal A, Kumar M, Mehta PK (2016) Microbial proteases in commercial applications. J Pharm Chem Biol Sci 4(3):365–374

Singh M, Chaudhary S, Sareen D (2017) Non-ribosomal peptide synthetases: identifying the cryptic gene clusters and decoding the natural product. J Biosci 42:175–187

Stierle A, Strobel G, Stierle D (1993) Taxol and taxane production by Taxomycesandreanae, an endophytic fungus of Pacific yew. Science 260(5105):214–216

Strobel GA (2003) Endophytes as sources of bioactive products. Microbes Infect 5(6):535–544

Strobel G (2006) Harnessing endophytes for industrial microbiology. Curr Opin Microbiol 9(3):240–244

Strobel GA, Miller RV, Martinez-Miller C, Condron MM, Teplow DB, Hess WM (1999) Cryptocandin, a potent antimycotic from the endophytic fungus Cryptosporiopsis cf. quercina. Microbiology 145(8):1919–1926

Su Z, Zeng Y, Li X, Perumal AB, Zhu J, Lu X, Dai M, Liu X, Lin F (2021) The endophytic fungus Piriformospora indica-assisted alleviation of cadmium in tobacco. J Fungi 7(8):675

Sunitha VH, Ramesha A, Savitha J, Srinivas C (2012) Amylase production by endophytic fungi Cylindrocephalumsp. Isolated from medicinal plant Alpinia calcarata (Haw.) Roscoe. Braz J Microbiol 43(3):1213

Tang D, Wang G, Zhou JM (2017) Receptor kinases in plant-pathogen interactions: more than pattern recognition. Plant Cell 29(4):618–637

Tian W, Hou C, Ren Z, Wang C, Zhao F, Dahlbeck D, Hu S, Zhang L, Niu QI, Li L, Staskawicz BJ (2019) A calmodulin-gated calcium channel links pathogen patterns to plant immunity. Nature 572(7767):131–135

Tian C, Gao H, Peng XP, Li G, Lou HX (2021) Fusidic acid derivatives from the endophytic fungus Acremonium pilosum F47. J Asian Nat Prod Res 23(12):1148–1155

Tiodar ED, Văcar CL, Podar D (2021) Phytoremediation and microorganisms-assisted phytoremediation of mercury-contaminated soils: challenges and perspectives. Int J Environ Res Public Health 18(5):2435

Toghueo RM, Zabalgogeazcoa I, de Aldana BV, Boyom FF (2017) Enzymatic activity of endophytic fungi from the medicinal plants Terminalia catappa, Terminalia mantaly and Cananga odorata. S Afr J Bot 109:146–153

Traving SJ, Thygesen UH, Riemann L, Stedmon CA (2015) A model of extracellular enzymes in free-living microbes: which strategy pays off? Appl Environ Microbiol 81(21):7385–7393

Van Der Burgh AM, Postma J, Robatzek S, Joosten MH (2019) Kinase activity of SOBIR1 and BAK1 is required for immune signalling. Mol Plant Pathol 20(3):410–422

Vasundhara M, Kumar A, Reddy MS (2016) Molecular approaches to screen bioactive compounds from endophytic fungi. Front Microbiol 7:203522

Verma N, Thakur S, Bhatt AK (2012) Microbial lipases: industrial applications and properties (a review)

Vimal A, Kumar A (2017) Biotechnological production and practical application of L-asparaginase enzyme. Biotechnol Genet Eng Rev 33(1):40–61

Vionnie WC, Yusuf RZ, Oki T, Wu J, Saez B, Wang X, Cook C, Baryawno N, Ziller MJ, Lee E, Gu H (2016) Epigenetic memory underlies cell-autonomous heterogeneous behavior of hematopoietic stem cells. Cell 167(5):1310–1322

Vogl AE (1898) Mehl und die anderenMehlprodukte der Cerealien und Leguminosen. Nahrungsm Unters Hyg Warenk 12:25–29

Wang L, Han X, Zhu G, Wang Y, Chairoungdua A, Piyachaturawat P, Zhu W (2018) Polyketides from the endophytic fungus Cladosporium sp. isolated from the mangrove plant Excoecariaagallocha. Front Chem 6:344

Wang X, Jiang D, Shi Q, Ren G, Liu C (2022) Microbial degradation of aristolochic acid I by endophytic fungus Ah-Fs-1 of Asarum heterotropoides. Front Microbiol 13:917117

Wen J, Okyere SK, Wang S, Wang J, Xie L, Ran Y, Hu Y (2022) Endophytic fungi: an effective alternative source of plant-derived bioactive compounds for pharmacological studies. J Fungi 8(2):205

Wiewióra B, Żurek G, Pańka D (2015) Is the vertical transmission of Neotyphodiumlolii in perennial ryegrass the only possible way to the spread of endophytes? PLoS One 10(2):e0117231

Wu SH, Zhao LX, Chen YW, Huang R, Miao CP, Wang J (2011) Sesquiterpenoids from the endophytic fungus Trichoderma sp. PR-35 of Paeonia delavayi. Chem Biodivers 8(9):1717–1723

Wu SH, Huang R, Miao CP, Chen YW (2013) Two new steroids from an endophytic fungus Phomopsis sp. Chem Biodivers 10(7):1276–1283

Xie XG, Huang CY, Fu WQ, Dai CC (2016) Potential of endophytic fungus Phomopsis liquidambari for transformation and degradation of recalcitrant pollutant sinapic acid. Fungal Biol 120(3):402–413

Xu XH, Su ZZ, Wang C, Kubicek CP, Feng XX, Mao LJ, Wang JY, Chen C, Lin FC, Zhang CL (2014) The rice endophyte harpophoraoryzae genome reveals evolution from a pathogen to a mutualistic endophyte. Sci Rep 4(1):5783

Xu Z, Tian J, Gan L, Tian Y (2020) Discovery of the endophytic fungi from Polygonum cuspidatum and biotransformation of resveratrol to pterostillbene by the endophyte Penicillium sp. F5. Appl Biochem Microbiol 56:313–320

Yang DH, Hettenhausen C, Baldwin IT, Wu J (2011) The multifaceted function of BAK1/SERK3: plant immunity to pathogens and responses 1 to insect herbivores. Plant Signal Behav 6(9):1322–1324

Yu X, Feng B, He P, Shan L (2017) From chaos to harmony: responses and signaling upon microbial pattern recognition. Annu Rev Phytopathol 55:109–137

Yuan P, Jauregui E, Du L, Tanaka K, Poovaiah BW (2017) Calcium signatures and signaling events orchestrate plant–microbe interactions. Curr Opin Plant Biol 38:173–183

Zhai YJ, Huo GM, Zhang Q, Li D, Wang DC, Qi JZ, Han WB, Gao JM (2020) Phaeosphaones: tyrosinase inhibitory thiodiketopiperazines from an endophytic Phaeosphaeriafuckelii. J Nat Prod 83(5):1592–1597

Zhang Y, Han T, Ming Q, Wu L, Rahman K, Qin L (2012) Alkaloids produced by endophytic fungi: a review. Nat Prod Commun 7(7):1934578X1200700742

Zhang Q, Gao B, Xiao Y, Yang H, Wang Y, Du L, Zhu D (2020) Purification and characterization of a novel β-glucuronidase precisely converts glycyrrhizin to glycyrrhetinic acid 3-O-mono-β-D-glucuronide from plant endophytic Chaetomium globosum DX-THS3. Int J Biol Macromol 159:782–792

Zhao S, Li J, Yang X (2023) Secondary metabolites of Alternaria: a comprehensive review of chemical diversity and pharmacological properties. Front Microbiol 13:1085666

Diversity of Endophytes from Medicinal Plants and Their Effect in Mitigating the Agricultural Problems

4

Nandita Jana, Hiran Kanti Santra, and Debdulal Banerjee

Abstract

Endophytes are hidden microorganisms that live symbiotically and asymptomatically in the intracellular compartments of green plants, and their co-existence gives a beneficial effect on hosts' survival, growth, and metabolism. Endophytes are ubiquitous in occurrence and are reported from almost every plant studied till date. Endophytic fungi, bacteria, and actinobacteria constitute the rich pool of plants' internal ecosystem and open up a new domain of Endophyte Biology. The innumerable occurrence of endophytes has made them a potent biotechnological tool to solve environmental as well as industrial/pharmaceutical problems. Though each and every type of plant holds endophytes, here the focus has been given on medicinal plants only. Medicinal plants contain phytochemicals that have curative properties, which have been used to treat diseases from time immemorial and are an invaluable source of medicines. Endophytic colonization increases levels of phytohormones, modulates the antioxidant system, improves nutrient uptake, and also ameliorates the effects of environmental stress by increasing productivity and resisting plant pathogens by strengthening the immune system. They build more sustainable and environmentally friendly agri-

N. Jana
Centre for Life Sciences, Vidyasagar University, Midnapore, West Bengal, India

H. K. Santra
Microbiology and Microbial Biotechnology Laboratory, Department of Botany and Forestry, Vidyasagar University, Midnapore, West Bengal, India

D. Banerjee (✉)
Centre for Life Sciences, Vidyasagar University, Midnapore, West Bengal, India

Microbiology and Microbial Biotechnology Laboratory, Department of Botany and Forestry, Vidyasagar University, Midnapore, West Bengal, India
e-mail: db@mail.vidyasagar.ac.in

K. A. Abd-Elsalam, S. F. AbuQamar (eds.), *Fungal Endophytes Volume I*,
https://doi.org/10.1007/978-981-97-7312-1_4

cultural practices. Biotic and abiotic stress/problems like increased temperature, low precipitation, high soil salinity, and improper use of synthetic chemicals reduce soil fertility. These environmental factors and plant pathogens affect the growth of agricultural crops. This needs to be tackled immediately, and endophytes along with their metabolites play a major role here. In this chapter, we enumerate the diversity of endophytes from medicinal plants and their potential to reduce abiotic and biotic stress in agricultural crop plants.

Keywords

Endophytes · Medicinal plants · Biotic and abiotic stress · Phytohormones · Secondary metabolites

4.1 Introduction

Endophytic fungi are eukaryotic microorganisms; they have been identified as the most widespread inhabitants that live in plant tissues and are beneficial to the host plant. The term "endophytic" fungi were first coined by De Barry in 1886, derived from the Greek words "endo" meaning "within" and "phyte" meaning "plant", to describe the symbiotic relationship between plants and microorganisms (Stone et al., 2000). The majority of recorded plant–endophyte partnerships are mutualistic, which means that both parties gain from the association. Endophytes are chiefly bacteria (Aguiar-Pulido et al. 2016) fungi (Gupta et al. 2020), and archaea (Moissl-Eichinger et al. 2018), and they spend a major part of their life cycle within the plant cell asymptomatically (Zimmerman and Vitousek 2012).

Various endophytes are being widely used to have plant–endophyte interactions that promote plant growth as bio-fertilizers in agriculture. Endophytic fungi act as natural chemical synthesizers (Owen and Hundley 2004); these natural products or chemicals serve as the main repository for the discovery of novel compounds with a variety of actions, such as antibiotic drugs, antioxidants, and anticancerous, and perform a crucial role in the agricultural practices, medicine, and aesthetic sectors (Tan and Zou 2001). An example of intriguing and economically significant interactions of endophytes with plants is seen in clavicipitaceous endophytes and grasses (Pérez et al. 2020). In the USA, Australia, and New Zealand in particular, *Epichole* sp. has been effectively cultivated and commercialized for agricultural application (Hume et al. 2016).

Endophytes derived from medicinal plants that create unique bioactive substances with a variety of functions and effects that are significant for the pharmaceutical industry might be used to address the demand for fresh medicinal drugs to cure diseases in humans (Gouda et al. 2016). Enzymes generated by endophytes include proteases, cellulases, glucoamylases, lactases, lipases, pectinases, and tannases (Tiwari 2015). Taxol is a plant alkaloid that is produced by endophytic fungi and is used as an anti-cancerous medicine (El-Sayed et al. 2020). Endophytes produce a wide range of bioactive secondary metabolites. The endophyte is able to produce a novel phenolic compound with strong antibacterial and antifungal activity, generating reactive oxygen species (ROS) (Yuan et al. 2016). Endophytic bacterial strains and fungi (Khan et al. 2015) produce essential plant growth hormones such as IAA (Indole Acetic Acid), Gas (Gibberelic acids), and cytokinin.

Endophytes attract an intriguing commitment to agricultural productivity because they improve plant growth, enhance nutrient uptake, control or resist plant diseases (Zhao et al. 2015), and increase abiotic stress tolerance (Zhang et al. 2022) in host plant. Endophytes enhance drought tolerance (Bacon and Hill 1996) and induce salt tolerance in crop plants (Baltruschat et al. 2008). Bacterial and fungal endophytes play a role in bioremediation and prevent contamination by breaking down toxic herbicides, organic pollutants, xenobiotic mercury, and heavy metals present in soil (Mello et al. 2020). Various endophytes are capable of synthesizing both silver and gold nanoparticles (Ganesan et al. 2020), playing an important role in biological control in the plant body through pathogen destruction (Naik 2020).

4.2 Endophytes from Medicinal Plant

4.2.1 Medicinal Plant

Traditional medicine has made considerable use of medicinal plants to boost immunity and human health. Due to their ability to synthesize a variety of therapeutic natural compounds, they have additionally been shown to contain a broad spectrum of pharmacological activities (Akerele et al. 1991). Medicinal plants were traditionally used to treat various diseases. Numerous studies have shown that medicinal plants produce beneficial metabolites; nevertheless, the synthesis of several of these compounds has also been documented from the endophytic microbiota, demonstrating its importance. Studies on endophytes revealed their significant contribution to the quality and amount of metabolites produced by medicinal plants (Berg et al. 2014). This is a result of their bioactive chemicals, which can play a key protective function for plants against a variety of biotic and abiotic challenges. In many countries' economies, medicinal plants are also quite significant. They are a vital component of human healthcare, a source of revenue for rural residents, and a source of supply of raw materials for industry (Table 4.1).

4.3 Endopytes Ameliorate Biotic Stress in Agriculturally Valuable Plants

According to Madani et al. (2019), live pathogens such as viruses, bacteria, fungi, nematodes, insects, arachnids, and weeds can produce biotic stress in plants. Nematodes, which feed on plant tissues, are the main offenders behind diseases transmitted by soil that cause nutritional deficiencies in plants, stunted development, and wilting (Osman et al. 2020). According to Pallas and García (2011), viruses have the ability to cause both local and systemic damage that leads to sickness including stunting and chlorosis. In contrast, insects and mites harm plants by

Table 4.1 A brief representation of potent endophytes isolated from different medicinal plants

Name of medicinal plant	Family	Isolated endophytes	References
Aegle marmelos	Rutaceae	**Fungi** *Bartalinia robillardoides*, *Xylaria psidii*	Gangadevi and Muthumary (2008), Arora et al. (2016)
		Bacteria *Kosakonia cowanii*	Panigrahi and Rath (2019)
Aquilaria sinensis	Thymelaeaceae	**Fungi** *Nemania aquilariae*, *Nemania yunnanensis*	Tibpromma et al. (2021)
		Bacteria *Acinetobacter radioresistens*, *Bacillus* sp.	Bhore et al. (2013a, b)
Azadirachta indica	Meliaceae	**Fungi** *Curvularia lunata*, *Phomopsis oblanga*, *Cladosporium cladosporioides*	Chukwuemerie et al. (2022)
		Pestalotiopsis sp., *Trichoderma* sp. and *Aspergillus* sp.	Verma et al. (2007)
		Actinomycetes *Streptomyces* sp., *Microspora* sp., *Saccharomonospora* sp., *Nocardia* sp.	Verma et al. (2009)
Camellia sinensis	Theaceae	**Fungi** *Aspergillus niger*, *Penicillium sclerotiorum*, *Fusarium oxysporum*	Xie et al. (2020)
		Bacteria *Alcaligenes* sp., *Brevibacillus* sp., *Microbacterium* sp., *Bacillus* sp.	Hazarika et al. (2021)
		Actinomycetes *Streptomyces* sp., *Actinomadura* sp., *Kribbella* sp., *Nocardia* sp., *Kytococcus*, *Leifsonia*, *Microbacterium*, *Micromonospora* sp., *Mobilicoccus* sp., *Mycobacterium* sp., *Nocardiopsis* sp., *Piscicoccus* sp., and *Pseudonocardia* sp.	Shan et al. (2018)
Camptotheca acuminata	Comaceae	**Fungi** *Penicillum polonicu*, *Diaporthe* sp.	Wen et al. (2020), Liu et al. (2021)
		Actinobacteria *Blastococcus endophyticus* sp.	Zhu et al. (2013)
Catharanthus roseus	Apocynaceae	**Fungi** *Fusarium solani*, *Chaetomium funicola*, *Penicillium rugulosumola*	Linh et al. (2021)
		Bacteria *Microbacterium* sp.	Anjum and Chandra (2019)
		Actinomycetes *Microbispora catharanthi*	Klykleung et al. (2020)

(continued)

Table 4.1 (continued)

Name of medicinal plant	Family	Isolated endophytes	References
Corchorus olitorius (*jute*)	Malvaceae	**Fungi** *Grammothele lineata*	Das et al. (2017)
		Bacteria *Staphylococcus* sp., *Micrococcus* sp., *Ralstonia solanacearum*	Haidar et al. (2018)
Crocus sativus	Iridaceae	**Fungi** *Rhizopus oryzae*, *Aspergillus fumigatiaffinis*, and *Aspergillus niger*	Chamkhi et al. (2018)
		Bacteria *Bacillus* sp., *Paenibacillu selgii*, *Brevibacillus* sp., *Pseudomonas putida*, *Staphylococcus hominis*, and *Enterobacter cloacae*	Sharma et al. (2015)
		Actinomycetes *Streptomyces yangpuensis*	Tian et al. (2022)
Cynodon dactylon	Poaceae	**Fungi** *Curvularia tsudae*, *Aspergillus flavus*	Nischitha et al. (2020), Kalimuthu et al. (2022)
Cyperus rotundus	Cyperaceae	**Fungi** *Aspergillus terreus*, *Fusarium oxysporum*	Kusmiyati et al. (2021)
Dicerandra frutescens	Lamiaceae	*Phomopsis longicolla*	Wagenaar and Clardy (2001)
Glycine max	Leguminosae	**Fungi** *Cladosporium sphaerospermum*	Hamayun et al. (2009)
		Bacteria *Enterobacter* sp., *Acinetobacter* sp., *Pseudomonas* sp., *Ochrobactrum* sp., and *Bacillus* sp.	Zaho et al. (2018)
		Actinomycetes *Plantactinospora soyae* sp., *Streptomyces* sp., *Micromonospora* sp., *Nocardia* sp., *Actinomadura* sp., *Microbispora* sp., and *Actinoplanes* sp.	Guo et al. (2016), Dalal and Kulkarni (2014)
Mentha pulegium	Lamiaceae	**Fungi** *Penicillium griseofulvum*, *Stemphylium globuliferum*	Debbab et al. (2009)
Nymphaea nouchali	Nymphaeaceae	**Fungi** *Chaetomium globosum*, *Trichoderma ghanense*, *Phyllosticta capitalensis*, *Curvularia papendorfii*, *Endomelanconiopsis* sp.	Dissanayake et al. (2016)

(continued)

Table 4.1 (continued)

Name of medicinal plant	Family	Isolated endophytes	References
Oryza sativa	Poaceae	**Bacteria** *Pantoea* sp., *Methylobacterium* sp., *Sphingomonas* sp., *Rhizobium* sp., *Microbacterium* sp. and *Pseudomonas* sp.	Liu et al. (2019)
Osimum sp.	Lamiaceae	**Fungi** *Exserohilum rostrata*, *Aspergillus nidulans*, *Aspergillus fumigatus*, and *Aspergillus flavus*	Bagur et al. (2020), Sharaf et al. (2022)
		Bacteria *Enterobacter cloacae*	Panigrahi et al. (2020)
		Actinomycetes *Streptomyces* sp.	Ayswaria et al. (2022)
Piper aduncum	Piperaceae	**Fungi** *Xylaria* sp., *Penicillium* sp.	Oliveira et al. (2011)
Taxus baccata	Taxaceae	**Fungi** *Gliocladium* sp., *Paraconiothyrium variabile*	Sreekanth et al. (2011), Somjaipeng et al. (2015)
Taxus brevifolia	Taxaceae	**Fungi** *Taxomyces andreanae*, *Pestalotiopsis* sp.	Strobel et al. (1993), Pulici et al. (1996)
		Bacteria *Paenibacillus kribensis*	Islam et al. (2018)
Tectona grandis	Lamiaceae	**Fungi** *Alternaria* sp., *Colletotricum* sp., *Nigrospora* sp., *Phomposis* sp.	Chareprasert et al. (2006)
		Bacteria *Curtobacterium* sp., *Ochrobactrum* sp., and *Bacillus* spp.	Pinto et al. (2022)
Terminalia arjuna	Combretaceae	**Fungi** *Pestalotiopsis terminaliae*, *Chaetomella raphigera*, *Alternaria brassicicola*	Gangadevi and Muthumary (2009), Gill and Vasundhara (2019)
		Actinomycetes *Streptomyces* sp., *Microbispora* sp., *Micromonospora* sp.	Gohain et al. (2015)
		Bacteria *Cytobacillus firmus*	Sudarsan et al. (2021)
Vitex negundo	Verbenaceae	**Fungi** *Colletotrichum gloeosporioides*, *Phomopsis archeri*, *Aspergillus flavus*, *Nigrospora sphaerica*	Desale and Bodhankar (2013)
		Bacteria *Bacillus* sp.	Ali and Vora (2014)

(continued)

Table 4.1 (continued)

Name of medicinal plant	Family	Isolated endophytes	References
Vitis vinifera	Vitaceae	**Fungi** *Cladosporium cladosporioides* and *Cladosporium herberum*, *Didymella* sp., *Aspergillus* sp., *Aureobasidium* sp.	Kulišová et al. (2021), Nigris et al. (2018)
		Bacteria *Bacillus licheniformis*, *Burkholderia* sp.	Compant et al. (2005)
Rouwolfia serpentina	Apocynaceae	**Fungi** *Colletotricum gloeosporioides*, *Penicillium* sp. and *Aspergillus awamori*	Nath et al. (2015)
		Curvularia sp.	Kaaniche et al. (2019)
		Bacteria *Pseudomonas aeruginosa*	Pal and Paul (2020)
Emblica officinalis	Phyllanthaceae	**Fungi** *Phomopsis* sp. and *Xylaria* sp.	Nath et al. (2012)
		Nigrospora oryzae	Rathod et al. (2014), Gangwar et al. (2015)
		Actinomycetes *Streptomyces* sp., *Micromonospora* sp. and *Microbispora* sp., *Micrococcus* sp., *Mycobacterium* sp., *Kitasatospora* sp., *Promicromonospora* sp.	Gohain et al. (2019)
Aloe vera	Asphodelaceae	**Fungi** *Chaetomium globosum*, *Trichoderma harzanium*	Chowdhary and Sharma (2020)
		Bacteria *Proteobacteria*, *Firmicutes*, *Bacteroidetes* and *Actinobacteria*	Silva et al. (2020)
Zingiber officinale	Zingiberaceae	**Fungi** *Colletotrichum gloeosporoides* and *Glomerella cingulata*, *Curvularia affinis* and *Cochliobolus geniculatus*	Ginting et al. (2013)
		Bacteria *Enterobacter cloacae*, *Enterobacter hormaechei*, *Burkholderia vietnaminesis*, *Pantoea agglomerans*	Rohini et al. (2018)
		Actinimycetes *Nocardiopsis* sp.	Pujiyanto et al. (2012)

laying eggs or eating (penetrating as well as sucking). The two types of fungal parasites are necrotrophic (destroy host cells by secreting toxins) and biotrophic (feed on live host cells). They have the ability to cause vascular wilts, leaf spots, and cankers in plants (Sobiczewski et al. 2017). Other viruses and bacteria can be transmitted by insects (Schumann and D'Arcy 2006). To fight these stressors, plants have evolved a complex immune system (Saijo and Loo 2020). By creating physical barriers, chemical substances, proteins, and enzymes, plants protect themselves against biotic stress. They give plant components strength and stiffness, which protects them and enables them to withstand biotic pressures.

4.3.1 Protection from Fungal and Bacterial Pathogens

Pathogenic fungi significantly reduce crop yield and quality, causing major economic losses in agriculture. They produce mycotoxins that are hazardous to human and animal health, secreted by fungal pathogens. For the control of pathogenic fungi, synthetic chemical fungicides have become a staple in agriculture. However, just like other pesticides, fungicides can have unintended negative effects on the environment, they can harm fungi that are essential for crop health, like arbuscular mycorrhizae, whose damage can significantly worsen plant health (Wilkes et al. 2020; Vázquez et al. 2021). Additionally, fungicides might deliberately harm pests and also beneficial bacteria if they are not the intended target (Reiff et al. 2021). The rate of photosynthesis, chlorophyll content, trichomes and stomata of plant tissues, antioxidant enzyme activity, cell lignifications, and phytoalexin accumulation have all been shown to increase and subsequent metabolites SA (salicylic acid), JA (jasmonic acid), and ethylene secretion increase with endophyte colonization (Qin et al. 2021). In addition to these defence mechanisms, endophytes and pathogenic fungi may be excluded through competition (Oliva et al. 2021).

4.3.2 Protection from Leaf-Cutting Ants

Endophytic fungi, like *Trichoderma* sp., have the potential to protect their plant hosts by acting as bodyguards, and could therefore be used in a Trojan horse strategy to lessen the negative effects of leaf-cutting ants in crops (Rocha et al. 2017). Endophytes are harmful to the ants' fungal symbiont, and cutting and processing leaves containing endophytes may be effective for the ants. A wide variety of compounds are produced by endophytes, some of which are threatening to bacteria and other fungus. Leaf-cutting ants, a significant herbivore in plant ecosystems, particularly in the Neotropics, have been linked to host choices by leaf endophytes (Van Bael et al. 2011). Leafcutter ants cut and harvest approximately 2% of the total leaf biomass in forests, significantly impacting crop production and causing substantial agricultural losses, estimated at around $1 billion per year (Estrada et al. 2014). An endophytic fungus Colletotrichum tropicale colonized in Merrima umbellate plants restricted leaf-cutting ants from cutting and harvesting of leaf.Plants with a higher degree of colonization of C. tropicale are more resistant to ants than plants with less colonization of the endophytic fungi (Van Bael et al. 2009).

4.3.3 Protection from Nematodes

Nematodes are common soil pests that affect plant parts, including stems, leaves, roots, flowers, and seeds. Nematodes that parasitize plants can be generically categorized as endoparasitic or ectoparasitic depending on where they feed. The root-knot caused by *Meloidogyne* spp. and cyst nematodes caused by *Heterodera* sp. and *Globodera* spp. are the most significant worms in terms of crop losses, according to Jones et al. (2013). Nematodes cause sores on plant parts where they feed and take nutrients, which allow secondary pathogens like fungi, bacteria, and viruses to infect the plant (Caboni et al. 2016). Chemical-based nematicides have been used to prevent nematodes from existing and proliferating. Chemical treatments, like other pesticides, might have an unwanted effect that harms a number of advantageous microbial communities in the local rhizosphere (Kumar et al. 2019). There are many reports that endophyte colonization can control nematode invasion. After being inoculated with the endophytic fungi *Fusarium redolens*, *Phialemoniopsis cornearis*, and *Macrophomina pseudophaseolina* the medicinal plant *Coleus forskohlii* produces the secondary metabolite forskolin because they increase the expression of diterpene synthases, which improves tolerance to *M. incognita* (Mastan et al. 2019).

Many fungal genera have been reported as having inhibitory effects on *Meloidogyne* species, including *Acremonium* sp., *Arthrobotrys* sp., *Chaetomium* sp., *Cladosporium* sp., *Diaporthe* sp., *Epiccocum* sp., *Fusarium* sp., *Gibellulopsis* sp., *Melanconium* sp., *Metacordyceps* sp., *Monacrosporium* sp., *Neotyphodium* sp., *Phialemonium* sp., *Phyllosticta* sp., *Piriformospora* sp., *Purpureocillium* sp., and *Trichoderma* sp. (Kumar and Dara 2021). It has been demonstrated that compounds produced by the nematode-invading strains *Alternaria* sp., *Fusarium* sp., *Cladosporium* sp., *Clonostachys* sp., *Phyllosticta* sp., *Piriformospora* sp., *Chaetomium* sp., and *Trichoderma* sp. alter the chemical composition of host plant metabolites or increase their production and induce increased resistance (Zhou et al. 2020).

4.3.4 Protection from Mites and Insects

Phytophagous mites play a significant role as pests in agriculture and horticulture worldwide. They cause harm by feeding on plants and can transmit viruses, resulting in reduced photosynthetic capacity, overall plant health, yield, and market value. Mite pests exhibit a remarkable ability to infest a wide range of hosts. For example, the two-spotted spider mite (*Tetranychus urticae*), considered an economically important mite species, has been documented on more than 1000 plant species spanning 130 different plant families, causing substantial production losses in economically relevant crops. The citrus red mite (*Panonychus citri*) and European red mite (*P. ulmi*) infect cucurbits, beans, hops, grapes, apples, and strawberries (Knapp et al. 2020).

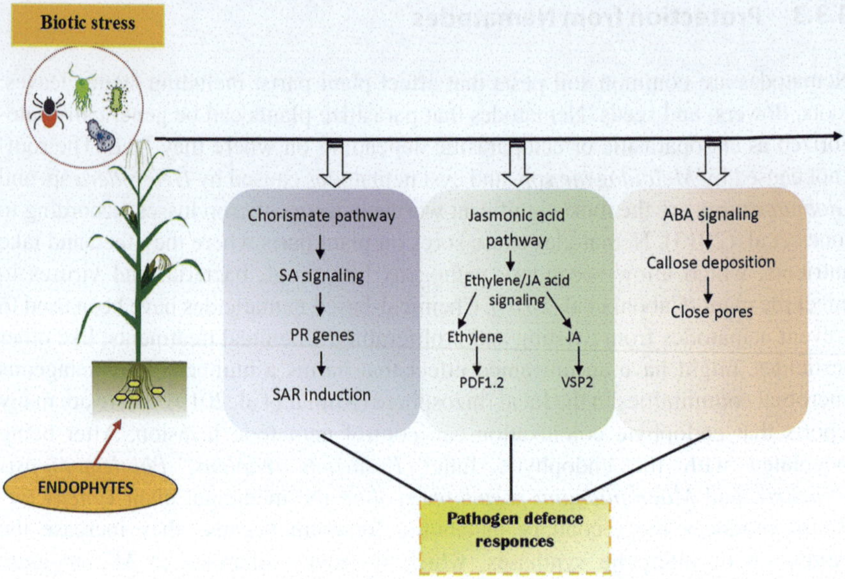

Fig. 4.1 The schematic representation of the response in endophyte colonized plant tissues under biotic stress. The critical cellular signaling molecules associated with defense against pathogenic attack get activated and cumulatively shield the host plant. SA, JA, PR, SAR, and ABA stand for salicylic acid, jasmonic acid, pathogenesis-related protein, systemic acquired resistance, and abscisic acid respectively

Fusarium solani Colonization of endophytic fungi Fusarium solani, Beauveria bassiana, Metarhizium brunneum, and M. robertsii protects tomato plants (Solanum lycopersicum) against two-spotted spider mite (Tetranychusurticae) (Pappas et al. 2018; Rasool et al. 2021). *Metarhizium robertsii* (ESALQ 1622) and *Beauveria bassiana* (ESALQ 3375) are two entomopathogenic fungal isolates that have been used to inoculate strawberry roots and have been shown to have an impact on *Tetranychus urticae*, a naturally occurring arthropod pest (Canassa et al. 2020). Either *B. bassiana* or *Purpureocillium lilacinum* were shown to have foliar endophyte colonization that controlled the aphid (*Aphis gossypii*) infection in field experiments (Castillo Lopez et al. 2014) (Fig. 4.1 and Table 4.2).

4.4 Endophytes in Ameliorating Abiotic Stress

Drought, salinity, heavy metal stress, low/high-temperature acidic conditions, excess light, and nutrient starvation are some of the major abiotic stresses. Because plants are immobile, they try to withstand abiotic conditions including drought, salt, and high temperatures, hazardous metals in the soil, including cadmium, arsenate, and aluminumthat limit plant production in agriculture and endanger food security. According to Zhang et al. (2022), these stresses disrupt plant growth and development and lower agricultural production.

Table 4.2 Endophytes promote biotic stress tolerance of plant

	Host plant	Application in biotic stress control	References
Endophytic fungi			
Alternaria tenuissima	*Cornus officinalis*	Against *Colletotrichum gloeosporioides*, *Alternaria alternate* and *Botryosphaeria dothidea*	Zaho et al. (2020)
Aspergillus neoniger	*Ficus carica*	Controls pathogen activity (*Penicillium avelaneum*, *P. notatum*, and *Aspergillus terreus*)	Abdou et al. (2021)
Aspergillus, *Penicillium*, and *Trichoderma*	*Eucalyptus benthamii*	Target *Botrytis cinerea*	Sbravatti Júnior et al. (2013)
Beauveria bassiana, *Metarhizium anisopliae*, *Isaria fumosorosea*	*Capsicum annum*	Control green peach aphid (*Myzuspersicae*)	Mantzoukas and Lagogiannis (2019)
Colletotrichum acutatum	*Angelica sinensis*	Antimicrobial, antibiofilm, antioxidant, antimalarial, and anti-proliferative properties	Yehia (2023)
Colletotrichum gloeosporioides	*Artemisia mongolica*	Produce colletotric acid, antimicrobial activity	Zou et al. (2000)
Daldinia cf. *concentric*	*Olea europaea*	Nematicidal activity, against *Meloidogyne javanica*	Liarzi et al. (2016)
Epicoccum nigrum	Sugarcane	Biocontrol activity, inhibits pathogens, such as *Sclerotinia sclerotiorum*, *Pythium*, and *Monilinia* spp. in sunflower, cotton, and apple respectively	Fávaro and Sebastianes (2012)
Fusarium oxysporum sensulato	*Musa* spp.	Antifungal activity, against *Pratylenchus goodeyi*	Mwaura et al. (2010)
Fusarium solani sensulato	*Vitis labrusca*	Against *Botrytis* sp.	Brum (2008)
Hypoxylon pulicicidum		Produces apantropical, insecticidal effect	Bills et al. (2012).
Metarhizium brunneum	Melon (*Cucumis melo*)	Leaf worm (*Spodoptera littoralis*)	Miranda-Fuentes et al. (2021)
Muscodor vitigenus	*Paullina paullinioides*	Controls insect attack	Daisy et al. (2002)
Nodulisporium sp.	*Bontia daphnoides*	Produces nodulisporic acids, exhibit insecticidal activity against the blowfly larvae	Strobel and Daisy (2003)
Phomopsis liquidambaris	Peanut	Against leaf spot and root rot causing pathogens	Zhang et al. (2020a)
Sarocladium strictum	*Cynanchum acutum*	Against insect pest, *Spodoptera littoralis*	El-Sayed et al. (2020)

(continued)

Table 4.2 (continued)

	Host plant	Application in biotic stress control	References
Sarocladiumzeae	Wheat	Implicated in biocontrol activity against *Fusarium* sp. (head blight disease)	Kemp et al. (2020)
Trichophyton sp.	*Symphytum officinale*	Against *Sclerotinia sclerotiorum*	Rocha et al. (2009)
Xylariales sp.	*Distylium chinense*	Against *Clavibacter michiganensis*, *Xanthomonas citri* pv. phaseoli var. fuscans, and *Pseudomonas syringae* pv. *lachrymans*	Zhang et al. (2020b)
Endophytic actinomycetes			
Streptomyces sp.	Soybean	Against blight disease caused by *Xanthomonas campestris* pv. glycines	Mingma et al. (2014)
Streptomyces sp.	*Lathyrus odoratus*	Against powdery mildew *Oidium* sp.	Sangmanee et al. (2009)
Streptomyces sp.	*Nicotiana tabacum*	Against brown spot *Alternaria* spp.	Gao et al. (2014)
Streptomyces sp. MBFA-172	Strawberry	Biocontrol activity, against *Glomerella cingulata*	Marian et al. (2020)
Streptomyces viridodiasticus	*Lactuca sativa*	Against basal drop disease *Sclerotinia minor*	El-Tarabily et al. (2000)
Endophytic bacteria			
Achromobacter xylosoxidans	Rice	Inhibits growth of *Magnaporthe oryzae* which cause rice blast	Joe et al. (2012)
B. amyloliquefaciens, *B. subtilis*, *B. pasteurii*, *B. cereus*, *B. pumilus*, *B. mycoides*, and *B. sphaericu*	Sugar beet, tobacco, Arabidopsis	Against crown-rotting fungal pathogen, nematodes, and stem-blight fungal pathogen	Kloepper et al. (2004)
Bacillus pumilus strain SE34	Pea	Produces phenolic compound, resistance to *Fusarium oxysporum* sp.	Benhamou et al. (1996)
Bacillus sp.	Olive	Antagonistic activity, against *Xylella fastidiosa*	Zicca et al. (2020)
Burkholderia cepacia (B3) and *Pseudomonas aeruginosa* (*P3*)	Oil palm	Inhibit basal stem rot caused by *Ganoderma boninense*	Sapak et al. (2008)
P. aeruginosa and *P. monteilii*	Ginger	Antimicrobial activity, against *Pythium myriotylum* and *Phyllosticta zingiberi*	Chen et al. (2014)
Pseudomonas aeruginosa PW09	Cucumber	Defence *Sclerotium rolfsii* infection	Pandey et al. (2012)
Pseudomonas fluorescens (Pf1) and *endophytic Bacillus* spp. (*EPB22*)	Banana	Produce pathogenesis-related (PR) proteins, defense-related proteins Protect from banana bunchy top virus (BBTV)	Harish et al. (2009)

(continued)

Table 4.2 (continued)

	Host plant	Application in biotic stress control	References
Pseudomonas fluorescens, *Burkholderia cepacia*	Eggplant	Produce 2,4-diacetylphloroglucinol, inhibit wilt caused by *Ralstonia solanacearum*	Ramesh et al. (2009)
Pseudomonas monteilii and *Glomus fasciculatum*	*Coleus forskohlii*	Inhibit root rot (*Fusarium chlamydosporum*) and wilt (*Ralstonia solanacearum*)	Singh et al. (2013)
Pseudomonas putida, *Bacillus megaterium* (*B. megaterium EU071712*)	Black pepper	Against *Phytophthora capsici* causing foot rot disease	Aravind et al. (2009)

According to FAO data, salinity or water logging has caused damage to 20% of irrigated land in developing countries. Nearly 1 billion people are in danger, and approximately 250 million people have been directly impacted by desertification. Forty-five percent of the world's land surface has been damaged by drought and 20% of cultivated land by salt stress (Rojas 2020). The main abiotic stressors, their harmful effects on plants, and the part that different endophytes play in abiotic stress tolerance are further detailed.

4.4.1 Endophytes Alleviate Drought Stress in Plant

Azospirillum lipoferum, a plant growth-promoting bacterium (PGPB), was inoculated into maize (*Zea mays* L.) under drought-stress conditions. *Azospirillum lipoferum* promoted maize growth by producing phytohormones such as ABA (Abscisic acid), IAA, and GAs. The findings indicate that ABA and GAs help host plants cope with drought stress (Cohen et al. 2009). Wheat, tomato, and watermelon plants are much more tolerant to drought when *Colletotrichum magna* and *C. protuberate* 4666D are present (Redman et al. 2001). *Phaseolus vulgaris* treated with *Rhizobium etli* was increase drought tolerance by increasing trehalose-6-phosphate synthase, which is up-regulated in response to stress and is important in carbon and nitrogen metabolism (Suarez et al. 2008). When *Populus deltoids* had inoculated with endophytes consortia under severe drought stress conditions, including *Rhodotorula graminis*, *Burkholderia vietnamiensis*, and *Rhizobium tropici*, the host plant's damage from reactive oxygen species was reduced (Khan et al. 2016). In water-stressed soybean plants, the effects of cotton stick-derived biochar in combination with PGPR strains (*Paraburkholderia phytofirmans* and *Bacillus* sp.) enhanced plant growth in dry, arid areas (Nawaz et al. 2023). According to Xu et al. (2017), *Piriformospora indica* may protect plants against the damaging effects of drought by increasing antioxidant enzyme activity, proline buildup, and the expression of genes relevant to drought, as well as reducing damage to membranes in maize plants. As demonstrated by Naveed et al. (2014), *Burkholderia phytofirmans* and *Enterobacter* sp. The synthesis of drought-responsive enzymes including peroxidase (POD), catalase (CAT), and superoxide dismutase (SOD) may be to blame for the enhanced drought tolerance.

4.4.2 Endophytes Alleviate Salt Stress in Plant

Piriformospora indica, one of the most extensively researched root endophytes, has been shown to stimulate development in a variety of plant systems when subjected to abiotic challenges, such as salt stress (Varma et al. 2012). Under salt stress, *P. indica* in Arabidopsis keeps the Na^+ and K^+ homeostasis stable (Arshad et al. 2017). Chickpeas are able to tolerate salt because of two bacterial endophytes, *Mesorhizobium ciceri* and *Bacillus subtilis*, which increase proline content and lower H_2O_2 concentrations (Egamberdieva et al. 2010). By boosting the activity of the essential enzyme for ethylene production, 1-aminocyclopropane-1-carboxylate deaminase, *Pseudomonesa fluorescens* and *P. migulae* reduce salt stress in tomato plants (Ali et al. 2014).

Aspergillus flavus CHS1-mediated salt tolerance in *Glycine max* L. was the enhancement of the host's endogenous hormone levels and antioxidant system (Asaf et al. 2018). According to Waqas et al. (2017), the presence of antioxidant enzymes at high concentrations dramatically decreased ROS-based negative effects such as lipid peroxidation, CAT activity, and peroxidase activities. IAA, siderophore production, nutrient solubilization, and ACC deaminase are just a few of the traits that rhizobial isolates (*Rhizobium phaseoli* S18 and *Rhizobium pusense* S6R2) and passenger endophytes (*Pantoea dispersa* YBB19B and *Bacillus tequilensis* NBB13) showed under salt-stressed conditions (Oviya et al. 2023) in groundnut.

Six weeks after inoculation, *Periconia macrospinosa*, *Cadophora* sp., and *Leptodontidium* sp. increased tomato shoot and root development under moderate and high salt stress conditions (Gaber et al. 2023). Plant growth-promoting bacterial endophytes Pseudomonas fluorescens YsS6 and P. migulae 8R6 with potent 1-aminocyclopropane-1- carboxylase (ACC) deaminase activity ensures the tomato plants' survival under salt stress (Ali et al. 2014).

4.4.3 Heavy Metal Stress

Fe, Mn, Zn, Cu, Mg, Mo, and Ni are considered essential micronutrients for normal plant growth, whereas Cd, Sb, Cr, Pb, As, Co, Ag, Se, and Hg are considered nonessential elements that reduce or inhibit plant growth (Rascio and Navari-Izzo 2011). The majority of the world's agricultural land is only mildly or moderately affected by metal toxicity. This is a result of anthropogenic activities, ongoing industrialization, and intense agricultural practices, including the prolonged use of fertilisers, the application of sludge, and poor irrigation (Passariello et al. 2002). Both vegetation and human health are severely impacted by these heavy metals. According to

Duruibe et al. (2007), heavy metals are nondegradable metallic elements with a density of more than 4 g/cm^3 that are hazardous at minute concentrations. Due to numerous human activities, the permissible limit of cadmium in agricultural soil (100 mg/kg) (Salt et al. 1995) is regularly exceeded.

According to Emamverdian et al. (2015), physical adaptations against metals in plants include thick cell walls, morphological features like a thick cuticle and biologically active tissues like trichomes plants exhibit avoidance behaviours or strategies to stop metal from entering their roots as first line of defence against metal toxicity (Viehweger 2014). If heavy metal ions are able to pass through biophysical barriers, plants will begin to develop tolerance mechanisms by inducing metallochaperones or chelators like nicotianamine, spermine, putrescine, organic acids, glutathione (Bricker et al. 2001), phytochelatins, and metallothioneins as well as cellular exudates like protons, heat shock proteins, specific amino acids like proline and histidine, and hormones like SA, JA, and ethylene (John et al. 2012) (Fig. 4.2 and Table 4.3).

4.5 Future Perspectives

Endophytic microorganisms potently alleviate both biotic and abiotic stresses in plants and opennew avenues for sustainable agricultural practices.

Understanding the precise mechanisms by which endophytes confer stress tolerance to their host plants is essential. Further research is needed to decipher the signaling pathways and effector molecules involved in the beneficial interactions between endophytes and plants. Advances in omics technologies (Crandall et al. 2020) including genomics, transcriptomics, and proteomics, will provide valuable insights into these complex interactions. Genetic modification of endophytes to enhance their stress-alleviating capabilities represents a promising avenue for future research. By manipulating the genetic makeup of endophytic strains, we can potentially enhance their ability to combat specific biotic and abiotic stresses. This approach could involve introducing genes responsible for the production of stress-responsive proteins or engineering endophytes to produce specific metabolites that alleviate stress in plants. Adopting a multi-strain and multi-trait approach can enhance the effectiveness of endophytes in stress alleviation. Utilizing a combination of endophytic strains with complementary stress-alleviating traits can provide synergistic effects and maximize the benefits for plants. This approach requires careful selection and screening of compatible endophytic strains with diverse stress-tolerance mechanisms. Moving beyond laboratory and greenhouse studies, it is crucial to evaluate the efficacy of endophytes in real-world agricultural systems. Assessing the long-term effects of endophyte application on crop productivity, sustainability, and ecological balance will be critical for their successful integration into agricultural practices.

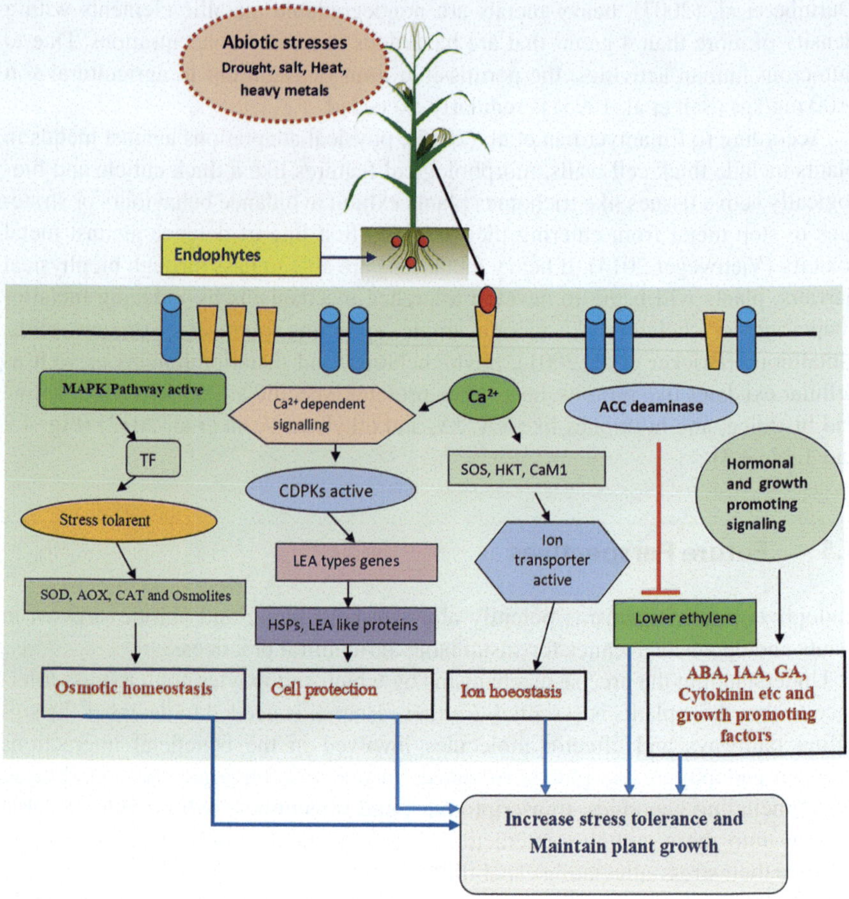

Fig. 4.2 Role of endophytic colonization in managing abiotic stress in plants. Colonization with endophytes activates different signaling pathways to tolerate abiotic stresses e.g. switches on antioxidant enzymes like SOD (Super Oxide Dismutase), AOX (Alternative Oxidase), and CAT (Catalase) to reduce ROS (Reactive Oxygen Species) stress-responsive proteins HSP (Heat Shock Protein), LEA (Late Embryogenesis Abundant) and protect plant cells. Also, activate SOS membrane protein and increase ion homeostasis. ACC deaminase signaling decreases ethylene production. Increase important plant hormone secretion (IAA, GA, Cytokinin, etc.) conducive to abiotic stress tolerance and maintain plant growth

4.6 Conclusions

In conclusion, endophytic microorganisms have demonstrated their immense potential in various aspects of plant health and productivity. Their ability to trigger numerous mechanisms within plants provides protection against both biotic and abiotic stresses, making them valuable tools for bioprospecting and genetic modification of plants. The secondary metabolites produced by endophytes exhibit significant

Table 4.3 Endophytes promote abiotic stress tolerance in plant

	Source	Application in abiotic stress control	References
Endophytic fungi			
Aspergillus ochraceus	*Hordeum vulgare*	Seawater stress	Badawy et al. (2021)
Chaetomium cupreum	*Miscanthus sinensis*	Stabilizing of metal (Al) stress	Haruma et al. (2019)
Cochliobolus sp.	*Ablemoschus esculentus*	Tolerance to salinity	Bibi et al. (2019)
Epichloë bromicola	*Hordeum vulgare*	Tolerance to salinity	Chen et al. (2019)
Fusarium oxysporum sensulato	*Oryza sativa*	Tolerance to salinity	Manasa et al. (2020)
Fusarium verticillioides and *Humicola* sp.	*Glycine max*	Tolerance to salinity	Radhakrishnan et al. (2015)
Paecilomyces formosus and *Penicillium funiculosum*	*Glycine max*	Alleviate metal (Ni, Cd, and Al) toxicity and heat	Bilal et al. (2020)
Penicillium citrinum, *Aureobasidium pullulans*, and *Dothideomycetes* sp.	*Citrus reticulata*	Tolerance to drought	Sadeghi et al. (2020)
Penicillium roqueforti	*Triticum aestivum*	Accumulate heavy metal (Ni, Cd, Cu, Zn, and Pb)	Zhu et al. (2018)
Phialophora mustea	*Lycopersicon esculentum*	Stabilize metal (Cd and Zn) stress	Ikram et al. (2018)
Phoma glomerata and *Penicillium* sp.	*Cucumis sativus*	Salinity	Waqas et al. (2017)
Piriformospora indica	*Solanum lycopersicum* and *Oryza sativa*	Alleviate salinity	Abdelaziz et al. (2019), Jogawat et al. (2013)
Piriformospora indica	*Zea mays*	Drought	Guler et al. (2016)
Piriformospora indica	*Medicago truncatula*	Salinity	Li et al. (2017)
Sinorhizobium melilotii	*Medicago sativa*	Saline stress	Gao et al. (2023)
Trametes hirsuta	*Triticum aestivum*	Metal (Pb)	Malik et al. (2020)
Trichoderma atroviride	*Zea mays*	Drought	Guler et al. (2016)
Trichoderma longibrachiatum	*Triticum aestivum*	Salinity	Zhang et al. (2016)
Yarrowia lipolytica	*Glycine max*	Salinity	Gul Jan et al. (2019)
Endophytic bacteria			
Pseudomonas pseudoalcaligenes along with *Bacillus pumilus*	*O. sativa*	Accumulates compatible solutes (i.e., glycine betaine) and reduces proline content (5%) to protect the host from osmotic stress	Jha et al. (2011)

(continued)

Table 4.3 (continued)

	Source	Application in abiotic stress control	References
Serratia marcescens strain SRM	Summer squash (*Cucurbita pepo*)	Influence seedling growth at cold temperatures. Solubilize P and produce IAA	Selvakumar et al. (2008)
Neotyphodium uncinatum	*Elymus dahuricus*	Antioxidative enzymes activities (AEA) and proline content increased	Zhang et al. (2012)
Staphylococcus arlettae	*Brassica juncea*	Arsenic reductase activity Increased soiled hydrogenase, phosphatase, and available phosphorus	Srivastava et al. (2013)
Pseudomonas koreensis AGB-1	*Miscanthus sinensis*	High tolerance to metal(loid) concentrations through extracellular sequestration, increased catalase and SOD activities in plants	Babu et al. (2015)
Pseudomonas aeruginosa PC5	*Prosopis cineraria*	Reducing the disease incidence under salt water irrigation	Al-Rashdi et al. (2023)

biological activity and hold great potential as valuable products. Exploring and characterizing endophytic microorganisms from unexplored plant species can lead to the discovery of new species capable of producing substances of interest. These substances can be used in biological control strategies against diseases, as elicitors of induced resistance, and in inducing tolerance to abiotic stresses. By acting as biological triggers, endophytes help plants adapt to adverse climatic conditions such as drought, heat, high salinity, nutrient deficiencies, and biotic stresses. As a result, they contribute to crop productivity and can be employed as eco-friendly approaches for sustainable agriculture and soil improvement.

References

Abdelaziz ME, Abdelsattar M, Abdeldaym EA, Atia MA, Mahmoud AWM, Saad MM, Hirt H (2019) Piriformospora indica alters Na^+/K^+ homeostasis, antioxidant enzymes and LeNHX1 expression of greenhouse tomato grown under salt stress. Sci Hortic 256:108532

Abdou R, Alqahtani AM, Attia GH (2021) Bioactive metabolites of *Aspergillus neoniger*, an endophyte of the medicinal plant *Ficus carica*. Indian J Pharm Sci 83(1):101–109

Aguiar-Pulido V, Huang W, Suarez-Ulloa V, Cickovski T, Mathee K, Narasimhan G (2016) Metagenomics, metatranscriptomics, and metabolomics approaches for microbiome analysis: supplementary issue: bioinformatics methods and applications for big metagenomics data. Evol Bioinform 12:15–16

Akerele O, Heywood V, Synge H (1991) Conservation of medicinal plants. Cambridge University Press

Ali MM, Vora D (2014) Potential of endophytic bacteria isolated from Vitex negundo L. to produce auxin. Res J Recent Sci 3:38–42

Ali S, Charles TC, Glick BR (2014) Amelioration of high salinity stress damage by plant growth-promoting bacterial endophytes that contain ACC deaminase. Plant Physiol Biochem 80:160–167

Al-Rashdi A, Al-Sadi AM, Al-Harrasi MMA, Al-Sabahi JN, Janke R, Velazhahan R (2023) The effect of NaCl on growth and volatile metabolites produced by antagonistic endophytic bacteria isolated from Prosopis cineraria. Australas Plant Pathol 52:1–8

Anjum N, Chandra R (2019) Endophytic bacteria of *Catharanthus roseus* as an alternative source of vindoline and application of response surface methodology to enhance its production. Arch Biol Sci 71(1):27–38

Aravind R, Kumar A, Eapen SJ, Ramana KV (2009) Endophytic bacterial flora in root and stem tissues of black pepper (*Piper nigrum L.*) genotype: isolation, identification and evaluation against Phytophthora capsici. Lett Appl Microbiol 48(1):58–64

Arora D, Sharma N, Singamaneni V, Sharma V, Kushwaha M, Abrol V et al (2016) Isolation and characterization of bioactive metabolites from Xylariapsidii, an endophytic fungus of the medicinal plant Aegle marmelos and their role in mitochondrial dependent apoptosis against pancreatic cancer cells. Phytomedicine 23(12):1312–1320

Arshad M, Gruber MY, Wall K, Hannoufa A (2017) An insight into microRNA156 role in salinity stress responses of alfalfa. Front Plant Sci 8:356

Asaf S, Hamayun M, Khan AL, Waqas M, Khan MA, Jan R et al (2018) Salt tolerance of Glycine max L. induced by endophytic fungus *Aspergillus flavus* CSH1, via regulating its endogenous hormones and antioxidative system. Plant Physiol Biochem 128:13–23

Ayswaria R, Alfred J, Gopalan G, Krishna R (2022) Endophytic Streptomyces isolated from *Ocimum tenuiflorum* and *Catharanthus roseus* and its efficacy test to develop antibacterial fabric for wound caring. J Textile Inst 114:1–12

Babu AG, Shea PJ, Sudhakar D, Jung IB, Oh BT (2015) Potential use of Pseudomonas koreensis AGB-1 in association with Miscanthus sinensis to remediate heavy metal(loid)-contaminated mining site soil. J Environ Manag 151:160–166

Bacon C, Hill N (1996) Symptomless grass endophytes: products of coevolutionary symbioses and their role in the ecological adaptations of grasses. In: Redlin SC, Carris LM (eds) Endophytic fungi in grasses and woody plants. APS Press, St. Paul, MN, pp 155–178

Badawy AA, Alotaibi MO, Abdelaziz AM, Osman MS, Khalil AM, Saleh AM et al (2021) Enhancement of seawater stress tolerance in barley by the endophytic fungus Aspergillus ochraceus. Metabolites 11(7):428

Bagur H, Poojari CC, Melappa G, Rangappa R, Chandrasekhar N, Somu P (2020) Biogenically synthesized silver nanoparticles using endophyte fungal extract of *Ocimum tenuiflorum* and evaluation of biomedical properties. J Clust Sci 31:1241–1255

Baltruschat H, Fodor J, Harrach BD, Niemczyk E, Barna B, Gullner G, Janeczko A, Kogel K-H, Schäfer P, Schwarczinger I, Zuccaro A, Skoczowski A (2008) Salt tolerance of barley induced by the root endophyte *Piriformospora indica* is associated with a strong increase in antioxidants. New Phytol 180(2):501–510

Benhamou N, Kloepper JW, Quadt-Hallman A, Tuzun S (1996) Induction of defense-related ultrastructural modifications in pea root tissues inoculated with endophytic bacteria. Plant Physiol 112(3):919–929

Berg G, Grube M, Schloter M, Smalla K (2014) Unraveling the plant microbiome: looking back and future perspectives. Front Microbiol 5:148

Bhore SJ, Preveena J, Kandasamy KI (2013a) Isolation and identification of bacterial endophytes from pharmaceutical agarwood-producing Aquilaria species. Pharm Res 5(2):134

Bhore SJ, Preveena J, Kandasamy KI (2013b) Isolation and identification of bacterial endophytes from pharmaceutical agarwood-producing Aquilaria species. Pharmacogn Res 5(2):134

Bibi N, Jan G, Jan FG, Hamayun M, Iqbal A, Hussain A et al (2019) *Cochliobolus* sp. acts as a biochemical modulator to alleviate salinity stress in okra plants. Plant Physiol Biochem 139:459–469

Bilal S, Shahzad R, Imran M, Jan R, Kim KM, Lee IJ (2020) Synergistic association of endophytic fungi enhances *Glycine max* L. resilience to combined abiotic stresses: heavy metals, high temperature and drought stress. Ind Crop Prod 143:111931

Bills GF, González-Menéndez V, Martín J, Platas G, Fournier J, Peršoh D, Stadler M (2012) *Hypoxylon pulicicidum* sp. Nov. (Ascomycota, Xylariales), a pantropical insecticide-producing endophyte. PLoS One 7(10):e46687

Bricker TJ, Pichtel J, Brown HJ, Simmons M (2001) Phytoextraction of Pb and Cd from a superfund soil: effects of amendments and croppings. J Environ Sci Health A 36(9):1597–1610

Brum MCP (2008) Fungos endofíticos de *Vitis labrusca* L. var. Niagara Rosada e o seu potencial biotecnológico. UMC, Mogi das Cruzes

Caboni P, Aissani N, Demurtas M, Ntalli N, Onnis V (2016) Nematicidal activity of acetophenones and chalcones against Meloidogyne incognita and structure–activity considerations. Pest Manag Sci 72(1):125–130

Canassa F, Esteca FC, Moral RA, Meyling NV, Klingen I, Delalibera I (2020) Root inoculation of strawberry with the entomopathogenic fungi *Metarhizium robertsii* and *Beauveria bassiana* reduces incidence of the two spotted spider mite and selected insect pests and plant diseases in the field. J Pest Sci 93(1):261–274

Castillo Lopez D, Zhu-Salzman K, Ek-Ramos MJ, Sword GA (2014) The entomopathogenic fungal endophytes *Purpureocillium lilacinum* (formerly *Paecilomyces lilacinus*) and *Beauveria bassiana* negatively affect cotton aphid reproduction under both greenhouse and field conditions. PLoS One 9(8):e103891

Chamkhi I, Sbabou L, Aurag J (2018) Endophytic fungi isolated from Crocus sativus L. (saffron) as a source of bioactive secondary metabolites. Pharmcogn J 10(6):1143–1148

Chareprasert S, Piapukiew J, Thienhirun S, Whalley AJ, Sihanonth P (2006) Endophytic fungi of teak leaves *Tectona grandis* L. and rain tree leaves Samanea saman Merr. World J Microbiol Biotechnol 22:481–486

Chen Y, Fan JB, Du L, Xu H, Zhang QH, He YQ (2014) The application of phosphate solubilizing endophyte *Pantoea dispersa* triggers the microbial community in red acidic soil. Appl Soil Ecol 84:235–244

Chen T, Li C, White JF, Nan Z (2019) Effect of the fungal endophyte *Epichloë bromicola* on polyamines in wild barley (*Hordeum brevisubulatum*) under salt stress. Plant Soil 436:29–48

Chowdhary K, Sharma S (2020) Plant growth promotion and biocontrol potential of fungal endophytes in the inflorescence of *Aloe vera* L. Proc Natl Acad Sci India B Biol Sci 90:1045–1055

Chukwuemerie OL, Bunu SJ, Iloh ES, Onwuzuluigbo CC, Onyegbule FA, Okoye FB (2022) Antimicrobial properties and characterization of secondary metabolites obtained from *Curvularia lunata*, an endophyte of *Azadirachta indica*. J Drug Deliv Therap 12(6):110–119

Cohen AC, Travaglia CN, Bottini R, Piccoli PN (2009) Participation of abscisic acid and gibberellins produced by endophytic *Azospirillum* in the alleviation of drought effects in maize. Botany 87(5):455–462

Compant S, Reiter B, Sessitsch A, Nowak J, Clément C, Ait Barka E (2005) Endophytic colonization of *Vitis vinifera* L. by plant growth-promoting bacterium *Burkholderia* sp. strain PsJN. Appl Environ Microbiol 71(4):1685–1693

Crandall SG, Gold KM, Jiménez-Gasco MDM, Filgueiras CC, Willett DS (2020) A multi-omics approach to solving problems in plant disease ecology. PLoS One 15(9):e0237975

Daisy BH, Strobel GA, Castillo U, Ezra D, Sears J, Weaver DK, Runyon JB (2002) Naphthalene, an insect repellent, is produced by *Muscodor vitigenus*, a novel endophytic fungus. Microbiology 148(11):3737–3741

Dalal JM, Kulkarni NS (2014) Antagonistic and plant growth promoting potentials of indigenous endophytic actinomycetes of soybean (*Glycine max* (L) Merril). CIBTech J Microbiol 3:1–12

Das A, Rahman MI, Ferdous AS, Amin A, Rahman MM, Nahar N et al (2017) An endophytic basidiomycete, *Grammothele lineata*, isolated from *Corchorus olitorius*, produces paclitaxel that shows cytotoxicity. PLoS One 12(6):e0178612

Debbab A, Aly AH, Edrada-Ebel R, Wray V, Müller WE, Totzke F et al (2009) Bioactive metabolites from the endophytic fungus *Stemphylium globuliferum* isolated from *Mentha pulegium*. J Nat Prod 72(4):626–631

Desale MG, Bodhankar MG (2013) Antimicrobial activity of endophytic fungi isolated from *Vitex negundo* Linn. Int J Curr Microbiol App Sci 2(12):389–395

Dissanayake RK, Ratnaweera PB, Williams DE, Wijayarathne CD, Wijesundera RL, Andersen RJ, de Silva ED (2016) Antimicrobial activities of endophytic fungi of the Sri Lankan aquatic plant *Nymphaea nouchali* and chaetoglobosin A and C, produced by the endophytic fungus *Chaetomium globosum*. Mycology 7(1):1–8

Duruibe JO, Ogwuegbu MOC, Egwurugwu JN (2007) Heavy metal pollution and human biotoxic effects. Int J Phys Sci 2(5):112–118

Egamberdieva D, Berg G, Lindström K, Räsänen LA (2010) Co-inoculation of *Pseudomonas* spp. with Rhizobium improves growth and symbiotic performance of fodder galega (*Galega orientalis* Lam.). Eur J Soil Biol 46:269–272

El-Sayed E-SR, Zaki AG, Ahmed AS, Ismaiel AA (2020) Production of the anticancer drug taxol by the endophytic fungus *Epicoccum nigrum* TXB502: enhanced production by gamma irradiation mutagenesis and immobilization technique. Appl Microbiol Biotechnol 104(16):6991–7003

El-Tarabily KA, Soliman MH, Nassar AH, Al-Hassani HA, Sivasithamparam K et al (2000) Biological control of Sclerotinia minor using a chitinolytic bacterium and actinomycetes. Plant Pathol 49:573–583

Emamverdian A, Ding Y, Mokhberdoran F, Xie Y (2015) Heavy metal stress and some mechanisms of plant defense response. Sci World J 2015:756120

Estrada C, Rojas EI, Wcislo WT, Van Bael SA (2014) Fungal endophyte effects on leaf chemistry alter the in vitro growth rates of leaf-cutting ants' fungal mutualist, *Leucocoprinus gongylophorus*. Fungal Ecol 8:37–45

Fávaro LCL, Sebastianes FLS, Araújo WL (2012) *Epicoccum nigrum* P16, a sugarcane endophyte, produces antifungal compounds and induces root growth. PLoS One 7:e36826

Gaber DA, Berthelot C, Blaudez D, Kovács GM, Franken P (2023) Impact of dark septate endophytes on salt stress alleviation of tomato plants. Front Microbiol 14:1124879

Ganesan V, Hariram M, Vivekanandhan S, Muthuramkumar S (2020) *Periconium* sp. (endophytic fungi) extract mediated sol-gel synthesis of ZnO nanoparticles for antimicrobial and antioxidant applications. Mater Sci Semicond Process 105:104739

Gangadevi V, Muthumary J (2008) Taxol, an anticancer drug produced by an endophytic fungus *Bartaliniaro billardoides* Tassi, isolated from a medicinal plant, Aegle marmelos Correa ex Roxb. World J Microbiol Biotechnol 24:717–724

Gangadevi V, Muthumary J (2009) Taxol production by *Pestalotiopsis terminaliae*, an endophytic fungus of *Terminalia arjuna* (arjun tree). Biotechnol Appl Biochem 52(1):9–15

Gangwar M, Kaur N, Saini P, Kalia A (2015) The diversity, plant growth promoting and antimicrobial activities of endophytic actinomycetes isolated from *Emblica officinalis* Gaertn. Int J Adv Res 3(4):1062–1071

Gao F, Wu Y, Wang M (2014) Identification and antifungal activity of an actinomycete strain against *Alternaria* spp. Span J Agric Res 12(4):1158–1165

Gao H, Yang D, Yang L, Han S, Liu G, Tang L et al (2023) Co-inoculation with Sinorhizobium meliloti and Enterobacter ludwigii improves the yield, nodulation, and quality of alfalfa (Medicago sativa L.) under saline-alkali environments. Ind Crop Prod 199:116818

Gill H, Vasundhara M (2019) Isolation of taxol producing endophytic fungus *Alternaria brassicicola* from non-Taxus medicinal plant *Terminalia arjuna*. World J Microbiol Biotechnol 35:1–8

Ginting RCB, Sukarno N, Widyastuti UTUT, Darusman LK, Kanaya S (2013) Diversity of endophytic fungi from red ginger (*Zingiber officinale* Rosc.) plant and their inhibitory effect to *Fusarium oxysporum* plant pathogenic fungi. HAYATI J Biosci 20(3):127–137

Gohain A, Gogoi A, Debnath R, Yadav A, Singh BP, Gupta VK et al (2015) Antimicrobial biosynthetic potential and genetic diversity of endophytic actinomycetes associated with medicinal plants. FEMS Microbiol Lett 362(19):fnv158

Gohain A, Sarma RK, Debnath R, Saikia J, Singh BP, Sarmah R, Saikia R (2019) Phylogenetic affiliation and antimicrobial effects of endophytic actinobacteria associated with medicinal plants: prevalence of polyketide synthase type II in antimicrobial strains. Folia Microbiol 64:481–496

Gouda S, Das G, Sen SK, Shin H-S, Patra JK (2016) Endophytes: a treasure house of bioactive compounds of medicinal importance. Front Microbiol 7:1538–1538

Gul Jan F, Hamayun M, Hussain A, Jan G, Iqbal A, Khan A, Lee IJ (2019) An endophytic isolate of the fungus Yarrowia lipolytica produces metabolites that ameliorate the negative impact of salt stress on the physiology of maize. BMC Microbiol 19(1):1–10

Guler NS, Pehlivan N, Karaoglu SA, Guzel S, Bozdeveci A (2016) *Trichoderma atroviride* ID20G inoculation ameliorates drought stress-induced damages by improving antioxidant defence in maize seedlings. Acta Physiol Plant 38:1–9

Guo X, Guan X, Liu C, Jia F, Li J, Li J et al (2016) *Plantactinospora soyae* sp. nov., an endophytic actinomycete isolated from soybean root [Glycine max (L.) Merr]. Int J Syst Evol Microbiol 66(7):2578–2584

Gupta S, Chaturvedi P, Kulkarni MG, Van Staden J (2020) A critical review on exploiting the pharmaceutical potential of plant endophytic fungi. Biotechnol Adv 39:107462

Haidar B, Ferdous M, Fatema B, Ferdous AS, Islam MR, Khan H (2018) Population diversity of bacterial endophytes from jute (*Corchorus olitorius*) and evaluation of their potential role as bioinoculants. Microbiol Res 208:43–53

Hamayun M, Afzal Khan S, Ahmad N, Tang DS, Kang SM, Na CI et al (2009) *Cladosporium sphaerospermum* as a new plant growth-promoting endophyte from the roots of *Glycine max* (L.) Merr. World J Microbiol Biotechnol 25:627–632

Harish S, Kavino M, Kumar N, Balasubramanian P, Samiyappan R (2009) Induction of defense-related proteins by mixtures of plant growth promoting endophytic bacteria against Banana bunchy top virus. Biol Control 51(1):16–25

Haruma T, Yamaji K, Ogawa K, Masuya H, Sekine Y, Kozai N (2019) Root-endophytic *Chaetomium cupreum* chemically enhances aluminium tolerance in *Miscanthus sinensis* via increasing the aluminium detoxicants, chlorogenic acid and oosporein. PLoS One 14(2):e0212644

Hazarika SN, Saikia K, Borah A, Thakur D (2021) Prospecting endophytic bacteria endowed with plant growth promoting potential isolated from Camellia sinensis. Front Microbiol 12:738058

Hume DE, Ryan GD, Gibert A, Helander M, Mirlohi A, Sabzalian MR (2016) Epichloë fungal endophytes for grassland ecosystems. In: Lichtfouse E (ed) Sustainable agriculture reviews. Springer, pp 233–305

Ikram M, Ali N, Jan G, Jan FG, Rahman IU, Iqbal A, Hamayun M (2018) IAA producing fungal endophyte *Penicillium roqueforti* Thom., enhances stress tolerance and nutrients uptake in wheat plants grown on heavy metal contaminated soils. PLoS One 13(11):e0208150

Islam N, Choi J, Baek KH (2018) Antibacterial activities of endophytic bacteria isolated from *Taxus brevifolia* against foodborne pathogenic bacteria. Foodborne Pathog Dis 15(5):269–276

Jha Y, Subramanian RB, Patel S (2011) Combination of endophytic and rhizospheric plant growth promoting rhizobacteria in *Oryza sativa* shows higher accumulation of osmoprotectant against saline stress. Acta Physiol Plant 33(3):797–802

Joe MM, Islam MR, Karthikeyan B, Bradeepa K, Sa T (2012) Resistance responses of rice to rice blast fungus after seed treatment with the endophytic *Achromobacter xylosoxidans* AUM54 strains. Crop Prot 42:141–148

Jogawat A, Saha S, Bakshi M, Dayaman V, Kumar M, Dua M (2013) *Piriformospora indica* rescues growth diminution of rice seedlings during high salt stress. Plant Signal Behav 8:e26891

John R, Ahmad P, Gadgil K, Sharma S (2012) Heavy metal toxicity: effect on plant growth, biochemical parameters and metal accumulation by *Brassica juncea* L. Int J Plant Prod 3(3):65–76

Jones JT, Haegeman A, Danchin EG, Gaur HS, Helder J, Jones MG et al (2013) Top 10 plant-parasitic nematodes in molecular plant pathology. Mol Plant Pathol 14:946–961. https://doi.org/10.1111/mpp.12057

Kaaniche F, Hamed A, Abdel-Razek AS, Wibberg D, Abdissa N, El Euch IZ et al (2019) Bioactive secondary metabolites from new endophytic fungus *Curvularia* sp. isolated from *Rauwolfia macrophylla*. PLoS One 14(6):e0217627

Kalimuthu AK, Pavadai P, Panneerselvam T, Babkiewicz E, Pijanowska J, Mrówka P et al (2022) Cytotoxic potential of bioactive compounds from *Aspergillus flavus*, an endophytic fungus isolated from *Cynodon dactylon*, against breast cancer: experimental and computational approach. Molecules 27(24):8814

Kemp ND, Vaughan MM, McCormick SP, Brown JA, Bakker MG (2020) *Sarocladium zeae* is a systemic endophyte of wheat and an effective biocontrol agent against Fusarium head blight. Biol Control 149:104329

Khan AL, Hussain J, Al-Harrasi A, Al-Rawahi A, Lee I-J (2015) Endophytic fungi: resource for gibberellins and crop abiotic stress resistance. Crit Rev Biotechnol 35(1):62–74

Khan Z, Rho H, Firrincieli A, Shang HH, Luna V, Masciarelli O (2016) Growth enhancement and drought tolerance of hybrid poplar upon inoculation with endophyte consortia. Curr Plant 6:38

Kloepper JW, Ryu CM, Zhang S (2004) Induced systemic resistance and promotion of plant growth by Bacillus spp. Phytopathology 94(11):1259–1266

Klykleung N, Yuki M, Kudo T, Ohkuma M, Phongsopitanun W, Pittayakhajonwut P, Tanasupawat S (2020) Microbispora catharanthi sp. nov., a novel endophytic actinomycete isolated from the root of *Catharanthus roseus*. Int J Syst Evol Microbiol 70(2):964–970

Knapp M, Palevsky E, Rapisarda C (2020) Insect and mite pests. In: Integrated pest and disease management in greenhouse crops. Springer, pp 101–146

Kulišová M, Vrublevskaya M, Lovecká P, Vrchotová B, Stránská M, Kolařík M, Kolouchová I (2021) Fungal endophytes of *Vitis vinifera*—plant growth promotion factors. Agriculture 11(12):1250

Kumar KK, Sridhar J, Murali-Baskaran RK, Senthil-Nathan S, Kaushal P, Dara SK, Arthurs S (2019) Microbial biopesticides for insect pest management in India: current status and future prospects. J Invertebr Pathol 165:74–81

Kumar KK, Dara SK (2021) Fungal and bacterial endophytes as microbial control agents for plant-parasitic nematodes. Int J Environ Res Public Health 18(8):4269

Kusmiyati N, Wicaksono ST, Maknuna D (2021) Identification of endophytic fungi in nutgrass (*Cyperus rotundus* L.) as solubilizing phosphate and indole-3-acetic acid producers. Biogenesis Jurnal Ilmiah Biologi 9(1):93–101

Li L, Li L, Wang X, Zhu P, Wu H, Qi S (2017) Plant growth-promoting endophyte *Piriformospora indica* alleviates salinity stress in *Medicago truncatula*. Plant Physiol Biochem 119:211–223

Liarzi O, Bucki P, Miyara SB, Ezra D (2016) Bioactive volatiles from an endophytic *Daldinia* cf. *concentrica* isolate affects the viability of the plant parasitic nematode *Meloidogyne javanica*. PLoS One 11(12):e0168437

Linh TM, Mai NC, Hoe PT, Ngoc NT, Thao PTH, Ban NK, Van NT (2021) Development of a cell suspension culture system for promoting alkaloid and vinca alkaloid biosynthesis using endophytic fungi isolated from local *Catharanthus roseus*. Plants 10(4):672

Liu Y, Xu P, Yang F, Li M, Yan H, Li N et al (2019) Composition and diversity of endophytic bacterial community in seeds of super hybrid rice 'Shenliangyou 5814' (*Oryza sativa* L.) and its parental lines. Plant Growth Regul 87:257–266

Liu Y, Cheng L, Shen Y (2021) Two new nonenolides from *Diaporthe* sp. SXZ-19, an endophytic fungus of *Camptotheca acuminata*. Chem Biodivers 18(5):e2001055

Madani B, Mirshekari A, Imahori Y (2019) Physiological responses to stress. In: Yahia EM (ed) Postharvest physiology and biochemistry of fruits and vegetables. Woodhead Publishing, UK, pp 405–423

Malik A, Butt TA, Naqvi STA, Yousaf S, Qureshi MK, Zafar MI et al (2020) Lead tolerant endophyte *Trametes hirsuta* improved the growth and lead accumulation in the vegetative parts of *Triticum aestivum* L. Heliyon 6(7):e04188

Manasa KM, Vasanthakumari MM, Nataraja KN, Shaanker RU (2020) Endophytic fungi of salt adapted *Ipomea pes-caprae* LR Br: their possible role in inducing salinity tolerance in paddy (*Oryza sativa* L.). Curr Sci 118(9):00113891

Mantzoukas S, Lagogiannis I (2019) Endophytic colonization of pepper (*Capsicum annum*) controls aphids (*Myzus persicae* Sulzer). Appl Sci 9(11):2239. https://doi.org/10.3390/app9112239

Marian M, Ohno T, Suzuki H, Kitamura H, Kuroda K, Shimizu M (2020) A novel strain of endophytic Streptomyces for the biocontrol of strawberry anthracnose caused by *Glomerella cingulata*. Microbiol Res 234:126428

Mastan A, Bharadwaj RKB, Kushwaha RK, Babu CSV (2019) Functional fungal endophytes in *Coleus forskohlii* regulate labdane diterpene biosynthesis for elevated forskolin accumulation in roots. Microb Ecol 78:914–926. https://doi.org/10.1007/s00248-019-01376-w

Mello IS, Targanski S, Pietro-Souza W, Stachack FFF, Terezo AJ, Soares MA (2020) Endophytic bacteria stimulate mercury phytoremediation by modulating its bioaccumulation and volatilization. Ecotoxicol Environ Saf 202:110818

Mingma R, Pathom-aree W, Trakulnaleamsai S, Thamchaipenet A, Duangmal K (2014) Isolation of rhizospheric and roots endophytic actinomycetes from Leguminosae plant and their activities to inhibit soybean pathogen, *Xanthomonas campestris* pv. glycine. World J Microbiol Biotechnol 30(1):271–280

Miranda-Fuentes P, Yousef-Yousef M, Valverde-Garcia P, Rodríguez-Gómez IM, Garrido-Jurado I, Quesada-Moraga E (2021) Entomopathogenic fungal endophyte-mediated tritrophic interactions between Spodoptera littoralis and its parasitoid Hyposoter didymator. J Pest Sci 94(3):933–945. https://doi.org/10.1007/s10340-020-01306-7

Moissl-Eichinger C, Pausan M, Taffner J, Berg G, Bang C, Schmitz RA (2018) Archaea are interactive components of complex microbiomes. Trends Microbiol 26:70–85

Mwaura P, Dubois T, Losenge T, Coyne D, Kahangi E (2010) Effect of endophytic *Fusarium oxysporum* on paralysis and mortality of Pratylenchus goodeyi. Afr J Biotechnol 9(8):1130–1134

Naik BS (2020) Nanoparticles from endophytic fungi and their efficacy in biological control. In: Devarajan T, Jeyabalan S, Ram P (eds) Nanotechnology for food, agriculture, and environment. Springer, pp 161–179

Nath A, Raghunatha P, Joshi SR (2012) Diversity and biological activities of endophytic fungi of *Emblica officinalis*, an ethnomedicinal plant of India. Mycobiology 40(1):8–13

Nath A, Chattopadhyay A, Joshi SR (2015) Biological activity of endophytic fungi of *Rauwolfia serpentina* Benth: an ethnomedicinal plant used in folk medicines in Northeast India. Proc Natl Acad Sci India Sect B Biol Sci 85:233–240

Naveed M, Mitter B, Reichenauer TG, Wieczorek K, Sessitsch A (2014) Increased drought stress resilience of maize through endophytic colonization by *Burkholderia phytofirmans* PsJN and *Enterobacter* sp. FD17. Environ Exp Bot 97:30–39

Nawaz F, Rafeeq R, Majeed S, Ismail MS, Ahsan M, Ahmad KS et al (2023) Biochar amendment in combination with endophytic bacteria stimulates photosynthetic activity and antioxidant enzymes to improve soybean yield under drought stress. J Soil Sci Plant Nutr 23(1):746–760

Nigris S, Baldan E, Tondello A, Zanella F, Vitulo N, Favaro G et al (2018) Biocontrol traits of *Bacillus licheniformis* GL174, a culturable endophyte of *Vitis vinifera* cv. Glera. BMC Microbiol 18:1–16

Nischitha R, Vasanthkumari MM, Kumaraswamy BE, Shivanna MB (2020) Antimicrobial and antioxidant activities and chemical profiling of *Curvularia tsudae* endophytic in *Cynodon dactylon* (L.) Pers. 3 Biotech 10:1–12

Oliva J, Ridley M, Redondo MA, Caballol M (2021) Competitive exclusion amongst endophytes determines shoot blight severity on pine. Funct Ecol 35:239

Oliveira CM, Regasini LO, Silva GH, Pfenning LH, Young MC, Berlinck RG et al (2011) Dihydroisocoumarins produced by *Xylaria* sp. and *Penicillium* sp., endophytic fungi associated with *Piper aduncum* and *Alibertia macrophylla*. Phytochem Lett 4(2):93–96

Osman HA, Ameen HH, Mohamed M, Elkelany US (2020) Efficacy of integrated microorganisms in controlling root-knot nematode *Meloidogyne javanica* infecting peanut plants under field conditions. Bull Natl Res Centre 44:1–10. https://doi.org/10.1186/s42269-020-00366-0

Oviya G, Rangasamy A, Ariyan M, Krishnamoorthy R, Senthilkumar M, Gopal NO et al (2023) Halotolerant nodule rhizobial and passenger endophytes alleviates salinity stress in groundnut (*Arachis hypogaea* L.). J Plant Growth Regul 42:1–16

Owen NL, Hundley N (2004) Endophytes—the chemical synthesizers inside plants. Sci Prog 87:79–99

Pallas V, García JA (2011) How do plant viruses induce disease? Interactions and interference with host components. J Gen Virol 92:2691–2705. https://doi.org/10.1099/vir.0.034603-0

Pal A, Paul AK (2020) In Vitro antimicrobial activity screening of bacteria endophytic to ethnomedicinal plant *Rauvolfia serpentina* (L.) Benth ex Kurz. J Appl Biotechnol Rep 7(3):176–184

Pandey PK, Yadav SK, Singh A, Sarma BK, Mishra A, Singh HB (2012) Cross-species alleviation of biotic and abiotic stresses by the endophyte *Pseudomonas aeruginosa* PW09. J Phytopathol 160(10):532–539

Panigrahi S, Rath CC (2019) Condition optimization for phosphate solubilization by *Kosakonia cowanii* MK834804, an endophytic bacterium isolated from *Aegle marmelos*. Int J Curr Microbiol App Sci 8:2823–2835

Panigrahi S, Mohanty S, Rath CC (2020) Characterization of endophytic bacteria *Enterobacter cloacae* MG00145 isolated from *Ocimum sanctum* with indole acetic acid (IAA) production and plant growth promoting capabilities against selected crops. S Afr J Bot 134:17–26

Pappas ML, Liapoura M, Papantoniou D, Avramidou M, Kavroulakis N, Weinhold A et al (2018) The beneficial endophytic fungus *Fusarium solani* strain K alters tomato responses against spider mites to the benefit of the plant. Front Plant Sci 9:1603

Passariello B, Giuliano V, Quaresima S, Barbaro M, Caroli S, Forte G, Iavicoli I (2002) Evaluation of the environmental contamination at an abandoned mining site. Microchem J 73(1–2):245–250

Pérez LI, Gundel PE, Zabalgogeazcoa I, Omacini M (2020) An ecological framework for understanding the roles of Epichloë endophytes on plant defenses against fungal diseases. Fungal Biol Rev 34(3):115–125

Pinto MDS, Inocente LB, Oliveira PND, Silva KJP, Carrer H (2022) Plant growth-promoting (PGP) traits of endophytic bacteria from in vitro cultivated *Tectona grandis* Lf. Forests 13(10):1539

Pujiyanto S, Lestari YULIN, Suwanto A, Budiarti S, Darusman LK (2012) Alpha-glucosidase inhibitor activity and characterization of endophytic actinomycetes isolated from some Indonesian diabetic medicinal plants. Int J Pharm Pharm Sci 4(1):327–333

Pulici M, Sugawara F, Koshino H, Uzawa J, Yoshida S, Lobkovsky E, Clardy J (1996) Pestalotiopsins A and B: new caryophyllenes from an endophytic fungus of *Taxus brevifolia*. J Org Chem 61(6):2122–2124

Qin X, Zhao X, Huang S, Deng J, Li X, Luo Z, Zhang Y (2021) Pest management via endophytic colonization of tobacco seedlings by the insect fungal pathogen *Beauveria bassiana*. Pest Manag Sci 77(4):2007–2018

Radhakrishnan R, Khan AL, Kang SM, Lee IJ (2015) A comparative study of phosphate solubilization and the host plant growth promotion ability of *Fusarium verticillioides* RK01 and *Humicola* sp. KNU01 under salt stress. Ann Microbiol 65(1):585–593

Ramesh R, Joshi AA, Ghanekar MP (2009) Pseudomonads: major antagonistic endophytic bacteria to suppress bacterial wilt pathogen *Ralstonia solanacearum* in the eggplant (*Solanum melongena* L.). World J Microbiol Biotechnol 25(1):47–55

Rascio N, Navari-Izzo F (2011) Heavy metal hyperaccumulating plants: how and why do they do it? and what makes them so interesting? Plant Sci 180(2):169–181

Rasool S, Cárdenas PD, Pattison DI, Jensen B, Meyling NV (2021) Isolate-specific effect of entomopathogenic endophytic fungi on population growth of two-spotted spider mite (*Tetranychus urticae Koch*) and levels of steroidal glycoalkaloids in tomato. J Chem Ecol 47:476–488

Rathod DP, Dar MA, Gade AK, Rai MK, Baba G (2014) Griseofulvin producing endophytic *Nigrospora oryzae* from Indian *Emblica officinalis* Gaertn: a new report. Austin J Biotechnol Bioeng 1(3):1–5

Redman RS, Dunigan DD, Rodriguez RJ (2001) Fungal symbiosis from mutualism to parasitism: who controls the outcome, host or invader? New Phytol 151(3):705–716

Reiff JM, Ehringer M, Hoffmann C, Entling MH (2021) Fungicide reduction favors the control of phytophagous mites under both organic and conventional viticulture. Agric Ecosyst Environ 305:107172

Rocha R, Luz DED, Engels C, Pileggi SAV, Jaccoud Filho DDS, Matiello RR, Pileggi M (2009) Selection of endophytic fungi from comfrey (*Symphytum officinale* L.) for in vitro biological control of the phytopathogen *Sclerotinia sclerotiorum* (Lib.). Braz J Microbiol 40:73–78

Rocha SL, Evans HC, Jorge VL, Cardoso LA, Pereira FS, Rocha FB et al (2017) Recognition of endophytic Trichoderma species by leaf-cutting ants and their potential in a Trojan-horse management strategy. R Soc Open Sci 4(4):160628

Rohini S, Aswani R, Kannan M, Sylas VP, Radhakrishnan EK (2018) Culturable endophytic bacteria of ginger rhizome and their remarkable multi-trait plant growth-promoting features. Curr Microbiol 75:505–511

Rojas O (2020) Agricultural extreme drought assessment at global level using the FAO-Agricultural Stress Index System (ASIS). Weather Clim Extremes 27:100184

Sadeghi F, Samsampour D, Seyahooei MA, Bagheri A, Soltani J (2020) Fungal endophytes alleviate drought-induced oxidative stress in mandarin (Citrus reticulata L.): toward regulating the ascorbate–glutathione cycle. Sci Hortic 261:108991

Saijo Y, Loo EPI (2020) Plant immunity in signal integration between biotic and abiotic stress responses. New Phytol 225:87–104. https://doi.org/10.1111/nph.15989

Salt DE, Prince RC, Pickering IJ, Raskin I (1995) Mechanisms of cadmium mobility and accumulation in Indian mustard. Plant Physiol 109(4):1427–1433

Sangmanee P, Bhromsiri A, Akarapisan A (2009) The potential of endophytic actinomycetes, (*Streptomyces* sp.) for the biocontrol of powdery mildew disease in sweet pea (*Pisum sativum*). Asian J Food Agro Ind 93:e8

Sapak Z, Meon S, Ahmad Z (2008) Effect of endophytic bacteria on growth and suppression of Ganoderma infection in oil palm. Int J Agric Biol 10:127–132

Sbravatti Júnior JA, Auer CG, Pimentel IC, dos Santos AF, Schultz B (2013) In vitro selection of endophytes for biological control of *Botrytis cinerea* in *Eucalyptus benthamii*. Floresta 43(1):145–151

Schumann GL, D'Arcy CJ (2006) Essential plant pathology. APS Press, London

Selvakumar G, Mohan M, Kundu S, Gupta AD, Joshi P, Nazim S, Gupta HS (2008) Cold tolerance and plant growth promotion potential of *Serratia marcescens* strain SRM (MTCC 8708) isolated from flowers of summer squash (*Cucurbita pepo*). Lett Appl Microbiol 46(2):171–175

Shan W, Zhou Y, Liu H, Yu X (2018) Endophytic actinomycetes from tea plants (*Camellia sinensis*): isolation, abundance, antimicrobial, and plant-growth-promoting activities. Biomed Res Int 2018:1470305

Sharaf MH, Abdelaziz AM, Kalaba MH, Radwan AA, Hashem AH (2022) Antimicrobial, antioxidant, cytotoxic activities and phytochemical analysis of fungal endophytes isolated from *Ocimum basilicum*. Appl Biochem Biotechnol 194:1271–1289

Sharma T, Kaul S, Dhar MK (2015) Diversity of culturable bacterial endophytes of saffron in Kashmir, India. SpringerPlus 4(1):1–13

Silva CFD, Vitorino LC, Mendonça MAC, Araújo WL, Dourado MN, Albuquerque LC et al (2020) Screening of plant growth-promoting endophytic bacteria from the roots of the medicinal plant *Aloe vera*. S Afr J Bot 134:3–16

Singh R, Soni SK, Kalra A (2013) Synergy between *Glomus fasciculatum* and a beneficial Pseudomonas in reducing root diseases and improving yield and forskolin content in *Coleus forskohlii* Briq. under organic field conditions. Mycorrhiza 23(1):35–44

Sobiczewski P, Iakimova E, Mikicínski A, Wêgrzynowicz-Lesiak E, Dyki B (2017) Necrotrophic behaviour of *Erwinia amylovora* in apple and tobacco leaf tissue. Plant Pathol 66:842–855. https://doi.org/10.1111/ppa.12631

Somjaipeng S, Medina A, Kwaśna H, Ortiz JO, Magan N (2015) Isolation, identification, and ecology of growth and taxol production by an endophytic strain of Paraconiothyrium variabile from English yew trees (Taxus baccata). Fungal Biol 119(11):1022–1031

Sreekanth D, Sushim GK, Syed A, Khan BM, Ahmad A (2011) Molecular and morphological characterization of a taxol-producing endophytic fungus, *Gliocladium* sp., from *Taxus baccata*. Mycobiology 39(3):151–157

Srivastava S, Verma PC, Chaudhry V, Singh N, Abhilash PC, Kumar KV, Singh N (2013) Influence of inoculation of arsenic-resistant *Staphylococcus arlettae* on growth and arsenic uptake in *Brassica juncea* (L.) Czern. Var. R-46. J Hazard Mater 262:1039–1047

Stone JK, Bacon CW, White JF Jr (2000) An overview of endophytic microbes: endophytism defined. Microbial Endophytes:17–44

Strobel G, Daisy B (2003) Bioprospecting for microbial endophytes and their natural products. Microbiol Mol Biol Rev 67(4):491–502

Strobel G, Stierle A, Stierle D, Hess WM (1993) *Taxomyces andreanae*, a proposed new taxon for a bulbilliferous hyphomycete associated with Pacific yew (*Taxus brevifolia*). Mycotaxon 47:71–80

Suarez R, Wong A, Ramirez M, Barraza A, Orozco MC, Cevallos MA (2008) Improvement of drought tolerance and grain yield in common bean by overexpressing trehalose-6-phosphate synthase in rhizobia. Mol Plant Microbe Interact 21:958–966

Sudarsan S, Kumar Shankar M, Kumar Belagal Motatis A, Shankar S, Krishnappa D, Mohan CD et al (2021) Green synthesis of silver nanoparticles by *Cytobacillus firmus* isolated from the stem bark of *Terminalia arjuna* and their antimicrobial activity. Biomolecules 11(2):259

Tan RX, Zou WX (2001) Endophytes: a rich source of functional metabolites. Nat Prod Rep 18:448–459

Tian L, Hu S, Wang X, Guo Y, Huang L, Wang L, Li W (2022) Antagonism of rhizosphere *Streptomyces yangpuensis* CM253 against the pathogenic fungi causing corm rot in saffron (*Crocus sativus* L.). Pathogens 11(10):1195

Tibpromma S, Zhang L, Karunarathna SC, Du TY, Phukhamsakda C, Rachakunta M et al (2021) Volatile constituents of endophytic fungi isolated from *Aquilaria sinensis* with descriptions of two new species of Nemania. Life 11(4):363

Tiwari K (2015) The future products: endophytic fungal metabolites. J Biodivers Bioprospect Dev 2(1):1000145

Van Bael SA, Fernández-Marín H, Valencia MC, Rojas EI, Wcislo WT, Herre EA (2009) Two fungal symbioses collide: endophytic fungi are not welcome in leaf-cutting ant gardens. Proc Biol Sci 276:2419

Van Bael SA, Estrada C, Wcislo WT (2011) Fungal-fungal interactions in leaf-cutting ant agriculture. Psyche 2011:617478

Varma A, Bakshi M, Lou B, Hartmann A, Oelmueller R (2012) *Piriformospora indica*: a novel plant growth-promoting mycorrhizal fungus. Agric Res 1:117–131

Vázquez MB, Moreno MV, Amodeo MR, Bianchinotti MV (2021) Effects of glyphosate on soil fungal communities: a field study. Rev Argent Microbiol 53(4):349–358

Verma VC, Gond SK, Kumar A, Kharwar RN, Strobel G (2007) The endophytic mycoflora of bark, leaf, and stem tissues of Azadirachta indica A. Juss (neem) from Varanasi (India). Microb Ecol 54:119–125

Verma VC, Gond SK, Kumar A, Mishra A, Kharwar RN, Gange AC (2009) Endophytic actinomycetes from *Azadirachta indica* A. Juss.: isolation, diversity, and anti-microbial activity. Microb Ecol 57:749–756

Viehweger K (2014) How plants cope with heavy metals. Botan Stud 55(1):35

Wagenaar MM, Clardy J (2001) Dicerandrols, new antibiotic and cytotoxic dimers produced by the fungus Phomopsis longicolla isolated from an endangered mint. J Nat Prod 64(8):1006–1009

Waqas A, Afsheen Z, Zarina A, Beena N (2017) Effect of salinization and alkalizationon seed germination and early seedling growth of leguminous species *Vigna mungo* L. Plant Sci J 6(1):1–11

Wen Y, Lv Y, Hao J, Chen H, Huang Y, Liu C et al (2020) Two new compounds of *Penicillium polonicum*, an endophytic fungus from *Camptotheca acuminata* Decne. Nat Prod Res 34(13):1879–1883

Wilkes TI, Warner DJ, Davies KG, Edmonds-Brown V (2020) Tillage, glyphosate and beneficial arbuscular mycorrhizal fungi: optimising crop management for plant–fungal symbiosis. Agriculture 10(11):520

Xie H, Feng X, Wang M, Wang Y, Kumar Awasthi M, Xu P (2020) Implications of endophytic microbiota in *Camellia sinensis*: a review on current understanding and future insights. Bioengineered 11(1):1001–1015

Xu L, Wang A, Wang J, Wei Q, Zhang W (2017) *Piriformospora indica* confers drought tolerance on *Zea mays* L. through enhancement of antioxidant activity and expression of drought-related genes. Crop J 5:251–258

Yehia RS (2023) Multi-function of a new bioactive secondary metabolite derived from endophytic fungus *Colletotrichum acutatum* of *Angelica sinensis*. J Microbiol Biotechnol 33(6):1–17

Yuan J, Sun K, Deng-Wang M-Y, Dai C-C (2016) The mechanism of ethylene signaling induced by endophytic fungus *Gilmaniella* sp. AL12 mediating sesquiterpenoids biosynthesis in Atractylodes lancea. Front Plant Sci 7:361

Zhang X, Li C, Nan Z (2012) Effects of cadmium stress on seed germination and seedling growth of *Elymus dahuricus* infected with the Neotyphodium endophyte. Sci China Life Sci 55(9):793–799

Zhang S, Gan Y, Xu B (2016) Application of plant-growth-promoting fungi *Trichoderma longibrachiatum* T6 enhances tolerance of wheat to salt stress through improvement of antioxidative defense system and gene expression. Front Plant Sci 7:1405

Zhang FM, He W, Wu CY, Sun K, Zhang W, Dai CC (2020a) *Phomopsis liquidambaris* inoculation induces resistance in peanut to leaf spot and root rot. BioControl 65:475–488

Zhang XQ, Qu HR, Bao SS, Deng ZS, Guo ZY (2020b) Secondary metabolites from the endophytic fungus *Xylariales* sp. and their antimicrobial activity. Chem Nat Compd 56:530–532

Zhang H, Zhu J, Gong Z, Zhu JK (2022) Abiotic stress responses in plants. Nat Rev Genet 23(2):104–119

Zhao L, Xu Y, Lai XH, Shan C, Deng Z, Ji Y (2015) Screening and characterization of endophytic Bacillus and Paenibacillus strains from medicinal plant *Lonicera japonica* for use as potential plant growth promoters. Braz J Microbiol 46(4):977–989

Zhou W, Verma VC, Wheeler TA, Woodward JE, Starr JL, Sword GA (2020) Tapping into the cotton fungal phytobiome for novel nematode biological control tools. Phytobiomes J 4(1):19–26

Zhu WY, Zhang JL, Qin YL, Xiong ZJ, Zhang DF, Klenk HP et al (2013) Blastococcus endophyticus sp. nov., an actinobacterium isolated from *Camptotheca acuminata*. Int J Syst Evol Microbiol 63(Pt_9):3269–3273

Zhu L, Li T, Wang C, Zhang X, Xu L, Xu R, Zhao Z (2018) The effects of dark septate endophyte (DSE) inoculation on tomato seedlings under Zn and Cd stress. Environ Sci Pollut Res 25(34):35232–35241

Zicca S, De Bellis P, Masiello M, Saponari M, Saldarelli P, Boscia D, Sisto A (2020) Antagonistic activity of olive endophytic bacteria and of *Bacillus* spp. strains against *Xylella fastidiosa*. Microbiol Res 236:126467

Zimmerman NB, Vitousek PM (2012) Fungal endophyte communities reflect environmental structuring across a Hawaiian landscape. Proc Natl Acad Sci 109:13022–13027

Zou WX, Meng JC, Lu H et al (2000) Metabolites of *Colletotrichum gloeosporioides*, an endophytic fungus in *Artemisia mongolica*. J Nat Prod 63(11):1529–1530

Foliar Endophytic Fungi: Interactions and Importance

5

Rupshikha Nath, Jayeeta Mitra, Sibashish Kityania, Rajat Nath, Deepa Nath, and Anupam Das Talukdar

Abstract

Foliar endophytic fungi are a group of fungi that reside within the tissues of healthy leaves of plants without causing any apparent disease symptoms. They form a symbiotic relationship with the plant, where they colonize the internal spaces of the leaf tissues without triggering any apparent negative effects. They derive benefits from the host plant, such as protection, nutrient availability, and enhanced tolerance to environmental stresses, while the plant may benefit from improved growth and defense mechanisms. Foliar endophytic fungi initially colonize the outer tissues of the leaves, such as the epidermis and the mesophyll. They then penetrate the leaf through natural openings or by directly penetrating the cell walls. Once inside, they proliferate and colonize the intercellular spaces. Foliar endophytic fungi have a significant role in plant ecology. They can enhance plant growth and fitness by promoting nutrient uptake, modulating plant hormone levels, and facilitating stress tolerance. They can also contribute to the defense mechanisms of plants by producing bioactive compounds that deter herbivores and pathogens. Foliar endophytic fungi are known to produce a wide array of secondary metabolites, including alkaloids, terpenoids, phenolics, saponins, and antimicrobial compounds. These metabolites can have potential applications in medicine, agriculture, and industry. Some studies suggest that endophytic foliar fungi can induce genetic and phenotypic changes in host plants

R. Nath · S. Kityania · R. Nath · A. D. Talukdar (✉)
Department of Life Science and Bioinformatics, Assam University, Silchar, Assam, India
e-mail: anupam@bioinfoaus.ac.in

J. Mitra
Department of Botany, Arunachal University of Studies, Namsai, Arunachal Pradesh, India

D. Nath
Department of Botany, Gurucharan College, Silchar, Assam, India

K. A. Abd-Elsalam, S. F. AbuQamar (eds.), *Fungal Endophytes Volume I*,
https://doi.org/10.1007/978-981-97-7312-1_5

despite being asymptomatic in nature. Moreover, another alteration in plant physiology has also been observed. Additionally, studies on foliar endophytic fungi have gained considerable attention due to their potential agricultural and biotechnological applications. Researchers are investigating their role in improving crop productivity, enhancing stress tolerance, and discovering novel bioactive compounds. This chapter focuses on the synergistic and ecological interactions of foliar endophytic fungi as well as their pharmacological and industrial usage.

Keywords

Endophytic fungi · Secondary metabolites · Plant hormones · Stress tolerance · Antimicrobial properties

5.1 Introduction

The intricate interplay between endophytic fungi and their host plants involves a complex web of interactions, with evidence suggesting their coexistence since the earliest appearance of higher plants on Earth (Redecker et al. 2000). Fungi that dwell inside plant tissues without damaging the host and frequently offer advantages are known as endophytic fungi. They are distributed in all plant tissues, including leaves, stems, seeds, roots, and fruits (Breen 1994), and offer valuable benefits to their hosts, such as secondary metabolites and organic acids, while also aiding in defense against pathogens by secreting enzymes (Khan et al. 2016). Moreover, they provide a variety of advantages, including accelerated plant development, bioremediation, the production of bioactive chemicals, the induction of systemic resistance, and increased nutrient availability. They are crucial to the growth (Kim et al. 2017; Li et al. 2012b), physiological functions, endurance to stress (Jia et al. 2020), and carbon sequestration of plants (Iqbal et al. 2012). Endophytic fungi are numerous in forest environments and can significantly influence the genetic and phenotypic expression of their hosts (Mejía et al. 2014). While infectious bacteria have received the majority of attention in the ecology of microorganisms that live on leaves, beneficial endophytic fungi in leaves are increasingly receiving greater attention (Jia et al. 2020). Numerous plant components, including roots, leaves, seeds, and flowers, have been examined for endophytic fungus. Both forest trees and herbaceous plants have been shown to have them in their leaves. In the leaves of numerous tree species across the world, several types of endophytic fungus have been discovered. Endophytic fungi's capacity to colonize leaves may be connected to their capacity to adapt to the defense compounds released by plants. These fungi cause internal, asymptomatic infections in plant tissues and are spread horizontally (Herre et al. 2007; Rodriguez et al. 2009). They affect the severity of disease, plant development, herbivore sensitivity, photosynthetic rates, drought tolerance, and nutrient metabolism in host plants, among other functional traits (Arnold and Engelbrecht 2007; Arnold et al. 2003; Bae et al. 2009; Christian et al. 2017; Estrada et al. 2015; Mejía et al. 2014). Endophytes have the ability to alter how host genes involved in defense, cell wall formation, photosynthesis, and nitrogen metabolism are expressed. Nitrogen intake and distribution inside plants can be impacted by the number and kind of endophytes present (Mejía et al. 2014). Beneficial endophytes can boost a plant's ability to absorb nitrogen as a whole, and nitrogen can be supplied differently to various endophyte-colonized leaves (Zanne et al. 2006). Additionally, the distribution of nitrogen among leaves may be impacted by the simultaneous colonization of endophytes and diseases. Some research suggests that endophyte colonization might alter how plants absorb nutrients and affect how N is distributed within plant tissues while a disease is attacking (Christian et al. 2019). Endophytic fungi have been found to produce plant hormones, such as indole-3-acetic acid (IAA) and gibberellins (GAs), which can influence plant growth and development (Manici et al. 2015). These hormones might be involved in the mutualistic interactions between plants and fungus. Various fungal species may produce these hormones in various amounts, and studies have shown that some endophytic fungi can stimulate plant development in the presence of abiotic stress (Khan et al. 2015). In addition, certain endophytic fungi generate cytokinin's that have an impact on plant cell division and other physiological processes (Nicoletti and Fiorentino 2015; Vadassery et al. 2008). Moreover, knowing how endophytic fungi interact with plants may be useful for farming and agricultural production. By generating hormones, obtaining nutrients, and secreting stress-adaptor metabolites, they have a favorable impact on the host plant. Plants give these fungus shelter and food in exchange (Igiehon et al. 2021; Odelade and Babalola 2019). Diverse bioactive substances, such as alkaloids, flavonoids, steroids, terpenoids, and phenolic compounds, can be produced by endophytic fungus. These substances have pharmacological effects such as antibacterial, antioxidant, antidiabetic, antimalarial, and anticancer capabilities (Wen et al. 2022). They have co-evolved distinctive routes to synthesize complex secondary metabolites, and they are a very varied and adaptable microbial population found in different plant species (Wang et al. 2022). When desirable active ingredients for pharmaceutical usage are hard to come by or difficult to synthesize from conventional plant sources, endophytic fungus provide a viable option (Wen et al. 2022). This mutualistic relationship holds immense potential in various industries, agriculture, pharmaceuticals, and environmental applications. Endophytic fungi stand out as prodigious producers of secondary metabolites with potential applications in pharmaceuticals, serving as sources of antibiotics and anticancer, antioxidant, and anti-parasitic compounds. Additionally, they harbor a wide range of enzymes, making them valuable resources for various industries, including food, textile, leather, agriculture, beverages, and human health (Mishra et al. 2017).

5.2 The Synergism

Foliar endophytic fungus (FEFs) generates toxins that are essential for the biocontrol of insects. Numerous poisons, such as the sesquiterpene lactone heptelidic acid (HA) (Calhoun et al. 1992) and the anthraquinone rugulosin (Miller et al. 2002), have been identified from FEFs in various tree species. These poisons demonstrate their toxicity to forest pests including the economically significant insect pest *Choristoneura fumiferana* (Sumarah et al. 2010). However, the biosynthetic routes of these poisons are still poorly known, making it difficult to use them in practice for pest management in forestry and agriculture. In order to advance the use of toxins as biocontrol agents, it may be necessary to analyze important enzymes involved in toxin formation. Pest activity and leaf damage can be significantly impacted by foliar endophytic fungus (FEFs) (Jia et al. 2020). By limiting pest activity and raising insect mortality, the poisons generated by FEFs in tree leaves can prevent leaf damage (Jaber and Ownley 2018). To treat a variety of pests, including the European corn borer (Bing and Lewis 1992; Wagner and Lewis 2000), red scale bug (Asensio et al. 2005), stem weevil (Biswas et al. 2013), and banana weevil (Akello et al. 2008), FEFs have been utilized as entomopathogenic fungi. FEFs can also modify the mutualism between farmed fungus and leaf-cutting ants, which can change ant behavior and leaf consumption (Bittleston et al. 2011). Additionally, FEFs modify insect ovipositional preferences and have an impact on leafminer productivity and death (Wilson and Faeth 2001). FEFs may also affect insect reproduction, as was the case when *Glomerella cingulata* affected the production of *Chelymorpha alternans*, a beetle, and its progeny production (Van Bael et al. 2009).

5.2.1 Promoting Plant Growth

Many foliar endophytic fungi (FEFs), including plant growth-promoting rhizobacteria (PGPR) (Stringlis et al. 2018) and root endophytic fungi, have the ability to promote plant growth (Khan et al. 2016). It has been demonstrated that FEFs increase photosynthesis, stimulate vegetative development, and create indole acetic acids (IAA) that support plant growth. For instance, IAA is produced by FEFs such *Preussia* sp. and *Pestalotiopsis neglecta*, which have a favorable effect on the growth of tomato and *Boswellia sacra* plants, respectively (Hoffman et al. 2013). Furthermore, FEFs like *Colletotrichum tropicale* have been shown to affect *Theobroma cacao's* nitrogen intake and distribution, indicating their involvement in fostering plant development (Christian et al. 2019).

5.2.2 Preventing Herbivory

The experiments demonstrate that foliar endophytes in forb leaves can influence insect feeding and growth rates. The effects of inoculating different endophytes on insects were observed, with some endophytes reducing growth of generalist insects while benefiting specialist insects. The potential of endophytes as insect control agents in agricultural systems warrants further investigation using powerful methods like metabolomics (Gange et al. 2012). The symbiotic relationship between woody plants and foliar endophytic fungi (FEF) provides a valuable experimental system for studying the effects of endophytes on host genetic and phenotypic expression (Friesen et al. 2011). While FEF species are present in all surveyed plants under natural conditions, they generally have minimal visible effects on their hosts (Rodriguez et al. 2009). However, experiments have shown that foliar endophytes often enhance host defenses against pathogens and herbivores (Arnold et al. 2003; Mejía et al. 2008). Studies on *Theobroma cacao*, the source of cacao beans, have revealed that certain FEF species isolated from *T. cacao* can reduce leaf and pod damage caused by pathogens and herbivores (Arnold et al. 2003; Krauss et al. 2010; Mejía et al. 2008). Moreover, research using gene expression analyses indicated that FEF can significantly influence various aspects of the host's physiology, metabolism, anatomy, and resistance to pathogens and herbivores (Mejía et al. 2014).

5.2.3 Conferring Abiotic Stress Tolerance in Plants

Abiotic stressors, such as osmotic stress, salt, drought, heat, cold, and heavy metals, have a negative impact on plant development and metabolism, which results in substantial losses in agricultural productivity (Boyer 1982; Khare and Arora 2014; Wang et al. 2003). Plants undergo molecular, cellular, and physiological changes in response to these stimuli, which aid in their ability to adapt (Phurailatpam and Mishra 2020). The most common factor limiting agricultural productivity is drought stress (Saranga et al. 2001), although salt stress has an impact on plant development. While cold stress results in membrane degradation and cellular dehydration (Balestrasse et al. 2010; Mishra et al. 2019b), high temperatures interfere with photosynthesis and membrane stability (Schlau-Cohen et al. 2015). Collectively, these abiotic stressors cause up to a 70% (Boyer 1982) reduction in agricultural productivity. Phytohormones, enzyme activity, ROS scavenging, and osmolyte synthesis are a few examples of the processes involved in plant adaptation (Hasanuzzaman 2020).

5.2.3.1 Drought

Endophytes encourage the growth of roots, improve nutrient and water uptake, and create phytohormones including IAA, cytokinins, and gibberellins that support plant growth in water-scarce environments (Dimkpa et al. 2009). Additionally, endophytes create volatiles such 2*R*,3*R* butanediol that trigger systemic stress responses (Choudhary et al. 2008; Heil and Silva Bueno 2007; Niinemets 2010). In order to aid plant growth under drought stress, endophytes also help to osmotic adjustment by secreting osmolytes such proline, quaternary ammonium compounds, and polyamines (Vurukonda et al. 2016). Their actions are affected by genes associated with drought, and they can control how many stress-related genes are expressed in the host plant, increasing drought tolerance (Hasanuzzaman 2020).

5.2.3.2 Salinity

During high-salinity stress, endophytes attached to plants enhance biomass, shoot growth, chlorophyll content, and photosynthetic activity (Khan et al. 2011). Through the production of antioxidants and the control of phytohormones, they give salt resistance. Endophytes improve ROS scavenging, preventing plants from suffering from oxidative stress under abiotic stress conditions. It has been demonstrated that fungal endophytes, including *Aspergillus fumigatus*, promote the elongation of aerial parts of plants growing in salty environments and increase the plants' resistance to salt stress (Khan et al. 2011). By reducing the production of different antioxidants such tocopherol, ascorbate, catalase, superoxide dismutase, and glutathione reductase, endophytes impart salt resistance to plants (Rouhier et al. 2008).

5.2.3.3 Temperature

Due to sessile nature of plants, plants have to endure a range of variable temperatures. Plants survive in the range of temperature between 60 and 75 °F (Clay and Schardl 2002; Marchin et al. 2022). Temperature above or below this range causes stress in plants. It has been found that application of microbes can help in mitigating heat stress to a large extent. Microbes confer thermotolerance to plants. Bacteria has been known to release exopolysaccharide when they are exposed to heat stress. These exopolysaccharides contain 97% water; thus, water remains available to the plants during heat stress and they can easily mitigate the stress periods.

Endophytes have a symbiotic relationship with plants and they help the plants to withstand abiotic stress such as drought, salt, and heat. It was found that endophytic microbe SA187 helped in mitigating heat stress. In an experimental setup, *Arabidopsis thaliana* and wheat plant were subjected to heat stress and the result showed that the groups that were inoculated with Enterobacter SA 187 could mitigate heat stress more efficiently than untreated plants. The treated plants could mitigate temperature as high as 44 °C. The endophytes not only induced thermopriming in plants they also helped in increasing the Agronomic traits. The inoculated members showed 10–14% increase in height, 40% increase in grain yield, and 12% increase in seed weighty. The overall result showed that Endophytic group SA187 play a beneficial role in mitigating heat tolerance. In a separate study it was found that *Septoglomus deserticola* and *S. constrictu* conferred heat stress to tomato seedlings. Endophyte-inoculated tomato seedlings showed decreased oxidative stress and minimized level of reactive oxygen species. It was found that there was a marked improvement in the cellular performance, stomatal conductance, and leaf water content (Marchin et al. 2022). Soybean plant showed thermoregulation activity when they were inoculated with *B. cereus* SA1 (Bisht et al. 2020).

5.2.3.4 Heavy Metal

Heavy metal toxicity is a major abiotic stress factor causing significant crop loss (25–80%) due to its detrimental effects on root system development (Singh et al. 2021). Plants counteract heavy metal toxicity through molecular and biochemical mechanisms like antioxidant enzyme activity, hormonal regulation, metal chelators, and transporters (Singh et al. 2021). Endophytic fungi play a vital role in managing heavy metal stress in plants by binding and absorbing heavy metals through their unique cell walls. Metal-tolerant microbe–plant interactions enable metal transfer and absorption, promoting bioavailability and phytoextraction (Rajkumar et al. 2010; Sessitsch et al. 2013). Endophytes like *Aureobasidium pullulans* and

Gaeumannomyces cylindrosporus enhance plant tolerance to heavy metal stress by increasing antioxidant defense enzymes, regulating soil enzymatic activities, and improving growth parameters like chlorophyll content and seedling biomass in cucumber and maize plants, respectively (Ali et al. 2019; Yihui et al. 2017).

5.3 Ecological Role

In forest ecosystems, the decomposition of litter has a significant ecological impact on nutrient cycling and forest production. Foliar endophytic fungi (FEFs), which function as saprotrophs and influence the quality and quantity of plant litter, can have a variety of effects on the breakdown of litter (Jia et al. 2020; Saikkonen et al. 2015). While certain FEFs have been shown to slow down the decomposition of litter, others are highly degrading and produce extracellular enzymes to quicken the process. FEFs can linger in leaf litter for different amounts of time, and their presence might affect the processes of microbial colonization and breakdown in the future (Jia et al. 2020). The overall function of FEFs in the breakdown of litter is intricate and varies depending on the particular fungi involved and the surrounding circumstances. Moreover, endophytes establish information transfer through long-term synergy, leading to the production of common metabolites and enzymes (Strobel 2002). They colonize host tissues despite host resistance and secrete enzymes to degrade metabolites (Sunitha et al. 2013). Enzymes shed light on their functional roles and genetic recombination with hosts. Studies explore the ecological and evolutionary aspects of endophytes' symbiotic role and their influence on host fitness (Sunitha et al. 2013).

5.3.1 Role of Endophytic Fungi for Sustainable Agriculture

Foliar endophytic fungus are essential for sustainable agriculture because they encourage crop development, improve nutrient absorption, and create symbiotic relationships with plants. A plant pathogen called Sclerotinia sclerotiorum in sunflowers (Pieckenstain et al. 2001), phytoplasma in apple trees (Musetti et al. 2011), are just a few examples of the pathogens that Epicoccum nigrum, an endophytic fungus, can biocontrol (Chhipa and Deshmukh 2019) by raising root system biomass and generating bioactive secondary metabolites, it serves as a natural defense mechanism. *E. nigrum* is a viable contender for organic farming methods (Fávaro et al. 2012). *Trichoderma* species are endophytic plant symbionts that use strategies including antibiosis, mycoparasitism, and nutrient competition to successfully manage plant diseases (Sharma et al. 2019; Singh et al. 2011; Talapatra et al. 2017). In addition to fostering plant growth, they offer resistance to abiotic stressors and possess antifungal and antibacterial qualities. Trichoderma is a useful tool for sustainable agricultural practices, since it offers a safe substitute for toxic agrochemicals and has demonstrated encouraging outcomes in crop experiments (Chhipa and Deshmukh 2019; López et al. 2019). Fungal endophytes help crops adapt to biotic and abiotic stress factors, such as climate change (Rodriguez et al. 2008), by providing protection, bioremediation, and biocontrol mechanisms. Crop productivity is significantly impacted by abiotic factors such as soil salinity, drought, water scarcity, and rising temperatures (Rodriguez et al. 2008). Through induced systemic resistance (ISR), bioremediation, and biocontrol processes, fungal endophytes provide protection and help plants acclimatize to these stresses (Rastegari et al. 2020; Yadav 2019). Fungal endophytes promote agricultural sustainability by mitigating soil salinity's negative effects on plant metabolism, crop production, and environmental health. By 2050, 50% of arable land will be affected by salt accumulation (Parande et al. 2013). Fungal endophytes have the potential to improve crop health and productivity by minimizing damage, boosting photosynthesis, and collecting metabolites, endophytes increase crop resistance in conditions of high temperatures, drought, and heavy metal stress, which will support sustainable agricultural practices. Moreover, advanced research and agro-waste-grown bioagents also contribute to eco-friendly soil fertility enhancement (Singh et al. 2021).

5.3.2 Role of Endophytic Fungi for Controlling Nematodes

Endophytes have been found to effectively control nematodes and thus they help in mitigating biotic stress (Seleiman et al. 2021). A similar kind of result was found by West et al. where endophyte-infected *Lolium arundinaceum* showed tolerance to nematode attack (West et al. 1988). Endophytic fungi are effective in protecting plants from nematodes. West et al. (1988) observed that endophyte-infected *Lolium arundinaceum* gained tolerance to drought and nematode infection, and Elmi and Robbins (2000) and other posterior works confirmed those premises relating both stresses (Elmi and Robbins 2000).

5.3.3 Phytohormone Production Ability of Endophytic Fungi

Plants that have symbiotic association with endophytes are able to absorb phosphorus, vital metabolites, and plant hormones. It has been found that the solubilizing capacity of phosphate also gets enriched (Firáková et al. 2007). Endophytes such as Penicillium, Piriformo, Sebacina vermifera, Colletotrichum, and Spora indica have been shown to generate specific metabolites and enzymes that aid in plant growth during stressful situations. (Redman et al. 2011). Gibberellic acid is released by *Cladosporium sphaerospermum* in rice and soybean plants, which helps in increasing the growth of these plants (Hamayun et al. 2010). Germination rate was found to be increased in plants containing *Pestalotiopsis microspora*, which secreted pestalotin. This metabolite acted as gibberellin (Li et al. 2018). It was found that *Fusarium tricinctum* and *Alternaria alternate* increased plant development since they secrete indole acetic acid derivative (Khan et al. 2014). The endophytes of *P. indica* helped in the absorption of phosphorus and nitrogen (Johnson et al. 2014).

5.3.4 Production of Enzymes and Their Uses

Some endophytic fungi produce diverse enzymes that target specific components of plant tissues, reflecting a co-evolutionary adaptation. This enzyme diversity benefits both fungi and host plants, aiding in defense against pathogens and nutrient availability. Foliar endophytic fungi produce valuable enzymes like cellulases, xylanases, amylases, and pectinases, with applications in various industries (Mishra et al. 2019a, b). Studying these fungi offers potential for novel enzyme discovery and sustainable bioprocessing, reducing reliance on chemicals.

Xylanases play a crucial role in cellulose and paper industries, aiding in the breakdown of xylan and lignin in the pulp structure (Paice et al. 1992). Endophytic fungi play a crucial role in producing xylanase enzymes with diverse properties and functionalities. These enzymes are effective in breaking down xylan, a component of plant cell walls, and have various biotechnological applications. Researchers have isolated thermotolerant xylanase from an endophytic fungal strain (*Aspergillus terreus*) residing in a native Brazilian plant (Sorgatto et al. 2012). Furthermore, foliar endophytic fungi from Thai medicinal plants (*Croton oblongifolius*) demonstrated the ability to produce xylanase enzymes.

Amylases are used in the paper industry to size the paper, resulting in various paper grades dependent on starch viscosity (Godfrey and West 1996). Several endophytic fungi have been isolated from various plants and studied for their amylolytic activity. A strain from *Alpinia calcarata* displayed the highest amylolytic activity with optimal conditions at 30 °C and pH 7.0, using maltose as the best carbon source and sodium nitrate as the best nitrogen source (Sunitha et al. 2012). Another study revealed α-amylase activity in an endophytic isolate (*Presumina minima*) from a native Australian plant with optimal conditions at 25 °C and pH 9.0 (Zaferanloo et al. 2014).

In textiles, cellulases are used for biostoning, replacing the harsh pumice stone in stone washing of jeans (Mishra et al. 2019a). In the pulp and paper industry, cellulases save energy during mechanical pulping and aid in deinking waste papers. Endophytic fungi have been studied for cellulase production, showing potential for oil accumulation and cellulase production simultaneously. Some endophytic fungi (*Periconia* sp.) produce thermotolerant β-glucosidase with applications in lignocellulosic substrate treatment for biofuels and chemicals (Harnpicharnchai et al. 2009). Other endophytic fungal isolates (e.g. *Fusiccoccum* sp.) have also demonstrated cellulase activity, indicating their biotechnological potential (Kanokratana et al. 2008).

Pectinolytic enzymes are used in wine processing to improve wine characteristics by maximizing juice yield, enhancing extraction, facilitating filtration, and intensifying color and flavor. In tea and coffee processing, pectinases break down pectins in the leaves, accelerating fermentation and acting as antifoaming agents (Mishra et al. 2019a).

Lipases have diverse applications in various industries. In the food industry, they are used to synthesize flavors and remove fat from dairy and meat products. In detergents, lipases act on greasy substances and improve their efficacy. In the pulp and paper industry, lipases help remove pitch from wood, making pulp less sticky and improving paper production. In leather processing, lipases remove fat from hides and skin. Moreover, lipases play a crucial role in biodiesel production by breaking down triacyl glycerides into fatty acids (Mishra et al. 2019a). In the second study, lipases from *Cercospora kikuchii*, associated with *Tithonia diversifolia*, were studied. Lipase production was highest on 2% soybean oil, yielding 9 U/mL. Spray drying with β-cyclodextrin maintained 100% activity after 8 months (Costa-Silva et al. 2011). *Lasiodiplodia* sp. was identified as the best lipase producer, with an activity of 108 U/mL (Venkatesagowda et al. 2012). Moreover, L-asparaginases, obtained from endophytic fungi, have potential as anti-tumor drugs, especially in leukemia treatment (Batool et al. 2016). Several endophytic fungi have been screened for their ability to produce L-asparaginase, which is also valuable in the food industry to reduce acrylamide formation in baked and fried foods (Nagarajan et al. 2014).

Endophytic fungi play a pivotal role in detoxifying hazardous substances and are harnessed for waste treatment and environmental cleanup due to their secretion of diverse extracellular enzymes (Mishra et al. 2017). They adapt fungal metabolic pathways under environmental stress, breaking down intricate polymers into simpler compounds (Li et al. 2012a). They also contribute to global mineral cycling and engage in biotransformations (Abourashed and Hufford 1996), particularly noteworthy, *Pestalotiopsis microspora* effectively degrades polyester polyurethane under both aerobic and anaerobic conditions (Russell et al. 2011).

5.4 Secondary Metabolites Production (Noble Source of Bioactive Compound Production)

The secondary metabolites that foliar endophytic fungus create are an important source of bioactive substances. These metabolites are not essential for the fungus's basic metabolic processes but often have significant pharmaceutical and agricultural applications. Alkaloids, terpenoids, flavonoids, xanthones, and camptothecins are just a few of the prominent secondary metabolites that foliar endophytes create. Foliar endophytic fungi are now a significant and prospective source of bioactive chemical synthesis as a result of their discovery, which has created new opportunities for medicine research and plant protection techniques. Some of the examples are discussed below:

Some endophytic fungi have the capacity to create volatile antibiotics that may be utilized as flavorings (Abrahão et al. 2013) and in mycofumigation (Park et al. 2010; Suwannarach et al. 2013). Many endophytic fungi produce terpenes like eremophilanes, which have the potential to be phytotoxic (Isaka et al. 2010). The content of essential oils in plants can also be altered by endophytic fungi, which may affect crop resilience to pests (Nicoletti and Fiorentino 2015).

Terpenoid substances are present in a variety of plants and essential oils, including sesquiterpene lactones, ginkgolides (Cui et al. 2012), agathic acid (Yuan et al. 2009), limonoids (Dos Santos and Rodrigues-Fo 2002), and toosendanin (Wang et al. 2007). It has been discovered that some endophytic fungi create these substances as fermentation byproducts, suggesting their contribution to the composition of the combination. These terpenoid substances have a wide range of bioactive qualities which made them attractive candidates for a number of uses, including the production of drugs and the management of insects (Nicoletti and Fiorentino 2015).

It has been shown that endophytic fungi create a wide range of secondary metabolites, including coumarins and their derivatives. These substances, including bergapten, umbelliferone, and mellein, have a variety of advantageous qualities, including antioxidant, anti-inflammatory, anticoagulant, antibacterial, anticancer, antihypertensive, and antihyperglycemic actions (Pal et al. 2011; Venugopala et al. 2013). Some of these substances, which were previously solely known from plants, have lately been described as metabolites of endophytic fungus, raising the possibility of their use in pharmaceuticals (Umashankar et al. 2014).

Numerous flavonoids, including kaempferol (Huang et al. 2014), quercetin, genkwanin, (+/−)-catechin, tectochrysin, and dimethylgalangin, have been discovered to be produced by endophytic fungus (Cheng et al. 2013). Other flavonoids produced by endophytic fungi include luteolin (Zhao et al. 2014), cajanol (Zhao et al. 2013), apigenin (Gao et al. 2012), vitexin (Zhou et al. 2011), and rotenone (Hu et al. 2005). One more notable metabolite is 2′,4′-dihydroxy-6′-methoxy-3′,5′-dimethylchalcone from *Ceriporia lacerata* (Wang et al. 2013) and *Silymarin*, a bioactive extract from milk thistle (Davis-Searles et al. 2005). The synthesis of these various flavonoids by endophytic fungus emphasizes their potential for use in pharmaceuticals and their function in interactions between plants and microbes (Nicoletti and Fiorentino 2015).

Xanthones, which may be derived from both plants and fungi and have structural similarities to flavonoids, have use in medicine. (Negi et al. 2010). It has been discovered that endophytic fungi create xanthones such as 1,7-dihydroxyxanthone, which is well-known for its antimalarial activities (Huang et al. 2012). Emodin, physcion, and questinol are examples of anthraquinones, which are commonly used as colors and have antibacterial characteristics in plants. They are also present in endophytic fungi (Marinho et al. 2005, 2013; Wang et al. 2008). The antidepressant and anticancer compound hypericin has been discovered in an endophytic strain (Kusari et al. 2008, 2009). Tanshinones, which have antioxidant and anticancer properties and are present in *Salvia* species, have also been isolated from endophytic strains (Ma et al. 2011). A number of pharmacologically active substances, including shikonin (Movahhedin et al. 2014) and lapachol (Govindappa 2014; Sadananda et al. 2011), have been retrieved from endophytic strains. These results point to the potential for medicinal uses for these endophytic fungi-produced chemicals.

For delicate forest ecosystems, biocontrol of forest diseases brought on by fungi like *Diplodia corticola* (Cimmino et al. 2016; Fernandes et al. 2014) and *Hymenoscyphus* (Schlegel et al. 2018) species is crucial. The terpenoids, alkaloids, and polyketides that are produced by foliar endophytic fungi (FEFs) have potent inhibitory effects on pathogenic bacteria (Masi et al. 2018; Mousa and Raizada 2013). FEFs isolated from *Pinus strobes* (Sumarah et al. 2011) and *Populus*

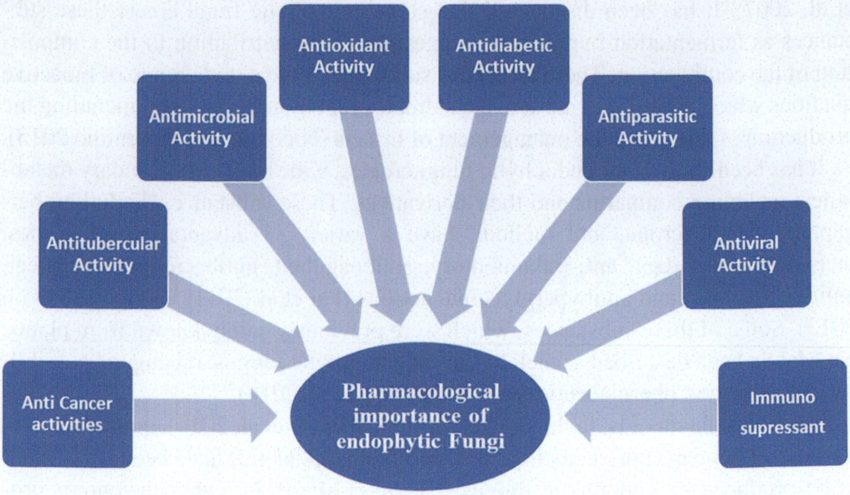

Fig. 5.1 Different potential of endophytic fungi in pharmacology

trichocarpa (McMullin et al. 2018) trees have shown antifungal activity and the ability to prevent disease. Additionally, several psychrophilic FEFs in Cupressaceae trees have antibacterial properties that help them withstand freezing and defend against freeze–thaw damage (Hosseyni Moghaddam and Soltani 2014).

5.5 Pharmacological Importance

The overall pharmacological activities of endophytic fungi are shown in Figure 5.1.

5.5.1 Anticancer Activities of Endophytic Fungi

Chemotherapy for cancer must overcome difficulties including toxicity and drug resistance (Bhadresha et al. 2022). Endophytic fungi produce natural anticancer compounds that have shown promising results against various cancers, including Kaposi's sarcoma, prostate, and breast cancer (Dos Santos et al. 2022; Tiwari and Bae 2022). These compounds induce apoptosis and inhibit cancer progression. Specific compounds like paclitaxel, podophyllotoxin, camptothecin, vinblastine, and hypericin have been isolated from endophytic fungi and exhibit cytotoxic effects against cancer cells (Kousar et al. 2022). Recent research has unveiled a plethora of bioactive compounds from fungal endophytes with promising anticancer properties. These novel compounds exhibit cytotoxic effects against various cancer cell lines, presenting potential as valuable assets for cancer therapy (Kousar et al. 2022). Examples include fusarubin and anhydrofusarubin from *Cladosporium* sp., which slow the spread of cancer by causing metastatic cells to undergo apoptosis and stopping the cell cycle (de Wit 2016). Various compounds act through various mechanisms, such as apoptosis induction and inhibition of key cellular processes, providing new opportunities for cancer treatment. Many endophytic fungi were found to possess anticancer activities and anticancer compounds isolated from many of the sources listed in Table 5.1 and structure shown in Fig. 5.2.

5.5.2 Antitubercular Activity

Tuberculosis has created such a big problem that elimination of TB has been enlisted among one of the Sustainable Development Goals of 2030. *Mycobacterium tuberculosis* is the causative agent of Tuberculosis. It has been found that synthetic drugs lose their efficacy against Mycobacterium in a short period of time. *Phomopsis* species isolated from *Garcinia* has been found to secrete several secondary metabolites such as Tenuazonic acid and 3-Nitropropanoic acid. These metabolites alter the pathway of isocitrate lyase enzyme and thus act as an effective biocontrol agent against *Mycobacterium tuberculosis* (Jena et al. 2015). Two other bioactive metabolites named Phomoenamide and Phomonitroester secreted by *Phomopsis* species isolated from *Garcinia* are also capable of inhibiting the growth of *Mycobacterium tuberculosis*. It has also been found that *Diaporthe* sp., an endophyte found in *Pandanas amaryllifolius*, secretes Benzopyran and Diaportheone A and B; both these compounds have been found to be effective against Tuberculosis (Chepkirui and Stadler 2017).

5.5.3 Antimicrobial Activity

Salmonella thypi, *Escherichia coli*, *Bacillus subtilis*, and *Streptococcus aureus* were all susceptible to the antibacterial effects of endophytic fungi extracted from *P. canescens* leaves. After experimental analysis it was observed in particular that seven extracts had high antibacterial capabilities against *Streptococcus aureus*, while five

Table 5.1 Few examples of endophytic fungi, their host, and isolated bioactive compounds

Sl. No.	Fungi	Host plant	Bioactive compound	Reference
1.	*Altenaria alternata*	*Passiflora incarnata*	Chrysin	Shaffique et al. (2022)
2.	*Alternaria alternata*	*Miquelia dentate*	Camptothecine	Mukhtar et al. (2023)
3.	*Phomopsis* sp. and *Fomitopsis* sp. *Aspergillus flavus*	*Solanum nigrum*	Podophyllotoxin	Jha and Sit (2022)
4.	*Fusarium oxysporum*	*Sabina recurva*	Podophyllotoxin	Ali et al. (2009)
5.	*Fusarium oxysporum*	*Catharanthus roseus*	Vinblastine	Abd El-Daim et al. (2014)

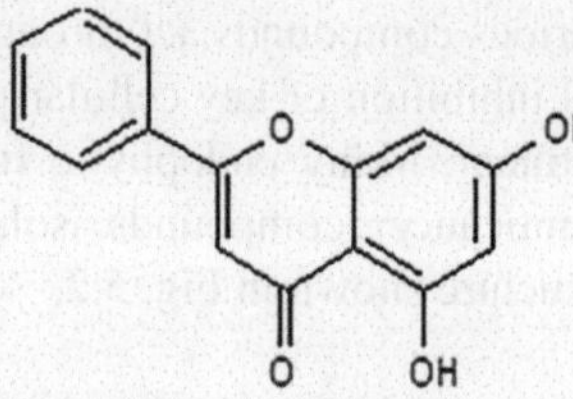

1.

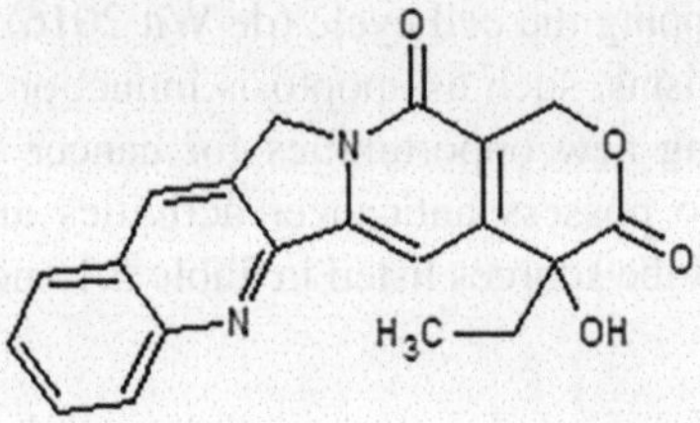

2.

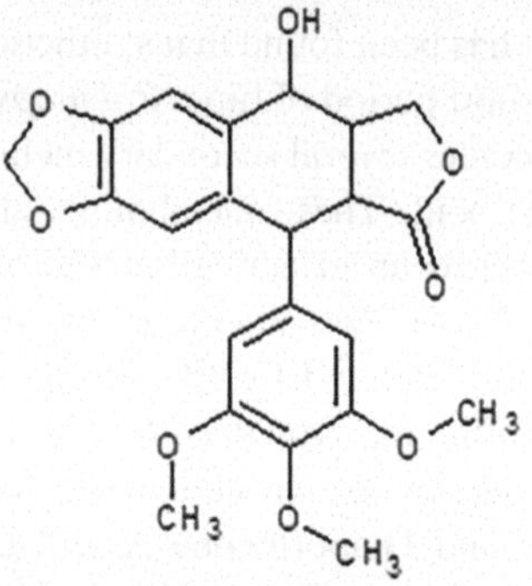

3.

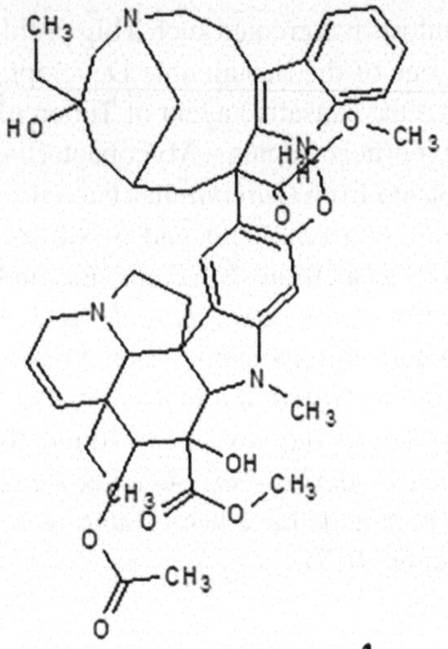

4.

Fig. 5.2 Chemical structure of some bioactive compounds (1) Chrysin, (2) Camptothecine, (3) Podophyllotoxin, (4) Vinblastine

extracts showed good antibacterial activity against *Escherichia coli*. The majority of the endophytic fungal extracts also showed potent antibacterial activity against *Bacillus subtilis* and *Salmonella typhi*. According to the observations, endophytic fungi found on *P. canescens* leaves may be sources of bioactive substances with potential for use in medicine (Elfita et al. 2022).

5.5.4 Antioxidant Activity

Like mystical inhabitants of the plant kingdom, endophytic fungi harbor many gifts; one of them is crafting antioxidant secondary metabolites from within, adding a touch of enchantment to nature's arsenal. In order to identify possible antioxidant sources, some researchers examine the antioxidant activity of secondary metabolites from *S. samarangense* and its endophytes (Budiono et al. 2019). They concluded that endophytic fungi are interesting sources for novel antioxidant chemicals as they can generate secondary compounds that have greater biological activity over their host plants. The *Lasiodiplodia venezuelensis* strain CBS 129753 (known as BJA-1) showed the highest homology and is therefore a promising candidate for further research as a source of bioactive chemicals (Budiono et al. 2019).

5.5.5 Antidiabetic Activity

Plant-dwelling endophytic fungus represent a prospective source of bioactive compounds. Some study suggested that mango, neem, and jambolan plants have historically been used to treat diabetes (Jaiswal et al. 2012). One such finding involved the collection of mature and healthy leaves from *Azadirachta indica*, *Syzygium cumini*, and *Mangifera indica* at the biopark of Bahauddin Zakariya University of Multan. The leaves were meticulously gathered, sterilized, and processed within 24 h. Fungal endophytes were isolated from the plant samples using modified methods. The fungi were morphologically characterized under a microscope. Subsequently, the endophytes were screened for their ability to produce antidiabetic metabolites through an alpha-amylase inhibitory assay. Statistical analysis was applied to the data, and significant differences were determined (Khan et al. 2019). A total of eight fungal endophytes were successfully isolated and identified, comprising *Aspergillus* sp., *Penicillium* sp., and *Alternaria* sp. These endophytes exhibited metabolic activity after 10 days at 30 °C. Notably, the extract from *Alternaria* sp. (from *Syzygium cumini* L.) displayed the highest alpha-amylase inhibition of 62%, outperforming the other fungal strain activity. This finding suggests that these endophytic fungi hold promise as potential tools for managing hyperglycemia and diabetes (Khan et al. 2019).

5.5.6 Antiparasitic Activity

Parasites are those organisms that depend on other organisms for their food and survival (Muhammed et al. 2019). The microorganisms that fall in the range of parasites are diverse, ranging from protozoans to helminths. *Plasmodium* the causative agent of Malaria resulted in death of about 3.3 billion people (Ateba et al. 2018). A census of the amount of people dying due to Malaria was conducted in 2016 and it was found that over 405,000 people died that year due to this disease and 228 million new cases were recorded in 91 countries (Muhammed et al. 2019). Fungal endophytes have been found to play a very effective role as medicine, and therefore pharmaceutical industries are always in look out for new fungal endophytes. *Vanilla albindia* harbors *Phomopsis archeri* which synthesizes phomoxanthones A-C. Phomoxanthone AC has been found to be effective against malaria (Aharwal et al. 2016). Another plant *Piptadenia adiantoides* carried endophyte *Cochliobolus* spp., which produced Cochlioquonone which was found effective against Leishmania (Wang et al. 2017). *Mycosphaerella* spp. produces Cercosporins, a secondary metabolite which have been found to be effective against a variety of parasites such as *Plasmodium falciparum*, *Trypanosoma cruzi*, and *Leishmania donovania* (Wang et al. 2017). Another secondary metabolite epoxycytochalasin H has been found to be effective against Malarial Parasite especially *Plasmodium falciparum* (Ferreira et al. 2017).

5.5.7 Antiviral Activity

Some endophytes have been found which showed bioactive activity against virus. So far very less amount of knowledge has been obtained in this field, but it shows a lot promise for future (Adeleke and Babalola 2021). Fungal endophytes secreting secondary metabolites such as cyclosporine U, sequoiatones C–F, Cytonic acid, and podophyllotoxin have been found to have effective antiviral activities (Adeleke and Babalola 2021).

5.5.8 Immunosupressive Properties of Endophytes

Immunosuppressive properties refer to the ability of a substance to suppress or dampen the immune system's response. In some cases, compounds derived from fungi have been found to possess immunosuppressive properties, but these studies have predominantly focused on fungi associated with medicinal or terrestrial sources rather than foliar endophytic fungi. Some experimental study suggested that three novel xanthone derivatives discovered by chemical analysis of the endophytic fungus *Aspergillus sydowii* discovered from host liverwort *Scapania ciliata* S. Lac may have immunosuppressive effects. These are sydoxanthone A, a xanthone derivative with ester groups, sydoxanthone B, and 13-*O*-acetylsydowinin B. The study assumed that fungus endophytes could contain immunosuppressive substances with potential therapeutic uses (Song et al. 2013). However, the potential immunosuppressive properties of foliar endophytic fungi would require dedicated research and investigations to identify and characterize specific compounds responsible for such effects.

5.6 Role of Medicinal Bioactive Compounds in Industries

Fungal endophytes are an excellent source of bioactive chemicals. They have been investigated for drug development since the isolation of Taxol and penicillin from plants. (Strobel et al. 1996). Bioactive compounds from endophytic fungi can be produced in vitro by using certain phyto-components to stimulate the synthesis of bioactive chemicals in endophytic fungus. By using this method, antimicrobial substances with good antimicrobial activity against *Fusarium oxysporum* were produced, including 5-hydroxy-2,3-dihydroxymethyl-7-methoxychromone, 5-hydroxy-3-acetoxymethyl-2-methyl-7-methoxychromone, and 5,7-dihydroxy-3-hydroxymethyl-2-methylchromone (Hu et al. 2020). These results point to a possible approach for identifying novel medication candidates in these fungi, stimulating endophyte metabolism (Wen et al. 2022). KB3.1 cells were moderately cytotoxic to *Neurospora udagawae* udagawanones A–B's significant antifungal efficacy against *Rhodoturula glutinis*. Against tumour cell lines, nigeriapyrones A and B (Liu et al. 2013a) showed strong antifungal activity. Significant antifungal activity was shown by ficipyrones A–B against *Gibberella zeae* (Liu et al. 2013b).

Pharmaceutical chemists are developing effective, safe, and low-toxic medications from natural sources, including endophytic fungi. They are exploring new bioactive substances and understanding their mechanisms of action using biological, pharmacodynamic, biochemical, bioinformatics, and pre-clinical techniques. Endophytes produce antibacterial compounds effective against antibiotic-resistant infections. To fully comprehend the impact of certain elements on the metabolism and synthesis of bioactive compounds by endophytic fungus, more investigation is required (Wen et al. 2022).

5.7 Future Perspectives

The time-consuming isolation and purification procedure is one of the current difficulties in using endophytic fungus for industrial uses because of the possibility of contamination (Thirumalanadhuni et al. 2021). Isolation is hampered by the difficulty of growing certain fungus on sterile mycelia. Low levels of cytotoxic substances generated by endophytes don't match industrial requirements, and the extinction of host plants in danger raises questions (Adeleke and Babalola 2021; Kharwar et al. 2011). The development of novel medications and the growth of endophytes that are uncultivable are made possible by molecular biology advances (Fadiji and Babalola 2020; Giddings and Newman 2022). Nanotechnology and metagenomics have the ability to increase metabolite synthesis and find novel bioactive chemicals (Yadav et al. 2022). The use of genomic and metabolomics methods in the future will help us comprehend how endophytic fungi interact with their host plants (Fadiji and Babalola 2020). Endophytic fungi produce low-toxic secondary metabolites from host plants, converting active compounds into more potent derivatives. Advances in fermentation culture, extraction, and structure identification enable the discovery of new valuable compounds. Challenges like limited yield, gene silencing, and low expression require addressing. Techniques like OSMAC and "omics" can enhance metabolite production (Wen et al. 2022).

5.8 Conclusion

Folar endophytic fungi play an important role in plant health, defense systems, and secondary metabolite synthesis. These fungi have enormous medicinal and agricultural potential, having antiviral, anticancer, antibacterial, antioxidant, and immunosuppressive properties. They also show promising biocontrol capabilities against pathogenic fungus in forest habitats, adding to environmentally friendly disease management strategies. Folar endophytic fungi may also be useful in the treatment of cancer, diabetes, antibiotic infections, parasitic disorders, and viral infections. More study is required to completely comprehend their interactions with host plants as well as their various biological activities. Understanding and utilizing the potential of these fungi might result in substantial advances in a variety of sectors, as well as contribute to long-term solutions for human and environmental health.

References

Abd El-Daim IA, Bejai S, Meijer J (2014) Improved heat stress tolerance of wheat seedlings by bacterial seed treatment. Plant Soil 379:337–350

Abourashed EA, Hufford CD (1996) Microbial transformation of artemether. J Nat Prod 59:251–253

Abrahão MR, Molina G, Pastore GM (2013) Endophytes: recent developments in biotechnology and the potential for flavor production. Food Res Int 52:367–372

Adeleke BS, Babalola OO (2021) Pharmacological potential of fungal endophytes associated with medicinal plants: a review. J Fungi 7:147

Aharwal RP, Kumar S, Sandhu SS (2016) Endophytic mycoflora as a source of biotherapeutic compounds for disease treatment. J Appl Pharm Sci 6:242–254

Akello J, Dubois T, Coyne D, Kyamanywa S (2008) Endophytic Beauveria bassiana in banana (Musa spp.) reduces banana weevil (Cosmopolites sordidus) fitness and damage. Crop Prot 27:1437–1441. https://doi.org/10.1016/j.cropro.2008.07.003

Ali SZ, Sandhya V, Grover M, Kishore N, Rao LV, Venkateswarlu B (2009) Pseudomonas sp. strain AKM-P6 enhances tolerance of sorghum seedlings to elevated temperatures. Biol Fertil Soils 46:45–55

Ali A, Bilal S, Khan AL, Mabood F, Al-Harrasi A, Lee I-J (2019) Endophytic Aureobasidium pullulans BSS6 assisted developments in phytoremediation potentials of Cucumis sativus under Cd and Pb stress. J Plant Interact 14:303–313

Arnold AE, Engelbrecht BM (2007) Fungal endophytes nearly double minimum leaf conductance in seedlings of a neotropical tree species. J Trop Ecol 23:369–372

Arnold AE, Mejía LC, Kyllo D, Rojas EI, Maynard Z, Robbins N, Herre EA (2003) Fungal endophytes limit pathogen damage in a tropical tree. Proc Natl Acad Sci 100:15649–15654

Asensio L, Lopez-Llorca LV, López-Jiménez JA (2005) Use of light, scanning electron microscopy and bioassays to evaluate parasitism by entomopathogenic fungi of the red scale insect of palms (Phoenicococcus marlatti Ckll., 1899). Micron 36:169–175. https://doi.org/10.1016/j.micron.2004.09.004

Ateba JE, Toghueo RM, Awantu AF, Mba'ning BM, Gohlke S, Sahal D, Rodrigues-Filho E, Tsamo E, Boyom FF, Sewald N (2018) Antiplasmodial properties and cytotoxicity of endophytic fungi from Symphonia globulifera (Clusiaceae). J Fungi 4:70

Bae H, Sicher RC, Kim MS, Kim S-H, Strem MD, Melnick RL, Bailey BA (2009) The beneficial endophyte Trichoderma hamatum isolate DIS 219b promotes growth and delays the onset of the drought response in Theobroma cacao. J Exp Bot 60:3279–3295

Balestrasse KB, Tomaro ML, Batlle A, Noriega GO (2010) The role of 5-aminolevulinic acid in the response to cold stress in soybean plants. Phytochemistry 71:2038–2045

Batool T, Makky EA, Jalal M, Yusoff MM (2016) A comprehensive review on L-asparaginase and its applications. Appl Biochem Biotechnol 178:900–923

Bhadresha KP, Jain NK, Rawal RM (2022) Assessing the protective effect of Moringa oleifera extract against bone metastasis: an in vitro simulated digestion approach. Nutr Cancer 74:1023–1036

Bing LA, Lewis LC (1992) Endophytic Beauveria bassiana (Balsamo) Vuillemin in corn: the influence of the plant growth stage and Ostrinia nubilalis (Hübner). Biocontrol Sci Tech 2:39–47. https://doi.org/10.1080/09583159209355216

Bisht N, Mishra SK, Chauhan PS (2020) Bacillus amyloliquefaciens inoculation alters physiology of rice (Oryza sativa L. var. IR-36) through modulating carbohydrate metabolism to mitigate stress induced by nutrient starvation. Int J Biol Macromol 143:937–951

Biswas C, Dey P, Satpathy S, Satya P, Mahapatra BS (2013) Endophytic colonization of white jute (Corchorus capsularis) plants by different Beauveria bassiana strains for managing stem weevil (Apion corchori). Phytoparasitica 41:17–21. https://doi.org/10.1007/s12600-012-0257-x

Bittleston LS, Brockmann F, Wcislo W, Van Bael SA (2011) Endophytic fungi reduce leaf-cutting ant damage to seedlings. Biol Lett 7:30–32. https://doi.org/10.1098/rsbl.2010.0456

Boyer JS (1982) Plant productivity and environment. Science 218:443–448

Breen J (1994) Acremonium endophyte interactions with enhanced plant resistance to insects. Annu Rev Entomol 39:401–423

Budiono B, Elfita E, Muharni M, Yohandini H, Widjajanti H (2019) Antioxidant activity of Syzygium samarangense L. and their endophytic fungi. Molekul 14:48–55

Calhoun LA, Findlay JA, Miller JD, Whitney NJ (1992) Metabolites toxic to spruce budworm from balsam fir needle endophytes. Mycol Res 96:281–286

Cheng M-J, Wu M-D, Chen J-J, Hsieh S-Y, Yuan G-F, Chen I-S, Chang C-W (2013) Secondary metabolites from the endophytic fungus of Annulohypoxylon ilanense. Chem Nat Compd 49:523–525

Chepkirui C, Stadler M (2017) The genus Diaporthe: a rich source of diverse and bioactive metabolites. Mycol Prog 16:477–494

Chhipa H, Deshmukh S (2019) Fungal endophytes: rising tools in sustainable agriculture production. Endophytes and secondary metabolites. Springer International Publishing AG, Cham, pp 1–24

Choudhary DK, Johri BN, Prakash A (2008) Volatiles as priming agents that initiate plant growth and defence responses. Curr Sci 94:595–604

Christian N, Herre EA, Mejia LC, Clay K (2017) Exposure to the leaf litter microbiome of healthy adults protects seedlings from pathogen damage. Proc R Soc B Biol Sci 284:20170641

Christian N, Herre EA, Clay K (2019) Foliar endophytic fungi alter patterns of nitrogen uptake and distribution in Theobroma cacao. New Phytol 222:1573–1583

Cimmino A, Maddau L, Masi M, Evidente M, Linaldeddu BT, Evidente A (2016) Further secondary metabolites produced by Diplodia corticola, a fungal pathogen involved in cork oak decline. Tetrahedron 72:6788–6793. https://doi.org/10.1016/j.tet.2016.09.008

Clay K, Schardl C (2002) Evolutionary origins and ecological consequences of endophyte symbiosis with grasses. Am Nat 160:S99–S127

Costa-Silva TA, Nogueira MA, Fernandes Souza CR, Oliveira WP, Said S (2011) Lipase production by endophytic fungus Cercospora kikuchii: stability of enzymatic activity after spray drying in the presence of carbohydrates. Dry Technol 29:1112–1119

Cui Y, Yi D, Bai X, Sun B, Zhao Y, Zhang Y (2012) Ginkgolide B produced endophytic fungus (Fusarium oxysporum) isolated from Ginkgo biloba. Fitoterapia 83:913–920

Davis-Searles PR, Nakanishi Y, Kim N-C, Graf TN, Oberlies NH, Wani MC, Wall ME, Agarwal R, Kroll DJ (2005) Milk thistle and prostate cancer: differential effects of pure flavonolignans from Silybum marianum on antiproliferative end points in human prostate carcinoma cells. Cancer Res 65:4448–4457

de Wit PJ (2016) Cladosporium fulvum effectors: weapons in the arms race with tomato. Annu Rev Phytopathol 54:1–23

Dimkpa C, Weinand T, Asch F (2009) Plant–rhizobacteria interactions alleviate abiotic stress conditions. Plant Cell Environ 32:1682–1694

Dos Santos RMG, Rodrigues-Fo E (2002) Meroterpenes from Penicillium sp found in association with Melia azedarach. Phytochemistry 61:907–912

Dos Santos I, Mohamed T, Borges L, Nahas HA, Abdel-Azeem M, Carvalho C, Bezerra J, Gupta V, Xavier-Santos S, Mossa M (2022) The global tendency in the research of biological activity in endophytic fungi: a scientometric analysis. Curr Res Environ Appl Mycol 12:1–14

Elfita E, Oktiansyah R, Mardiyanto M, Widjajanti H, Setiawan A (2022) Antibacterial and antioxidant activity of endophytic fungi isolated from Peronema canescens leaves. Biodiversitas 23:241213

Elmi W, Robbins K (2000) Endophyte effects on reproduction of a root-knot nematode (Meloidogyne marylandi) and osmotic adjustment in tall fescue. Grass Forage Sci 55:166–172

Estrada C, Degner EC, Rojas EI, Wcislo WT, Van Bael SA (2015) The role of endophyte diversity in protecting plants from defoliation by leaf-cutting ants. Curr Sci 109:55–61

Fadiji AE, Babalola OO (2020) Elucidating mechanisms of endophytes used in plant protection and other bioactivities with multifunctional prospects. Front Bioeng Biotechnol 8:467

Fávaro LCDL, Sebastianes FLDS, Araújo WL (2012) Epicoccum nigrum P16, a sugarcane endophyte, produces antifungal compounds and induces root growth. PLoS One 7:e36826

Fernandes I, Alves A, Correia A, Devreese B, Esteves AC (2014) Secretome analysis identifies potential virulence factors of Diplodia corticola, a fungal pathogen involved in cork oak (Quercus suber) decline. Fungal Biol 118:516–523. https://doi.org/10.1016/j.funbio.2014.04.006

Ferreira MC, Cantrell CL, Wedge DE, Gonçalves VN, Jacob MR, Khan S, Rosa CA, Rosa LH (2017) Antimycobacterial and antimalarial activities of endophytic fungi associated with the ancient and narrowly endemic neotropical plant Vellozia gigantea from Brazil. Mem Inst Oswaldo Cruz 112:692–697

Firáková S, Šturdíková M, Múčková M (2007) Bioactive secondary metabolites produced by microorganisms associated with plants. Biologia 62:251–257

Friesen ML, Porter SS, Stark SC, von Wettberg EJ, Sachs JL, Martinez-Romero E (2011) Microbially mediated plant functional traits. Annu Rev Ecol Evol Syst 42:23–46

Gange AC, Eschen R, Wearn JA, Thawer A, Sutton BC (2012) Differential effects of foliar endophytic fungi on insect herbivores attacking a herbaceous plant. Oecologia 168:1023–1031. https://doi.org/10.1007/s00442-011-2151-5

Gao Y, Zhao J, Zu Y, Fu Y, Liang L, Luo M, Wang W, Efferth T (2012) Antioxidant properties, superoxide dismutase and glutathione reductase activities in HepG2 cells with a fungal endophyte producing apigenin from pigeon pea [Cajanus cajan (L.) Millsp.]. Food Res Int 49:147–152

Giddings L-A, Newman DJ (2022) Extremophilic fungi from marine environments: underexplored sources of antitumor, anti-infective and other biologically active agents. Mar Drugs 20:62

Godfrey T, West S (1996) Textiles. In: Industrial enzymology. Macmillan Press, London

Govindappa M (2014) First report of anti-cancer agent, lapachol producing endophyte, Aspergillus niger of Tabebuia argentea and its in vitro cytotoxicity assays. Bangladesh J Pharmacol 9:129–139

Hamayun M, Khan SA, Khan AL, Rehman G, Kim Y-H, Iqbal I, Hussain J, Sohn E-Y, Lee I-J (2010) Gibberellin production and plant growth promotion from pure cultures of Cladosporium sp. MH-6 isolated from cucumber (Cucumis sativus L.). Mycologia 102:989–995

Harnpicharnchai P, Champreda V, Sornlake W, Eurwilaichitr L (2009) A thermotolerant β-glucosidase isolated from an endophytic fungi, Periconia sp., with a possible use for biomass conversion to sugars. Protein Expr Purif 67:61–69

Hasanuzzaman M (2020) Plant ecophysiology and adaptation under climate change: mechanisms and perspectives II: Mechanisms of adaptation and stress amelioration. Springer Nature

Heil M, Silva Bueno JC (2007) Within-plant signaling by volatiles leads to induction and priming of an indirect plant defense in nature. Proc Natl Acad Sci 104:5467–5472

Herre EA, Mejía LC, Kyllo DA, Rojas E, Maynard Z, Butler A, Van Bael SA (2007) Ecological implications of anti-pathogen effects of tropical fungal endophytes and mycorrhizae. Ecology 88:550–558

Hoffman MT, Gunatilaka MK, Wijeratne K, Gunatilaka L, Arnold AE (2013) Endohyphal bacterium enhances production of indole-3-acetic acid by a foliar fungal endophyte. PLoS One 8:e73132

Hosseyni Moghaddam MS, Soltani J (2014) Psychrophilic endophytic fungi with biological activity inhabit Cupressaceae plant family. Symbiosis 63:79–86. https://doi.org/10.1007/s13199-014-0290-2

Hu M, Zhong G, Sun ZT, Sh G, Liu H, Liu X (2005) Insecticidal activities of secondary metabolites of endophytic Pencillium sp. in Derris elliptica Benth. J Appl Entomol 129:413–417

Hu Z, Wu Z, Su Q, Li M, Wu S, Meng R, Ding W, Li C (2020) Metabolites with phytopathogenic fungi inhibitory activities from the mangrove endophytic fungus Botryosphaeria ramose. Bioorg Chem 104:104300

Huang Z, Yang J, Cai X, She Z, Lin Y (2012) A new furanocoumarin from the mangrove endophytic fungus Penicillium sp. (ZH16). Nat Prod Res 26:1291–1295

Huang J-X, Zhang J, Zhang X-R, Zhang K, Zhang X, He X-R (2014) Mucor fragilis as a novel source of the key pharmaceutical agents podophyllotoxin and kaempferol. Pharm Biol 52:1237–1243

Igiehon NO, Babalola OO, Cheseto X, Torto B (2021) Effects of rhizobia and arbuscular mycorrhizal fungi on yield, size distribution and fatty acid of soybean seeds grown under drought stress. Microbiol Res 242:126640

Iqbal J, Siegrist JA, Nelson JA, McCulley RL (2012) Fungal endophyte infection increases carbon sequestration potential of southeastern USA tall fescue stands. Soil Biol Biochem 44:81–92

Isaka M, Chinthanom P, Boonruangprapa T, Rungjindamai N, Pinruan U (2010) Eremophilane-type sesquiterpenes from the fungus Xylaria sp. BCC 21097. J Nat Prod 73:683–687

Jaber LR, Ownley BH (2018) Can we use entomopathogenic fungi as endophytes for dual biological control of insect pests and plant pathogens? Biol Control 116:36–45. https://doi.org/10.1016/j.biocontrol.2017.01.018

Jaiswal N, Srivastava S, Bhatia V, Mishra A, Sonkar A, Narender T, Srivastava A, Tamrakar A (2012) Inhibition of alpha-glucosidase by Acacia nilotica prevents hyperglycemia along with improvement of diabetic complications via aldose reductase inhibition. J Diabet Metab 6

Jena L, Deshmukh S, Waghmare P, Kumar S, Harinath BC (2015) Study of mechanism of interaction of truncated isoniazid–nicotinamide adenine dinucleotide adduct against multiple enzymes of Mycobacterium tuberculosis by a computational approach. Int J Mycobacteriol 4:276–283

Jha AK, Sit N (2022) Extraction of bioactive compounds from plant materials using combination of various novel methods: a review. Trends Food Sci Technol 119:579–591. https://doi.org/10.1016/j.tifs.2021.11.019

Jia Q, Qu J, Mu H, Sun H, Wu C (2020) Foliar endophytic fungi: diversity in species and functions in forest ecosystems. Symbiosis 80:103–132

Johnson JM, Alex T, Oelmüller R (2014) Piriformospora indica: the versatile and multifunctional root endophytic fungus for enhanced yield and tolerance to biotic and abiotic stress in crop plants. J Trop Agric 52:103–122

Kanokratana P, Chantasingh D, Champreda V, Tanapongpipat S, Pootanakit K, Eurwilaichitr L (2008) Identification and expression of cellobiohydrolase (CBHI) gene from an endophytic fungus, Fusicoccum sp. (BCC4124) in Pichia pastoris. Protein Expr Purif 58:148–153

Khan AL, Hamayun M, Kim Y-H, Kang S-M, Lee I-J (2011) Ameliorative symbiosis of endophyte (Penicillium funiculosum LHL06) under salt stress elevated plant growth of Glycine max L. Plant Physiol Biochem 49:852–861

Khan N, Khan M, Asgher M, Fatma M, Masood A, Syeed S (2014) Salinity tolerance in plants: revisiting the role of sulfur metabolites. J Plant Biochem Physiol 2:2

Khan AL, Hussain J, Al-Harrasi A, Al-Rawahi A, Lee I-J (2015) Endophytic fungi: resource for gibberellins and crop abiotic stress resistance. Crit Rev Biotechnol 35:62–74

Khan AL, Al-Harrasi A, Al-Rawahi A, Al-Farsi Z, Al-Mamari A, Waqas M, Asaf S, Elyassi A, Mabood F, Shin J-H (2016) Endophytic fungi from frankincense tree improves host growth and produces extracellular enzymes and indole acetic acid. PLoS One 11:e0158207

Khan R, Naqvi STQ, Fatima N, Muhammad SA (2019) 30. Study of antidiabetic activities of endophytic fungi isolated from plants. Pure Appl Biol 8:1287–1295

Khare E, Arora NK (2014) Effects of soil environment on field efficacy of microbial inoculants. In: Plant microbes symbiosis: applied facets. Springer, pp 353–381

Kharwar RN, Mishra A, Gond SK, Stierle A, Stierle D (2011) Anticancer compounds derived from fungal endophytes: their importance and future challenges. Nat Prod Rep 28:1208–1228

Kim D, Abdelaziz ME, Ntui VO, Guo X, Al-Babili S (2017) Colonization by the endophyte Piriformospora indica leads to early flowering in Arabidopsis thaliana likely by triggering gibberellin biosynthesis. Biochem Biophys Res Commun 490:1162–1167. https://doi.org/10.1016/j.bbrc.2017.06.169

Kousar R, Naeem M, Jamaludin MI, Arshad A, Shamsuri AN, Ansari N, Akhtar S, Hazafa A, Uddin J, Khan A (2022) Exploring the anticancer activities of novel bioactive compounds derived from endophytic fungi: mechanisms of action, current challenges and future perspectives. Am J Cancer Res 12:2897

Krauss U, Hidalgo E, Bateman R, Adonijah V, Arroyo C, García J, Crozier J, Brown NA, ten Hoopen GM, Holmes KA (2010) Improving the formulation and timing of application of endophytic biocontrol and chemical agents against frosty pod rot (Moniliophthora roreri) in cocoa (Theobroma cacao). Biol Control 54:230–240

Kusari S, Lamshöft M, Zühlke S, Spiteller M (2008) An endophytic fungus from Hypericum perforatum that produces hypericin. J Nat Prod 71:159–162

Kusari S, Zühlke S, Kosuth J, Cellarova E, Spiteller M (2009) Light-independent metabolomics of endophytic Thielavia subthermophila provides insight into microbial hypericin biosynthesis. J Nat Prod 72:1825–1835

Li H-Y, Wei D-Q, Shen M, Zhou Z-P (2012a) Endophytes and their role in phytoremediation. Fungal Divers 54:11–18

Li X, Ren A, Han R, Yin L, Wei M, Gao Y (2012b) Endophyte-mediated effects on the growth and physiology of Achnatherum sibiricum are conditional on both N and P availability. PLoS One 7:e48010

Li S-J, Zhang X, Wang X-H, Zhao C-Q (2018) Novel natural compounds from endophytic fungi with anticancer activity. Eur J Med Chem 156:316–343

Liu D, Li XM, Li CS, Wang BG (2013a) Nigerasterols A and B, antiproliferative sterols from the mangrove-derived endophytic fungus Aspergillus niger MA-132. Helv Chim Acta 96:1055–1061

Liu S, Liu X, Guo L, Che Y, Liu L (2013b) 2H-Pyran-2-one and 2H-Furan-2-one derivatives from the plant endophytic fungus Pestalotiopsis fici. Chem Biodivers 10:2007–2013

López AC, Alvarenga AE, Zapata PD, Luna MF, Villalba LL (2019) Trichoderma spp. from Misiones, Argentina: effective fungi to promote plant growth of the regional crop Ilex paraguariensis St. Hil. Mycology 10:210–221

Ma C, Jiang D, Wei X (2011) Mutation breeding of Emericella foeniculicola TR21 for improved production of tanshinone IIA. Process Biochem 46:2059–2063

Manici L, Kelderer M, Caputo F, Mazzola M (2015) Auxin-mediated relationships between apple plants and root inhabiting fungi: impact on root pathogens and potentialities of growth-promoting populations. Plant Pathol 64:843–851

Marchin RM, Backes D, Ossola A, Leishman MR, Tjoelker MG, Ellsworth DS (2022) Extreme heat increases stomatal conductance and drought-induced mortality risk in vulnerable plant species. Glob Chang Biol 28:1133–1146

Marinho AM, Rodrigues-Filho E, Moitinho MLR, Santos LS (2005) Biologically active polyketides produced by Penicillium janthinellum isolated as an endophytic fungus from fruits of Melia azedarach. J Braz Chem Soc 16:280–283

Marinho AM, Marinho PS, Santos LS, Rodrigues Filho E, Ferreira IC (2013) Active polyketides isolated from Penicillium herquei. An Acad Bras Cienc 85:909–912

Masi M, Maddau L, Linaldeddu BT, Scanu B, Evidente A, Cimmino A (2018) Bioactive metabolites from pathogenic and endophytic fungi of forest trees. Curr Med Chem 25:208–252

McMullin DR, Nguyen HDT, Daly GJ, Menard BS, Miller JD (2018) Detection of foliar endophytes and their metabolites in Picea and Pinus seedling needles. Fungal Ecol 31:1–8. https://doi.org/10.1016/j.funeco.2017.09.003

Mejía LC, Rojas EI, Maynard Z, Van Bael S, Arnold AE, Hebbar P, Samuels GJ, Robbins N, Herre EA (2008) Endophytic fungi as biocontrol agents of Theobroma cacao pathogens. Biol Control 46:4–14

Mejía LC, Herre EA, Sparks JP, Winter K, García MN, Van Bael SA, Stitt J, Shi Z, Zhang Y, Guiltinan MJ (2014) Pervasive effects of a dominant foliar endophytic fungus on host genetic and phenotypic expression in a tropical tree. Front Microbiol 5:479

Miller JD, Mackenzie S, Mark F, Adams GW, Findlay JA (2002) Needles of white spruce inoculated with rugulosin-producing endophytes contain rugulosin reducing spruce budworm growth rate. Mycol Res 106:471–479

Mishra VK, Passari AK, Chandra P, Leo VV, Kumar B, Uthandi S, Thankappan S, Gupta VK, Singh BP (2017) Determination and production of antimicrobial compounds by Aspergillus clavatonanicus strain MJ31, an endophytic fungus from Mirabilis jalapa L. using UPLC-ESI-MS/MS and TD-GC-MS analysis. PLoS One 12:e0186234

Mishra R, Kushveer J, Revanthbabu P, Sarma VV (2019a) Endophytic fungi and their enzymatic potential. In: Advances in endophytic fungal research: present status and future challenges. Springer, pp 283–337

Mishra S, Gupta D, Ranjan R (2019b) Molecular approaches for enhancing abiotic stress tolerance in plants. Approaches for enhancing abiotic stress tolerance in plants. CRC Press, pp 405–422

Mousa WK, Raizada MN (2013) The diversity of anti-microbial secondary metabolites produced by fungal endophytes: an interdisciplinary perspective. Front Microbiol 4:65. https://doi.org/10.3389/fmicb.2013.00065

Movahhedin N, Albadry M, Hamann M (2014) Isolation and characterization of cytotxic compounds from endophytes of an endangered American cactus, Mammillaria hahniana. Planta Med 80:PD136

Muhammed D, Dada E, Iyaji F, Abraham O, Chijioke N (2019) Investigation of biochemical parameters of Plasmodium berghei infected mice after administration of ethanolic leaf extract of Eucalyptus citriodora. Int J Pathogen Res 2:1–9

Mukhtar T, Ali F, Rafique M, Ali J, Afridi MS, Smith D, Mehmood S, Amna, Souleimanov A, Jellani G (2023) Biochemical characterization and potential of Bacillus safensis strain SCAL1 to mitigate heat stress in Solanum lycopersicum L. J Plant Growth Regul 42:523–538

Musetti R, Grisan S, Polizzotto R, Martini M, Paduano C, Osler R (2011) Interactions between 'Candidatus Phytoplasma mali' and the apple endophyte Epicoccum nigrum in Catharanthus roseus plants. J Appl Microbiol 110:746–756

Nagarajan A, Thirunavukkarasu N, Suryanarayanan T, Gummadi SN (2014) Screening and isolation of novel glutaminase free L-asparaginase from fungal endophytes. Res J Microbiol 9:163

Negi JS, Singh P, Joshi GP, Rawat MS, Bisht VK (2010) Chemical constituents of Asparagus. Pharmacogn Rev 4:215–220. https://doi.org/10.4103/0973-7847.70921

Nicoletti R, Fiorentino A (2015) Plant bioactive metabolites and drugs produced by endophytic fungi of Spermatophyta. Agriculture 5:918–970

Niinemets Ü (2010) Mild versus severe stress and BVOCs: thresholds, priming and consequences. Trends Plant Sci 15:145–153

Odelade KA, Babalola OO (2019) Bacteria, fungi and archaea domains in rhizospheric soil and their effects in enhancing agricultural productivity. Int J Environ Res Public Health 16:3873

Paice M, Gurnagul N, Page D, Jurasek L (1992) Mechanism of hemicellulose-directed prebleaching of kraft pulps. Enzym Microb Technol 14:272–276

Pal S, Chatare V, Pal M (2011) Isocoumarin and its derivatives: an overview on their synthesis and applications. Curr Org Chem 15:782–800

Parande S, Zamani GR, Zahan MHS, Ghaderi M (2013) Effects of silicon application on the yield and component of yield in the common bean (Phaseolus vulgaris) under salinity stress. Int J Agron Plant Product 4:1574–1579

Park MS, Ahn J, Choi GJ, Choi YH, Jang KS, Kim J-C (2010) Potential of the volatile-producing fungus Nodulisporium sp. CF016 for the control of postharvest diseases of apple. Plant Pathol J 26:253–259

Phurailatpam L, Mishra S (2020) Role of plant endophytes in conferring abiotic stress tolerance. In: Plant ecophysiology and adaptation under climate change: mechanisms and perspectives II: Mechanisms of adaptation and stress amelioration, pp 603–628

Pieckenstain FL, Bazzalo ME, Roberts AM, Ugalde RA (2001) Epicoccum purpurascens for biocontrol of Sclerotinia head rot of sunflower. Mycol Res 105:77–84

Rajkumar M, Ae N, Prasad MNV, Freitas H (2010) Potential of siderophore-producing bacteria for improving heavy metal phytoextraction. Trends Biotechnol 28:142–149

Rastegari AA, Yadav AN, Yadav N (2020) New and future developments in microbial biotechnology and bioengineering: trends of microbial biotechnology for sustainable agriculture and biomedicine systems: diversity and functional perspectives. Elsevier

Redecker D, Kodner R, Graham LE (2000) Glomalean fungi from the Ordovician. Science 289:1920–1921

Redman RS, Kim YO, Woodward CJ, Greer C, Espino L, Doty SL, Rodriguez RJ (2011) Increased fitness of rice plants to abiotic stress via habitat adapted symbiosis: a strategy for mitigating impacts of climate change. PLoS One 6:e14823

Rodriguez RJ, Henson J, Van Volkenburgh E, Hoy M, Wright L, Beckwith F, Kim Y-O, Redman RS (2008) Stress tolerance in plants via habitat-adapted symbiosis. ISME J 2:404–416

Rodriguez R, White J Jr, Arnold A, Redman ARA (2009) Fungal endophytes: diversity and functional roles. New Phytol 182:314–330

Rouhier N, San Koh C, Gelhaye E, Corbier C, Favier F, Didierjean C, Jacquot J-P (2008) Redox based anti-oxidant systems in plants: biochemical and structural analyses. Biochim Biophys Acta (BBA) General Subjects 1780:1249–1260

Russell JR, Huang J, Anand P, Kucera K, Sandoval AG, Dantzler KW, Hickman D, Jee J, Kimovec FM, Koppstein D (2011) Biodegradation of polyester polyurethane by endophytic fungi. Appl Environ Microbiol 77:6076–6084

Sadananda T, Nirupama R, Chaithra K, Govindappa M, Chandrappa C, Vinay Raghavendra B (2011) Antimicrobial and antioxidant activities of endophytes from Tabebuia argentea and identification of anticancer agent (lapachol). J Med Plants Res 5:3643–3652

Saikkonen K, Mikola J, Helander M (2015) Endophytic phyllosphere fungi and nutrient cycling in terrestrial ecosystems. Curr Sci 109:121–126

Saranga Y, Menz M, Jiang C-X, Wright RJ, Yakir D, Paterson AH (2001) Genomic dissection of genotype × environment interactions conferring adaptation of cotton to arid conditions. Genome Res 11:1988–1995

Schlau-Cohen GS, Yang H-Y, Krüger TP, Xu P, Gwizdala M, Van Grondelle R, Croce R, Moerner W (2015) Single-molecule identification of quenched and unquenched states of LHCII. J Phys Chem Lett 6:860–867

Schlegel M, Queloz V, Sieber TN (2018) The endophytic mycobiome of European ash and sycamore maple leaves—geographic patterns, host specificity and influence of ash dieback. Front Microbiol 9:2345. https://doi.org/10.3389/fmicb.2018.02345

Seleiman MF, Al-Suhaibani N, Ali N, Akmal M, Alotaibi M, Refay Y, Dindaroglu T, Abdul-Wajid HH, Battaglia ML (2021) Drought stress impacts on plants and different approaches to alleviate its adverse effects. Plants 10:259

Sessitsch A, Kuffner M, Kidd P, Vangronsveld J, Wenzel WW, Fallmann K, Puschenreiter M (2013) The role of plant-associated bacteria in the mobilization and phytoextraction of trace elements in contaminated soils. Soil Biol Biochem 60:182–194

Shaffique S, Khan MA, Imran M, Kang S-M, Park Y-S, Wani SH, Lee I-J (2022) Research progress in the field of microbial mitigation of drought stress in plants. Front Plant Sci 13:870626

Sharma S, Kour D, Rana KL, Dhiman A, Thakur S, Thakur P, Thakur S, Thakur N, Sudheer S, Yadav N (2019) Trichoderma: biodiversity, ecological significances, and industrial applications. In: Recent advancement in white biotechnology through fungi, Diversity and enzymes perspectives, vol 1. Springer, pp 85–120

Singh LP, Gill SS, Tuteja N (2011) Unraveling the role of fungal symbionts in plant abiotic stress tolerance. Plant Signal Behav 6:175–191

Singh N, Singh A, Dahiya P (2021) Plant growth-promoting endophytic fungi from different habitats and their potential applications in agriculture. In: Recent trends in mycological research. Springer Nature, pp 69–87
Song X-Q, Zhang X, Han Q-J, Li X-B, Li G, Li R-J, Jiao Y, Zhou J-C, Lou H-X (2013) Xanthone derivatives from Aspergillus sydowii, an endophytic fungus from the liverwort Scapania ciliata S. Lac and their immunosuppressive activities. Phytochem Lett 6:318–321
Sorgatto M, Guimarães N, Zanoelo F, Marques M, Peixoto-Nogueira S, Giannesi G (2012) Purification and characterization of an extracellular xylanase produced by the endophytic fungus, Aspergillus terreus, grown in submerged fermentation. Afr J Biotechnol 11:8076–8084
Stringlis IA, Yu K, Feussner K, de Jonge R, Van Bentum S, Van Verk MC, Berendsen RL, Bakker PAHM, Feussner I, Pieterse CMJ (2018) MYB72-dependent coumarin exudation shapes root microbiome assembly to promote plant health. Proc Natl Acad Sci 115:E5213–E5222. https://doi.org/10.1073/pnas.1722335115
Strobel GA (2002) Rainforest endophytes and bioactive products. Crit Rev Biotechnol 22:315–333
Strobel G, Yang X, Sears J, Kramer R, Sidhu RS, Hess WM (1996) Taxol from Pestalotiopsis microspora, an endophytic fungus of Taxus wallachiana. Microbiology 142:435–440. https://doi.org/10.1099/13500872-142-2-435
Sumarah MW, Puniani E, Sørensen D, Blackwell BA, Miller JD (2010) Secondary metabolites from anti-insect extracts of endophytic fungi isolated from Picea rubens. Phytochemistry 71:760–765
Sumarah MW, Kesting JR, Sørensen D, Miller JD (2011) Antifungal metabolites from fungal endophytes of Pinus strobus. Phytochemistry 72:1833–1837
Sunitha V, Ramesha A, Savitha J, Srinivas C (2012) Amylase production by endophytic fungi Cylindrocephalum sp. isolated from medicinal plant Alpinia calcarata (Haw.) Roscoe. Braz J Microbiol 43:1213
Sunitha V, Nirmala Devi D, Srinivas C (2013) Extracellular enzymatic activity of endophytic fungal strains isolated from medicinal plants. World J Agric Sci 9:1–9
Suwannarach N, Kumla J, Bussaban B, Nuangmek W, Matsui K, Lumyong S (2013) Biofumigation with the endophytic fungus Nodulisporium spp. CMU-UPE34 to control postharvest decay of citrus fruit. Crop Prot 45:63–70
Talapatra K, Das AR, Saha A, Das P (2017) In vitro antagonistic activity of a root endophytic fungus towards plant pathogenic fungi. J Appl Biol Biotechnol 5:068–071
Thirumalanadhuni V, Yerraguravagari LL, Palempalli UMD (2021) Endophytic microflora: the fountainhead of anticancer metabolites—a systematic review. In: Recent developments in applied microbiology and biochemistry. Elsevier, pp 13–20
Tiwari P, Bae H (2022) Endophytic fungi: key insights, emerging prospects, and challenges in natural product drug discovery. Microorganisms 10. https://doi.org/10.3390/microorganisms10020360
Umashankar T, Govindappa M, Ramachandra Y (2014) In vitro antioxidant and antimicrobial activity of partially purified coumarins from fungal endophytes of Crotalaria pallida. Int J Curr Microbiol Appl Sci 3:58–72
Vadassery J, Ritter C, Venus Y, Camehl I, Varma A, Shahollari B, Novák O, Strnad M, Ludwig-Müller J, Oelmüller R (2008) The role of auxins and cytokinins in the mutualistic interaction between Arabidopsis and Piriformospora indica. Mol Plant-Microbe Interact 21:1371–1383
Van Bael SA, Valencia MC, Rojas EI, Gómez N, Windsor DM, Herre EA (2009) Effects of foliar endophytic fungi on the preference and performance of the leaf beetle Chelymorpha alternans in Panama. Biotropica 41:221–225. https://doi.org/10.1111/j.1744-7429.2008.00476.x
Venkatesagowda B, Ponugupaty E, Barbosa AM, Dekker RF (2012) Diversity of plant oil seed-associated fungi isolated from seven oil-bearing seeds and their potential for the production of lipolytic enzymes. World J Microbiol Biotechnol 28:71–80
Venugopala KN, Rashmi V, Odhav B (2013) Review on natural coumarin lead compounds for their pharmacological activity. Biomed Res Int 2013:963248. https://doi.org/10.1155/2013/963248
Vurukonda SSKP, Vardharajula S, Shrivastava M, SkZ A (2016) Enhancement of drought stress tolerance in crops by plant growth promoting rhizobacteria. Microbiol Res 184:13–24

Wagner BL, Lewis LC (2000) Colonization of corn, *Zea mays*, by the entomopathogenic fungus *Beauveria bassiana*. Appl Environ Microbiol 66:3468–3473. https://doi.org/10.1128/AEM.66.8.3468-3473.2000

Wang W, Barak T, Vinocur B, Shoseyov O, Altman A (2003) Abiotic resistance and chaperones: possible physiological role of SP1, a stable and stabilizing protein from Populus. In: Plant biotechnology 2002 and beyond: Proceedings of the 10 th IAPTC&B Congress June 23–28, 2002 Orlando, Florida, USA. Springer, pp 439–443

Wang Q, Fu Y, Gao J, Wang Y, Li X, Zhang A (2007) Preliminary isolation and screening of the endophytic fungi from Melia azedarach L. Acta Agric Boreali-Occident Sin 16:224–227

Wang F, Hou Z, Wang C, Li P, Shi D (2008) Bioactive metabolites from Penicillium sp., an endophytic fungus residing in Hopea hainanensis. World J Microbiol Biotechnol 24:2143–2147

Wang J, Yao L-y, Lu Y-h (2013) Ceriporia lacerata DMC1106, a new endophytic fungus: isolation, identification, and optimal medium for 2′,4′-dihydroxy-6′-methoxy-3′,5′-dimethylchalcone production. Biotechnol Bioprocess Eng 18:669–678

Wang L-W, Wang J-L, Chen J, Chen J-J, Shen J-W, Feng X-X, Kubicek CP, Lin F-C, Zhang C-L, Chen F-Y (2017) A novel derivative of (−)mycousnine produced by the endophytic fungus Mycosphaerella nawae, exhibits high and selective immunosuppressive activity on T cells. Front Microbiol 8:1251

Wang S-S, Liu J-M, Sun J, Huang Y-T, Jin N, Li M-M, Liang Y-T, Fan B, Wang F-Z (2022) Analysis of endophytic bacterial diversity from different Dendrobium stems and discovery of an endophyte produced dendrobine-type sesquiterpenoid alkaloids. Front Microbiol 12:775665

Wen J, Okyere SK, Wang S, Wang J, Xie L, Ran Y, Hu Y (2022) Endophytic fungi: an effective alternative source of plant-derived bioactive compounds for pharmacological studies. J Fungi 8:205

West C, Izekor E, Oosterhuis D, Robbins R (1988) The effect of Acremonium coenophialum on the growth and nematode infestation of tall fescue. Plant Soil 112:3–6

Wilson D, Faeth SH (2001) Do fungal endophytes result in selection for leafminer ovipositional preference? Ecology 82:1097–1111. https://doi.org/10.1890/0012-9658(2001)082[1097:DFERIS]2.0.CO;2

Yadav A (2019) Endophytic fungi for plant growth promotion and adaptation under abiotic stress conditions. Acta Sci Agric 3:91–93

Yadav AN, Kour D, Kaur T, Devi R, Yadav A (2022) Endophytic fungal communities and their biotechnological implications for agro-environmental sustainability. Folia Microbiol 67:203–232

Yihui B, Zhouying X, Yurong Y, Zhang H, Hui C, Ming T (2017) Effect of dark septate endophytic fungus Gaeumannomyces cylindrosporus on plant growth, photosynthesis and Pb tolerance of maize (Zea mays L.). Pedosphere 27:283–292

Yuan L, Zhao PJ, Ma J, Lu CH, Shen YM (2009) Labdane and tetranorlabdane diterpenoids from Botryosphaeria sp. MHF, an endophytic fungus of Maytenus hookeri. Helv Chim Acta 92:1118–1125

Zaferanloo B, Bhattacharjee S, Ghorbani MM, Mahon PJ, Palombo EA (2014) Amylase production by Preussia minima, a fungus of endophytic origin: optimization of fermentation conditions and analysis of fungal secretome by LC-MS. BMC Microbiol 14:1–12

Zanne AE, Lower SS, Cardon ZG, Orians CM (2006) 15N partitioning in tomato: vascular constraints versus tissue demand. Funct Plant Biol 33:457–464

Zhao J, Li C, Wang W, Zhao C, Luo M, Mu F, Fu Y, Zu Y, Yao M (2013) Hypocrea lixii, novel endophytic fungi producing anticancer agent cajanol, isolated from pigeon pea (Cajanus cajan [L.] Millsp.). J Appl Microbiol 115:102–113

Zhao J, Ma D, Luo M, Wang W, Zhao C, Zu Y, Fu Y, Wink M (2014) In vitro antioxidant activities and antioxidant enzyme activities in HepG2 cells and main active compounds of endophytic fungus from pigeon pea [Cajanus cajan (L.) Millsp.]. Food Res Int 56:243–251

Zhou S-L, Zhou S-L, Wang M-X, Chen S-L (2011) Two compounds from the endophytic Colletotrichum sp. of Ginkgo biloba. Nat Prod Commun 6:1934578X1100600821

Marine Algal-Derived Endophytic Fungi: Secondary Metabolites and Applications

6

Rajagopal Kalyanaraman, Banu Narasimhan, Nivetha Venkatesan, and Meenambiga Setti Sudharsan

Abstract

Marine algae, including red, brown, and green algae, are a major group of organisms in the marine ecosystem. They play a crucial role in nutrient recycling, regulation, and stabilization of the ecosystem, making them an important area of study. Additionally, these algae have economic value and potential as a source of biofuels, further highlighting the need for research on these organisms and their endomycobiomes—the fungi that live within them. Endomycobiomes have the potential to enhance the growth and development of marine algae, aid in nutrient uptake and recycling, and improve the host's defense against pathogens. However, there has been limited focus on these endophytic fungi, with most studies instead focusing on bacterial microbiomes. As a result, little is known about the biology, ecology, and biotechnological potential of fungal endophytes in seaweed microbiomes. Seaweeds are essential components of the marine ecosystem and hold significant commercial value in the global market. Fungal endophytes isolated from marine hosts are a rich source of secondary metabolites, enzymes, and bioactive compounds, making them a promising area of research. Given the significance of these organisms, it is crucial to expand our understanding of fungal endophytes in seaweeds. Further research in this area has the potential to uncover valuable insights and applications, making it a worthwhile area to explore.

R. Kalyanaraman
Department of Botany, Ramakrishna Mission Vivekananda College, Chennai, India

B. Narasimhan
Department of Botany, Bharathi Women's College, George Town, Chennai, India

N. Venkatesan · M. S. Sudharsan (✉)
Department of Bioengineering, School of Engineering, Vels Institute of Science, Technology and Advanced Studies (VISTAS), Chennai, India
e-mail: meenambiga.se@velsuniv.ac.in

K. A. Abd-Elsalam, S. F. AbuQamar (eds.), *Fungal Endophytes Volume I*,
https://doi.org/10.1007/978-981-97-7312-1_6

Keywords

Marine ecosystem · Seaweeds · Green · Brown · Red algae · Fungal endophytes · Bioactive compounds

6.1 Introduction

Microorganisms are ubiquitous and have been identified in all ecosystems worldwide. Current scientific knowledge encompasses only a limited number of microbial species; research suggests that less than 5% of fungi and merely 1% of bacterial species have been characterized to date. Consequently, several million microbial species remain unidentified and concealed within various ecosystems (Felício et al. 2015). The marine ecosystem represents a vast, underexplored reservoir of fungal diversity; however, only a few thousand fungal species have been documented, with only a limited number being newly identified (Devadatha et al. 2018). Fungal endophytes have been detected in every plant species examined, and it is estimated that millions of endophytic fungi may exist within plants globally (Rampadarath et al. 2018). In this chapter, the term "endophytes" refers to the symptomless fungal endosymbionts residing within marine macroalgae. Given the demonstrated capacity of terrestrial endophytic fungi to synthesize a diverse array of bioactive compounds, ongoing research aims to isolate and evaluate microbes, particularly fungal endophytes from various habitats and unique environments, to discover new and novel metabolites. One such ecological niche is represented by the microorganisms associated with marine endophytic fungi (Vazquez-Rodriguez et al. 2015; Felício et al. 2015). Currently, fungal endophytes isolated from marine hosts are garnering increased attention as a promising reservoir of natural products with significant potential (Kjer et al. 2010). Jones et al. (2008) reported that fungal endophytes from the marine algae *Drechslera dematioidea* synthesized several bioactive compounds, mainly sesquiterpenes. The marine ecosystem is an enormously diverse reservoir of living organisms, and across the array of organisms there is an untapped pool of structurally unique natural compounds. Nearly 70% of the earth's surface is covered with water and it encompasses 500,000 biological species divided into different phyla. The total coastline of the world is about 312,000 km and a volume of 137 × 106 cubic kilometers making it the largest ecosystem on earth (Trindade et al. 2015; Kumar et al. 2015; Zainuddin et al. 2010; Rajulu et al. 2022).

The chapter examines secondary metabolites produced by endophytic fungi derived from marine algae. It discusses their potential uses in pharmaceuticals, agriculture, and environmental remediation. The chapter offers a comprehensive overview of their chemical diversity, biological activities, and potential benefits for human health and sustainable development.

6.2 Endophyte Circumscription

Endophytes are organisms that infect a host, but do not produce any symptoms or cause harm within plant tissues. This definition excludes pathogenic fungi. Some researchers have expanded this definition to include all organisms that reside in different plant organs for at least part of their life and can colonize plant tissues without causing deceptive harm to the host (Carroll 1986; Petrini 1991). This definition would encompass epiphytic fungi and latent pathogens that exist as endophytes during certain stages of their life cycle. However, it is important to note that some endophytic fungi can become pathogenic if their hosts are susceptible.

The presence of fungal endophytes in plant tissues was first discovered over 80 years ago in *Lolium* grass (Sampson 1935). However, modern research on fungal endophytes did not truly begin until Bernstein and Carroll (1977) described their occurrence in needles of *Pseudotsuga menziesii*. The study of fungal endophytes not only sheds light on the diversity of the fungal kingdom, but also highlights their potential usefulness. This is because many fungal endophytes have the ability to synthesize new secondary metabolites or bioactive compounds with agricultural, pharmaceutical, and industrial applications.

6.3 Ecology of Fungal Endophytes

Fungal endophytes are ubiquitous. They have been isolated from such diverse groups of plants as marine as marine algae, hydrophytes, crop plants, mosses, and ferns, coniferous trees, angiosperms, forest trees and grasses, etc. (Cubit 1976; Petrini 1986; Carroll and Carroll 1978; Johnson and Whitney 1992; Clay 1990; White Jr 1988; Suryanarayanan and Rajagopal 1997). Studies on the species composition of fungal endophyte groups of many plants have shown that a huge number of fungal taxa could be isolated from a single host plant species. Yet generally one or a few fungal endophytic species dominate in a host plant (Petrini 1991; Suryanarayanan et al. 1998; Petrini et al. 1989). The assemblage and distribution of fungal endophytes in a plant species is predisposed by age of the host plant, biotic and abiotic factors. A number of fungal endophyte species can be isolated from a host decline with extreme climatic conditions (Suryanarayanan et al. 1998; Sieber et al. 1991).

Numerous studies on fungal endophytes have revealed that the colonization frequency of these organisms increases with the age, surface area, and chemistry of the tissue in which they reside (Espinosa-Garcia and Langenheim 1990; Petrini and Carroll 1981). Endophytes that colonize trees are known to be host specific. Petrini and Fisher (1988) and Petrini (1991) reported that concurrent sampling of *Pinus* and *Fagus* growing in a similar area showed that the fungal endophyte assemblages of the two plant species are dissimilar. Likewise, the fungal endophyte community of a host is also affected by air quality. Fungal endophytes inhabit a unique ecological niche. Although they may not be exposed to the external environment like epiphytic or phylloplane fungi, fungal endophytes must overcome the host's defense mechanisms. This suggests that their life strategies are expected to differ from those of other fungi.

Several marine-derived fungi, which are isolated from different marine hosts like sponges, fish, algae, and mangroves, have been attracted due to their ability to synthesize compounds with high biological activities (Di Camillo et al. 2017; Zainuddin et al. 2010; Bajpai 2016; Giddings et al. 2015; Sahoo et al. 2021). Zhang et al. (2009) reported that many new compounds were discovered from marine-derived

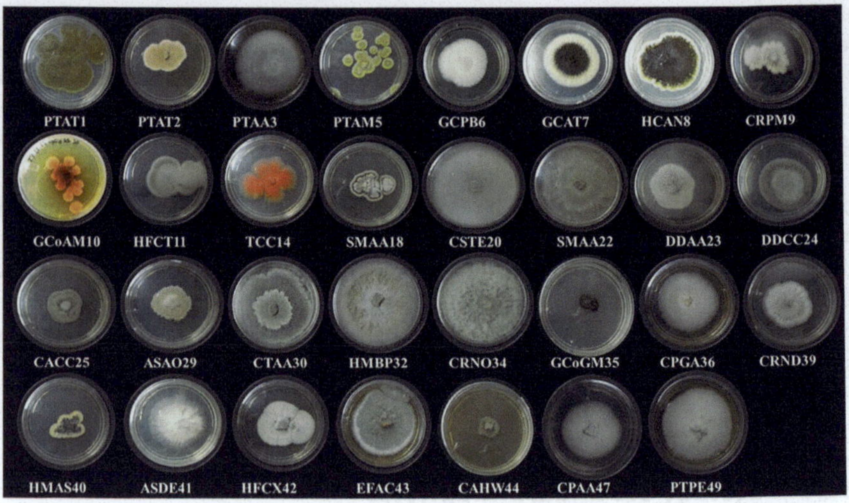

Fig. 6.1 Endophytic fungus isolated from water-based brown, green, and red algae have antibacterial, antifungal, and cytotoxic activity, as well as antioxidant capabilities. These fungi include *Phoma maricula*, *Daldinia eschscholtzii*, *Phoma moricola*, and *Alternaria alternata* (Sahoo et al. 2021)

fungi particularly by epiphytic or endophytes (Zhang et al. 2009). Even though terrestrial fungi have a remarkable mycodiversity of life, the greatest mycodiversity is in the oceans and seas, eventually in the deep sea, ocean floors, and coral reefs (Chanda et al. 2010). The living conditions like nutrition, salinity in the water, higher pressure, variations in the temperature, and interaction among bacteria, fungi, plants, and other organisms makes the genetic and metabolic diversity of marine fungi (Manivasagan et al. 2013). Additionally, marine fungi survive in close proximity with other marine organisms which have poor defense mechanisms, thus depending on the defense through microflora associated with them for their survival. Thus, marine-derived fungi have been known as reservoirs of novel biologically active metabolites (Samuel et al. 2011; Wahl et al. 2012). Recently, 31 different endophytic fungi were isolated from 18 marine macroalgae by Sahoo and his co-workers (2021) (Fig. 6.1).

6.4 Taxonomy of Fungal Endophytes

The fungal endophytes of aerial plant organs primarily belong to the Hyphomycete, Coelomycete, and Ascomycetes, along with their anamorphic conidial forms. However, a few Basidiomycetes and Oomycetes have also been isolated as fungal endophytes. According to Petrini (1986), the fungal endophyte collection of a plant could be classified into two broad categories—the nonspecific, epiphytic genera (e.g., *Alternaria, Aspergillus, Fusarium, Curvularia, Cladosporium*) and the true or exclusive fungal endophytes (e.g., *Phoma, Pestalotiopsis, Phyllosticta, Phomopsis*). However, Bills (1996) reported that any attempt to isolate fungal endophytes from a plant organ certainly yield quite a few non-sporulating or sterile fungi. So, the absence of any sporulating structures like conidia or fruit bodies in such non-sporulating forms prevents their unambiguous identification. Hence, mycologists use morphological culture characteristics of these fungal endophytes to differentiate them from one another (Dobranic et al. 1995; Suryanarayanan et al. 1998). Fungal endophytes include a highly diverse ecological and taxonomic group. Bender et al. (2016), Strobel and Daisy (2003), and Vinit et al. (2018) reported that every ecosystem with the highest diversity is the one with the greatest number of fungal endophytes and the most varied microorganisms.

6.5 Physiology of Fungal Endophytes

Fungal endophytes represent a significant group of organisms that inhabit the tissues of plants, occupying both intercellular and intracellular spaces. The presence of these organisms confers several advantages to the host plant. In recent decades, extensive research has been conducted on fungal endophytes, revealing their biodiversity, broad host range, and complex interactions with host plants as well as their associations with other plant microbiomes. Due to their strong association with host plants, fungal endophytes are known to synthesize various secondary metabolites, enzymes, and hormones. Several studies have indicated that the interaction between plants and fungal endophytes can lead to the production of novel and beneficial bioactive compounds. The metabolites produced by either the plant or the fungal endophytes have applications in modern medicine, agriculture, and industry (Alam et al. 2021). Sieber et al. (1991) and Pugh (1973) reported that extracellular enzymes such as amylases, esterases, cellulases, proteases, and pectinases are synthesized by fungal endophytes. Apart from enzymes fungal endophyte produces plant growth promoting substances like IAA. Carroll and Petrini (1983) studied the ability of fungal endophytes to overcome or metabolize phenolics and other host defense phytochemicals to enter and to survive in the host tissues. Suryanarayanan and Rajagopal (1998) reported that fungal endophytes of neem produced compounds that interfere with plant cell division.

Curiosity in fungal endophytes studies has vastly increased with the discovery of the billion-dollar anticancer drug paclitaxel (taxol), a diterpenoid used in the treatment of breast and ovarian cancers produced by the fungal endophyte *Taxomyces andreanae* from the host *Taxus brevifolia*. The World Health Organization reported that the number of new cases of cancer patients could increase drastically between 2008 and 2030, particularly in low-income countries compared to high-income countries. This frightening fact made scientists and mycologists discover new resources to synthesize the anticancer compound (taxol).

Fungal endophytes can produce a variety of compounds, likely influenced by their internal environment and the interactions between the host and the endophytic fungi (Tenguria et al. 2011). Various ecosystems, including oceans, deep-sea environments, mangrove forests, coastal vegetation, marine algae, and sponges, have been identified as ecological niches that harbor significant mycodiversity of endophytic fungi (Kalaiselvam 2015). These endophytes are notably prevalent in marine sources and environments, particularly within sponges, exhibiting a higher ratio of novel compounds compared to those previously identified in algicolous fungi. Fungal endophytes have been documented across a wide range of taxa, including mosses, ferns, gymnosperms, and angiosperms, found in temperate and boreal forests, as well as arctic, alpine, and xerophytic habitats (Stone et al. 2004). The majority of the endophytes that have been studied primarily belong to the Ascomycota and its anamorphs, as well as Coelomycetes, Basidiomycota, and Zygomycota, which have been identified as endophytic fungi. Among the various genus isolated as fungal endophytes, some of the common genera are *Alternaria*, *Acremonium*, *Bipolaris*, *Drechslera*, *Curvularia*, *Nodulisporium*, *Cladosporium*, *Epicoccum*, *Fusarium*, *Pestalotiopsis*, *Chaetomium*, *Talaromyces*, *Phoma*, and *Phomopsis*. Liu et al. (2010) stated that even though some of the genera are common to both temperate and tropical regions (e.g., in the temperate areas, *Phomopsis* and *Phoma*), Xylariaceae and Phyllosticta prevail as endophytes in the tropics. Fungal endophytes produce new and biologically active metabolites, namely phenols, alkaloids, steroids, terpenoids, lactones, and isocoumarins (Uzma et al. 2018). Bioactive compounds produced by fungal endophytes have antibacterial, antifungal, antiviral, antioxidant, cytotoxic, anticancer, and immunosuppressive activities (Bano et al. 2016).

The bioactive compounds synthesized by fungal endophytes generally initiate from biosynthetic pathways and include benzopyranones, tetralones, cytochalasins, enniatins, isoprenoids, polyketide and amino acid, terpenoids, xanthones, quinines, steroids, phenols, isocoumarins (Zhang et al. 2009; Zeilinger et al. 2016). Several fungal endophytes belonging to the following genera of fungal endophytes (i.e., *Aspergillus*, *Alternaria*, *Acremonium*, *Monochaetia*, *Mucor*, *Ozonium*, *Papulaspora*, *Periconia*, *Pestalotia*, *Botryodiplodia*, *Botrytis*, *Pithomyces*, *Taxomyces*, *Cladosporium*, *Ectostroma*, *Fusarium*, *Metarhizium*, *Pestalotiopsis*, *Phyllosticta*, and *Tubercularia*) capable of synthesizing paclitaxel and its analogs, for instance, 10-deacetylbaccatin III and baccatin III among others (Zhao et al. 2010; Manivasagan et al. 2013; Hu et al. 2016).

6.6 Biotechnological Potential of Fungal Endophytes

Fungal endophytes inhabit unique and often extreme environments and are increasingly recognized as a group of biotypes that may serve as reliable sources of novel metabolites with potential applications in biotechnology and agriculture (Bills and Polishook 1992). Numerous fungal endophytes have been shown to produce antibiotic compounds in culture that exhibit activity against both human and plant pathogens. Fisher et al. (1984a, b) demonstrated that over 30% of the fungal endophytes they analyzed possess antimicrobial properties. Additionally, a fungal endophyte isolated from *Vaccinium* sp. was found to produce a broad-spectrum antibiotic (Fisher et al. 1984b). Fungal endophyte such as *Aureobasidium pullulans* and *Epicoccum purpurascens* predisposed seed germination by synthesizing phytohormones (Petrini et al. 1993). In advanced countries, fungal endophytes are regarded as biocontrol agents against fungal pathogens of plant (Clay 1989; Petrini 1991). Fungal endophytes associations with the host confer utmost potential for biocontrol programs because of their strong association with host (Clay 1989). Dewan and

Sivasithamparam (1989) described that a fungal endophyte isolated from wheat offers substantial protection to the host from infection by "take all" fungal pathogen. Even some of the metabolites produced by fungal endophytes of conifer needles produce secondary metabolites which ward off insect pests. The best instance for fungal endophyte-mediated antagonism towards an insect pest is that of Douglas fir. The needles of this conifer contain *Rhabdocline parkeri* as an endophyte, and this fungal endophyte controls the gall midge, *Contarinia* sp. Therefore, fungal endophytes are potential organisms for biocontrol studies (Carroll and Carroll 1978; Dorworth and Callan 1996).

They are readily available selectable vector components of the plant-fungus system and are perfect for experimental manipulations. Fungal endophytes epitomize natural genomes conferred with useful qualities that could be identified and introduced into host plants for required traits. On the other hand, fungal endophytes could be used as vectors of biological gene delivery into host plants (Petrini et al. 1993). Fungal endophytes are existence explored by medical, pharmaceutical, and agricultural industries as they signify an unexploited pool of bioactive and secondary metabolites (Monaghan et al. 1995; Dreyfuss and Chapela 1994). Li et al. (1998) stated that a fungal endophyte residing in the bark of *Taxus wallichiana* growing in Himalayas synthesizes an anticancer drug taxol. Merck Research Laboratories, USA, works together with Costa Rica in a project to screen fungal endophytes from different forest trees for novel and new antibiotics (Bills and Polishook 1994; Bunders et al. 1996).

The agricultural industry and pharmaceutical companies have a continuous demand for bioactive compounds. To address this need, it is crucial to identify precise sources of such compounds. While these compounds can exhibit a wide variety of structures, they have not evolved to be as suitable for specific receptors as many natural products have. This situation necessitates a higher screening throughput for products derived from combinatorial chemistry, particularly those sourced from marine organisms, especially fungi, which remain largely untapped for the discovery of new, novel, and bioactive secondary metabolites (Lu and Shen 2004). The diversity and distribution of endophytic mycoflora vary with the host plant and induce changes in the host at physiological, biochemical, ecological, and genetic levels. These changes, induced by endophytic fungi, enable the host to survive in harsh environments, such as marine ecosystems because both biotic and abiotic factors control the distribution of marine fungi. Hence, marine fungi, including both epiphytic and endophytic species, have adapted to extreme environments, indicating that they are promising sources of natural products. Endophytic microbes may include actinomycetes, bacteria, fungi, mycoplasma, or yeast. These organisms inhabit the internal tissues of healthy plants without causing any symptoms or diseases in the host. Among the diverse array of endophytic organisms, endophytic fungi can significantly influence the ability of hosts to survive in extreme ecological niches, resist diseases, overcome abiotic stress, and facilitate nutrient recycling. Several marine substrates, such as seagrasses, mangrove plants, algae, and marine sponges, serve as natural and inexhaustible reservoirs of endophytic fungi (Deshmukh et al. 2018).

Despite the extensive research conducted on the symptomless endosymbionts of plants, known as endophytes, their biology, physiology, and precise roles within their host organisms remain inadequately understood. Recent studies have, however, established that these endophytes exhibit significant diversity and represent a promising source of novel bioactive compounds. Nonetheless, there exists a notable scarcity of information concerning the endophytic fungi associated with marine hosts. In recent years, there has been a growing interest in investigating the endophytic mycodiversity of marine plants and algae, as well as screening these endophytic fungi for bioactive compounds. For several decades, the treatment of a wide spectrum of diseases has relied on natural products. Consequently, numerous pharmaceutical companies and research institutions are increasingly focusing on a diverse array of lead compounds with potential bioactivity derived from medicinal plants, particularly those sourced from marine microbiomes, including bacteria, fungi, and algae (Lauritano et al. 2016). The products of marine ecosystems have become a focal point of global research, especially in the fields of biological and medical sciences, due to the richness and diversity of marine plants, algae, and sponges. The resilience of organisms within marine ecosystems enables them to withstand harsh environmental conditions, such as acidification, salinity, chemical pollution, and climate change. Over the past decade, several novel bioactive compounds derived from marine endophytic fungi have been isolated and reported, demonstrating specific clinical targets (Marris 2006). Thus, endophytic organisms from marine ecosystems are poised to serve as a sustainable source of therapeutically significant bioactive secondary metabolites (Deshmukh et al. 2018). *Paecilomyces variotii* an endophytic fungus isolated from the marine algae produced indole derivatives which showed cytotoxic effect cancer cell lines. Physcion extracted from an endophytic fungus *Microsporum* sp. from a red alga induced apoptosis in HeLa cells (Wijesekara et al. 2014). Likewise, *Aspergillus wentii* isolated from the *Sargassum* sp. synthesized Wentilactone B which showed cytotoxicity activity against MDA-MB-231, HeLa and HepG2, cell lines at IC50 17 μM (Zhang et al. 2013).

The diversity and composition of endophytic mycoflora vary depending on the host plant, as well as its anatomical, chemical, physiological, genetic, and ecological conditions (Wang et al. 2022). Endophytic mycoflora play a crucial role in helping host plants survive in harsh environments, such as halophilic conditions found in marine ecosystems. The distribution of both epiphytic and endophytic mycoflora is influenced by biotic and abiotic factors within these marine ecosystems. Consequently, marine endophytic fungi that have acclimatized to extreme environments represent valuable and novel sources of natural products.

6.7 Fungal Endophytes of Seaweeds

Marine macroalgae, including green algae, brown algae, and red algae, form essential components of benthic ecosystems and are among the most productive organisms in marine environments. Endophytic fungi are a group of symbiotic organisms that colonize the intercellular spaces of plants in natural systems. These fungi reside within living organisms without causing any visible symptoms. In nature, nearly all plants harbor endophytes. Numerous studies have investigated the fungal endophytes of seaweeds to explore their diversity, ecology, and potential metabolites (see Tables 6.1, 6.2, and 6.3). Recent research on the diversity of fungal endophytes in

Table 6.1 List of fungal endophytes of green algae

Name of the green algae	Name of the endophytic fungi	References
Enteromorpha	*Aspergillus unguis*	Kamat et al. (2020a, b)
	Perenniporia sp.	
	Hypoxylon fragiforme	
	Aspergillus sp.	
Chaetomorpha sp.	*Colletotrichum gloeosporioides*	
	Aspergillus niger	
	Penicillium cinnamopurpureum	
	Daldinia eschscholtzii	
	Chaetomium globosum	
	Hypoxylon fragiforme	
	Pestalotiopsis sp.	
	Aspergillus nidulans	
	Aspergillus unguis	
Ulva sp.	*Pleosporales* sp.	
Chaetomorpha antennina	*Penicillium chrysogenum*	Parthasarathy et al. (2020)
Chaetomorpha media	*Chaetomium globosum*	Kamat et al. (2020a, b)
Caulerpa racemosa	*Corynascus sepedonium*	Ramya et al. (2023)
Ceratophyllum submersum	*Talaromyces aurantiacus*	
	Amesia atrobrunnea	
	Microascus gracilis	
	Curvularia platzii	
	Chaetomium perlucidum	
Ulva sp.	*Ascochyta saliconiae*	Sarasan et al. (2017)
Blidingia minima	*Penicillium* sp.	
Boergesenia forbesii	*Aspergillus flavus*	Rajulu et al. (2022)
	A. niger	
	A. terreus	
	Aspergillus sp.	
	Aureobasidium pullulans	
	Fusarium sp.	
	Nigrospora oryzae	
	Nodulisporium sp.	
	Penicillium digitatum	
	P. oxalicum	
	Talaromyces sp.	
	Xylaria sp.	
Halimeda opuntia	*Alternaria* sp.	
	Aspergillus niger	
	Curvularia sp.	
	Lasiodiplodia theobromae	
	Nodulisporium sp.	
	Talaromyces sp.	
	Trichoderma yunnanense	
	Xylaria sp.	

(Continued)

Table 6.1 continued

Name of the green algae	Name of the endophytic fungi	References
Valonia utricularis	*Chaetomium* sp.	Flewelling et al. (2015)
Enteromorpha sp.	*Coniothyrium cereale*	
	Wardomyces anomalus	
Ulva linza	*Alternaria* sp.	
Ulva fasciata	*Aspergillus niger*	
	A. terreus	
	Aspergillus sp.	
	Chaetomium sp.	
	Curvularia lunata	
	Curvularia sp.	
	Paecilomyces sp.	
Ulva lactuca	*Aspergillus niger*	
	A. terreus	
	Aspergillus sp.	
	Chaetomium sp.	
	Cladosporium sp.	
	Eurotium sp.	
	Leotiomyceta sp.	
	Nigrospora sp.	
	Penicillium chrysogenum	
	Penicillium sp.	
	Pseuderuotium bakeri	
	Stilbella fimetaria	
Ulva intestinalis	*Aspergillus* sp.	
	Bionectria ochroleuca	
	Cordyceps sp.	
	Leptosphaeria sp.	
	Penicillium sp.	
Ulva pertusa	*Chaetomium globosum*	
Ulva sp.	*Penicillium* sp.	
	Dendryphiella salina	
Sphingomorpha arcta	*Penicillium soppii*	
	P. spinulosum	
	Penicillium sp.	
Codium fragile	*Aspergillus versicolor*	
	Gibberella zeae	
	Trichoderma longibrachiatum	
Caulerpa racemosa	*Aspergillus niger*	
	A. terreus	
	Aspergillus sp.	
	Aureobasidium pullulans	
	Chaetomium sp.	
	Fusarium sp.	
	Mucor sp.	
	Penicillium sp.	
	Phoma sp.	
	Pyrenochaeta sp.	

Name of the green algae	Name of the endophytic fungi	References
Caulerpa scalpelliformis	*Aspergillus niger*	
	A. terreus	
	Aspergillus sp.	
	Curvularia sp.	
	Paecilomyces sp.	
	Penicillium sp.	
Caulerpa sertularioides	*Aspergillus niger*	
	A. terreus	
	Aspergillus sp.	
	Chaetomium sp.	
	Cladosporium sp.	
	Fusarium sp.	
	Myrothecium sp.	
	Penicillium sp.	
	Phialophora sp.	
Halimeda macroloba	*Aspergillus niger*	
	A. terreus	
	Aspergillus sp.	
	Paecilomyces sp.	
	Penicillium sp.	
	Trichoderma sp.	
Prasiola crispa	*Mastodia tessellata*	

Table 6.2 List of fungal endophytes of brown algae

Name of the brown algae	Name of the fungal endophyte	References
Ascophyllum nodosum	*Aspergillus fumigatus*	Flewelling et al. (2015)
	Aspergillus sp.	
	Cladosporium sp.	
	Dendryphiella salina	
	Lichtheimia corymbifera	
	Mycosphaerella ascophylli	
Cystoseira sp.	*Haloguignardia irritans*	
Halidrys sp.	*Haloguignardia irritans*	
Colpomenia sinuosa	*Aspergillus niger*	
Cladostephus spongiosus	*Acremonium* sp.	
Dictyota dichotoma	*Aspergillus niger*	
	A. terreus	
	Trichoderma sp.	
Fucus sp.	*Dendryphiella arenaria*	
	Phoma sp.	

(Continued)

Table 6.2 continued

Name of the brown algae	Name of the fungal endophyte	References
Fucus vesiculosus	*Alternaria* sp.	
	Aspergillus fumigatus	
	Chalara sp.	
	Coniothyrium sp.	
	Epicoccum sp.	
	Phoma sp.	
	Penicillium sp.	
	Dendryphiella salina	
	Tolypocladium inflatum	
Fucus sp.	*Arthrinium* sp.	
Fucus serratus	*Aspergillus fumigatus*	
	Coniothyrium sp.	
	Penicillium sp.	
Fucus spiralis	*Penicillium* sp.	
Laminaria digitata	*Dendryphiella salina*	
Lobophora variegata var. *indica*	*Aspergillus niger*	
	A. terreus	
	Chaetomium sp.	
	Emericella nidulans	
	Nigrospora sp.	
Padina gymnospora	*Aspergillus terreus*	
	Chaetomium sp.	
	Cladosporium sp.	
	Penicillium sp.	
	Phoma sp.	
	Trichoderma sp.	
Padina tetrastromatica	*Acremoniella* sp.	
	Ascotricha sp.	
	Aspergillus niger	
	A. terreus	
	Aspergillus sp.	
	Chaetomium sp.	
	Monodictys sp.	
	Paecilomyces sp.	
	Penicillium sp.	
	Phialophora sp.	
Sargassum fusiforme	*Aspergillus wentii*	
S. horneri	*Pestalotiopsis* sp.	
S. kjellmaniamum	*Aspergillus ochraceus*	
S. palladium	*Penicillium chrysogenum*	
S. thunbergii	*Aspergillus versicolor*	
	Eurotium cristatum	
	Penicillium glabrum	

Name of the brown algae	Name of the fungal endophyte	References
Sargassum sp.	*Aspergillus wentii*	
	Alternaria sp.	
	Aspergillus niger	
	A. terreus	
	Colletotrichum sp.	
	Curvularia lunata	
	Drechslera sp.	
	Helicosporium sp.	
	Nigrospora sp.	
	Oidiodendron sp.	
	Penicillium sp.	
	Taeniolella sp.	
	Varicosporium sp.	
S. ilicifolium	*Aspergillus terreus*	
	Corollospora sp.	
	Dendryphiella arenaria	
S. wightii	*Aspergillus niger*	
	A. terreus	
	Aspergillus sp.	
	Cladosporium sp.	
	Nigrospora sp.	
	Paecilomyces sp.	
	Penicillium sp.	
	Yeast sp.	
Saccharina latissima	*Aspergillus* sp.	
	Penicillium chrysogenum	
	P. soppii	
Stoechospermum marginatum	*Alternaria* sp.	
	Aspergillus terreus	
	Chaetomium sp.	
	Emericella nidulans	
	Nigrospora sp.	
Turbinaria decurrens	*Alternaria* sp.	
	Aspergillus terreus	
	Chaetomium sp.	
	Curvularia sp.	
	Drechslera sp.	
	Emericella nidulans	
	Nigrospora sp.	
	Phialophora sp.	
	Pseudogymnoascus	
Turbinaria conoides	*Aspergillus terreus*	
	Aspergillus sp.	
	Chaetomium sp.	
	Cladosporium sp.	
	Emericella nidulans	
	Phoma sp.	

(Continued)

Table 6.2 continued

Name of the brown algae	Name of the fungal endophyte	References
Turbinaria sp.	*Alternaria* sp.	
	Aspergillus terreus	
	Aspergillus sp.	
	Chaetomium sp.	
	Cladosporium sp.	
	Colletotrichum sp.	
	Curvularia lunata	
	C. tuberculata	
	Curvularia sp.	
	Drechslera papendorfii	
	Drechslera sp.	
	Emericella nidulans	
	Fusarium sp.	
	Monodictys sp.	
	Paecilomyces sp.	
	Phaeotrichoconis sp.	
	Varicosporium sp.	
Undaria pinnatifida	*Guignardia* sp.	
Xiphophora gladiata	*Penicillium* sp.	
Padina sp.	*Trichoderma harzianum*	Handayani et al. (2019)
Sargassum sp.	*Aspergillus wentii*	
S. kjellmanianum	*Aspergillus ochraceus*	
Colpomenia sinuosa	*Aspergillus niger*	
Sargassum oligocystum	*Aspergillus* sp.	Zainee et al. (2019)
Sargassum polycystum	*Aspergillus* sp.	
Sargassum wightii	*Trichoderma harzianum*	Thirunavukkarasu et al. (2015)
Dictyota dichotoma	*Cladosporium uredinicola*	Vasavi and Umamaheswari (2018)

Name of the brown algae	Name of the fungal endophyte	References
Colpomenia sinuosa	*Aspergillus niger*	Rajulu et al. (2022)
	Aspergillus sp.	
	Cladosporium sp.	
	Fusarium sp.	
	Lasiodiplodia theobromae	
	Mucor sp.	
	Nodulisporium sp.	
	Penicillium oxalicum	
	Penicillium sp.	
	Pestalotiopsis sp.	
	Talaromyces sp.	
	Torulomyces sp.	
	Trichoderma reesei	
Padina gymnospora	*Aspergillus giganteus*	
	A. nidulans	
	Aspergillus sp.	
	Colletotrichum sp.	
	Fusarium sp.	
	Nodulisporium sp.	
	Penicillium oxalicum	
	Trichoderma lixii	
	Trichoderma reesei	
	Xylaria sp.	
Padina pavonica	*Aspergillus niger*	
	Aspergillus sp.	
	Aureobasidium pullulans	
	Eupenicillium sp.	
	Fusarium sp.	
	Mucor sp.	
	Paecilomyces sp.	
	Penicillium digitatum	
	Penicillium sp.	
	Pestalotiopsis sp.	
	Trichoderma crissum	
	T. yunnanense	
	Xylaria sp.	
Sargassum ilicifolium	*Arthrinium* sp.	
	Aspergillus giganteus	
	A. niger	
	Aspergillus sp.	
	Cladosporium sp.	
	Lasiodiplodia theobromae	
	Paecilomyces sp.	
	Penicillium digitatum	
	Talaromyces sp.	
	Trichoderma pleuroticola	
	T. yunnanense	
	Xylaria sp.	

(Continued)

Table 6.2 continued

Name of the brown algae	Name of the fungal endophyte	References
Sargassum polycystum	*Aspergillus fumigatus*	
	A. niger	
	Aspergillus sp.	
	Fusarium sp.	
	Nodulisporium sp.	
	Penicillium purpurogenum	
	Penicillium sp.	
	Trichoderma pleuroticola	
	Xylaria sp.	
Sargassum wightii	*Alternaria* sp.	
	Aspergillus flavus	
	A. niger	
	Fusarium sp.	
	Penicillium digitatum	
	Torulomyces sp.	
	Trichoderma lixii	
	T. yunnanense	
	Xylaria sp.	
Sargassum sp.	*Aspergillus fumigatus*	
	A. niger	
	Aspergillus sp.	
	Penicillium digitatum	
	Penicillium sp.	
	Talaromyces sp.	
	Torulomyces sp.	
	Trichoderma yunnanense	
	Xylaria sp.	
Turbinaria conoides	*Aspergillus clavatus*	
	A. nidulans	
	A. terreus	
	Aspergillus sp.	
	Eupenicillium sp.	
	Fusarium sp.	
	Mucor sp.	
	Penicillium sp.	
	Pestalotiopsis sp.	
	Talaromyces sp.	
Turbinaria decurrens	*Aspergillus giganteus*	
	A. niger	
	A. terreus	
	Aspergillus sp.	
	Nodulisporium sp.	
	Penicillium digitatum	
	P. oxalicum	
	Pestalotiopsis sp.	
	Talaromyces sp.	
	Torulomyces sp.	
	Trichoderma reesei	
	T. yunnanense	

Table 6.3 List of endophytic fungi of red algae

Name of the red algae	Name of the endophytic fungi	References
Chondrus crispus	*Alternaria metachromatica*	Bonthond et al. (2022)
	Metschnikowia reukaufii	
	Acremonium exuviarum	
	Stemphylium vesicarium	
	Rhodotorula mucilaginosa	
Mastocarpus stellatus	*Didymella glomerata*	
	D. exigua	
	Mucor laxorrhizus	
Acanthophora spicifera	*Aspergillus giganteus*	Rajulu et al. (2022)
	A. niger	
	A. terreus	
	Aureobasidium pullulans	
	Lasiodiplodia theobromae	
	Nodulisporium sp.	
	Penicillium digitatum	
	P. oxalicum	
	Pestalotiopsis sp.	
	Trichoderma atrobrunneum	
	Xylaria sp.	
Dichotomaria obtusata	*Aspergillus terreus*	
	Eupenicillium sp.	
	Fusarium sp.	
	Mucor sp.	
	Nodulisporium sp.	
	Paecilomyces sp.	
	Penicillium digitatum	
	Pestalotiopsis sp.	
	Phoma sp.	
	Talaromyces sp.	
	Trichoderma lixii	
	T. yunnanense	
	Xylaria sp.	
Gracilaria edulis	*Aspergillus* sp.	
	Nodulisporium sp.	
	Penicillium oxalicum	
	Trichoderma crassum	
Gracilaria lantaensis	*Aspergillus nidulans*	
	A. niger	
	Aspergillus sp.	
	Eupenicillium sp.	
	Fusarium sp.	
	Lasiodiplodia theobromae	
	Penicillium digitatum	
	Pestalotiopsis sp.	
	Trichoderma crassum	
	T. inhamatum	
	Xylaria sp.	

(continued)

Table 6.3 (continued)

Name of the red algae	Name of the endophytic fungi	References
Gracilaria salicornia	*Alternaria* sp.	
	Aspergillus fumigatus	
	A. niger	
	A. terreus	
	Fusarium sp.	
	Nodulisporium sp.	
	Penicillium digitatum	
	Pestalotiopsis sp.	
	Trichoderma crassum	
	T. parareesei	
	Xylaria sp.	
Hypnea valentiae	*Aspergillus niger*	
	Eupenicillium sp.	
	Fusarium sp.	
	Nodulisporium sp.	
	Penicillium digitatum	
	P. oxalicum	
	Penicillium sp.	
	Pestalotiopsis sp.	
	Talaromyces sp.	
	Trichoderma reesei	
	Xylaria sp.	
Gymnogongrus flabelliformis	*Aspergillus wentii*	Sarasan et al. (2017)
Heterosiphonia japonica	*Aspergillus oryzae*	
Laurencia sp.	*Penicillium chrysogenum*	
Polysiphonia sp.	*Apiospora montagnei*	
Asparagopsis taxiformis	*Annulohypoxylon stygium*	Medina et al. (2019a, b)
	A. yungensis	
Bostrychia tenella	*Trichoderma atroviride*	Felício et al. 2015
	Nigrospora oryzae	
	Phomopsis sp.	
	Penicillium decaturense	
	P. waksmanni	
	Acremonium sp.	
	A. implicatum	
	Eurotium sp.	
	Xylaria sp.	

Name of the red algae	Name of the endophytic fungi	References
Gracilaria sp.	*Aspergillus aegyptiacus*	Baghazadeh Daryaii et al. (2021)
	A. chevalieri	
	A. terreus	
	Penicillium chrysogenum	
Champia parvula	*Aspergillus terreus*	
	A. puniceus	
	Penicillium sp.	
Hypnea sp.	*Aspergillus terreus*	
Gracilaria foliifera	*Aspergillus carlsbadensis*	
Gracilariopsis persica	*Penicillium chrysogenum*	
Grateloupia lithophila	*Penicillium chrysogenum*	
Acanthophora spicifera	*Aspergillus sydowii*	Flewelling et al. (2015)
Apophlaea sp.	*Mycosphaerella apophlaeae*	
Bostrychia radicans	*Phomopsis longicolla*	
Ceramium sp.	*Chaetomium* sp.	
	Phaeosphaeria spartinae	
Chondrus crispus	*Aspergillus* sp.	
	Penicillium crustosum	
	Lautitia danica	
Chondrus ocellatus	*Penicillium echinulatum*	
Corallina officinalis	*Aspergillus flavus*	
Digenea simplex	*Dendryphiella arenaria*	
Dilsea carnosa	*Mycaureola dilseae*	
Devaleraea ramentacea	*Aureobasidium pullulans*	
	Botrytis sp.	
	Cladosporium sp.	
	Coniothyrium sp.	
	Penicillium decumbens	
	Trametes versicolor	
Gelidiella acerosa	*Alternaria* sp.	
	Aspergillus niger	
	A. terreus	
	A. versicolor	
	Nigrospora sp.	
	Penicillium sp.	
	Phoma sp.	

(Continued)

Table 6.3 (continued)

Name of the red algae	Name of the endophytic fungi	References
Gelidiella crassa	*Trichoderma* sp.	
Gracilaria edulis	*Aspergillus terreus*	
	Nigrospora sp.	
Gracilaria sp.	*Aphanocladium* sp.	
	Aspergillus niger	
	A. terreus	
	Monilinia sp.	
	Paecilomyces sp.	
	Yeast sp.	
	Daldinia eschscholzii	
Grateloupia lithophila	*Aspergillus niger*	
	A. terreus	
	Aspergillus sp.	
	Chaetomium sp.	
	Cladosporium sp.	
	Emericella nidulans	
	Nigrospora sp.	
Gymnogongrus flabelliformis	*Aspergillus wentii*	
Halymenia sp.	*Aspergillus niger*	
	A. terreus	
	Aspergillus sp.	
	Chaetomium sp.	
	Cladosporium sp.	
	Emericella nidulans	
	Nigrospora sp.	
Heterosiphonia japonica	*Aspergillus oryzae*	
Hypnea musciformis	*Dendryphiella arenaria*	
Laurencia ceylanica	*Aspergillus terreus*	
Laurencia similis	*Exophiala oligosperma*	
Laurencia sp.	*Alternaria alternata*	
	Penicillium chrysogenum	
Liagora viscida	*Drechslera dematioidea*	
Mastocarpus stellatus	*Penicillium decumbens*	
	Penicillium sp.	
Palmaria palmata	*Botrytis* sp.	
	Helicomyces sp.	
	Hypoxylon sp.	
	Penicillium chrysogenum	
	P. crustosum	
	P. decumbens	

Name of the red algae	Name of the endophytic fungi	References
Porphyra umbilicalis	*Aspergillus sydowii*	
Porphyra sp.	*Microdochium* sp.	
	Rhodotorula mucilaginosa	
Plocamium cartilagineum	*Acremonium* sp.	
	Cladosporium sp.	
	Metschnikowia sp.	
	Penicillium biourgeianum	
Plocamium sp.	*Acremonium* sp.	
	Nodulisporium sp.	
Polysiphonia sp.	*Colletotrichum* sp.	
	Geniculosporium sp.	
Polysiphonia lanosa	*Aureobasidium pullulans*	
	Botryotinia fuckeliana	
	Phaeosphaeria sp.	
Polysiphonia violacea	*Apiospora montagnei*	
Polysiphonia urceolata	*Chaetomium globosum*	
Portieria hornemannii	*Aspergillus terreus*	
	Aspergillus sp.	
	Cladosporium sp.	
	Emericella nidulans	
	Memnoniella sp.	
	Penicillium sp.	
	Trimmatostroma sp.	
	Yeast sp.	

seaweeds from the Andaman and Nicobar Islands indicates that Eurotiomycetes, which are less common in terrestrial plants, are predominant in seaweeds (Rajulu et al. 2022).

6.8 Secondary Metabolites of Marine Algae-Derived Endophytic Fungi

Endophytes are microorganisms that inhabit the tissues of plants without causing visible symptoms. These organisms are ubiquitous in nature, found in all plant species. Certain endophytes synthesize and release physiologically active compounds known as secondary metabolites, which can enhance the host plant's ability to resist bacterial, fungal, and other pest-related diseases (Fig. 6.2). Endophytes are primarily sourced from aquatic plants and algae (Gouda et al. 2016). Research has demonstrated that endophytic fungi derived from marine algae are consistent producers of novel secondary metabolites characterized by remarkable scaffold diversity, structural complexity, and significant bioactivity (Gao and Zhang 2023). Marine algae, which inhabit dynamic ecosystems, exhibit adaptations to both recurrent and

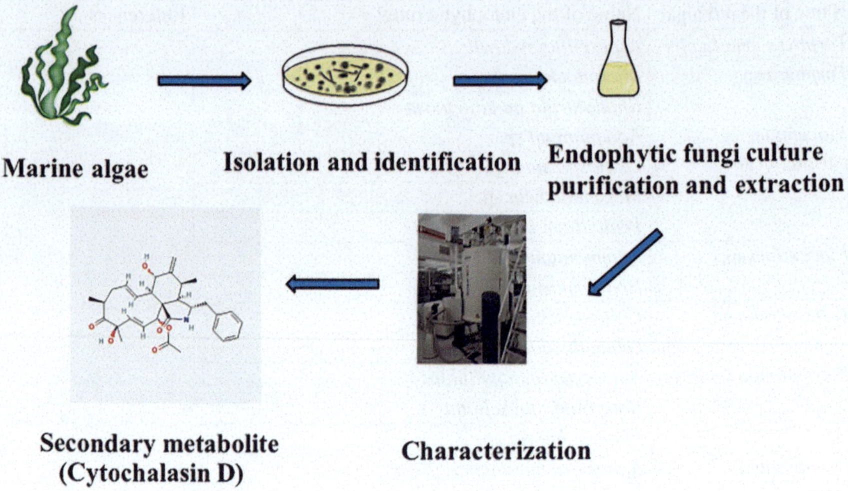

Fig. 6.2 Secondary metabolites of marine algae-derived endophytic fungi

sporadic environmental stressors, including drought, high salinity, low oxygen levels, nutrient limitation, excessive light, and other factors that may induce stress in endophytes. Such stressors can stimulate the production of bioactive secondary metabolites that bolster the host's defense mechanisms. Consequently, numerous chemical investigations into algal-derived endophytic fungal natural products have been conducted over the past few decades. These studies have led to the isolation and identification of several novel compounds, which exhibit a range of biological activities, including anticancer, antibiotic, antiviral, antioxidative, and kinase inhibitory properties (Zhang et al. 2016).

The relationship between the host and its endophytic fungi exhibits symbiotic characteristics, as the endophytic resident regularly receives nutrients from the host. In return, it actively promotes algal development and enhances the host's fitness by producing specific beneficial metabolites (Bamisile et al. 2018). This interaction between fungi and sea algae stimulates the synthesis of metabolites with potential therapeutic value. Consequently, several species of macroalgae have been globally investigated in relation to their associated fungal populations, which encompass various genera (Teixeira et al. 2019). Among these metabolites, the most isolated chemical classes include alkaloids, lactones, peptides, polyketides, and terpenes.

6.8.1 Anticancer Compounds

Steroids are a class of constitutive bioactive elements that are widely present in mycoendophytes. A novel steroidal derivative, 3,11-dihydroxy ergosta-8,24(28)-dien-7-one, and a rare anticancer norsteroid molecule, 7-nor-ergosterolide, have been isolated from the endophytic fungus *Aspergillus ochraceus* living in the macroalga *Sargassum kjellmanianum*. Seven-nor-ergosterolide has been shown to have anticancer properties against the human cancer cell lines NCI-H460, SMMC-7721, and SW1990. In contrast, 3β,11α-dihydroxy ergosta-8,24(28)-dien-7-one only slightly inhibits the proliferation of SMMC-7721 cells (Cui et al. 2010a, b) *Chaetomium* appears to be a repository for anticancer substances such cytoglobosins C and D (Cui et al. 2010a, b). An endophytic fungus called *Penicillium chrysogenum* QEN 24S was isolated from an unnamed maritime red algae species of the genus *Laurencia*. It produced penicitide A, which had a mild cytotoxic activity against a cell line from a human liver cancer with an IC50 of 32 μg/mL (Gao et al. 2010). As cytotoxic agents, norditerpenoids have garnered a lot of interest. As predicted, a number of new norditerpenoids (asperolides A and B and wentilactones A and B) from *Aspergillus wentii* EN-48, an algicolous endophyte of *Sargassum* sp., shown cytotoxicity against a number of human tumor cell lines. Similar outcomes with *N*-hydroxy-2-pyridine, an alkaloid of *Penicillium* sp. of the brown alga *Xiphophora gladiata* (IC50 1.8 g/mL), were seen in the murine leukemia bioassay (P388) (Sun et al. 2012). Cytochalasin D, a toxin produced by the fungus *Xylaria* sp., was discovered from the Brazilian coastal seaweed *Bostrychia tenella*, which had sporadic action against the cancer cells SF-295 and HCT-8 (Felício et al. 2015). Recently, Kojic acid (5-Hydroxy-2-(hydroxymethyl)-4*H*-pyran-4-one), a molecule isolated from *Aspergillus* sp., shown cytotoxicity against A-431 (human skin carcinoma) (Taritla 2023). Xylarichalasin A from the *Xylaria* cf. *curta* exhibited superior cytotoxicity to the medication cisplatin against the SMMC-7721 and MCF-7 cell lines (Wang et al. 2019a). Many important anticancer compounds isolated from algal endophytes are shown in Fig. 6.3.

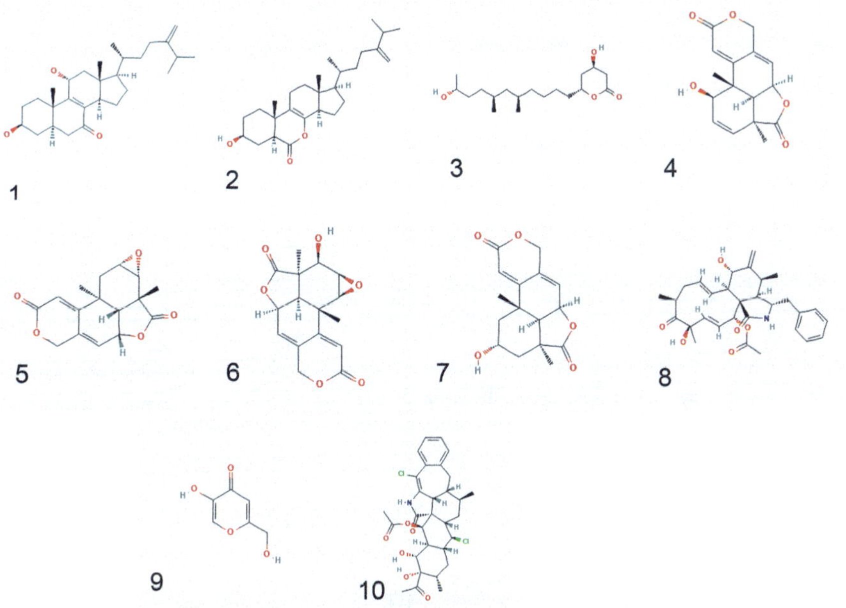

Fig. 6.3 Anticancer compounds from algal endophytes. (1) 3,11-dihydroxy ergosta-8,24(28)-dien-7-one, (2) 7-nor-ergosterolide, (3) penicitide A, (4) asperolides A, (5) asperolides B, (6) wentilactones A, (7) wentilactones B, (8) cytochalasin D, (9) (5-Hydroxy-2-(hydroxymethyl)-4*H*-pyran-4-one), (10) Xylarichalasin A

6.8.2 Antibacterial Compounds

The new chamiside A was discovered from *Chaetomium nigricolor* and had antibacterial action against *S. aureus*. It included a novel 6/6/5 fused tricyclic core skeleton holding a benzene ring and a rare seven-membered lactone (Wang et al. 2019b) Penicitroamide was obtained from the culture of *Penicillium* sp., and it exhibited antibacterial action against *Erwinia carotovora* and *Sclerotium rolfsii*. Its structure has a bicyclo(3.2.1)octane core with a high degree of carbonylization (Feng et al. 2016). The first instance of a ketal-bridged ring component in a natural sesquiterpene structure of the degraded bisabolane type is Aspergoterpenin A, which was isolated from *Aspergillus versicolor*. Aspergoterpenin A exhibited antibacterial action against *Erwinia carotovora* subsp. *carotovora* (Guo et al. 2018) violaceol I, violaceol II, and diorcinol was obtained from the culture filtrate of fungus *Aspergillus tabacinus* SFC20160407-M11 revealed a wide range of antibacterial activity, with minimum inhibitory concentrations ranging from 6.3 to 200 μg/mL (Nguyen et al. 2022). *Talaromyces* sp. BTBU20213036, a marine-derived fungus, was used to isolate four new polyketide-derived oligophenalenone dimers, bacillisporins K and L, rugulosin D, macrosporusone D, as well as four known compounds, bacillisporin B, macrosporusone D, rugulosin A, and penicillide. These substances were investigated for their antibacterial and antifungal properties against the bacterial strains *Staphylococcus aureus* (G+), *Escherichia coli* (G−), and *Candida albicans* (F). With minimal inhibitory concentrations ranging from 0.195 to 100 μg/mL, these substances demonstrated potential inhibitory effects against *S. aureus* (Song et al. 2022) (Fig. 6.4).

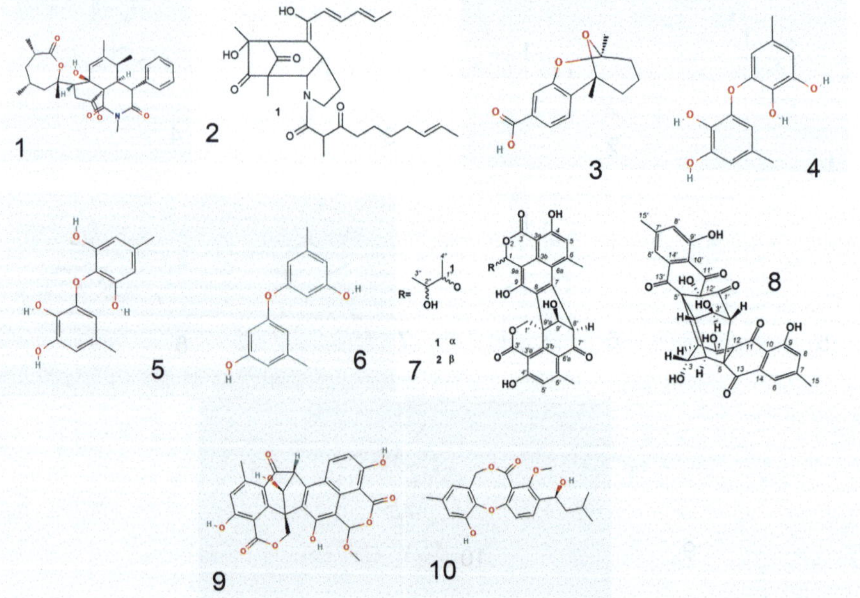

Fig. 6.4 Antibacterial compounds from algal endophytes (1) chamiside A, (2) Penicitroamide, (3) Aspergoterpenin A, (4) violaceol I, (5) violaceol II, (6) diorcinol, (7) bacillisporins K and L, (8) rugulosin D, (9) macrosporusone D, (10) penicillide

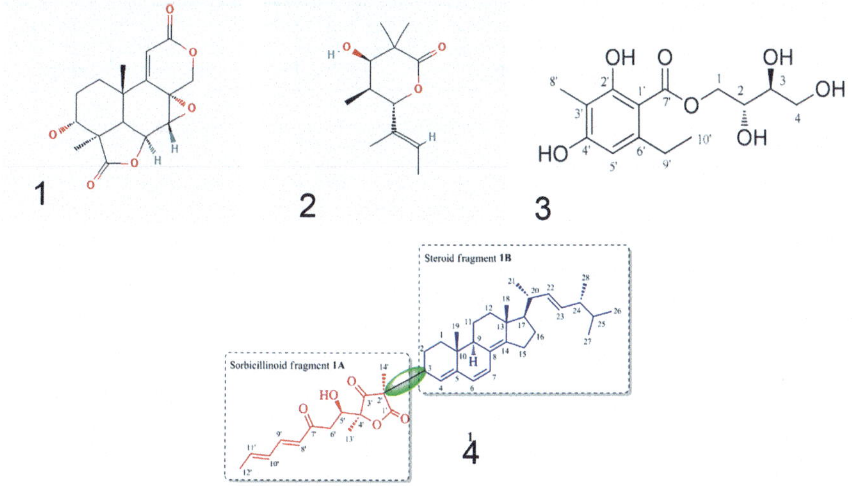

Fig. 6.5 Antifungal compounds from algal endophytes (1) tetranorditerpenoid, (2) helicascolide A, (3) penicierythritols A, (4) tennessenoid A

6.8.3 Antifungal Compounds

A tetranorditerpenoid derivative from the internal mycobiota of Sargassum species, *A. wentii* EN-48, is very intriguing due to both its powerful fungicidal and anticancer characteristics. The specific substance was discovered to have significant antifungal activity against *C. albicans*, with an MIC value of 16 μg/mL (Sun et al. 2012). According to Tarman and his team, helicascolide C that accumulated in the broth culture of endophytic *Daldinia eschscholzii* displays fungistatic properties (Tarman et al. 2012). Penicierythritols A was isolated from the marine algal-derived endophytic fungus *Penicillium chrysogenum* XNM-12. With an MIC value of 8 μg/mL, Penicierythritols A has mild antifungal activity against the plant pathogen *Alternaria alternata* (Xu et al. 2020). Tennessenoid A was discovered from the inner tissue of marine red algae, identified as a novel steroid-sorbicillinoid adduct through a C–C link that was rarely seen in natural products. *Panteria* showed broad-spectrum activities with inhibition zone diameters ranging from 2 to 7 mm against *Sclerotium rolfsii* Sacc., *Fusarium oxysporum* (Schl.) f.sp. *cucumerinum* Owen, *Coniella diplodiella* Petrak et Sydow, *Physalospora piricola* Nose., *Fusarium graminearum* Schw., and *Alternaria mali* rob (Zhao et al. 2022). The structures of antifungal compounds are shown in Fig. 6.5.

6.8.4 Antioxidant Compounds

Numerous studies do show that endosymbionts have strong antioxidant capability. The antioxidant activity of five new polyhydroxylated hydroanthraquinone derivatives, namely 8-hydroxyconiothyrinone B, 8,11-dihydroxyconiothyrinone B, 4*R*,8-dihydroxyconiothyrinone B, 4*S*,8-dihydroxyconiothyrinone B, and

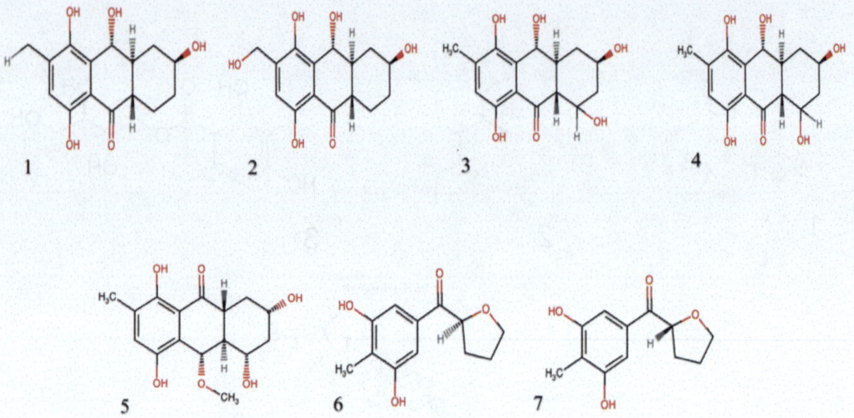

Fig. 6.6 Antioxidant compounds from algal endophytes (1) 8-hydroxyconiothyrinone B, (2) 8,11-dihydroxyconiothyrinone B, (3) 4*R*,8-dihydroxyconiothyrinone B, (4) 4*S*,8-dihydroxyconiothyrinone B, (5) 4*S*,8-dihydroxy-10-*O*-methyldendryol E, (6) (−)-1*S*-myrothecol, (7) (+)-1*R*-myrothecol

4*S*,8-dihydroxy-10-*O*-methyldendryol E isolated from the inner tissue of the marine red alga *Laurencia okamurai*. A novel aromatic bisabolene-type sesquiterpenoid (7*S*-8*S*)-8-hydroxy sydowic acid was extracted and identified from the marine red alga *Symphyocladia latiuscula* by the endophytic fungus strain *Aspergillus sydowii* EN-434. The DPPH-free radical scavenging activity of (7*S*-8*S*)-8-hydroxy sydowic acid was strongly shown, with an IC50 value of 113.5 μmol/L. The secondary metabolites produced by marine fungus of various origins are discussed by Saravanakumar et al. (2020). It is suggested that a number of marine fungi that have been isolated from brown, red, and green algae, soft corals, and red, blue, and black sponges may have therapeutic benefits. It has been demonstrated that *Penicillium flavigenum*-derived fungal metabolites have antioxidant and anti-proliferative properties. (−)-1*S*-myrothecol and (+)-1*R*-myrothecol, a pair of 2-benzoyl tetrahydrofuran enantiomers found from deep-sea-derived fungus *Myrothecium* sp. showed antioxidant activity in the oxygen radical absorbance capacity and 2,2-azino-bis(3-ethylbenzothiazoline-6-sulfonic acid) tests (Lu et al. 2020). The chemical structures of different antioxidant compounds from algal endophytes are shown in Fig. 6.6.

6.8.5 Other Bioactive Compounds

Three chloramphenicol derivatives were isolated from marine alga-derived endophytic fungus, *Acremonium vitellinum*. Compounds 1–3 were tested for their insecticidal effects against the cotton bollworm, *Helicoverpa armigera* (Fig. 6.7). With an LC50 value of 0.56 ± 0.03 mg/mL (as opposed to matrine's LC50 value of

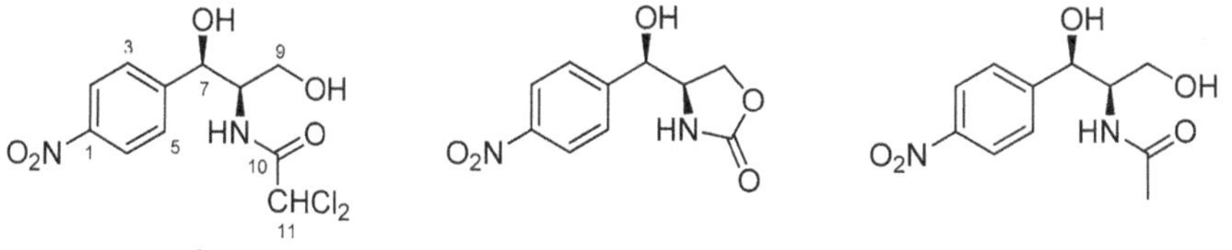

1 2 3

Fig. 6.7 The isolated chemical structures of the natural compounds 1–3 from *Acremonium vitellinum*

0.24 ± 0.01 mg/mL), the natural chemical 2 demonstrated significant efficacy against *H. armigera* (Chen et al. 2018) Rebeca et al. isolated five new botryane sesquiterpenes from the marine red alga *Asparagopsis taxiformis*. Among those compounds, nemenonediol A exhibited moderate acetylcholinesterase inhibition whereas (+)-(2*R*,4*S*,5*R*,8*S*)-4-deacetyl-5-hydroxy-botryenalol inhibited both acetylcholinesterase and butyrylcholinesterase (Medina et al. 2019a, b). One monoterpene derivative penicimonoterpene was obtained from endophytic fungus *Penicillium chrysogenum* QEN-24S, isolated from an unknown maritime red algal species of the genus *Laurencia*, showed strong resistance to the pathogen *A. brassicae* (Gao et al. 2010). The compound methyl(*Z*)-3-(3,4-dihydroxyphenyl)-2-formamidoacrylate discovered from the marine algal-derived endophytic fungi *Penicillium oxalicum* EN-290 demonstrated efficacy against *Nitzschia closterium*, a species responsible for toxic algal blooms, with inhibition zones of 20, 16, and 10 mm at 20, 10, and 5 mg/mL, respectively (Li et al. 2015). The fungus *Halosphaeriaceae* sp. OUPS-135D-4, which was acquired from the marine algae *Sargassum thunbergii*, produced Halosmysin A, a novel 14-membered macrodiolide with an unorthodox skeleton, demonstrated to be extremely effective against human HL-60, murine HL-60, and L1210 leukemia cell lines, with IC50 values ranging from 2.2 ± 3.1 to 11.7 ± 2.8 μM (Yamada et al. 2020). The structures of other bioactive compounds are shown in Fig. 6.8.

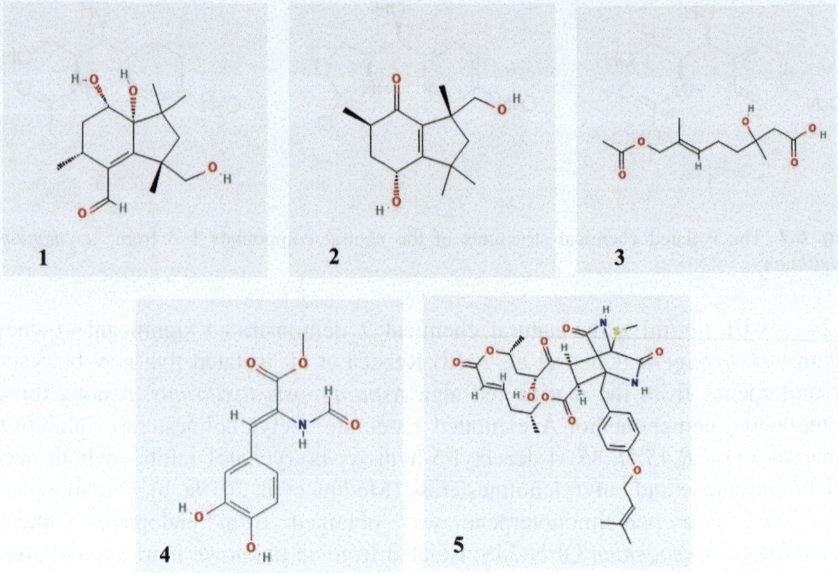

Fig. 6.8 Other bioactive compounds from algal endophytes (1) (+)-(2*R*,4*S*,5*R*,8*S*)-4-deacetyl-5-hydroxy-botryenalol, (2) nemenonediol A, (3) penicimonoterpene, (4) methyl(Z)-3-(3,4-dihydroxyphenyl)-2-formamidoacrylate, (5) Halosmysin A

6.9 Conclusion

There is an urgent need for new and novel sustainable compounds that have potential bioactivity because of the emergence of life-threatening new infections and diseases caused by bacteria, fungi, mycoplasma, and viruses. Several bacteria have already developed resistance against existing antibacterial compounds, becoming multidrug-resistant bacteria. Currently, certain microbial by-products appear to have a unique and novel mechanism of action. Hence, we are focusing on microorganisms that survive in unparalleled biotopes. In this context, microorganisms growing in unique and unexplored/underexplored ecological niches, such as endophyte biotypes and several other minor marine phyla, are presently considered as a research source of bioactive drugs from the marine ecosystem (El-Gendy et al. 2017; Handayani et al. 2019).

In summary, marine endophytic fungi generated from algae have shown to be a valuable source of secondary metabolites with a wide range of uses. Numerous bioactive substances, such as antioxidants, antivirals, antibacterials, and anticancer agents, can be produced by this fungus. The enormous potential of these metabolites in a variety of industries, including pharmaceuticals, agriculture, and environmental remediation, has been revealed by scientists via intensive research and discovery. Marine algae are a fascinating field of study for the identification of new natural products because of their special biological niche, which provides a large reservoir of endophytic fungi. Deeper research and comprehension of endophytic fungi generated from marine algae could lead to novel and long-lasting solutions for problems related to agriculture, human health, and the environment.

References

Alam B, Lǐ J, Gě Q, Khan MA, Gōng J, Mehmood S et al (2021) Endophytic fungi: from symbiosis to secondary metabolite communications or vice versa? Front Plant Sci 12:3060

Baghazadeh Daryaii L, Samsampour D, Bagheri A, Sohrabipour J (2021) Macroalgae-derived fungal endophytes promoted the growth of Mexican lime seedlings under heat stress in greenhouse condition. J Phycol Res 5(2):691–712

Bajpai VK (2016) Antimicrobial secondary metabolites from marine fungi: a mini review. NISCAIR-CSIR

Bamisile BS, Dash CK, Akutse KS, Keppanan R, Wang L (2018) Fungal endophytes: beyond herbivore management. Front Microbiol 9:544

Bano N, Rizvi IF, Sharma N, Siddiqui MH, Khan MA, Akhtar S (2016) Production of bioactive secondary metabolites from endophytic fungi. Int Res J Eng Tech 3(6):1859–1866

Bender SF, Wagg C, van der Heijden MG (2016) An underground revolution: biodiversity and soil ecological engineering for agricultural sustainability. Trends Ecol Evol 31(6):440–452

Bernstein ME, Carroll GC (1977) Internal fungi in old-growth Douglas fir foliage. Can J Bot 55(6):644–653

Bills GF (1996) Isolation and analysis of endophytic fungal communities from woody plants. In: Endophytic fungi in grasses and woody plants: systematics, ecology, and evolution. APS Press, pp 31–65

Bills GF, Polishook JD (1992) Recovery of endophytic fungi from Chamaecyparis thyoides. Sydowia 44(1):1–12

Bills GF, Polishook JD (1994) Abundance and diversity of microfungi in leaf litter of a lowland rain forest in Costa Rica. Mycologia 86:187–198

Bonthond G, Barilo A, Allen RJ, Cunliffe M, Krueger-Hadfield SA (2022) Fungal endophytes vary by species, tissue type, and life cycle stage in intertidal macroalgae. J Phycol 58(2):330–342

Bunders J, Haverkort B, Hienstra W (1996) Biotechnology: building on farmer's knowledge. Macmillan Education Ltd., London

Carroll GC (1986) The biology of endophytism in plants with particular reference to woody perennials. In: Microbiology of the phyllosphere. Cambridge University Press, pp 203–222

Carroll GC, Carroll FE (1978) Studies on the incidence of coniferous needle endophytes in the Pacific Northwest. Can J Bot 56(24):3034–3043

Carroll G, Petrini O (1983) Patterns of substrate utilization by some fungal endophytes from coniferous foliage. Mycologia 75:53–63

Chanda S, Dave R, Kaneria M, Nagani K (2010) Seaweeds: a novel, untapped source of drugs from sea to combat infectious diseases. In: Current research, technology and education topics in applied microbiology and microbial biotechnology, vol 1. FORMATEX Research Center, pp 473–480

Chen D, Zhang P, Liu T, Wang XF, Li ZX, Li W, Wang FL (2018) Insecticidal activities of chloramphenicol derivatives isolated from a marine alga-derived endophytic fungus, Acremonium vitellinum, against the cotton bollworm, Helicoverpa armigera (Hübner) (Lepidoptera: Noctuidae). Molecules 23(11):2995

Clay K (1989) Clavicipitaceous endophytes of grasses: their potential as biocontrol agents. Mycol Res 92(1):1–12

Clay K (1990) Fungal endophytes of grasses. Annu Rev Ecol Syst 21(1):275–297

Cubit JD (1976) Interactions of seasonally changing physical factors and grazing affecting high intertidal communities on a rocky shore. PhD Thesis, Department of Biology, Oregon State University, USA

Cui CM, Li XM, Li CS, Proksch P, Wang BG (2010a) Cytoglobosins A–G, cytochalasans from a marine-derived endophytic fungus, Chaetomium globosum QEN-14. J Nat Prod 73(4):729–733

Cui CM, Li XM, Meng L, Li CS, Huang CG, Wang BG (2010b) 7-Nor-ergosterolide, a pentalactone-containing norsteroid and related steroids from the marine-derived endophytic Aspergillus ochraceus EN-31. J Nat Prod 73(11):1780–1784

Deshmukh SK, Prakash V, Ranjan N (2018) Marine fungi: a source of potential anticancer compounds. Front Microbiol 8:2536

Devadatha B, Sarma VV, Jeewon R, Jones EG (2018) Morosphaeria muthupetensis sp. nov. (Morosphaeriaceae) from India: morphological characterization and multigene phylogenetic inference. Bot Mar 61(4):395–405

Dewan MM, Sivasithamparam K (1989) Growth promotion of rotation crop species by a sterile fungus from wheat and effect of soil temperature and water potential on its suppression of take-all. Mycol Res 93(2):156–160

Di Camillo CG, Cerrano C, Romagnoli T, Calcinai B (2017) Living inside a sponge skeleton: the association of a sponge, a macroalga and a diatom. Symbiosis 71:185–198

Dobranic JK, Johnson JA, Alikhan QR (1995) Isolation of endophytic fungi from eastern larch (Larix laricina) leaves from Now Brunswick, Canada. Can J Microbial 41:194–198. https://doi.org/10.5897/AJMR11.163

Dorworth DE, Callan BE (1996) Manipulation of endophytic fungi to promote their utility as vegetation biocontrol agents. In: Endophytic fungi in grasses and woody plants: systematics, ecology, and evolution. APS Press, pp 209–216

Dreyfuss MM, Chapela IH (1994) Potential of fungi in the discovery of novel, low-molecular weight pharmaceuticals. In: Discovery of novel natural products with therapeutic potential. Butterworth-Heinemann, pp 49–80

El-Gendy MMAA, Al-Zahrani SHM, El-Bondkly AMA (2017) Construction of potent recombinant strain through intergeneric protoplast fusion in endophytic fungi for anticancerous enzymes production using rice straw. Appl Biochem Biotechnol 183:30–50

Espinosa-Garcia FJ, Langenheim JH (1990) The endophytic fungal community in leaves of a coastal redwood population diversity and spatial patterns. New Phytol 116(1):89–97

Felício RD, Pavão GB, Oliveira ALLD, Erbert C, Conti R, Pupo MT et al (2015) Antibacterial, antifungal and cytotoxic activities exhibited by endophytic fungi from the Brazilian marine red alga Bostrychia tenella (Ceramiales). Rev Bras 25:641–650

Feng ZW, Lv MM, Li XS, Zhang L, Liu CX, Guo ZY et al (2016) Penicitroamide, an antimicrobial metabolite with high carbonylization from the endophytic fungus Penicillium sp. (No. 24). Molecules 21(11):1438

Fisher PJ, Anson AE, Petrini O (1984a) Antibiotic activity of some endophytic fungi from ericaceous plants. Bot Helv 94(2):249–253

Fisher PJ, Anson AE, Petrini O (1984b) Novel antibiotic activity of an endophytic Cryptosporiopsis sp. isolated from Vaccinium myrtillus. Trans Br Mycol Soc 83(1):145–148

Flewelling AJ, Currie J, Gray CA, Johnson JA (2015) Endophytes from marine macroalgae: promising sources of novel natural products. Curr Sci 109:88–111

Gao LW, Zhang P (2023) An update on chemistry and bioactivities of secondary metabolites from the marine algal-derived endophytic fungi. Phytochem Rev 22:587–614

Gao SS, Li XM, Du FY, Li CS, Proksch P, Wang BG (2010) Secondary metabolites from a marine-derived endophytic fungus Penicillium chrysogenum QEN-24S. Mar Drugs 9(1):59–70

Giddings LA, Newman DJ, Giddings LA, Newman DJ (2015) Bioactive compounds from marine extremophiles. Springer International Publishing, pp 1–124

Gouda S, Das G, Sen SK, Shin HS, Patra JK (2016) Endophytes: a treasure house of bioactive compounds of medicinal importance. Front Microbiol 7:1538

Guo ZY, Tan MH, Liu CX, Lv MM, Deng ZS, Cao F et al (2018) Aspergoterpenins A–D: four new antimicrobial bisabolane sesquiterpenoid derivatives from an endophytic fungus Aspergillus versicolor. Molecules 23(6):1291

Handayani D, Ananda N, Artasasta MA, Ruslan R, Fadriyanti O, Tallei TE (2019) Antimicrobial activity screening of endophytic fungi extracts isolated from brown algae Padina sp. J Appl Pharm Sci 9(3):009–013

Hu X, Li W, Yuan M, Li C, Liu S, Jiang C et al (2016) Homoharringtonine production by endophytic fungus isolated from Cephalotaxus hainanensis Li. World J Microbiol Biotechnol 32:1–9

Johnson JA, Whitney NJ (1992) Isolation of fungal endophytes from black spruce (Picea mariana) dormant buds and needles from New Brunswick, Canada. Can J Bot 70(9):1754–1757

Jones EG, Stanley SJ, Pinruan U (2008) Marine endophyte sources of new chemical natural products: a review. Bot Mar. https://doi.org/10.1515/BOT.2008.028

Kalaiselvam M (2015) Marine fungal diversity and bioprospecting. In: Springer handbook of marine biotechnology. Springer, pp 13–25

Kamat S, Kumari M, Sajna KV, Jayabaskaran C (2020a) Endophytic fungus, Chaetomium globosum, associated with marine green alga, a new source of Chrysin. Sci Rep 10(1):18726

Kamat S, Kumari M, Taritla S, Jayabaskaran C (2020b) Endophytic fungi of marine alga from Konkan coast, India—a rich source of bioactive material. Front Mar Sci 7:31

Kjer J, Debbab A, Aly AH, Proksch P (2010) Methods for isolation of marine-derived endophytic fungi and their bioactive secondary products. Nat Protoc 5(3):479–490

Kumar KA, Gousia SK, Latha JNL (2015) Evaluation of biological activity of secondary metabolites of Neurospora crassa from Machilipatnam Sea Water. Res J Microbiol 10(8):377

Lauritano C, Andersen JH, Hansen E, Albrigtsen M, Escalera L, Esposito F et al (2016) Bioactivity screening of microalgae for antioxidant, anti-inflammatory, anticancer, anti-diabetes, and antibacterial activities. Front Mar Sci 3:68

Li JY, Sidhu RS, Bollon ART, Strobel GA (1998) Stimulation of taxol production in liquid cultures of Pestalotiopsis microspora. Mycol Res 102(4):461–464

Li X, Li XM, Zhang P, Wang BG (2015) A new phenolic enamide and a new meroterpenoid from marine alga-derived endophytic fungus Penicillium oxalicum EN-290. J Asian Nat Prod Res 17(12):1204–1212

Liu AR, Chen SC, Lin XM, Wu SY, Xu T, Cai FM, Jeewon R (2010) Endophytic Pestalotiopsis species associated with plants of palmae, Rhizophoraceae, Planchonellae and Podocarpaceae in Hainan, China. Afr J Microbiol Res 4(24):2661–2669

Lu C, Shen Y (2004) Harnessing the potential of chemical defenses from antimicrobial activities. BioEssays 26(7):808–813

Lu X, He J, Wu Y, Du N, Li X, Ju J et al (2020) Isolation and characterization of new anti-inflammatory and antioxidant components from deep marine-derived fungus Myrothecium sp. Bzo-l062. Mar Drugs 18(12):597

Manivasagan P, Venkatesan J, Sivakumar K, Kim SK (2013) RETRACTED: Marine actinobacterial metabolites: current status and future perspectives. Microbiol Res 168(6):311–332

Marris E (2006) Marine natural products: drugs from the deep. Nature 443(7114):904–906

Medina RP, Araujo AR, Andersen RJ, Soares MA, Silva FDA, Silva DH (2019a) Aromatic compounds produced by endophytic fungi isolated from red alga Asparagopsis taxiformis-Falkenbergia stage. Nat Prod Res 33(3):443–446

Medina RP, Araujo AR, Batista JM Jr, Cardoso CL, Seidl C, Vilela AF et al (2019b) Botryane terpenoids produced by Nemania bipapillata, an endophytic fungus isolated from red alga Asparagopsis taxiformis-Falkenbergia stage. Sci Rep 9(1):12318

Monaghan RL, Polishook JD, Pecore VJ, Bills GF, Nallin-Omstead M, Streicher SL (1995) Discovery of novel secondary metabolites from fungi—is it really a random walk through a random forest? Can J Bot 73(S1):925–931

Nguyen MV, Han JW, Kim H, Choi GJ (2022) Phenyl ethers from the marine-derived fungus Aspergillus tabacinus and their antimicrobial activity against plant pathogenic fungi and bacteria. ACS Omega 7(37):33273–33279

Parthasarathy R, Chandrika M, Rao HY, Kamalraj S, Jayabaskaran C, Pugazhendhi A (2020) Molecular profiling of marine endophytic fungi from green algae: assessment of antibacterial and anticancer activities. Process Biochem 96:11–20

Petrini O (1986) Taxonomy of endophytic fungi of aerial plant tissues. In: Microbiology of the phyllosphere. Cambridge University Press, pp 175–187
Petrini O (1991) Fungal endophytes of tree leaves. In: Microbial ecology of leaves. New York, Springer New York, pp 179–197
Petrini O, Carroll G (1981) Endophytic fungi in foliage of some Cupressaceae in Oregon. Can J Bot 59(5):629–636
Petrini O, Fisher PJ (1988) A comparative study of fungal endophytes in xylem and whole stem of Pinus sylvestris and Fagus sylvatica. Trans Br Mycol Soc 91(2):233–238
Petrini LE, Petrini O, Laflamme G (1989) Recovery of endophytes of Abies balsamea from needles and galls of Paradiplosis tumifex. Phytoprotection 70(2):97–103
Petrini O, Sieber TN, Toti L, Viret O (1993) Ecology, metabolite production, and substrate utilization in endophytic fungi. Nat Toxins 1(3):185–196
Pugh GJF (1973) Saprophytic fungi and seeds. In: Heydecker W (ed) Seed ecology. Butterworths, London, UK, pp 337–345
Rajulu MG, Rajamani T, Murali TS, Suryanarayanan TS, Minj D (2022) The fungal endobiome of seaweeds of the Andaman Islands, India. Curr Sci 123(12):1508
Rampadarath S, Puchooa D, Jeewon R, Bandhoa K (2018) Diversity, seasonal variation and antibacterial activity of endophytic fungi associated with the genus Jatropha in Mauritius. J Biotechnol Biomater 8(1):1–8
Ramya AK, Sethumadhavan K, Devika R (2023) Biodiversity of endophytic fungi from marine algae and its phylogenetic studies. J Survey Fish Sci 10:2260–2268
Sahoo S, Subban K, Chelliah J (2021) Diversity of marine macro-algicolous endophytic fungi and cytotoxic potential of Biscogniauxia petrensis metabolites against cancer cell lines. Front Microbiol 12:650177
Sampson K (1935) The presence and absence of an endophytic fungus in Lolium temulentum and L. perenne. Trans Br Mycol Soc 19(4):337–343
Samuel P, Prince L, Prabakaran P (2011) Antibacterial activity of marine derived fungi collected from South East Coast of Tamilnadu, India. J Microbiol Biotechnol Res 1(4):86–94
Sarasan M, Puthumana J, Job N, Han J, Lee JS, Philip R (2017) Marine algicolous endophytic fungi—a promising drug resource of the era. J Microbiol Biotechnol 27(6):1039–1052
Saravanakumar K, Rajendren N, Kathiresan K, Wang MH (2020) Medicinal drug-related bioactive agents from marine fungi. In: Encyclopedia of marine biotechnology, vol 4, pp 2173–2190
Sieber TN, Sieber-Canavesi F, Petrini O, Ekramoddoullah AKM, Dorworth CE (1991) Characterization of Canadian and European Melanconium from some Alnus species by morphological, cultural, and biochemical studies. Can J Bot 69(10):2170–2176
Song F, Dong Y, Wei S, Zhang X, Zhang K, Xu X (2022) New antibacterial secondary metabolites from a marine-derived Talaromyces sp. strain BTBU20213036. Antibiotics 11(2):222
Stone JK, Polishook JD, White JF (2004) Endophytic fungi. Biodivers Fungi 241:270
Strobel G, Daisy B (2003) Bioprospecting for microbial endophytes and their natural products. Microbiol Mol Biol Rev 67(4):491–502
Sun HF, Li XM, Meng L, Cui CM, Gao SS, Li CS et al (2012) Asperolides A–C, tetranorlabdane diterpenoids from the marine alga-derived endophytic fungus Aspergillus wentii EN-48. J Nat Prod 75(2):148–152
Suryanarayanan TS, Rajagopal K (1997) Occurrence of fungal endophytes in forage grasses of South India. In: Neotyphodium/grass interactions. Springer US, Boston, MA, pp 81–83
Suryanarayanan TS, Rajagopal K (1998) Fungal endophytes in leaves of some south Indian tree species. In: Proceedings of the Asia–Pacific mycological conference on biodiversity and biotechnology, Hua-Hin, Thailand, pp 252–256
Suryanarayanan TS, Kumaresan V, Johnson JA (1998) Foliar fungal endophytes from two species of the mangrove Rhizophora. Can J Microbiol 44(10):1003–1006
Taritla S (2023) Isolation, structure elucidation and increasing anticancer efficacy of an anti-cancer secondary metabolite from a marine-derived endophytic fungus, Aspergillus species. Doctoral dissertation

Tarman K, Palm GJ, Porzel A, Merzweiler K, Arnold N, Wessjohann LA et al (2012) Helicascolide C, a new lactone from an Indonesian marine algicolous strain of Daldinia eschscholzii (Xylariaceae, Ascomycota). Phytochem Lett 5(1):83–86
Teixeira TR, Santos GSD, Armstrong L, Colepicolo P, Debonsi HM (2019) Antitumor potential of seaweed derived-endophytic fungi. Antibiotics 8(4):205
Tenguria RK, Khan FN, Quereshi S (2011) Endophytes—mines of pharmacological therapeutics. World J Sci Technol 1(5):127–149
Thirunavukkarasu N, Jahnes B, Broadstock A, Rajulu MG, Murali TS, Gopalan V, Suryanarayanan TS (2015) Screening marine-derived endophytic fungi for xylan-degrading enzymes. Curr Sci 109:112–120
Trindade M, Van Zyl LJ, Navarro-Fernández J, Abd Elrazak A (2015) Targeted metagenomics as a tool to tap into marine natural product diversity for the discovery and production of drug candidates. Front Microbiol 6:890
Uzma F, Mohan CD, Hashem A, Konappa NM, Rangappa S, Kamath PV et al (2018) Endophytic fungi—alternative sources of cytotoxic compounds: a review. Front Pharmacol 9:309
Vazquez-Rodriguez S, Joao Matos M, Borges F, Uriarte E, Santana L (2015) Bioactive coumarins from marine sources: origin, structural features and pharmacological properties. Curr Top Med Chem 15(17):1755–1766
Vinit K, Doilom M, Wanasinghe DN, Bhat DJ, Brahmanage RS, Jeewon R et al (2018) Phylogenetic placement of Akanthomyces muscarius, a new endophyte record from Nypa fruticans in Thailand. Curr Res Environ Appl Mycol 8(3):404–417
Wahl M, Goecke F, Labes A, Dobretsov S, Weinberger F (2012) The second skin: ecological role of epibiotic biofilms on marine organisms. Front Microbiol 3:292
Wang WX, Lei X, Yang YL, Li ZH, Ai HL, Li J et al (2019a) Xylarichalasin A, a halogenated hexacyclic cytochalasan from the fungus Xylaria cf. curta. Org Lett 21(17):6957–6960
Wang HH, Li G, Qiao YN, Sun Y, Peng XP, Lou HX (2019b) Chamiside A, a cytochalasan with a tricyclic core skeleton from the endophytic fungus Chaetomium nigricolor F5. Org Lett 21(9):3319–3322
Wang R, Zhang Q, Ju M, Yan S, Zhang Q, Gu P (2022) The endophytic fungi diversity, community structure, and ecological function prediction of Sophora alopecuroides in Ningxia, China. Microorganisms 10(11):2099
White JF Jr (1988) Endophyte-host associations in forage grasses. XI. A proposal concerning origin and evolution. Mycologia 80:442–446
Wijesekara I, Zhang C, Van Ta Q, Vo TS, Li YX, Kim SK (2014) Physcion from marine-derived fungus Microsporum sp. induces apoptosis in human cervical carcinoma HeLa cells. Microbiol Res 169(4):255–261
Xu K, Wei XL, Xue L, Zhang ZF, Zhang P (2020) Antimicrobial meroterpenoids and erythritol derivatives isolated from the marine-algal-derived endophytic fungus Penicillium chrysogenum XNM-12. Mar Drugs 18(11):578
Yamada T, Kogure H, Kataoka M, Kikuchi T, Hirano T (2020) Halosmysin A, a novel 14-membered macrodiolide isolated from the marine-algae-derived fungus Halosphaeriaceae sp. Mar Drugs 18(6):320
Zainee NFA, Ibrahim N, Ismail A, Taip ME, Ismail A (2019) Diversity of endophytic fungi isolated from the leaves, stipes and receptacles of the Malaysian marine brown algae, Sargassum (Phaeophyta, Sargassaceae). Malay Nat J 71(2):139–148
Zainuddin N, Alias SA, Lee CW, Ebel R, Othman NA, Mukhtar MR, Awang K (2010) Antimicrobial activities of marine fungi from Malaysia. Bot Mar. https://doi.org/10.1515/bot.2010.067
Zeilinger S, Gruber S, Bansal R, Mukherjee PK (2016) Secondary metabolism in Trichoderma—chemistry meets genomics. Fungal Biol Rev 30(2):74–90
Zhang Y, Mu J, Feng Y, Kang Y, Zhang J, Gu PJ et al (2009) Broad-spectrum antimicrobial epiphytic and endophytic fungi from marine organisms: isolation, bioassay and taxonomy. Mar Drugs 7(2):97–112

Zhang Z, Miao L, Lv C, Sun H, Wei S, Wang B et al (2013) Wentilactone B induces G2/M phase arrest and apoptosis via the Ras/Raf/MAPK signaling pathway in human hepatoma SMMC-7721 cells. Cell Death Dis 4(6):e657–e657

Zhang P, Li X, Wang BG (2016) Secondary metabolites from the marine algal-derived endophytic fungi: chemical diversity and biological activity. Planta Med 82(09/10):832–842

Zhao J, Zhou L, Wang J, Shan T, Zhong L, Liu X, Gao X (2010) Endophytic fungi for producing bioactive compounds originally from their host plants. In: Current research, technology and education topics in applied microbiology and microbial biotechnology, vol 1. Formatex Research Center, pp 567–576

Zhao DL, Wang HS, Gao LW, Zhang P (2022) Tennessenoid A, an unprecedented steroid—sorbicillinoid adduct from the marine-derived endophyte of Aspergillus sp. strain 1022LEF. Front Mar Sci 9:923128

Fungal Endophytes: A Potent Bioactive Metabolites Reservoir

7

Nahdhoit Ahamada Rachid and Nihal Doğruöz Güngör

Abstract

Scientists emphasize the enormous potential of new sources of bioactive compounds to combat global diseases. One of the most important of these sources is the endophytic fungi. They are defined as fungi that asymptomatically colonize intracellular spaces within plant tissues. The biodiversity of these fungi, which have a symbiotic relationship with the plant, is impressive. In addition, the bioactive substances produced by these endophytic fungi are shaped by the characteristics of the plant with which they are in contact. Fungal endophytes produce beneficial and invaluable bioactive metabolic compounds with anticancer, antimicrobial, anti-inflammatory, insecticidal, antioxidant, etc. properties. Some of these bioactive compounds are terpenoids, phenolics, flavonoids, macrolides, alkaloids, steroids, polyketides, auxin, quinone, and indole acetic acid.

Endophytic fungi have also been found to be used in agriculture and industries other than medicine. The potential of natural bioactive compounds obtained from endophytic fungi to be produced in large quantities using biotechnological processes and to offer more economical alternatives are also their positive aspects. Enzymes, biocontrol, and plant growth-promoting substances produced by endophytic fungi have shown that they have biotechnology potential and have made these organisms more popular. This chapter resumes the potential use of endophytic fungal bioactive compounds for human welfare by highlighting the richness of these organisms in terms of natural biological products.

N. Ahamada Rachid
Institute of Graduate Studies in Sciences, Istanbul University, Istanbul, Turkey

N. Doğruöz Güngör (✉)
Department of Biology, Fundamental and Industrial Microbiology, Faculty of Science, Istanbul University, Istanbul, Turkey
e-mail: ndogruoz@istanbul.edu.tr

K. A. Abd-Elsalam, S. F. AbuQamar (eds.), *Fungal Endophytes Volume I*,
https://doi.org/10.1007/978-981-97-7312-1_7

Keywords

Fungal metabolites · Endophytic fungi · Bioactive compound · Natural products

7.1 Introduction

Plants, like other living organisms, establish close relationships with microorganisms. The term endophyte, first used by De Bary (1866), is defined as "an organism such as bacteria, fungi and similar groups which live in plant tissues without causing any apparent harm to the plant host" (De Bary 1866). Endophytes are found in almost all plant species, including plants in different ecosystems such as grasses, trees, marine plants, mosses, and algae (Mishra et al. 2014). The organisms found in the host plant are named bacterial endophytes, fungal endophytes, algal endophytes, etc. They provide many advantages to their hosts, such as increasing nutrient absorption, protecting against biotic and abiotic stresses, promoting their growth, and increasing resistance to some diseases (Nisa et al. 2015; Kusam et al. 2019).

Bioactive metabolites, these natural compounds exhibiting positive effects on human health, are produced by various organisms resulting from their metabolic activities. These metabolites are not directly related to the growth and development of the producing organism. They may contain compounds with biological activities such as antibiotic, antiviral, antiprotozoal, antitumor, antifungal, anti-inflammatory, antioxidant, and immunosuppressant (Karthikeyan et al. 2022; Caruso et al. 2022). Bioactive substances have been researched, for a long time, from traditional microbial sources such as soil and water. However, the decrease in new bioactive substance discoveries has directed natural product researchers to alternative sources. For the discovery of new compounds, researchers are searching for alternative sources of microorganisms in a wide spectrum, from different extreme environments to different organisms (Rachid and Güngör 2023; Aydoğdu et al. 2023). Thereby, we may have a chance to fight for both the discovery of new drugs and against the resistance developed against the used drugs (Sharma et al. 2019). Endophytes are one of these unique niches.

It is estimated that there are approximately 2.2–3.8 million species of fungi in the world, and the number of known fungal species is 148,000 (Zhou and May 2022). Fungi are an important group in terms of biodiversity and play a role in maintaining ecosystem functioning. In this context, it is of great importance to continue scientific research and fill the gaps in this group. Fungi inhabit a wide variety of ecological niches, including swamps, hot springs, caves, deserts, and aquatic environments such as freshwater and seas. They can form symbiotic associations with animals and plants (Blackwell 2011).

Compounds with a wide range of applications, including the first discovered antibiotic, have been discovered from the rich secondary metabolites of fungi. Fungal endophytes are an important resource that has been little studied and has great potential for the discovery of new bioactive compounds (Gakuubi et al. 2021). Throughout evolution, fungal endophytes have developed a wide range of interactions with their host plants (Gupta and Shukla 2020). The richness of this fungal diversity, the fact that every plant examined so far contains at least one fungal endophyte and even the fact that some plants contain hundreds of fungal strains (Faeth and Fagan 2002) have been determined in previous studies. In addition, certain endophytic fungi, considered promising sources of novel compounds, have been found to produce a wide range of metabolites similar to those produced by host plants, including alkaloids, saponins, steroids flavonoids, phenolic acids, peptides, terpenes, and many others (Singh et al. 2021).

This chapter underlines the potential of endophytic fungi as a potential source of new bioactive metabolites and their important role in the fight against human diseases. It also highlights the biotechnological applications of endophytic fungi in industry and agriculture.

7.2 Fungal Endophytes

Endophytic fungi enhance crop yield and growth by effectively tolerating biotic and abiotic stresses, including high temperature, salinity, and drought (Rodriguez et al. 2009). These fungi exhibit wide adaptability and can colonize various plant tissues such as leaves, roots, and stems, establishing a symbiotic relationship with the host (Ferdous et al. 2019). Importantly, endophytic fungi do not cause visible symptoms or harm to the host species. In the fungus-plant relationship, the fungus utilizes the carbon fixed by the associated plant through photosynthesis. In return, it supplies the plant with vital nutrients from the soil rhizosphere (Wu et al. 2012). This symbiotic relationship plays a crucial role in ecosystem stability. The structure of endophyte fungal groups such as richness, species diversity, and abundance are regulated by various abiotic and biotic factors such as plant types, plant-microbe interactions, biogeography, and soil conditions (Ferdous et al. 2019; Hashem et al. 2023). In addition, endophytic fungi are an important part of the biodiversity as they affect plant growth under adverse conditions. Fungal endophytes have been identified in various taxonomic plant groups living in various habitats worldwide (Kouipou Toghueo and Boyom 2019) (Fig. 7.1).

7.3 Fungal Endophyte-Derived Bioactive Metabolites

Endophytic fungi are a relatively unexplored group that can synthesize bioactive compounds, which can be used effectively against numerous pathogenic infections. They are a reliable source of biologically and chemically active compounds and hold strong potential for new drug discoveries.

Endophytes are exposed to biotic and abiotic stresses within the host plant and interact with both each other and the host plant, resulting in the production of various secondary metabolites (Wang et al. 2023c; Verma et al. 2022). Consequently, these metabolites enhance the plant's resistance to abiotic stress, disease attacks, insects, and mammalian herbivores (Cagigal 2016). Research on fungal endophytes

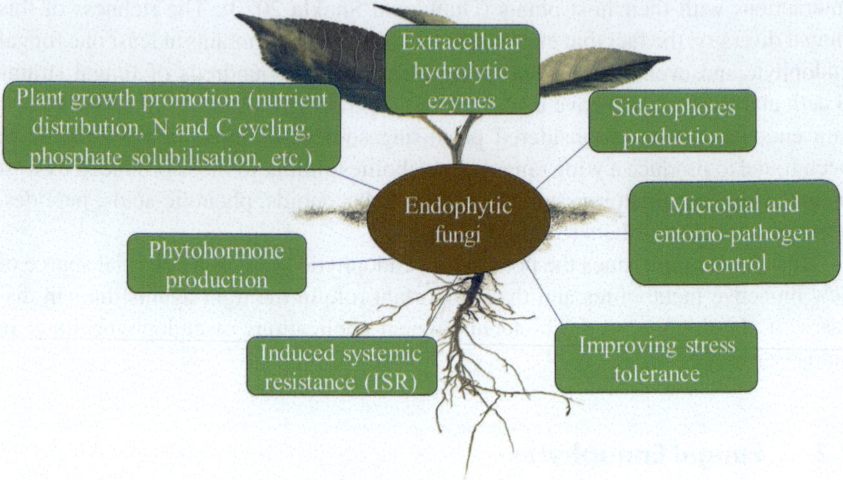

Fig. 7.1 Plant benefits of endophytic fungi

reveals that they produce a diverse range of bioactive natural compounds such as terpenoids, phenylpropanoids, isocoumarins, benzopyranones, lignans, alkaloids, steroids, quinones, xanthones, phenols, cytochalasins, furanoids, flavonoids, perylene, steroids, enniatines, and lactones (Nisa et al. 2015). These endophytic fungi containing abundant bioactive metabolites have demonstrated various biological activities, including antifungal, anticancer, antiviral, antioxidant, immunosuppressive, anti-inflammatory, antiparasitic, and antibacterial properties (Gandhi et al. 2015). Besides their medicinal applications, these compounds can also be utilized in agriculture, cosmetics, and food industries.

7.3.1 Endophytic Fungi as Antibacterial Sources

As previously shown, endophytes can act as a protector of the plant against pathogenic organisms. This property pushed scientists to research these valuable natural antimicrobial compounds. Some studies have shown that endophytic organisms can use some host-produced compounds and transform them for other bioactive properties (Ludwig-Müller 2015; Rustamova et al. 2020). Moreover, large quantities of antimicrobial and other biological compounds are found to be produced by endophytic fungi from medicinal plants (Wang et al. 2007; Ababutain et al. 2021). Search for these compounds from these endophytes is encouraged since it is thought that the medicinal features of these hosts can be derived from these endophytic organisms (Kusari et al. 2013).

Results of studies realized in this thematic, antibacterial compound from endophytic fungi isolated from medicinal plants encourage the multiplication of research in this field for using these endophytes as alternative sources of pharmaceutical compounds. They can help to avoid the long period and area required for plant growth and the high cost of the extraction and purification of antimicrobial compounds from plants. The medicinal properties of the plant species *Artemisia sieberi* are known from traditional medicine. Ababutain et al. (2021) showed that every fungal isolated from the lives of this plant has an antibacterial effect with the highest activity observed against Gram-positive bacteria. Isolates of *P. chrysogenum* had the highest antimicrobial activity against all the tested pathogens.

On the other hand, extracellular culture and extracts of the *Chaetomium globosum*, a fungus inhabiting the medicinal plant *Moringa oleifera*, were tested for their antibacterial activities. Some of the pathogens resisting the extracellular culture were sensitive to the organic extracts. The highest activity from the extracellular culture was observed against *K. pneumoniae* 1, *S. aureus* and Methicillin-Resistant and Biofilm-Producing *S. aureus* (MRSA), whereas the highest MIC value of chloroform extract was shown against *S. flexneri*, *K. pneumoniae* 2, and *S. typhimurium* 1. The authors noted the broad spectrum of the chloroform extract of this fungus. However, they have also shown the biosafety of the extracellular culture (Kaur and Arora 2020).

Infection is one of the major problems in agriculture (Liu et al. 2020; Li-Bin et al. 2021). In this context, research for alternative bactericidal from natural resources can increase the year yield of the agricultural sector. Studies of endophytic fungi for this aim have been multiplied with promising results. The study run by Chen et al. (2022b) discovered isobenzofuranones and isocumarins compounds from the kiwi endophytic fungus *Paraphaeosphaeria sporulosa*. These compounds exhibit antibacterial activity against the kiwi pathogen bacterium *Pseudomonas syringae* pv. *Actinidiae* (Psa). The identification of these compounds will help to control the canker disease caused by the previously cited pathogen, which hinders the development of the kiwifruit industry. On the other hand, an eremophilane sesquiterpenoid (compound **4**) and the cholesterol analogous (22*E*)-3β-hydroxy-26,27-bisnorcholesta-5,22-dien-24-one (compound **20**) isolated from the endophytic fungus *Septoria rudbeckiae* shown antibacterial activities against Psa with the respective minimal inhibitory concentration (MIC) values of 6.25 and 12.5 μM (Li-Bin et al. 2021).

7.3.2 Endophytic Fungi as Antifungal Sources

Fungi are one of the potential sources of bioactive compounds. However, the pathogenicity of these organisms is also worrying. They can cause disease in plants, humans, and animals. The important damages caused by pathogenic fungi include the decrease of agricultural yields for the plant fungal pathogens, which can directly impact different industries such as food industries (like the production of juice and fruit extracts) (Doehlemann et al. 2017). On the other hand, some *Candida* (yeast) species cause fungal infections (candidiasis) in humans (Willems et al. 2020). Like the other antimicrobials, the drug resistance mechanisms and alternative sources of antifungal are emerging and need to be studied.

Trichoderma species are endophytic fungi known for their multiple functions on the plants. Some of them have antifungal activity against plant pathogens. The

T. spirale T76-1 showed antimicrobial activity against two lettuce leaf spot fungi: *Corynespora cassiicola* and *Curvularia aeria* (Baiyee et al. 2019). The root rot in cocoyam and the stem rot in carnation are respectively caused by the fungal species *Pythium myriotylum* and *Sclerotium*, which are both controlled by the endophytic *Trichoderma asperellum* (Mbarga et al. 2012; Vinodkumar et al. 2017). Wonglom et al. (2020) demonstrated that the antifungal activity of *T. asperellum* T1 is mediated by the volatile organic compounds emitted by this endophyte. The use of *Trichoderma* species as biological controls for the protection of plants and enhancement of crop yield is previously well known (Vinodkumar et al. 2017; Vinale et al. 2008).

Fungal infections cause complications in immuno-compromised patients due to the difficulties of therapeutic decisions. Human mucormycoses are destructive invasive diseases caused by different fungi, including *Syncephalastrum*, *Rhizopus*, and *Mucor*. The control of these diseases was hard, especially in immuno-compromised patients. Recent studies run for fungal infection therapeutic solutions are directed at natural resources of fungicides, and the results obtained from endophytic fungi are promising (Pilmis et al. 2018; Talukdar et al. 2022; Hashem et al. 2022). Hashem et al. (2022) studied the antifungal activity of *Aspergillus terreus* isolated from *Moringa oleifera* against some mucormycoses-causing fungi. They observed the antifungal activity of the ethyl acetate crude extract with MIC values of 0.3125 mg/mL toward *Mucor racemosus*, 1.25 mg/mL against *Rhizopus orizae* and 2.5 mg/mL toward *Syncephalastrum racemosum*. In addition, the damage to the pathogen's structures, like their mycelia and cell membranes, and the increase of the membrane lipid peroxidation, which shows the destruction of the membranes, confirmed the fungicide function of the ethyl acetate crude extract of the endophytic *A. terreus*.

Plants of the family Cupressacae are known for their resistance against plant pathogens. It is thought that this resistance comes from the antimicrobial activities of the endophytes that they are hosting (Soltani et al. 2016). Erfandoust et al. (2020) studied the anti-Asperilli activities of some endophytic fungi inhabiting the Iranian Cupressacae plant family. Even if *Aspergillus* species are used in biotechnology due to their high metabolic rates, this fungal genus has pathogenic effects that can cause strong health problems to humans, animals, and plants. Their spores are reported to have allergic reactions such as asthma in humans and animals (Soltani 2016; Erfandoust et al. 2020). Erfandoust et al. (2020) observed antifungal activities of the cultures, volatile organic compounds (VOCs), and extracellular seconder metabolites (SMs) of studied endophytes against *A. niger* and *A. fumigatus*. The highest activities from the culture and VOCs were noted from the *T. koningii* CSE_{32} and *T.atroviride* JCE_{33} against the two tested human pathogens.

7.3.3 Antiviral Activity of Endophytic Fungi

A wide range of secondary metabolites from fungal endophytes show antiviral activities against human, animal, and plant viruses (Lacerda et al. 2022; Carrie et al. 2023). Viruses are particles that use the cellular machinery of host cells for reproduction. They have different structures and, thus, various mechanisms of infection. Some of the enveloped viruses are sensitive to chemical solvents. These solvents dissolve the lipids on the envelope, limiting or reducing their infection power. However, non-enveloped viruses and other viruses that are deprived of this lipid structure express a resistance toward these solvents. In this context, research for other compounds targeting other viral structures is imperative. In addition, drug resistance, the high mutation rate of viruses, and the side effects of antiretroviral treatments led to research of new and alternative natural compounds (Lacerda et al. 2022).

In the study run by Nzimande et al. (2022), three crude extracts of *Alternaria alternata*, an endophytic fungus from *Sclerocary birrea* and *Hyposis* plants, showed anti-HIV-1 activity, and the virus reproduction inhibition profile was maintained on PBMCs and CD4+ T cells. The GC-MS analysis of the purified extracts showed various compounds, including cyclotrisiloxane octamethyl, Propaninitrile, Coumarin, 3,4-dihydro-4,5,7-trimethyl-2-chromanone, and Pyrrolol[1,2-a]pyrazine-1,4-dione, hexahydro-3-(2-methyl propyl). A new 14-nordrimane-type sesquiterpenoid was characterized by Liu et al. (2019) for the first time from the endophytic fungus *Phoma* sp. isolated from the *Aconitum vilmorinianum* plant. This compound was observed with antiviral activity against influenza A virus H1N1 with IC_{50} value of 2.9 ± 0.64 μg/mL.

7.3.4 Antioxidant Activities of Endophytic Fungi

Antioxidants are compounds, including vitamins, minerals, and enzymes, that protect organisms against damage caused by reactive oxygen species (ROS) and oxygen-derived free radicals. Naturally occurring antioxidants such as superoxide dismutase, glutathione, and catalase, among others, help reduce the harmful effects of ROS and free radicals and strive to maintain the body's redox balance. However, when the production of ROS and free radicals exceeds the capacity of the body's natural antioxidant system, oxidative damage can occur, leading to various pathological effects such as diabetes, cancer, atherosclerosis, arthritis, DNA damage, and cellular degeneration (Pizzino et al. 2017). To alleviate the negative effects of these on the human body, exogenous antioxidants are needed. In recent years, there has been a significant increase in the search for natural antioxidants due to the growing demand for dietary supplements that do not contain synthetic antioxidants (Lourenço et al. 2019). Various antioxidant compounds have been obtained from endophytic fungi isolated from plants found in different regions of the world, making endophytic fungi among the most popular organisms for obtaining new antioxidant substances.

Rai et al. (2023) determined that the ethyl acetate extract of *Colletotrichum gloeosporioides* associated with *Oroxylum indicum* (L.) Kurz exhibited antioxidant and cytotoxic activity, and these bioactive compounds were produced by the fungus depending on the host. Elghaffar et al. (2022) identified that the ethyl acetate crude extract of endophytic *Alternaria alternata* from *Ziziphus spina-christi* plant contained terpenoids, tannins, glycosides, flavonoids, alkaloids, and phenols. Additionally, it was found that the extract showed no toxicity in the Vero normal cell line and possessed potential antioxidant activity. Tang et al. (2021) reported that the *Penicillium oxalicum* isolated from *Ligusticum chuanxiong* Hort could produce several high-value polyphenols, such as ferulic acid (1.17 μmol/g), alternariol (12.64 μmol/g), and hesperidin (36.06 μmol/g), making it a potential source for the pharmaceutical industry.

7.3.5 Anticancer Activity of Endophytic Fungi

Cancer is a global disease characterized by uncontrolled cell proliferation and the spread of cells to new tissues. According to the World Health Organization (WHO), there were 19.3 million new cancer cases worldwide in 2020, and this number is projected to exceed 30 million by 2040. Cancer treatment typically involves a combination of surgery, chemotherapy, and radiotherapy. However, these methods often come with significant side effects (Pierce et al. 2017; Nurgali et al. 2018). Additionally, there is a risk of cancer recurrence or the development of treatment resistance by the cells. Researchers are actively seeking new natural resources, with endophytic fungi being recognized as a significant source.

The presence of camptothecin, a potential anticancer compound, was identified through phytochemical screening of extracts from endophytic fungi *Fusarium* and *Alternaria alternata* species isolated from *Mappia foetida*. These extracts exhibited activity against non-small cell lung carcinoma (H1975), breast carcinoma (MCF-7), cervical carcinoma (HeLa), and hepatocellular carcinoma cell line (Hep G2) (Ravi et al. 2022). The GC-MS analysis of endophytic fungi *Chlonostachys* sp., *Fusarium solani*, and *Fusarium oxysporum* cultures isolated from *C. decandra* revealed 80 compounds. The study by Munshi et al. (2022) demonstrated the anticancer potential of these compounds against breast cancer cells through the estrogen receptor pathway. A study conducted by Duan et al. (2019) found that out of 154 fungal endophytes isolated from the *Distylium chinense* plant, 27 exhibited anticancer activity against the human papillary thyroid carcinoma cell line (IHH4) and human pancreatic adenocarcinoma cell line (CFPAC-1). Further exploration of endophytic fungi and their metabolites may provide potential solutions in addressing the increasing global burden of cancer (El-Ghonemy et al. 2023).

7.3.6 Endophytic Fungi as Insecticidal Sources

One important application of bioactive secondary metabolites produced by fungi is their insecticidal activity. Blundell et al. (2020) discovered that endophytic fungi in the rhizosphere of agricultural plants synthesize salicylic acid, which reduces the edibility of host plants and exhibits antagonistic effects on herbivorous insects, thereby protecting plant health. In a study conducted by Kaur et al. (2019), *Alternaria destruens* EGSimmons, isolated from *Calotropis gigantea* L. plant, exhibited significant α-glycosidase inhibitory activity and was investigated for its potential as an insecticide against *Spodoptera litura*. The study revealed that α-glucosidase inhibitors resulted in increased mortality, delayed development time, and induced deformities in adults. Hou (2020) found that asperpolyketide A, a newly synthesized polyketide derived from the endophytic fungus *Aspergillus oryzae*, demonstrated significant insecticidal properties.

Trichoderma hamatum, an endophyte isolated from the roots of *Brassica oleracea* var. acephala (Kale), produces insecticidal secondary metabolites that infect *Spodoptera littoralis* (Cotton leafworm) larvae through natural openings. The insecticidal metabolite responsible for this activity is rhizoferrin (Lana et al. 2023). Similarly, Bai et al. (2020) discovered that bioactive secondary metabolites produced by a *Penicillium* sp. JY246 derived from mangroves exhibited growth inhibition activities against newly hatched larvae of *Helicoverpa armigera* Hubner. The studies indicate that endophytic fungi have significant potential as a valuable source to produce environmentally friendly insecticides.

7.3.7 Anti-inflammatory Activities of Endophytic Fungus

Inflammation is a protective response that occurs after cell damage, with the goal of eliminating the cause of damage and safeguarding living tissues during the repair process (Karin and Clevers 2016). The immune system cells play a role in this response. Chronic inflammation is strongly linked to various diseases, such as cancer and atherosclerosis (Grivennikov et al. 2010; Chakrabarti et al. 2014). Endophytic fungi may contain a wide range of compounds that have the potential for use in the development of new drugs as alternatives to synthetic ones. Kim et al. (2023) found that Colletogloeopyrone A (1), monocillin II (5), and monocillin II glycoside (6), which were isolated from *Colletotrichum gloeosprioides* JS0419 obtained from the leaves of *Halofit Suaeda japonica*, demonstrated effective reduction in nitric oxide production without causing any cytotoxic effects.

A study showed that paraconion B, a compound newly identified from *Paraconiothyrium* sp. found in the *Artemisia selengensis* plant, effectively inhibited lipopolysaccharide-induced nitric oxide (NO) production in RAW 264.7 cells. The IC50 value for this inhibition was measured at 51.7 μM (Zeng et al. 2023). Another study by Lv et al. (2023) revealed that compound **6**, derived from metabolites of *Xylaria grammica* isolated from the Chinese medicinal plant *Sophora tonkinensis* Gagnep, demonstrated anti-inflammatory activity by inhibiting NO production in LPS-induced RAW 264.7 cells. The IC50 value for this inhibition was measured at 23.24 ± 1.3 μM.

7.3.8 Plant Growth-Promoting Features of Fungal Endophytes

The growing global population and the demand for reliable food sources require innovative improvements in plant growth and sustainable agricultural productivity.

The use of chemical fertilizers to increase crop yield has detrimental effects on soil fertility, water retention capacity, and the long-term viability of crop production. Additionally, these fertilizers have a negative impact on soil structure, ecosystem health, and all living organisms (Granada et al. 2018). Plants have evolved intricate defense mechanisms to withstand abiotic factors such as drought and biotic factors like nematodes and pathogens (Kour et al. 2020). Phytohormones are growth regulators that affect plant metabolism and play a significant role in defense mechanisms. They are produced by both plants and endophytic fungi, which obtain nutrients from plants and can also produce phytohormones and various metabolites (Gill et al. 2016). Fungal endophytes have been found to support plant growth through the synthesis of phytohormones, including gibberellins, cytokinins, and auxins.

They also enhance plant tolerance to environmental stresses, facilitate nutrient absorption, suppress plant pathogens, repel insects, and contribute to atmospheric nitrogen fixation (Omomowo and Babalola 2019; Adeleke and Babalola 2020; Fadiji and Babalola 2020). *Fusarium proliferatum*, isolated from the *Rhazya stretcha* plant, was tested for its ability to improve agronomic traits of sunflowers under drought stress. The analysis showed that this symbiotic relationship significantly enhanced the growth of the host and provided protection against drought stress induced by PEG (Seema et al. 2023). Asaf et al. (2023) demonstrated that the endophytic fungus *Curvularia lunata* SL1 strain established a symbiotic relationship with tomato plants exposed to Cadmium (Cd) stress. Under Cd stress, tomato seedling growth decreased by up to 42.8%, but after SL1 inoculation, the reduction was limited to 25%. SL1 inoculation was found to inhibit Cd uptake and translocation in plants while also increasing levels of abscisic acid, jasmonic acid, and ethylene during Cd stress (source). *Rhizopus oryzae* isolated from coriander roots was found to promote spinach growth through the production of indole-3-acetic acid (IAA) (Parvez et al. 2023a, b). *Aspergillus clavatus* Y2H0002, obtained from *Nymphoides peltate* plant, was discovered to enhance the growth of hots in freshwater by producing the plant growth hormone gibberellin (Nayak et al. 2020) (Table 7.1).

7.3.9 Enzymatic Activities of Endophytic Fungi and Their Applications

Enzymes are proteinaceous molecules that act as biological catalysts, specifically targeting and catalyzing biochemical reactions on specific substrates (Thapa et al. 2019). The global industrial enzyme market is projected to reach $8.7 billion by 2026, with a compound annual growth rate of 6.3% for 2021–2026 (BCC Research 2021). Enzymes can be synthesized from various biological sources, such as animals, plants, and microorganisms. The high growth rate and metabolic activities of microorganisms, along with their low development cost, diversity, distribution, and ease of genetic manipulation, make them favorable sources of industrial enzymes (Raveendran et al. 2018). Enzymes from microorganisms are renowned for their ability to function in extreme environmental conditions, such as low or high

Table 7.1 Biological compounds and applications areas of endophytic fungi

Endophytes	Plant hosts	Compounds	nature	Biological activities	Application areas	References
Trichoderma asperillum A-YMD-9-2	*Gracilaria verrucosa*	Hazardione (1) 3*S*-hydroxyharzianone (2) 4-Cadinen-11,12-diol (3) 4-Cadinen-11,13-diol (4)	Terpenoid Diterpenoid Sesquiterpenoids	Antibacterial	Pharmaceutical industry	Song et al. (2019)
Bipoloris sp.	Kiwi	Bipolarisorokin M4 Bipolarisorokin N5	Terpenoids	Antibacterial	Pharmaceutical industry	Yu et al. (2022)
Fusarium cicatricum	*Betula pendula* Roth	Betulin	Triterpenoids	Anticancer	Cosmetic and pharmaceutical industries	Nazari et al. (2023)
Penicillium citrinum						
Alternaria alternate		Betulinic acid		Antiviral		
Neofusicocum luteum	*Kigelia Africana* (Lam.)	C16-terpene dilactone	Terpene	Antibacterial	Pharmaceutical	Bodede et al. (2022)
Pharaphaespharia sp.	*Ginkgo biloba*	Paraphaone Paraphaterpene A, C, D, Aletrnariol methyl ether	Terpenoids	Antifungal	Biological control	Su et al. (2023)
		Paraphaterpene A (2), B (4), Paraconiothin D (3), Alternariol methyl ether (7)		Antifeedant		
Fusarium sporotrichioides	*Rauwolfia yunnanensis*	Compound **2**		Anticancer	Pharmaceutical	Wang et al. (2023b)
		Compounds **3** and **5**		Cytotoxic activity		

(continued)

Table 7.1 (continued)

Endophytes	Plant hosts	Compounds	nature	Biological activities	Application areas	References
Aplosporella javeedii	*Orychophragmus violaceus* (L.) O. E. Schul	Fusaproliferin (4)	Sesterterpenes	Cytotoxicity Antibacterial	Medicine, pharmaceutical industries	Gao et al. (2020)
		Mutolide (7)	Macrolide	Cytotoxicity		
				Antibacterial		
		(−)-Terpestacin (3)		Antibacterial		
		6,7,9,10-Tetrahydromutolide (6)	Macrolide	Antibacterial		
Aspergillus sp. 16-5c	*S. apetala*	DIKETOPIPERAZINE alkaloids aspergiamides A	Alkaloids	Anti-diabetic	Pharmaceutical	Ye et al. (2021)
Fusarium sambucinum TE-6L	*Nicotiana tabacum* L.	Pyrano[2,3-g] indole moieties, amoenamide C (1)	(PIA) Prenylated indole alkaloids	Antibacterial Antifungal	Pharmaceutical industries; biological control	Zhang et al. (2019)
		Sclerotiamide B (2)		Insecticidal	Biological control	
		Sclerotiamide (3)	Congeners	Antibacterial Antifungal	Pharmaceutical industries; biological control	
		Notoamide B (4)		Insecticidal	Biological control	
Nigrospora sphaerica	*Euphorbia hirta* L.	2,4-Ditert-butylphenol	Phenolic	Antioxidant	Pharmaceutical	Gautam et al. (2022)
Unidentified II	*Vanda cristata*	ND	ND	Auxin production Phosphate solubilization	Agriculture	Chand et al. (2020)
Agaricus bisporus		ND	ND			
Mycolepto discus		ND		Ammonia synthesis		

Endophytes	Plant hosts	Compounds	nature	Biological activities	Application areas	References
Acrocalymma sp.	*Sinomenium acutum*	Acrocalysterols B (2)	Steroids	Anticancer	Pharmaceutical	Yang et al. (2022)
Aspergillus nidulans	*Annelida Whitmania pigra Whitman*	(7Z,9Z,17Z)-,2α,3β-dihydroxypregna-7,9,17(20)-trien-18-al	Steroids	Anticancer	Pharmaceutical	Li et al. (2022)
Colletotrichum sp.	*Gannan navel orange*	Colletene A (1)	Sesquiterpeniods	ND	ND	Wang et al. (2023a)
		Compound **2** (3,4-dihydroxybenzyl alcohol (2))	Phenolic	Antibacterial	Pharmaceutical	
Alternaria sp. MG1	*Vitis quinquangularis*	Alternaritin B (2), Alternaritin C (3)	Polyketides	Anti-inflammation	Pharmaceutical	Tian et al. (2021)
Cytospora sp.	*Ceriops tagal*	Cytospomarin (2),	Polyketides	Antibacterial	Pharmaceutical	Wei et al. (2020)
Trichoderma afroharzianum	*Ficus elastica*	Compound **1**, compound **2**	Polyketides	Antifungal	Pharmaceutical	Ding et al. (2020)
Penicillium sp. BJR-P2	*Avicennia marinav* (Forsk.) Vierh	Penicillol B (2)	Polyketides	Anti-inflammation	Pharmaceutical	Chen et al. (2022a)
Didymella sp. SZMC 26648 strain	*Sophora flavescens*	IAA (indoleacetic acid)	Plant growth hormone	Plant growth hormone promoter	Agriculture	Turbat et al. (2020)
Penicillium sp.	*Artemisia annua*	Emodin (1) Physcion (2)	Quinone	Antimalarial	Pharmaceutical	Alhadrami et al. (2021)
Talaromyces sp.						
Aspergillus niger 58	*Terminalia catappa* L. (West Indian-Almond)	ND	ND	Antimalarial	Pharmaceutical	Kouipou Toghueo et al. (2021)
Paecilomyces sp. EJC01.1	*Schnella splendens*	Phomoxanthone A (1)	Xanthones	Anti-Leishmania	Pharmaceutical	Ramos et al. (2022)

temperatures, acidic or basic environments, and high pressure (Thapa et al. 2019). Endophytic fungi secrete secondary metabolites and enzymes to protect plant hosts against biotic and abiotic stresses. Therefore, studying the enzymatic activities of these fungi can contribute to the development of the enzyme production market and enhance industrial production activities. The thermostability, pH stability, and high-pressure adaptation are interesting features of enzymes produced by endophytic fungi (Bhadra et al. 2022).

7.3.9.1 Cellulases

Cellulose is a widely abundant biopolymer in nature and a key component of plant cell walls. It has various applications in biotechnology and industry, such as fuel production, soil fertilization, crop protection, pulp and paper manufacturing, animal feed, and wine production (Bhadra et al. 2022; Rana et al. 2019). However, cellulose must undergo enzymatic breakdown into glucose before it can be utilized. Cellulase enzymes are responsible for catalyzing this conversion. Different types of cellulases, such as cellobiases, cellobiohydrolases, oxidative cellulases, endonucleases, and cellulose phosphorylases, exist with distinct structures and mechanisms (Bhadra et al. 2022; Rana et al. 2019). Endophytic fungi, including species like *Penicillium*, *Pestalotipsis*, *Colletotrichum*, and *Phialocefala*, are recognized as potential sources of cellulase enzymes (Surono and Narisawa 2021; Li et al. 2021; Goukanapalle et al. 2020; Batista et al. 2022).

Strains of the dark septate endophytic fungus *Phialocephala fortinii* were screened by Surono and Narisawa (2021) for their cellulose hydrolysis activity. The study observed a high cellulolytic index from the *P. fortinii* III.Pi.I8 strain. In contrast, *Rhizopus* sp. and *Mucor* sp., isolated from the leaves of the medicinal plant *Ziziplus spina*, exhibited cellulose degradation percentages of 0.47325 and 0.511, respectively (Hawar 2022). Batista et al. (2022) evaluated hydrolytic enzyme production from 50 isolates obtained from *Euterpe precatoria*. The study revealed that 58.5% of the isolates exhibited cellulase enzyme production, with the highest cellulase index observed in *Guignardia* sp. L11, followed by *Colletotrichum* sp. S1. The authors recommend optimizing the cultivation conditions of these fungi to enhance enzymatic production and activity.

7.3.9.2 Phytase

Phytase is an enzyme that breaks down phytate, a stored organic phosphorous compound. This process releases soluble phosphate, which plays a crucial role in the growth of plants and the nutrition and digestion of humans and animals (Elliott et al. 2022; Mei et al. 2021). Phytase finds applications in various fields such as environmental bioremediation, human nutrition, animal feed and aquaculture, plant growth promotion, bread manufacturing, pulp and paper industries, and lower inositol production (Zhou et al. 2022). Microorganisms, specifically bacteria and fungi, are considered the primary source of phytases, although animals and plants can also produce them. Numerous studies have reported the presence of endophytic fungal phytases, both in ancient and recent research.

The *Rhizopus oryzae* isolated from water-stressed soil was found to have the ability to solubilize phosphate in various phosphorous compounds. This solubilization process is facilitated by the endophyte's production of phosphatase and phytase enzymes (Parvez et al. 2023a, b). Additionally, *Candida* sp. was identified as an endophyte in rice sprouts. Although extracellular phytase was not detected, the yeast demonstrated degradation of sodium phytate. The concentration of solubilized phosphate in the broth containing soybean meal and inoculated with the identified isolate increased. Additionally, the promotion of rice seedlings inoculated with the endophytic *Candida* sp. was observed. The yeast contributed to an increase in shoot fresh weight (80%), root fresh weight (60%), and shoot length (35%) (Zhu et al. 2019).

7.3.9.3 Lipase

The microbial lipase market is estimated at $425.0 million and projected to reach $590.2 million. This enzyme belongs to the serine hydrolases family EC 3.1.1.3 (triacylglycerol ester hydrolase) and catalyzes the hydrolysis of long-chain triglycerides, synthesizing smaller lipid chains including fatty acids, diacylglycerol, monoacylglycerol, and glycerol. Additionally, it can catalyze other reactions such as esterification, alcoholysis, and aminolysis (Chandra et al. 2020). It is commonly used in the food industry for producing cheese, bakery products, tea processing, and flavoring confectionery and cheese (Omar et al. 2016; Chandra et al. 2020).

Furthermore, lipase enzyme plays a significant role in various industries such as cosmetic and personal care products manufacturing, pulp and paper, leather degreasing, textile, and fat and oil industries (Xiaoyang et al. 2019; Chandra et al. 2020). It also has applications in environmental bioremediation for plastics and polymer biodegradation, as well as wastewater and effluent treatment (Yadav et al. 2019; Agobo 2017). Additionally, the medical sector has shown great interest in utilizing this enzyme for its antioxidant, antibiotic, diagnostic tool, and disease treatment properties, including cancer (Trono 2019). It plays a significant role in the manufacturing of detergents and the production of bioenergy/biodiesel (Kalita et al. 2022).

In a study conducted by Sopalun et al. (2021), it was found that 78% of the endophytic fungi isolated from Thai mangrove plants exhibited lipolytic activity. The highest enzymatic production was observed in the isolate Xml-02, identified as *Colletotrichum gloeosporioides*, with a production of 82.22 U/mL. Another study by Rocha et al. (2020) focused on the lipase activity of *Stemphylium lycopersici* and *Sordaria* sp. and their application in esterification reactions. The enzymatic activity of the *Sordaria* sp. extract was found to be 74 U/mL after 96 h and 86 U/mL after 120 h.

7.3.9.4 Amylase

Amylolytic enzymes are well-known enzymes and make up approximately 30% of the global enzyme production market (van der Maarel et al. 2002). These enzymes catalyze the hydrolysis of starch by breaking down the internal glycosidic bonds. They belong to the glycoside hydrolase family 13 (GH-13) (Sundarram and Murthy 2014; Farooq et al. 2021). Various types of amylases, such as endo-amylases, exo-amylases, and glycotransferases, have been identified from different sources. The most extensively studied types are α, β, and γ-amylases (Kamon et al. 2015; Gopinath et al. 2017). Most amylases are metalloenzymes that rely on calcium ions for their activities, stability, and structure (Bordbar et al. 2005; Sunitha et al. 2012).

Endophytic fungi from various plants have demonstrated amylase activities, making them potential alternative sources of industrial and laboratory-grade amylases. Examples include *A. niger*, *Mucor* sp., and *Rhizopus* sp., isolated from *Ziziphus spina* leaves (Hawar 2022); *Trichoderma* sp. and *Aspergillus* sp., isolated from the stem of *Baliospermum montanum* (Jagannath et al. 2021); *Penicillium* sp. L3 and *Guignardia* sp. S22, isolated from açai palm (Euterpe precatoria) (Batista et al. 2022); and *Aspergillus* sp. A9 and A36, isolated from the Amazonian *Myrcia guianensis* (Matias et al. 2021).

7.3.9.5 Protease

Proteases are essential for catalyzing protein and polypeptide hydrolysis, leading to the formation of oligopeptides or free amino acids (Mandal and Banerjee 2019). These enzymes hold approximately 60% of the enzyme market and have significant physiological, biotechnological, and industrial importance. Proteases play functional and structural roles in almost all living organisms, ranging from nutrition to defense mechanisms (Mandal and Banerjee 2019; Ningthoujam et al. 2009). Enzymes are widely utilized in various industrial sectors such as food, pharmaceuticals, leather, and laboratory research. These vital enzymes are synthesized by plants, animals, and microorganisms. Fungi are estimated to synthesize 60% of the industrial proteases, according to Mandal and Banerjee (2019) and Mishra et al. (2019). Compared to bacterial sources, this high percentage can be attributed to the ease of enzyme purification and the wide biochemical diversity of fungal proteases (Mandal and Banerjee 2019).

Industrial applications require enzymes that can function under various conditions, including different pH and temperature ranges. Therefore, it is crucial to search for fungal proteases and other enzymes from different sources to enhance industrial production and improve quality. Plants serve as an intriguing ecosystem for fungi, particularly endophytic fungi. These fungi, which reside within plants, produce a wide range of proteases with potential applications in various fields. Since plants exist in diverse ecological environments, they offer unique characteristics to their endophytic organisms. In a study conducted by Mathur et al. (2022), *Aspergillus niger* isolated from *Albizia lebbeck* (L.) Benth exhibited various enzymatic activities, including protease activity. The crude enzyme derived from this endophyte demonstrated the ability to effectively remove blood stains when combined with detergent. El-Khonezy et al. (2021) investigated the thermos-alkaline protease activity of another *Aspergillus* species isolated from *Ruprechita saicifolia* in a separate study. The results indicated that *Aspergillus ochraceus* BT21 produces a thermotolerant and alkaline-serine protease. The purified form of this enzyme demonstrated a broad range of activity at temperatures between 20 and 60 °C, as well as pH levels ranging from 5.00 to 11. The highest level of activity was observed at 50 °C and pH 8.00. These characteristics make this enzyme well-suited for industrial applications, particularly in the production of detergents.

7.3.9.6 Other Enzymes

Pectinase enzymes play a crucial role in breaking down pectin, a complex polymer found in plant cell walls (Moharram et al. 2022). There are various types of pectinase enzymes, including pectin acetyl esterases, arabinases, and polymethylgalacturonases (Adapa et al. 2014). These enzymes work together to biodegrade pectin. They hold a significant market share of 10% and find applications in food industries such as tea and coffee fermentation, juice preparation, paper, textile, and agriculture (Moharram et al. 2022; Bhadra et al. 2022). Studies on the source of these enzymes have identified numerous fungal producers. However, research on endophytic fungi in low-temperature environments is still limited and requires further development. *Trichoderma* sp., *Aspergillus* sp., *C. gloeosporioides*, *Cladosporium sphaerospermum*, *Mycelia sterilia*, and *Cladosporium* sp. are endophytic fungi known for their ability to produce pectinase enzymes (Shubha and Srinivas 2017; Garg et al. 2016; Bezerra et al. 2012; Uzma et al. 2016).

L-asparaginase, known for its therapeutic function, exhibits antitumor activity by reducing asparagine levels. Asparagine is a necessary molecule for the multiplication and development of leukemia cells in serum (Mishra et al. 2019). L-Asparaginase catalyzes the hydrolysis of asparagine into ammonia and L-aspartate (Singh et al. 2016; Bhadra et al. 2022). Fungal L-asparaginase is preferred over the prokaryotic form due to the multiple side effects associated with bacterial L-asparaginase (Udayan and Gnanadoss 2023; Mishra et al. 2019; Batool et al. 2016). This enzyme is reported in numerous endophytic fungi, such as *Aspergillus* sp., *Colletotrichum* sp., and *Fusarium* sp. (Udayan and Gnanadoss 2023).

Chitinase enzymes are utilized in the remediation of chitin, which is found in the exoskeletons of insects, shells of crustaceans, and cell walls of fungi. These enzymes have been shown to play a role in various functions within fungal cells, such as hyphal extension and ramification, spore formation and germination, cell division, and parasitism (Karlsson and Stenlid 2008). Additionally, chitinase enzymes are employed as a biocontrol method against plant fungal and insect pathogens and in wastewater remediation processes. Species such as *T. harzianum* TH 5-12, *T. atroviride* TA 2-2-1, and *T. harzianum* TH 10-2-2 (Dolatabad et al. 2017) have been identified as chitinase-producing endophytic fungi. Additionally, *Aureobasidium pullulans* and *Lasiodiplodia theobromae* (Govinda Rajulu et al. 2011) are also known for their ability to produce chitinase.

Xylanases are glycosidase enzymes that work alongside other enzymes to break down xylan, a polysaccharide found in hemicellulose. Xylan is the second most abundant polysaccharide in nature (Rana et al. 2019) and constitutes approximately one-third of all renewable organic carbon on Earth (Motta et al. 2013). Xylanases can be sourced from various organisms, including bacteria, protozoa, fungi, and gastropods. The xylanase enzyme derived from the endophytic fungal entomopathogen *Beaveria bassiana* has shown potential for wastepaper deinking (Amobonye et al. 2021). The *Talaromyces funiculosus* and *Cochliobolus* sp. isolated from South African plants exhibited xylanase activities of 0.20 and 0.58, respectively (Abdalla et al. 2020). Ratul et al. (2023) reported the xylanase activity of *A. niger* (32.00 ± 2.18 μg/mL), *P. crustosum* (28.00 ± 2.32 μg/mL), *P. chrysogenum* F5 (27.55 ± 1.22 μg/mL), and *Fusarium* sp. (25.25 ± 1.23 μg/mL) isolated from *Camellia sinensis*, all within a 30-min timeframe.

Lignin is widely acknowledged as the second most abundant polymer in nature, following cellulose. It is a major constituent of plant cell walls and is derived from waste materials generated by industries such as paper, pulp, textiles, and pharmaceuticals. Ligninolytic enzymes, along with other enzymes, are employed to degrade these polymers (Jelic et al. 2012; Cabana et al. 2007). These enzymes find extensive application in the food industry to improve product quality, production efficiency, and overall effectiveness. They are also used in the remediation of wastewater and environments contaminated by industrial wastes such as toxic dyes and pharmaceutical compounds (Kasonga et al. 2023; Melati et al. 2023; Rana et al. 2019). The white rot fungi are considered the earliest producers. Additionally, promising results have been found in the synthesis of ligninolytic enzymes by endophytic fungi. For instance, *Cladosporium tenuissimum* KSP *and Curvularia lunata* PP exhibited high ligninolytic activity (Melati et al. 2023). Kasonga et al. (2023) observed the production of lignin peroxidase and manganese peroxidase by the endophytic fungus *Rhizopus delemar* (Fig. 7.2).

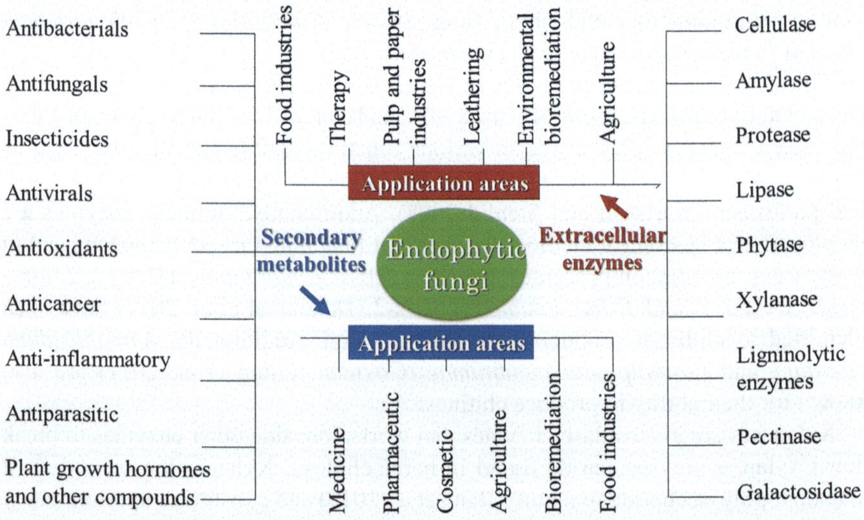

Fig. 7.2 Bioactive compounds and biotechnological/industrial applications of endophytic fungi

7.4 Conclusion

The search for alternative sources of antibiotics has become increasingly important due to the rise of antibiotic resistance. Endophytic fungi have emerged as a valuable source in recent years, as they possess the potential to produce bioactive metabolites with diverse bioactivities. Compounds derived from endophytic fungi, such as terpenoids, alkaloids, phenolic compounds, flavonoids, and steroid derivatives, have been discovered to exhibit antibacterial, antiviral, antitumor, anti-inflammatory, antimalarial, antifungal properties, among others.

Endophytic fungi play a crucial role in protecting plants from both biotic and abiotic stresses through their symbiotic relationship with the plant. This association allows the plant to effectively combat multiple stressors and enhance its overall yield. Additionally, the metabolic activities of these endophytes produce enzymes with varying stability and activity levels, making them valuable for various biotechnological and industrial applications. The endophytic fungus, isolated from various plants in diverse environments, synthesizes enzymes such as cellulase, pectinase, asparaginase, protease, lipase, and amylase. These enzymes have applications in several industries, including pulp and paper, pharmaceuticals, food, leather, and bioremediation. The research on endophytic fungi is limited in terms of purifying, understanding the mechanisms, and optimizing the screened compounds despite the discovery of numerous secondary metabolites and bioactive compounds such as enzymes. Additionally, there is limited literature on the biodiversity of these fungi and their metabolic activities.

References

Ababutain IM, Aldosary SK, Aljuraifani A, Alghamdi AI, Alabdalall AH, Al-Khaldi EM, Aldakeel SA, Almandil NB, Abdulazeez S, Borgio JF (2021) Identification and antibacterial characterization of endophytic fungi from Artemisia sieberi. Int J Microbiol 2021:1. https://doi.org/10.1155/2021/6651020

Abdalla MA, Aro AO, Gado D, Passari AK, Mishra VK, Singh BP, McGaw LJ (2020) Isolation of endophytic fungi from South African plants, and screening for their antimicrobial and extracellular enzymatic activities and presence of type I polyketide synthases. S Afr J Bot 134:336. https://doi.org/10.1016/j.sajb.2020.03.021

Adapa V, Ramya LN, Pulicherla KK, Rao KR (2014) Cold active pectinases: advancing the food industry to the next generation. Appl Biochem Biotechnol 172(5):2324–2337. https://doi.org/10.1007/s12010-013-0685-1

Adeleke BS, Babalola O (2020) The endosphere microbial communities, a great promise in agriculture. Int Microbiol 24:1–17. https://doi.org/10.1007/s10123-020-00140-2

Agobo KU, Arazu VA, Uzo K, Igwe CN (2017) Microbial lipases: a prospect for biotechnological industrial catalysis for green products: a review. Ferment Technol 6(144):2. https://doi.org/10.4172/2167-7972.1000144

Alhadrami HA, Sayed AM, El-Gendy AO, Shamikh YI, Gaber Y, Bakeer W, Sheirf NH, Attia EZ, Shaban GM, Khalifa BA, Ngwa CJ, Pradel G, Rateb ME, Hassan HM, Alkhalifah DHM, Abdelmohsen UR, Hozzein WN (2021) A metabolomic approach to target antimalarial metabolites in the Artemisia annua fungal endophytes. Sci Rep 11(1):2770. https://doi.org/10.1038/s41598-021-82201-8

Amobonye A, Bhagwat P, Singh S, Pillai S (2021) Beauveria bassiana xylanase: characterization and wastepaper deinking potential of a novel glycosyl hydrolase from an endophytic fungal entomopathogen. J Fungi (Basel, Switzerland) 7(8):668. https://doi.org/10.3390/jof7080668

Asaf S, Jan R, Khan MA, Lubna, Khan AL, Asif S, Bilal S, Ahmad W, Waqas M, Kim KM, Al-Harrasi A, Lee IJ (2023) Unraveling the mutualistic interaction between endophytic Curvularia lunata CSL1 and tomato to mitigate cadmium (Cd) toxicity via transcriptomic insights. Sci Total Environ 861:160542. https://doi.org/10.1016/j.scitotenv.2022.160542

Aydoğdu EÖA, Kesiktaş M, Şanli NÖ, Güngör ND, Sancar S, Yildiz T, Yemişken E (2023) Preliminary study on antimicrobial activities of skin mucus from by-catch of Elasmobranch species. Oceanol Hydrobiol Stud 52(2):137–146. https://doi.org/10.26881/oahs-2023.2.01

Bai M, Zheng CJ, Chen GY (2020) Study on bioactive secondary metabolites from a mangrove-derived fungus Penicillium sp. JY246. Chinese J Mar Drugs 39(02):11–18

Baiyee B, Pornsuriya C, Ito S, Sunpapao A (2019) Trichoderma spirale T76-1 displays biocontrol activity against leaf spot on lettuce (Lactuca sativa L.) caused by Corynespora cassiicola or Curvularia aeria. Biol Control 129:195–200. https://doi.org/10.1016/j.biocontrol.2018.10.018

Batista BN, Matias RR, Oliveira RLE, Albuquerque PM (2022) Hydrolytic enzyme production from açai palm (Euterpe precatoria) endophytic fungi and characterization of the amylolytic and cellulolytic extracts. World J Microbiol Biotechnol 38(2):30. https://doi.org/10.1007/s11274-021-03217-w

Batool T, Makky EA, Jalal M, Yusoff MM (2016) A comprehensive review on L-asparaginase and its applications. Appl Biochem Biotechnol 178(5):900–923. https://doi.org/10.1007/s12010-015-1917-3

BCC Research (2021). https://www.bccresearch.com/market-research/biotechnology/global-markets-for-enzymes-in-industrial-applications.html

Bezerra JDP, Santos MGS, Svedese VM, Lima DMM, Fernandes MJS, Paiva LM, Souza-Motta CM (2012) Richness of endophytic fungi isolated from Opuntia ficus-indica Mill. (Cactaceae) and preliminary screening for enzyme production. World J Microbiol Biotechnol 28:1989–1995. https://doi.org/10.1007/s11274-011-1001-2

Bhadra F, Gupta A, Vasundhara M, Reddy MS (2022) Endophytic fungi: a potential source of industrial enzyme producers. 3 Biotech 12(4):86. https://doi.org/10.1007/s13205-022-03145-y

Blackwell M (2011) The fungi: 1, 2, 3 … 5.1 million species? Am J Bot 98(3):426–438. https://doi.org/10.3732/ajb.1000298

Blundell R, Schmidt JE, Igwe A, Cheung AL, Vannette RL, Gaudin ACM, Casteel CL (2020) Organic management promotes natural pest control through altered plant resistance to insects. Nat Plants 6(5):483–491. https://doi.org/10.1038/s41477-020-0656-9

Bodede O, Kuali M, Prinsloo G, Moodley R, Govinden R (2022) Anti-Pseudomonas aeruginosa activity of a C16-terpene dilactone isolated from the endophytic fungus Neofusicoccum luteum of Kigelia africana (Lam.). Sci Rep 12(1):780. https://doi.org/10.1038/s41598-021-04747-x

Bordbar AK, Omidiyan K, Hosseinzadeh R (2005) Study on interaction of alpha-amylase from Bacillus subtilis with cetyl trimethylammonium bromide. Colloids Surf B Biointerfaces 40(1):67–71. https://doi.org/10.1016/j.colsurfb.2004.10.002

Cagigal EFM (2016) Dark septate endophytes (DSE) in polluted areas. In: Hughes E (ed) Endophytic Fungi: Diversity, Characterization and Biocontrol. Nova Science Publishers, New York, pp 125–146

Cabana H, Jones JP, Agathos S (2007) Elimination of endocrine disrupting chemicals using white rot fungi and their lignin modifying enzymes: a review. Eng Life Sci 7:429. https://doi.org/10.1002/ELSC.200700017

Carrie W, Zothanpuiab, Mehetre G, Dekaa P, Lalnunmawiia E, Singh BP (2023) Management of plant diseases using endophytes as biocontrol agents: present status and future prospects. In: Singh BP, Dubey AK (eds) Endophytic association: what, why and how. Elsevier, pp 367–390. https://doi.org/10.1016/B978-0-323-91245-7.00024-9

Caruso DJ, Palombo EA, Moulton SE, Zaferanloo B (2022) Exploring the promise of endophytic fungi: a review of novel antimicrobial compounds. Microorganisms 10(10):1990. https://doi.org/10.3390/microorganisms10101990

Chakrabarti S, Jahandideh F, Wu J (2014) Food-derived bioactive peptides on inflammation and oxidative stress. Biomed Res Int 2014:1

Chand K, Shah S, Sharma J, Paudel MR, Pant B (2020) Isolation, characterization, and plant growthpromoting activities of endophytic fungi from a wild orchid Vanda cristata. Plant Signal Behav 15(5):1744294. https://doi.org/10.1080/15592324.2020.1744294

Chandra P, Enespa, Singh R, Arora PK (2020) Microbial lipases and their industrial applications: a comprehensive review. Microb Cell Factories 19(1):169. https://doi.org/10.1186/s12934-020-01428-8

Chen C, Ye G, Tang J, Li J, Liu W, Wu L, Long Y (2022a) New polyketides from mangrove endophytic fungus Penicillium sp. BJR-P2 and their anti-inflammatory activity. Mar Drugs 20(9):583. https://doi.org/10.3390/md20090583

Chen Q, Yu JJ, He J, Feng T, Liu JK (2022b) Isobenzofuranones and isocoumarins from kiwi endophytic fungus Paraphaeosphaeria sporulosa and their antibacterial activity against Pseudomonas syringae pv. actinidiae. Phytochemistry 195:113050. https://doi.org/10.1016/j.phytochem.2021.113050

de Bary A (1866) Morphologie und physiologie pilze, flechten, und myxomyceten. In: Hofmeister's handbook of physiological botany, vol 2. Verlag Von Wilhelm Engelmann, Leipzig

Ding Z, Wang X, Kong FD, Huang HM, Zhao YN, Liu M, Wang ZP, Han J (2020) Overexpression of global regulator Talae1 leads to the discovery of new antifungal polyketides from endophytic fungus Trichoderma afroharzianum. Front Microbiol 11:622785. https://doi.org/10.3389/fmicb.2020.622785

Doehlemann G, Ökmen B, Zhu W, Sharon A (2017) Plant pathogenic fungi. Microbiol Spectr 5(1). https://doi.org/10.1128/microbiolspec.FUNK-0023-2016

Dolatabad HK, Javan-Nikkhah M, Shier WT (2017) Evaluation of antifungal, phosphate solubilisation, and siderophore and chitinase release activities of endophytic fungi from Pistacia vera. Mycol Progress 16:777–790. https://doi.org/10.1007/s11557-017-1315-z

Duan X, Xu F, Qin D, Gao T, Shen W, Zuo S, Yu B, Xu J, Peng Y, Dong J (2019) Diversity and bioactivities of fungal endophytes from Distylium chinense, a rare waterlogging tolerant plant endemic to the Three Gorges Reservoir. BMC Microbiol 19(1):278. https://doi.org/10.1186/s12866-019-1634-0

Elghaffar RYA, Amin BH, Hashem AH, Sehim AE (2022) Promising endophytic Alternaria alternata from leaves of Ziziphus spina-christi: phytochemical analyses, antimicrobial and antioxidant activities. Appl Biochem Biotechnol 194(9):3984–4001. https://doi.org/10.1007/s12010-022-03959-9

El-Ghonemy DH, Ali SA, Abdel-Megeed RM, Elshafei AM (2023) Therapeutic impact of purified Trichoderma viride L-asparaginase in murine model of liver cancer and in vitro Hep-G2 cell line. J Genet Eng Biotechnol 21(1):38. https://doi.org/10.1186/s43141-023-00493-x

El-Khonezy MI, Elgammal EW, Ahmed EF, Abd-Elaziz AM (2021) Detergent stable thiol-dependant alkaline protease produced from the endophytic fungus Aspergillus ochraceus BT21: Purification and kinetics. Biocatal Agric Biotechnol 35:102046. https://doi.org/10.1016/j.bcab.2021.102046

Erfandoust R, Habibipour R, Soltani J (2020) Antifungal activity of endophytic fungi from Cupressaceae against human pathogenic Aspergillus fumigatus and Aspergillus niger. J Mycol Med 30(3):100987. https://doi.org/10.1016/j.mycmed.2020.100987

Elliott H, Woods P, Green BD, Nugent AP (2022) Can sprouting reduce phytate and improve the nutritional composition and nutrient bioaccessibility in cereals and legumes? Nutr Bull 47(2):138–156. https://doi.org/10.1111/nbu.12549

Fadiji AE, Babalola O (2020) Elucidating mechanisms of endophytes used in plant protection and other bioactivities with multifunctional prospects. Front Bioeng Biotechnol 8. https://doi.org/10.3389/fbioe.2020.00467

Faeth SH, Fagan WF (2002) Fungal endophytes: common host plant symbionts but uncommon mutualists. Integr Comp Biol 42(2):360–368. https://doi.org/10.1093/icb/42.2.360

Farooq MA, Ali S, Hassan A, Tahir HM, Mumtaz S, Mumtaz S (2021) Biosynthesis and industrial applications of α-amylase: a review. Arch Microbiol 203(4):1281–1292. https://doi.org/10.1007/s00203-020-02128-y

Ferdous KJ, Afroz F, Islam MR, Mazid MA, Sohrab MH (2019) Isolated endophytic fungi from the plant Curcuma longa and their potential bioactivity—a review. Pharmacol Pharm 10(05):244–270. https://doi.org/10.4236/pp.2019.105021

Garg G, Singh A, Kaur A, Singh R, Kaur J, Mahajan R (2016) Microbial pectinases: an ecofriendly tool of nature for industries. 3. Biotech 6:1–13. https://doi.org/10.1007/s13205-016-0371-4

Gakuubi MM, Munusamy M, Liang ZX, Ng SB (2021) Fungal endophytes: A promising frontier for discovery of novel bioactive compounds. J Fungi (Basel, Switzerland) 7(10):786. https://doi.org/10.3390/jof7100786

Gandhi SG, Mahajan V, Bedi YS (2015) Changing trends in biotechnology of secondary metabolism in medicinal and aromatic plants. Planta 241(2):303–317. https://doi.org/10.1007/s00425-014-2232-x

Gautam VS, Singh A, Kumari P, Nishad JH, Kumar J, Yadav M et al (2022) Phenolic and flavonoid contents and antioxidant activity of an endophytic fungus Nigrospora sphaerica (EHL2), inhabiting the medicinal plant Euphorbia hirta (dudhi) L. Arch Microbiol 204(2):140. https://doi.org/10.1007/s00203-021-02650-7

Gao Y, Stuhldreier F, Schmitt L, Wesselborg S, Wang L, Müller WEG, Kalscheuer R, Guo Z, Zou K, Liu Z, Proksch P (2020) Sesterterpenes and macrolide derivatives from the endophytic fungus Aplosporella javeedii. Fitoterapia 146:104652. https://doi.org/10.1016/j.fitote.2020.104652

Gill SS, Gill R, Trivedi DK, Anjum NA, Sharma KK, Ansari MW, Ansari AA, Johri AK, Prasad R, Pereira E, Varma A, Tuteja N (2016) Piriformospora indica: potential and significance in plant stress tolerance. Front Microbiol 7:332. https://doi.org/10.3389/fmicb.2016.00332

Gopinath SC, Anbu P, Arshad MK, Lakshmipriya T, Voon CH, Hashim U, Chinni SV (2017) Biotechnological processes in microbial amylase production. Biomed Res Int 2017:1272193. https://doi.org/10.1155/2017/1272193

Goukanapalle PKR, Kanderi DK, Rajoji G, Shanthi Kumari BS, Bontha RR (2020) Optimization of cellulase production by a novel endophytic fungus Pestalotiopsis microspora TKBRR isolated from Thalakona forest. Cellulose 27:6299–6316. https://doi.org/10.1007/s10570-020-03220-8

Govinda Rajulu MB, Thirunavukkarasu N, Suryanarayanan TS, Ravishankar JP, El Gueddari NE, Moerschbacher BM (2011) Chitinolytic enzymes from endophytic fungi. Fungal Divers 47:43–53. https://doi.org/10.1007/s13225-010-0071-z

Granada CE, Passaglia LM, De Souza EM, Sperotto RA (2018) Is phosphate solubilization the forgotten child of plant growth-promoting rhizobacteria? Front Microbiol 9:2054. https://doi.org/10.3389/fmicb.2018.02054

Grivennikov SI, Greten FR, Karin M (2010) Immunity, inflammation, and cancer. Cell 140(6):883–899. https://doi.org/10.1016/j.cell.2010.01.025

Gupta M, Shukla KK (2020) Endophytic fungi: a treasure trove of novel bioactive compounds. In: Bioactive natural products in drug discovery. Springer, pp 427–449. https://doi.org/10.1007/978-981-15-1394-7_14

Hashem AH, Shehabeldine AM, Abdelaziz AM, Amin BH, Sharaf MH (2022) Antifungal activity of endophytic Aspergillus terreus extract against some fungi causing mucormycosis: ultrastructural study. Appl Biochem Biotechnol 194(8):3468–3482. https://doi.org/10.1007/s12010-022-03876-x

Hashem AH, Attia MS, Kandil EK, Fawzi MM, Abdelrahman AS, Khader MS, Abdelaziz AM (2023) Bioactive compounds and biomedical applications of endophytic fungi: a recent review. Microb Cell Factories 22(1):1–23. https://doi.org/10.1186/s12934-023-02118-x

Hawar SN (2022) Extracellular enzyme of endophytic fungi isolated from Ziziphus spina leaves as medicinal plant. Int J Biomater 2022:1. https://doi.org/10.1155/2022/2135927

Hou YT (2020) A new polyketide from endophytic fungus Aspergillus oryzae derived from Paris polyphylla var. yunnanensis. Chin Tradit Herb Drug 24;4393–4396

Jagannath S, Konappa N, Lokesh A, Bhuvaneshwari, Dasegowda T, Udayashankar AC, Chowdappa S, Cheluviah M, Satapute P, Jogaiah S (2021) Bioactive compounds guided diversity of endophytic fungi from Baliospermum montanum and their potential extracellular enzymes. Anal Biochem 614:114024. https://doi.org/10.1016/j.ab.2020.114024

Jelic A, Cruz-Morató C, Marco-Urrea E, Sarrà M, Perez S, Vicent T, Petrović M, Barcelo D (2012) Degradation of carbamazepine by Trametes versicolor in an air pulsed fluidized bed bioreactor and identification of intermediates. Water Res 46(4):955–964. https://doi.org/10.1016/j.watres.2011.11.063

Kalita P, Basumatary B, Saikia P, Das B, Basumatary S (2022) Biodiesel as renewable biofuel produced via enzyme-based catalyzed transesterification. Energy Nexus 6:100087. https://doi.org/10.1016/j.nexus.2022.100087

Kamon M, Sumitani J, Tani S, Kawaguchi T (2015) Characterization and gene cloning of a maltotriose-forming exo-amylase from Kitasatospora sp. MK-1785. Appl Microbiol Biotechnol 99(11):4743–4753. https://doi.org/10.1007/s00253-015-6396-5

Karin M, Clevers H (2016) Reparative inflammation takes charge of tissue regeneration. Nature 529(7586):307–315. https://doi.org/10.1038/nature17039

Karlsson M, Stenlid J (2008) Comparative evolutionary histories of the fungal chitinase gene family reveal non-random size expansions and contractions due to adaptive natural selection. Evol Bioinformatics Online 4:47–60. https://doi.org/10.4137/ebo.s604

Karthikeyan A, Joseph A, Nair BG (2022) Promising bioactive compounds from the marine environment and their potential effects on various diseases. J Genet Eng Biotechnol 20(1):14. https://doi.org/10.1186/s43141-021-00290-4

Kasonga TK, Kamika I, Ngole-Jeme VM (2023) Ligninolytic enzyme activity and removal efficiency of pharmaceuticals in a water matrix by fungus Rhizopus sp. isolated from cassava. Environ Technol 44(14):2157–2170. https://doi.org/10.1080/09593330.2021.2024885

Kaur N, Arora DS (2020) Prospecting the antimicrobial and antibiofilm potential of Chaetomium globosum an endophytic fungus from Moringa oleifera. AMB Express 10(1):206. https://doi.org/10.1186/s13568-020-01143-y

Kaur J, Sharma A, Sharma M, Kumari Manhas R, Kaur S, Kaur A (2019) Effect of α-glycosidase inhibitors from endophytic fungus Alternaria destruens on survival and development of insect pest Spodoptera litura Fab. and fungal phytopathogens. Sci Rep 9(1):11400. https://doi.org/10.1038/s41598-019-47838-6

Kim J, Baek JY, Bang S, Kim JY, Jin Y, Lee JW, Jang DS, Kang KS, Shim SH (2023) New anti-inflammatory β-resorcylic acid lactones derived from an endophytic fungus, Colletotrichum sp. ACS Omega 8(3):3530–3538. https://doi.org/10.1021/acsomega.2c07962

Kouipou Toghueo RM, Kemgne EAM, Sahal D, Yadav M, Kenou Kagho DU, Yang B, Baker BJ, Boyom FF (2021) Specialized antiplasmodial secondary metabolites from Aspergillus niger 58, an endophytic fungus from Terminalia catappa. J Ethnopharmacol 269:113672. https://doi.org/10.1016/j.jep.2020.113672

Kouipou Toghueo RM, Boyom FF (2019) Endophytic fungi from Terminalia species: a comprehensive review. J Fungi 5(2):43. https://doi.org/10.3390/jof5020043

Kour D, Rana KL, Yadav A, Yadav NR, Kumar M, Kumar V, Vyas P, Dhaliwal HS, Saxena AK (2020) Microbial biofertilizers: bioresources and eco-friendly technologies for agricultural and environmental sustainability. Biocatal Agric Biotechnol 23:101487. https://doi.org/10.1016/j.bcab.2019.101487

Kusam LR, Divjot K, Imran S, Neelam Y, Ajar NY, Vinod K, Bhim PS, Harcharan SD, Anil KS (2019) Biodiversity of endophytic fungi from diverse niches and their biotechnological applications advances in endophytic fungal. In: Advances in endophytic fungal research. Springer, Cham, pp 105–144. https://doi.org/10.1007/978-3-030-03589-1_6

Kusari P, Kusari S, Spiteller M, Kayser O (2013) Endophytic fungi harbored in Cannabis sativa L.: diversity and potential as biocontrol agents against host plant-specific phytopathogens. Fungal Divers 60:137–151. https://doi.org/10.1007/s13225-012-0216-3

Lacerda ÍCDS, Polonio JC, Golias HC (2022) Endophytic fungi as a source of antiviral compounds—a review. Chem Biodivers 19(6):e202100971. https://doi.org/10.1002/cbdv.202100971

Lana M, Simón O, Velasco P, Rodríguez VM, Caballero P, Poveda J (2023) First study on the root endophytic fungus Trichoderma hamatum as an entomopathogen: development of a fungal bio-insecticide against cotton leafworm (Spodoptera littoralis). Microbiol Res 270:127334. https://doi.org/10.1016/j.micres.2023.127334

Li H, Dou M, Wang X, Guo N, Kou P, Jiao J, Fu Y (2021) Optimization of cellulase production by a novel endophytic fungus Penicillium oxalicum R4 isolated from Taxus cuspidata. Sustainability 13(11):6006. https://doi.org/10.3390/su13116006

Li Q, Zheng Y, Fu A, Wei M, Kang X, Chen C, Zhu H, Zhang Y (2022) 30-norlanostane triterpenoids and steroid derivatives from the endophytic fungus Aspergillus nidulans. Phytochemistry 201:113257. https://doi.org/10.1016/j.phytochem.2022.113257

Li-Bin L, Xiao J, Zhang Q, Han R, Xu B, Yang SX et al (2021) Eremophilane sesquiterpenoids with antibacterial and anti-inflammatory activities from the endophytic fungus Septoria rudbeckiae. J Agric Food Chem 69(40):11878–11889. https://doi.org/10.1021/acs.jafc.1c04131

Liu SS, Jiang JX, Huang R, Wang YT, Jiang BG, Zheng KX, Wu SH (2019) A new antiviral 14-nordrimane sesquiterpenoid from an endophytic fungus Phoma sp. Phytochem Lett 29:75–78. https://doi.org/10.1016/j.phytol.2018.11.005

Liu HW, Ji QT, Ren GG, Wang F, Su F, Wang PY, Zhou X, Wu ZB, Li Z, Yang S (2020) Antibacterial functions and proposed modes of action of novel 1,2,3,4-tetrahydro-β-carboline derivatives that possess an attractive 1,3-diaminopropan-2-ol pattern against rice bacterial blight, kiwifruit bacterial canker, and citrus bacterial canker. J Agric Food Chem 68(45):12558–12568. https://doi.org/10.1021/acs.jafc.0c02528

Lourenço SC, Moldão-Martins M, Alves VD (2019) Antioxidants of natural plant origins: from sources to food industry applications. Molecules (Basel, Switzerland) 24(22):4132. https://doi.org/10.3390/molecules24224132

Ludwig-Müller J (2015) Plants and endophytes: equal partners in secondary metabolite production? Biotechnol Lett 37(7):1325–1334. https://doi.org/10.1007/s10529-015-1814-4

Lv LX, Mo TX, Liang M, Huang LL, Li BC, Qin XY, Huang SH, Li J, Xu WF, Yang RY (2023) Secondary metabolites from the endophytic fungus Xylaria grammica and their anti-inflammatory activities. Chem Nat Compd:1–3. https://doi.org/10.1007/s10600-023-03940-w

Mandal S, Banerjee D (2019) Proteases from endophytic fungi with potential industrial applications. In: Recent advancement in white biotechnology through fungi, Diversity and enzymes perspectives, vol 1. Springer, pp 319–359. https://doi.org/10.1007/978-3-030-10480-1_10

Mathur P, Ghosh U, Chetani R, Chaturvedi P, Sharma C, Bhatnagar P (2022) Enzymatic bioprospecting of endophytic Aspergillus niger isolated from Albizia lebbeck (L.) Benth. S Afr J Bot 148:580–587. https://doi.org/10.1016/j.sajb.2022.05.033

Matias RR, Sepúlveda AMG, Batista BN, de Lucena JMVM, Albuquerque PM (2021) Degradation of Staphylococcus aureus biofilm using hydrolytic enzymes produced by Amazonian endophytic fungi. Appl Biochem Biotechnol 193:2145–2161. https://doi.org/10.1007/s12010-021-03542-8

Mbarga JB, Ten Hoopen GM, Kuatú J, Adiobo A, Ngonkeu MEL, Ambang Z et al (2012) Trichoderma asperellum: a potential biocontrol agent for Pythium myriotylum, causal agent of cocoyam (Xanthosoma sagittifolium) root rot disease in Cameroon. Crop Prot 36:18–22

Mei C, Chretien RL, Amaradasa BS, He Y, Turner A, Lowman S (2021) Characterization of phosphate solubilizing bacterial endophytes and plant growth promotion in vitro and in greenhouse. Microorganisms 9(9):1935. https://doi.org/10.3390/microorganisms9091935

Melati I, Rahayu G, Effendi H, Henny C, Yanto DHY (2023) Biodecolorization of anthraquinone and azo dyes by dark septate endophytic fungi. Bioresour Technol Rep 22:101427. https://doi.org/10.15294/biosaintifika.v13i1.26148

Mishra Y, Singh A, Batra A, Sharma MM (2014) Understanding the biodiversity and biological applications of endophytic fungi: a review. J Microb Biochem Technol 2014:1–11. https://doi.org/10.4172/1948-5948.S8-004

Mishra R, Kushveer JS, Revanthbabu P, Sarma VV (2019) Endophytic Fungi and Their Enzymatic Potential. In: Singh B (ed) Advances in Endophytic Fungal Research. Fungal Biology. Springer, Cham. https://doi.org/10.1007/978-3-030-03589-1_14

Moharram AM, Zohri AA, Hesham AE, Abdel-Raheam HEF, Al-Ameen Maher M, Al-Bedak OA (2022) Production of cold-active pectinases by three novel Cladosporium species isolated from Egypt and application of the most active enzyme. Sci Rep 12(1):15599. https://doi.org/10.1038/s41598-022-19807-z

Motta FL, Andrade CC, Santana MH (2013) A review of xylanase production by the fermentation of xylan: classification, characterization and applications. InTech. https://doi.org/10.5772/53544

Munshi M, Zilani MNH, Islam MA, Biswas P, Das A, Afroz F, Hasan MN (2022) Novel compounds from endophytic fungi of Ceriops decandra inhibit breast cancer cell growth through estrogen receptor alpha in in-silico study. Inform Med Unlocked 32:101046. https://doi.org/10.1016/j.imu.2022.101046

Narisawa K (2021) The cellulolytic activity and symbiotic potential of dark septate endophytic fungus Phialocephala fortinii to promote non-mycorrhizal plants growth. IOP Conf Ser Earth Environ Sci 648(1):012165. https://doi.org/10.1088/1755-1315/648/1/012165

Nayak S, Samanta S, Mukherjee AK (2020) Beneficial role of Aspergillus sp. in agricultural soil and environment. In: Nayak SK, Mishra BB (eds) Frontiers in soil and environmental microbiology, vol 3, pp 17–36. https://doi.org/10.1201/9780429485794

Nazari J, Payamnoor V, Kavosi MR, Asadi J (2023) Increasing terpenes in bark endophytic fungi of Betula pendula roth as an anticancer potential source by cellulose nanofibers and sodium nitrate. S Afr J Bot 157:592. https://doi.org/10.1016/j.sajb.2023.04.032

Ningthoujam DS, Kshetri P, Sanasam S, Nimaichand S (2009) Screening, identification of best producers and optimization of extracellular proteases from moderately halophilic alkali thermotolerant indigenous actinomycetes. World Appl Sci J 7:907–916

Nisa H, Kamili AN, Nawchoo IA, Shafi S, Shameem N, Bandh SA (2015) Fungal endophytes as prolific source of phytochemicals and other bioactive natural products: a review. Microb Pathog 82:50–59. https://doi.org/10.1016/j.micpath.2015.04.001

Nurgali K, Jagoe RT, Abalo R (2018) Adverse effects of cancer chemotherapy: Anything new to improve tolerance and reduce sequelae? Front Pharmacol 9:245. https://doi.org/10.3389/fphar.2018.00245

Nzimande B, Kumalo HM, Ndlovu SI, Mkhwanazi NP (2022) Secondary metabolites produced by endophytic fungi, Alternaria alternata, as potential inhibitors of the human immunodeficiency virus. Front Genet 13:1077159. https://doi.org/10.3389/fgene.2022.1077159

Omar KA, Gounga ME, Liu R, Mlyuka E, Wang X (2016) Effects of microbial lipases on hydrolyzed milk fat at different time intervals in flavour development and oxidative stability. J Food Sci Technol 53(2):1035–1046. https://doi.org/10.1007/s13197-015-2158-8

Omomowo OI, Babalola OO (2019) Bacterial and fungal endophytes: tiny giants with immense beneficial potential for plant growth and sustainable agricultural productivity. Microorganisms 7(11):481. https://doi.org/10.3390/microorganisms7110481

Parvez M, e-Rana G, Hussain F, Khan M, Sajid H (2023a) Concurrent application of indole acetic acid and crude fungal extract from Rhizopus oryzae synergistically improved vegetative and physiochemical attributes in spinach. J Soil Sci Plant Nutr:1–12. https://doi.org/10.1007/s42729-023-01179-6

Parvez M, Hussain F, Khan M, Sajid H (2023b) Characterization of phosphate solubilizing fungal endophyte associated with roots of Coriandrum sativum L growing in water stressed soil. Symbiosis 89(1):83–94. https://doi.org/10.1007/s13199-022-00887-4

Pierce L, Hocking MC, Schwartz LA, Alderfer MA, Kazak AE, Barakat LP (2017) Caregiver distress and patient health-related quality of life: psychosocial screening during pediatric cancer treatment. Psycho-Oncology 26(10):1555–1561. https://doi.org/10.1002/pon.4171

Pilmis B, Alanio A, Lortholary O, Lanternier F (2018) Recent advances in the understanding and management of mucormycosis. F1000Research 7:F1000 Faculty Rev-1429. https://doi.org/10.12688/f1000research.15081.1

Pizzino G, Irrera N, Cucinotta M, Pallio G, Mannino F, Arcoraci V, Squadrito F, Altavilla D, Bitto A (2017) Oxidative stress: harms and benefits for human health. Oxidative Med Cell Longev 2017:8416763. https://doi.org/10.1155/2017/8416763

Rai N, Gupta P, Verma A, Tiwari RK, Madhukar P, Kamble SC et al (2023) Ethyl acetate extract of Colletotrichum gloeosporioides promotes cytotoxicity and apoptosis in human breast cancer cells. ACS Omega 8(4):3768–3784. https://doi.org/10.1021/acsomega.2c05746

Rachid NA, Güngör ND (2023) Screening of bioactive compounds for biomedical and industrial uses from actinobacteria isolated from the Parsik Cave (Turkey). Johnson Matthey Technol Rev 67(2):159–170. https://doi.org/10.1595/205651322X16482034395036

Ramos GDC, Silva-Silva JV, Watanabe LA, Siqueira JES, Almeida-Souza F, Calabrese KS, Marinho AMDR, Marinho PSB, Oliveira AS (2022) Phomoxanthone A, compound of endophytic fungi Paecilomyces sp. and its potential antimicrobial and antiparasitic. Antibiotics (Basel, Switzerland) 11(10):1332. https://doi.org/10.3390/antibiotics11101332

Rana KL, Kour D, Sheikh IA, Yadav NR, Yadav A, Kumar V, Singh BP, Dhaliwal HS, Saxena AK (2019) Biodiversity of endophytic fungi from diverse niches and their biotechnological applications. In: Advances in endophytic fungal research: present status and future challenges. Springer, pp 105–144

Ratul N, Sharma GD, Madhumita B (2023) Phosphate solubilization, siderophore production and extracellular enzyme production activities of endophytic fungi isolated from tea (Camellia sinensis) bushes of Assam, India. Res J Biotechnol 18:6. https://doi.org/10.25303/1806rjbt049057

Raveendran S, Parameswaran B, Ummalyma SB, Abraham A, Mathew AK, Madhavan A, Rebello S, Pandey A (2018) Applications of microbial enzymes in food industry. Food Technol Biotechnol 56(1):16–30. https://doi.org/10.17113/ftb.56.01.18.5491

Ravi P, Somu P, Acharya D, Gomez LA, Thathapudi JJ, Ramachandra YL, Rudraiah SB, Isaq M, Karua CS, Arifullah M, Poojari CC, Lee YR (2022) Isolation and phytochemical screening of endophytic fungi isolated from medicinal plant Mappia foetida and evaluation of its in vitro cytotoxicity in cancer. Appl Biochem Biotechnol 194(10):4570–4586. https://doi.org/10.1007/s12010-022-03929-1

Rocha KS, Queiroz MS, Gomes BS, Dallago R, de Souza RO, Guimarães DO et al (2020) Lipases of Endophytic Fungi Stemphylium lycopersici and Sordaria sp.: Application in the synthesis of solketal derived Monoacylglycerols. Enzym Microb Technol 142:109664. https://doi.org/10.1016/j.enzmictec.2020.109664

Rodriguez RJ, White JF Jr, Arnold AE, Redman RS (2009) Fungal endophytes: diversity and functional roles. New Phytol 182(2):314–330. https://doi.org/10.1111/j.1469-8137.2009.02773.x

Rustamova N, Bozorov K, Efferth T, Yili A (2020) Novel secondary metabolites from endophytic fungi: synthesis and biological properties. Phytochem Rev 19:425–448. https://doi.org/10.1007/s11101-020-09672-x

Seema N, Hamayun M, Hussain A, Shah M, Irshad M, Qadir M, Iqbal A, Alrefaei AF, Ali S (2023) Endophytic Fusarium proliferatum reprogrammed phytohormone production and antioxidant system of Oryza sativa under drought stress. Agronomy 13(3):873

Sharma N, Sharma V, Abrol V, Panghal A, Jaglan S (2019) An update on bioactive natural products from endophytic fungi of medicinal plants. In: Pharmaceuticals from microbes: impact on drug discovery. Springer, pp 121–153. https://doi.org/10.1007/978-3-030-04675-0_6

Shubha J, Srinivas C (2017) Diversity and extracellular enzymes of endophytic fungi associated with Cymbidium aloifolium L. Afr J Biotechnol 16(48):2248–2258. https://doi.org/10.13140/RG.2.2.28400.56326

Singh R, Kumar M, Mittal A, Mehta PK (2016) Microbial enzymes: industrial progress in 21st century. 3 Biotech 6(2):174. https://doi.org/10.1007/s13205-016-0485-8

Singh A, Singh DK, Kharwar RN, White JF, Gond SK (2021) Fungal endophytes as efficient sources of plant-derived bioactive compounds and their prospective applications in natural product drug discovery: insights, avenues, and challenges. Microorganisms 9(1):197. https://doi.org/10.3390/microorganisms9010197

Soltani J (2016) Secondary metabolite diversity of the genus Aspergillus: recent advances. In: New and future developments in microbial biotechnology and bioengineering. Elsevier, pp 275–292. https://doi.org/10.1016/B978-0-444-63505-1.00035-X

Soltani J, Zaheri-Shoja M, Hamzei J, Hosseyni-Moghaddam MS, Pakvaz S (2016) Diversity and bioactivity of bacterial endophyte community of Cupressaceae. For Pathol 46(4):353–361. https://doi.org/10.1111/efp.12270

Sopalun K, Laosripaiboon W, Wachirachaikarn A, Iamtham S (2021) Biological potential and chemical composition of bioactive compounds from endophytic fungi associated with thai mangrove plants. S Afr J Bot 141:66–76. https://doi.org/10.1016/j.sajb.2021.04.031

Song Y, Miao F, Liang X, Yin X, Ji N (2019) Harziane and cadinane terpenoids from the alga-endophytic fungus Trichoderma asperellum A-YMD-9-2. Phytochem Lett 32:38. https://doi.org/10.1016/J.PHYTOL.2019.05.001

Su S, Yang XQ, Yang YB, Ding ZT (2023) Antifungal and antifeedant terpenoids from Paraphaeosphaeria sp. cultured by extract of host Ginkgo biloba. Phytochemistry 210:113651. https://doi.org/10.1016/j.phytochem.2023.113651

Sundarram A, Murthy TPK (2014) α-Amylase production and applications: a review. J Appl Environ Microbiol 2(4):166–175

Sunitha VH, Ramesha A, Savitha J, Srinivas C (2012) Amylase production by endophytic fungi Cylindrocephalum sp. isolated from medicinal plant Alpinia calcarata (Haw.) Roscoe. Brazil J Microbiol 43(3):1213–1221. https://doi.org/10.1590/S1517-83822012000300049

Talukdar R, Paul S, Tayung K (2022) Antifungal drugs from endophytic microbes: present and future prospects. In: Bioresource utilization and management, Applications in therapeutics, biofuels, agriculture, and environmental science. Routledge, pp 3–37. https://doi.org/10.1201/9781003057826-2

Tang Z, Qin Y, Chen W, Zhao Z, Lin W, Xiao Y, Chen H, Liu Y, Chen H, Bu T, Li Q, Cai Y, Yao H, Wan Y (2021) Diversity, chemical constituents, and biological activities of endophytic fungi isolated from Ligusticum chuanxiong Hort. Front Microbiol 12:771000

Thapa S, Li H, OHair J, Bhatti S, Chen FC, Nasr KA, Johnson T, Zhou S (2019) Biochemical characteristics of microbial enzymes and their significance from industrial perspectives. Mol Biotechnol 61(8):579–601. https://doi.org/10.1007/s12033-019-00187-1

Tian LL, Ren H, Xi JM, Fang J, Zhang JZ, Wu QX (2021) Diverse anti-inflammation and anticancer polyketides isolated from the endophytic fungi Alternaria sp. MG1. Fitoterapia 153:105000. https://doi.org/10.1016/j.fitote.2021.105000

Trono D (2019) Recombinant enzymes in the food and pharmaceutical industries. In: Advances in enzyme technology. Elsevier. https://doi.org/10.1016/B978-0-444-64114-4.00013-3

Turbat A, Rakk D, Vigneshwari A, Kocsubé S, Thu H, Szepesi Á, Bakacsy L, Škrbić BD, Jigjiddorj EA, Vágvölgyi C, Szekeres A (2020) Characterization of the plant growth-promoting activities of endophytic fungi isolated from Sophora flavescens. Microorganisms 8(5):683. https://doi.org/10.3390/microorganisms8050683

Udayan E, Gnanadoss JJ (2023) Isolation and characterization of L-asparaginase producing endophytic fungi from medicinal plants of Rutaceae family. Biosci Biotechnol Res Asia 20:241. https://doi.org/10.13005/bbra/3085

Uzma F, Konappa N, Chowdappa S (2016) Diversity and extracellular enzyme activities of fungal endophytes isolated from medicinal plants of Western Ghats, Karnataka. Egypt J Basic Appl Sci 3:335–342. https://doi.org/10.1016/j.ejbas.2016.08.007

van der Maarel MJ, van der Veen B, Uitdehaag JC, Leemhuis H, Dijkhuizen L (2002) Properties and applications of starch-converting enzymes of the alpha-amylase family. J Biotechnol 94(2):137–155. https://doi.org/10.1016/s0168-1656(01)00407-2

Verma A, Shameem N, Jatav HS, Sathyanarayana E, Parray JA, Poczai P, Sayyed RZ (2022) Fungal endophytes to combat biotic and abiotic stresses for climate-smart and sustainable agriculture. Front Plant Sci 13:953836. https://doi.org/10.3389/fpls.2022.953836

Vinale F, Sivasithamparam K, Ghisalberti EL, Marra R, Woo SL, Lorito M (2008) Trichoderma–plant–pathogen interactions. Soil Biol Biochem 40(1):1–10. https://doi.org/10.1016/j.soilbio.2007.07.002

Vinodkumar S, Indumathi TR, Nakkeeran S (2017) Trichoderma asperellum (NVTA2) as a potential antagonist for the management of stem rot in carnation under protected cultivation. Biol Control 113:58–64. https://doi.org/10.1016/J.BIOCONTROL.2017.07.001

Wang FW, Jiao RH, Cheng AB, Tan SH, Song YC (2007) Antimicrobial potentials of endophytic fungi residing in Quercus variabilis and brefeldin A obtained from Cladosporium sp. World J Microbiol Biotechnol 23:79–83. https://doi.org/10.1007/s11274-006-9195-4

Wang H, Liu Z, Duan F, Chen Y, Qiu K, Xiong Q, Lin H, Zhang J, Tan H (2023a) Isolation, identification, and antibacterial evaluation of endophytic fungi from Gannan navel orange. Front Microbiol 14:1172629. https://doi.org/10.3389/fmicb.2023.1172629

Wang Y, Liu C, Yang Y, Li C, Pei Y (2023b) A new tetracyclic triterpenoid from endophytic fungus Fusarium sporotrichioide. Chin Herbal Med 16:231. https://doi.org/10.1016/j.chmed.2022.10.007

Wang Z, Wang L, Pan Y, Zheng X, Liang X, Sheng L, Zhang D, Sun Q, Wang Q (2023c) Research advances on endophytic fungi and their bioactive metabolites. Bioprocess Biosyst Eng 46(2):165–170. https://doi.org/10.1007/s00449-022-02840-7

Wei C, Deng Q, Sun M, Xu J (2020) Cytospyrone and cytospomarin: two new polyketides isolated from mangrove endophytic fungus, Cytospora sp. Molecules (Basel, Switzerland) 25(18):4224. https://doi.org/10.3390/molecules25184224

Willems HME, Ahmed SS, Liu J, Xu Z, Peters BM (2020) Vulvovaginal candidiasis: a current understanding and burning questions. J Fungi (Basel, Switzerland) 6(1):27. https://doi.org/10.3390/jof6010027

Wonglom P, Ito S, Sunpapao A (2020) Volatile organic compounds emitted from endophytic fungus Trichoderma asperellum T1 mediate antifungal activity, defense response and promote plant growth in lettuce (Lactuca sativa). Fungal Ecol 43:100867. https://doi.org/10.1016/j.funeco.2019.100867

Wu QS, He XH, Zou YN, He KP, Sun YH, Cao MQ (2012) Spatial distribution of glomalin-related soil protein and its relationships with root mycorrhization, soil aggregates, carbohydrates, activity of protease and β-glucosidase in the rhizosphere of Citrus unshiu. Soil Biol Biochem 45:181–183. https://doi.org/10.1016/j.envpol.2022.119703

Xiaoyang S, Shao-jun T, Li-fen Z, Jianchun X (2019) Effect of phospholipase A1-catalyzed degumming on oryzanol, tocopherols, and tocotrienols of dewaxed rice bran oil. J Chem 2019:1. https://doi.org/10.1155/2019/1608750

Yadav AN, Singh S, Mishra S, Gupta A (2019) Recent advancement in white biotechnology through fungi. Springer International Publishing, Cham, p 528

Yang T, Liu J, Lin LB, Hu JY, Wu GW, Fan PH, Duan DZ, Xiao J, Wang XL (2022) Acrocalysterols A and B, two new steroids from endophytic fungus Acrocalymma sp. Phytochem Lett 48:77–80. https://doi.org/10.1016/j.phytol.2022.02.003

Ye G, Huang C, Li J, Chen T, Tang J, Liu W, Long Y (2021) Isolation, structural characterization and antidiabetic activity of new diketopiperazine alkaloids from mangrove endophytic fungus aspergillus sp. 16-5c. Mar Drugs 19(7):402. https://doi.org/10.3390/md19070402

Yu JJ, Wei WK, Zhang Y, Cox RJ, He J, Liu JK, Feng T (2022) Terpenoids from kiwi endophytic fungus Bipolaris sp. and their antibacterial activity against Pseudomonas syringae pv. actinidiae. Front Chem 10:990734. https://doi.org/10.3389/fchem.2022.990734

Zeng Q, Cui H, Yao H, Yuan T (2023) Five sesquiterpenes from Paraconiothyrium sp. and their anti-inflammatory activity. Chem Biodivers:e202300477. https://doi.org/10.1002/cbdv.202300477

Zhang P, Yuan XL, Du YM, Zhang HB, Shen GM, Zhang ZF et al (2019) Angularly prenylated indole alkaloids with antimicrobial and insecticidal activities from an endophytic fungus Fusarium sambucinum TE-6L. J Agric Food Chem 67(43):11994–12001. https://doi.org/10.1021/acs.jafc.9b05827

Zhou Y, Anoopkumar AN, Tarafdar A, Madhavan A, Binoop M, Lakshmi NM et al (2022) Microbial engineering for the production and application of phytases to the treatment of the toxic pollutants: a review. Environ Pollut 308:119703. https://doi.org/10.1016/j.envpol.2022.119703

Zhou LW, May TW (2022) Fungal taxonomy: current status and research agendas for the interdisciplinary and globalisation era. Mycology 14(1):52–59. https://doi.org/10.1080/21501203.2022.2103194

Zhu A, Tan H, Cao L (2019) Isolation of phytase-producing yeasts from rice seedlings for prospective probiotic applications. 3 Biotech 9(6):216. https://doi.org/10.1007/s13205-019-1746-0

Bioprospecting Endophytic Fungi from *Fusarium* Genus as Sources of Bioactive Metabolites

8

Kaathambari Purushothaman, Esaimozhi Sivasankar, V. G. Sowmeya, and Mythili Sathiavelu

Abstract

The global landscape of health is dynamic. The emergence and re-emergence of diseases present great challenges that demand attention to find alternative solutions to combat health issues. Amidst these issues, the endophytic microbes are a beacon of hope that provides novel metabolites to combat the health threats leveraging humanity. Endophytic fungi are a great source of compounds that could serve as potential leads. The most diverse and highly reported genus of endophytic fungi is *Fusarium* spp. They produce diverse metabolites with varied biological activities, namely antimicrobial, anti-inflammatory, anticancer, antioxidant, and many more. This chapter focuses on the metabolites secreted by the endophytic *Fusarium* spp. and their biological activities.

Keywords

Endophytic fusaria · Bioactive metabolites · Antimicrobial

8.1 Introduction

In traditional medicine, plants have played a vital role. The bioactive components present in the plants have been largely exploited to cure several ailments. The quest to identify novel therapeutics has led to the exploration of microbial flora for their metabolites. Though plant compounds are highly active, the availability of the plants and the yield of the bioactive component are questionable. The demand for natural product-based drugs is ever-increasing due to the emergence of new

K. Purushothaman · E. Sivasankar · V. G. Sowmeya · M. Sathiavelu (✉)
School of Biosciences and Technology, VIT, Vellore, India
e-mail: smythili@vit.ac.in

K. A. Abd-Elsalam, S. F. AbuQamar (eds.), *Fungal Endophytes Volume I*,
https://doi.org/10.1007/978-981-97-7312-1_8

diseases. Thus, in recent years, the focus has shifted to the microorganisms that thrive within plants, termed endophytes. These can be beneficial to the plants by supporting growth without causing any symptomatic diseases. These non-pathogenic organisms secrete a broad range of compounds that are not yet explored completely (Gouda et al. 2016). These untapped resources may open new avenues for the treatment of diseases. Endophytes can also play a great role in boosting plant growth and help in tolerating stress. Since the endophytes are chemical powerhouses, they can even serve the role of fertilizers, thereby providing a sustainable solution to chemical fertilizers (Wang et al. 2023). Several bacteria and fungi are endophytic to some plants, rendering some beneficial effects on the host plant. The isolation of endophytic fungi has gained interest over the past decade. Endophytic fungal diversity is greatly attributed to geographical locations and the environmental variables, growth season, genotype of the host, and so on (Gakuubi et al. 2021).

They were capable of producing diverse bioactive metabolites such as steroids, alkaloids, phenols, terpenoids, isocoumarins, xanthones, quinones, peptides, and many others. This makes them prime candidates for unearthing novel drugs that could help mitigate diseases such as diabetes, cancer, antibiotic resistance, and many others (Adeleke and Babalola 2021; Tiwari and Bae 2022). Studies also suggest that endophytic fungi are capable of producing some bioactive metabolites inherent to their host plant. This trait is attributable to the close relationship of the microbe with the host plant (Hashem et al. 2023; Nisa et al. 2015). "One strain many compounds"—this approach makes fungal endophytes a great source for the biosynthesis of bioactive metabolites of pharmacological significance and commercialization (Singh et al. 2021). Many well-known medications have natural sources of origin, which has led to the establishment and recognition of the idea of bioprospecting natural products. The discovery that *Penicillium* sp. is capable of synthesizing penicillin has fueled the desire to study microbes for possible medical applications for decades (Fleming 1929). As a result, it was discovered that endophytic fungi originating from plants were capable of producing these kinds of beneficial compounds (Strobel and Daisy 2003; Strobel 2018). One of the most diverse and frequently reported genera of endophytic fungi is the *Fusarium* spp. The taxonomic classification of the fungi is given in Fig. 8.1. This filamentous fungus is a double-edged sword.

They are responsible for many diseases of the crops, leading to yield reduction, quality compromise, and so on. On the other hand, they produce an assorted array of bioactive metabolites that can act as life-saving drugs. Unquestionably, endophytic *Fusarium* species that reside in the tissues of host plants without exhibiting any disease symptoms have gradually demonstrated a remarkable ability to produce compounds that are officially approved for the treatment of several illnesses. The biosynthetic ability of the fungi belonging to *Fusarium* sp. makes them thrive in diverse ecological niches. The diverse groups of metabolites support plant growth by defending them from pathogens and pest attacks, and they can also help mitigate abiotic stress. They produce a plethora of compounds belonging to butenolides, sterols, alkaloids, phenalenones, terpenoids, cytochalasins, xanthones, cyclic peptides, depsipeptides, diphenyl ether, polyketides, anthraquinone derivatives, and

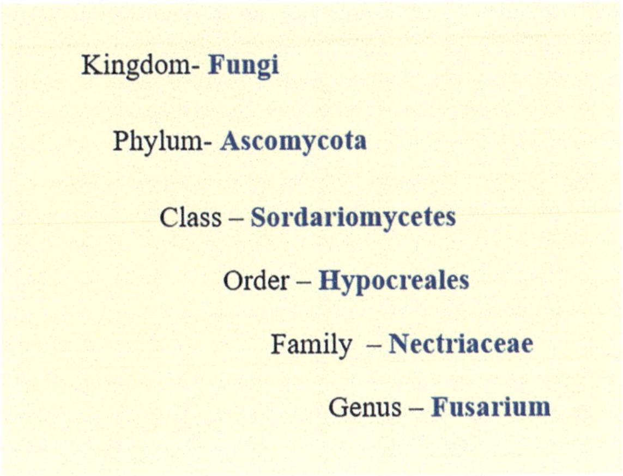

Fig. 8.1 Taxonomic classification of *Fusarium* species. There are at least 300 phylogenetically distinct species or species complexes in the filamentous fungal genus Fusarium, which is distributed across the world

many more that are capable of exerting various biological activities such as anticancer, antimicrobial, anti-inflammatory, and antioxidant activities (Ahmed et al. 2023; Wei and Wu 2020). Drugs like vincristine and vinblastine (Kumar et al. 2013a, b; Zhang et al. 2000; Tung et al. 2002); podophyllotoxin (Kour et al. 2008; Nadeem 2012), camptothecin and its analogs (Kusari et al. 2009; Ran et al. 2017; Shweta et al. 2010; Venugopalan et al. 2016), taxol (Elavarasi et al. 2012; Li et al. 2009; Xiong et al. 2013; Gurudatt et al. 2010), rohitukine (Kumara et al. 2014), quinine and cinchonidine (Hidayat 2016), ginkgolide B (Cui et al. 2012) have been identified as byproducts of various Fusarium species fermented in vitro. The main objective of this chapter is to explore the potential of endophytic fungi from the Fusarium genus as sources of bioactive metabolites through bioprospecting.

8.2 Antimicrobial Compounds

The rise and recurrence of multidrug-resistant (MDR) pathogens pose a challenge to the current healthcare system. Multidrug-resistant pathogen infections are notoriously difficult to treat, requiring the patient to have repeated courses of treatment using broad-spectrum antibiotics, which are also more toxic, expensive, and less effective (Lara et al. 2010). In the current era, new antimicrobial compounds are desperately needed to enter the pipeline for drug development to solve this problem. Numerous investigations have suggested that endophytes from the *Fusarium* genus may be a viable source of antibacterial chemicals. Research teams have looked into the efficacy of extracts obtained from endophytic fungus of the *Fusarium* genus against various bacterial and fungal diseases to identify possible avenues for developing novel antibiotics. The primary objective of these investigations is to pinpoint strong extracts that could be further investigated through chemical analysis to separate and describe possible active ingredients (Table 8.1).

These compounds might present a chance to produce the building blocks needed to build a library of semisynthetic substances, which would aid in hastening the

Table 8.1 Compounds produced by *Fusarium* spp. with antimicrobial activity

Species	Host	Metabolite	Biological activity	References
F. solani	*Taxus baccata*	Mixed volatile compounds	Antimicrobial	Tayung et al. (2011)
F. oxysporum	*Annona squamosa*	Polyketide	Antimicrobial	Tayung et al. (2011)
F. oxysporum	*Acianthera teres*	Crude	Antimicrobial	Tayung et al. (2011)
F. oxysporum	*Dracaena cambodiana*	Crude	Antimicrobial	Tayung et al. (2011)
F. oxysporum	*Discorea zingerberensis*	Crude	Antimicrobial	Tayung et al. (2011)
Fusarium sp. Dzf18 and *F. equiseti*	*Parvifolia Garcinia*	Crude	Antimicrobial (antibacterial)	Sim et al. (2010)
F. oxysporum	*Nothapodytes foetida*	Crude	Antimicrobial	Musavi and Balakrishnan (2013)
F. oxysporum	*C. odorata*	Crude	Antimicrobial	Toghueo et al. (2016)
F. oxysporum	*R. apiculate*	Crude	Antimicrobial	Moron et al. (2018)
F. oxysporum	*Candida albicans*	Crude	Antimicrobial (very active against yeast)	Moron et al. (2018)
F. lateritium	*Rhizophora mucronata*	Crude		Chowdhary et al. (2022)
Fusarium sp.	*Selaginella pallescens*	Pentaketide (2-methyl butyraldehyde-substituted-α-pyrone)	Antimicrobial	Brady and Clardy (2000)
F. oxysporum	Bark of *Cinnamomum kanehirae*	Beauvericin	Antimicrobial (suppresses growth of methicillin-resistant bacteria)	Chowdhary et al. (2022)
Fusarium sp. Ppf4	*P. polyphylla*	5alpha, 8alpha-epidioxyergosta-6, 22-dien-3beta-ol, butanedioic acid, 22-dien-3beta, 5alpha, 6beta, 7alpha-tetraol	Antimicrobial	Kaul et al. (2012)

Species	Host	Metabolite	Biological activity	References
Fusarium solani	*G. glabra*	3-*O*-methylfusarubin 3,6,9-Trihydroxy-7-methoxy-4,4-dimethyl-3,4-dihydro-1H-benzo[g]isochromene-5,10-dione Javanicin Fusarubin	Antibacterial	Shah et al. (2017)
F. solani	*C. alata*	Fusarubin Bostrycoidin Anhydrofusarubin	Antibacterial	Khan et al. (2018)
F. proliferatum	*Syzygium cordatum*	Ergosta-5,7,22-trien-3β-ol 9-*O*-methyl fusarubin Bostrycoidin-9-methyl ether 8-Hydroxy-5,6-dimethoxy-2-methyl-3-(2-oxo-propyl)-1,4-naphthoquinone Nectriafurone-8-methyl ether Bostrycoidin	Antibacterial	Dame et al. (2016)
F. oxysporum	*C. kanehirae*	Beauvericin	Antibacterial Acts against *Leishmania braziliensis*	Do Nascimento et al. (2012) and Wang and Xu (2012)
F. redolens Dzf2	*D. zingiberensis*	Crude	Antibacterial	Wang et al. (2011)
F. tricinctum	*Salicornia bigelovii*	Fusarielin B enniatin B sesquiterpenoid, fusartricin	Antimicrobial	Zhang et al. (2015)
F. equiseti	*Padina pavonica*	w-Hydroxyemodin cordycepin 17-Demethyl-2,11-dideoxy-rhizoxin and w-hydroxyemodin Cyclo (D-cis-Hyp-L-Leu)	Acts against *S. aureus* and *B. subtilis* *C. albicans* *B. megaterium*	Hawas et al. (2016)
Fusarium sp.	*Selaginella pallescens*	Antifungal agent, pentaketide, CR377	Acts against *C. albicans*	Brady and Clardy (2000)

(continued)

Table 8.1 (continued)

Species	Host	Metabolite	Biological activity	References
F. proliferatum AF-04	*Chinese onion (green)*	Indol-3-acetic acid Bassiatin Epicyclonerodiol oxide Fusaproliferin 5-*O*-methyljavanicin Anhydrojavanicin (1*R*,2*S*,3*R*)-3-((*S*)-2-hydroxy-6-methylhept-5-en-2-yl)-1,2-dimethylcyclopentan-1-ol Methyl indolyl-3-acetate Beauvericin 3β-Hydroxy-β-acorenol 5-*O*-methylsolaniol Methyl ether fusarubin	Antibacterial	Tang et al. (2022)
Fusarium sp. IFB-121	*Quercus variabilis*	(2*S*,2′*R*,3*R*,3′*E*,4*E*,8*E*)-1-*O*-beta-D-glucopyranosyl-2-*N*-(2′-hydroxy-3′-octadecenoyl)-3-hydroxy-9-methyl-4,8-sphingadienine Fusaruside	Antibacterial	Shu et al. (2004)
Fusarium sp.	*Mentha longifolia* L.	Fusaripeptide A	Antifungal	Ibrahim et al. (2018)
F. chlamydosporium	*Anvillea garcinii*	Fusarithioamide A	Antibacterial	Ibrahim et al. (2016a)
F. tricinctum	*Hordeum sativum*	Enniatins Q, B1, B2, B, A, A1	Antibacterial	Zaher et al. (2015)
Fusarium sp.	*Mentha longifolia*	Integracides G and F	Antileishmanial activity	Ibrahim et al. (2016b)

search for novel antimycobacterial medications. To accelerate the search for novel drugs to treat infectious diseases and numerous other neglected tropical diseases, these natural compounds can offer new scaffolds. Analysis of active and extremely complex crude extracts may lead to the purification of unidentified compounds with a wide to narrow spectrum of activity. With bioguided analysis, dozens of unidentified compounds with remarkable activity profiles were found among the crude metabolites of *Fusarium* spp. It has been reported numerous times that this effect-related analytical methodology is the best method for investigating natural compounds for potential medication candidates. This method is considered effective for examining complex mixtures to find certain parts (Weller 2012).

The most serious parasite-related illness is malaria, which has been estimated to have caused 219 million cases of morbidity and 435,000 deaths globally in 2017 (World Health Organization 2018). The ongoing emergence of parasite resistance to antimalarial medications creates a hurdle to the global elimination of malaria, although the infection is treatable (World Health Organization 2018). There is an urgent need for novel antiplasmodial medicines to fill the antimalarial drug development research pipeline. Historically, the investigation of natural materials has produced antimalarial medications. Thus, the identification of such lead compounds may result from the screening of natural sources. Fusaripeptide A has been reported to have a strong IC50 value of 0.34 μM for an antiplasmodial effect against *P. falciparum* (D6 clone) (Ibrahim et al. 2018). This substance, which operates in the nanomolar range, can be an excellent spark to start looking for novel anti-malarial drugs.

Though the extracts obtained from many *Fusarium* species showed strong potency against resistant *P. falciparum* strains, not many compounds derived from *Fusarium* spp. have been analyzed for antiplasmodial activity. This finding implies that investigating these species would present an opportunity to find substances that could treat an infection induced by resistant strains (Ateba et al. 2018). Additionally, it was observed that extracts from *Fusarium* sp. N240, isolated from *C. odorata*, was effective against both susceptible and resistant strains of *P. falciparum* (Toghueo et al. 2018). Similar results were obtained after investigating the ethyl acetate extract obtained from *Fusarium* sp. AMst1 endophyte of *Annona muricata* against *P. falciparum* Pf3D7, PfDd2, and PfIndo. Additionally, it was discovered that this extract prevented the ring from moving into the trophozoite stage (Toghueo et al. 2019). *P. falciparum* 3D7 was significantly inhibited by the extract from the endophytic fungus *Fusarium* sp. of a marine alga (Kaushik et al. 2014). These findings show that more thorough bio-guided research is required to thoroughly examine the metabolome of these fungi in search of novel antiplasmodial drugs (Toghueo 2020).

8.3 Anticancerous Compounds

The global rise in cancer is said to be 20 million cases, with 9.7 million deaths in 2022 (Bray et al. 2024). Developing anticancerous substances from endophytic fusarium fungi gives hope in drug development. In-vitro production of taxol, an anticancer metabolite by *Fusarium solani* has been reported as an endophyte from various yew (Chakravarthi et al. 2008). Taxol produced by *Fusarium culmorum* SVJM072 showed effective activity against human cancer cell lines. This taxol was identified based on spore production mechanism, fungal culture morphology, spore characteristics, and molecular characterization (Sonaimuthu et al. 2010) (Table 8.2).

Table 8.2 Anticancer compounds produced by endophytic *Fusarium* species

Species	Host	Metabolite	Activity	Reference
Fusarium sp. 2ST2	Mangrove	Fusarritsetins Two new 3-decalinoyltetramic acid (3DTA)	Anticancer	Chen et al. (2022)
Fusarium proliferatum AED3	Mangrove plant *Ardipsia elliptica*	Beauvericin	Anticancer (against T47D cancer cell line)	Handayani et al. (2021)
Fusarium solani	*Camptotheca* trees	CPT same as plant	Anticancer (inhibits cell growth and induces apoptosis on vero cells)	Chen et al. (2022)
Fusarium solani		Taxol	Anticancer	Chakravarthi et al. (2008)
Fusarium culmorum SVJM072		Taxol	Anticancer (against human cancer cell lines)	Handayani et al. (2021)
Fusarium solani	*Podophyllum hexandrum* roots	Podophyllotoxin	Precursor of some anticancer drugs like teniposide, etoposide, and etopohos phosphate	Nadeem (2012)
Fusarium redolens	Camptothecin	Taxol	Anticancer	Hoque et al. (2022)
Fusarium oxysporum	*Catharanthus roseus* inner bark	3β,5α-Dihydroxy-ergosta-7,22-dien-6-one, *p*-hydroxy benzaldehyde (3), 3-(*R*)-7-butyl-6,8-dihydroxy-3-pent-11-enylisochroman-1-one, vincristine, 3β,5α,9α-trihydroxy-ergosta-7,22-dien-6-one, beauvericin	Lead compound for anticancer drug development	Hoque et al. (2022)
Fusarium sp.	*Camptotheca acuminata* inner bark	9-Methoxycamptothecin, 10-hydroxycamptothecin	Anticancer (potent inhibitor of topoisomerase-1 enzyme)	Kusari et al. (2009)

(continued)

Table 8.2 (continued)

Species	Host	Metabolite	Activity	Reference
Fusarium solani	*Taxus chinensis*	Taxol	Anticancer	Tayung et al. (2011)
Fusarium oxysporum	*Nothapodytes nimmoniana*	Camptothecin	Anticancer	Tayung et al. (2011)
Fusarium oxysporum	*Annona squamosa*	Polyketide	Anticancer	Tayung et al. (2011)
Fusarium oxysporum	*Taxus cuspidate*	Taxol	Anticancer	Tayung et al. (2011)
Fusarium oxysporum	Leaves of *Rhizomphoru annamalayana*	Taxol	Anticancer	Elavarasi et al. (2012)
Fusarium oxysporum	*Cinnamomum kanehirae* bark	Beauvericin	Prostate cancer, human pancreatic carcinoma, lung carcinoma epithelial	Wang et al. (2011)
Fusarium oxysporum	*Ephedra fasciculata* root	Beauvericin	CNS glioma, metastatic prostate cancer, breast cancer cell line, fibroblast like fetal lung cell, human hepatocellular liver carcinoma	Ivanova et al. (2006), Zhan et al. (2007)
Fusarium oxysporum EPH2RAA	*Cylindropuntia echinocarpus* stem	Beauvericin	Pancreatic carcinoma, non-small cell lung cancer, CNS glioma, metastatic prostate cancer, breast cancer cell lines	Zhan et al. (2007)
Fusarium oxysporum CECIS	*Cylindropuntia echinocarpus* stem	Bikaverin	Pancreatic carcinoma, non-small cell lung cancer, breast cancer cell lines, CNS glioma, Ehrlich ascites carcinoma, sarcoma	Fuska et al. (1975), Zhan et al. (2007)
Fusarium solani	*Taxus mairei*	Taxol	Anticancer	Tayung et al. (2011)
Fusarium solani	*Apodytes dimidiate*	Campothecin	Anticancer	Tayung et al. (2011)

8.4 Anti-inflammatory Compounds

The Inflammatory response occurs as a defensive response against any injury or infection to the tissues. However, chronic inflammation can lead to the progression of varied diseases such as atherosclerosis, chronic obstructive pulmonary disease, inflammatory bowel disease, autoimmune diseases, cancer, glomerulonephritis, neurodegenerative disorders, and so on. Therefore, it is essential to suppress the excessive production of proinflammatory cytokines (Chen et al. 2018b). Recently, the search for novel compounds of microbial origin has been of great interest. Several studies have explored the potential of endophytic *Fusarium* spp. for use as an anti-inflammatory agent. *Fusarium* endophytes isolated from *Garcinia gummigutta, G. morella, G. indica*, and *G. xanthochymus* were found to have anti-inflammatory activity. Ethyl acetate extracts of the fungus at a concentration of 100 μg were able to inhibit human COX-2 and LOX-lipoxygenase (Kumar et al. 2013c). Chlamydosterol A and B produced by *Fusarium chlamydosporum*, an endophyte of *Anvillea garcinii* was capable of inhibiting 5-LOX with an IC50 value of about 3.06 μM (Al-Rabia et al. 2021). The roots of *Mentha longifolia* harbor *Fusarium* sp. as an endophyte that produces new ergosterol derivatives Fusaristerols B, C, and D. Fusaristerol B showed remarkable inhibitory activity against 5-LOX with an IC50 value of 3.61 μM (Khayat et al. 2019). Fusariumin A, B, asterpenoid A, agahtic acid, guiganardone, and trametenolic acid were produced by *Fusarium* sp. YD-2 obtained from *Santalum album.* Asterpenoid A exhibited anti-inflammatory activity by downregulating nitric oxide production at an IC50 value of 1.6 μM in LPS-activated macrophage cell lines (RAW 264.7) (Yan et al. 2018). Fusarubin, 8(Z)-lucilactaene, 4(Z) lucilactaene, javinicin, solaniol, 9-desmethylherbarine, NG391, and NG393 were isolated from *Fusarium* sp. QF001 derived from *Scutellaria baicalensis.* All the compounds inhibited the production of NO, TNF-α, and IL-6 in macrophages stimulated by LPS (Maharjan et al. 2020). Fusopoltide A from *Fusarium solani* effectively inhibited COX-2 (Chen et al. 2018a). Studies indicate the potency of compounds obtained from *Fusarium* sp.

8.5 Antioxidants

Recently, antioxidants have gained attention. These days, there is increased interest in the topic of free radical chemistry. Reactive oxygen and nitrogen species, or free radicals, are produced by the body's many physiological activities. Unregulated production of free radicals damages DNA, proteins, enzymes, and membrane lipids, resulting in oxidative stress and eventually cell death. Numerous degenerative human diseases, including Alzheimer's disease, Parkinson's disease, atherosclerosis, cancer, aging, diabetes mellitus, and inflammatory diseases, are brought on by these reactive oxygen species—ROS. Consuming adequate exogenous antioxidants can improve defense against free radicals. An antioxidant is a molecule with good stability that gives a free radical on the rampage of an electron to stop the chain reaction before any other important molecules are harmed. Dietary antioxidants such as vitamin C, E, and polyphenolic substances, are considered to be the most beneficial nutrients for preventing disorders linked to oxidative stress. A trustworthy and plentiful source of new antioxidant molecules is fungal endophytes. To investigate the possible impacts of fungal endophytes, numerous studies were carried out on their antiviral, anticancer, antidiabetic, and antibacterial properties; however, very few researchers have focused on their antioxidant capacity (Rajamanikyam et al. 2017). Endophytic *Fusarium* spp. has also been reported to produce several antioxidant molecules.

Taxol, a potent antioxidant is extracted from various *Fusarium* endophytes (Chakravarthi et al. 2008; Deng et al. 2009; Elavarasi et al. 2012; Garyali 2013; Li et al. 2008, 2009; Strobel et al. 1996; Xiong et al. 2013; Xu et al. 2006). Another antioxidant camptothecin, which is also an anticancerous agent, has been extracted

Table 8.3 Antioxidants secreted by *Fusarium* species via in vitro fermentation

Fungi	Host plant	Antioxidant	Reference
F. lateritium	*Rhizophora mucronate*	Methanol extract	Hamzah et al. (2018)
F. oxysporum	Flower of *Dendrobium lindleyi*	Ethyl acetate extract	Bungtongdee et al. (2019)
F. tricinctum		Crude metabolites	Vasundhara et al. (2016).
F. proliferatum, *F. oxysporum* (ERP-10), and *F. solani* (ERP-07)	*Cajanus cajan*	Cajaninstilbene acid (natural antioxidant)	Zhao et al. (2012)
F. oxysporum	*Curcuma amada*	1,5-Pentanediol, 2,3-pentanediol (antioxidant and antiaging)	Tiwari et al. (2014)
Fusarium spp.		Islandic acid-II methyl ester (ability to restore fertility) Cladobotrin V (*E*)-3-(5-ethyl-4-methoxy2-oxo-2*H*-pyran-6-yl) acrylic acid	Shiono et al. (2015)
F. oxysporum, *F. incarnatum*, *F. solani*	*C. calisaya*	Quinine and cinchonidine	Hidayat (2016)
F. solani	*A. dimidiata* *C. acuminata*	10-Hydroxycamptothecin	Kusari et al. (2009), Shweta et al. (2010)
F. solani	*A. dimidiata* *C. acuminata*	9-Methoxycamptothecin	Kusari et al. (2009), Shweta et al. (2010)

(continued)

Table 8.3 (continued)

Fungi	Host plant	Antioxidant	Reference
F. oxysporum, *F. solani*	*J. recurva* *P. hexandrum*	Podophyllotoxin	Kour et al. (2008), Nadeem (2012)
F. oxysporum *F. solani*	*D. binectariferum* *D. binectariferum*	Rohitukine	Kumara et al. (2014)
F. oxysporum	*C. roseus*	Vinblastine	Kumar et al. (2013b)
F. solani	*C. roseus*	Vinblastine and vincristine	Kumar et al. (2013a)

from Fusarium endophytes, namely, *F. solani* and *F. sacchari* (Gurudatt et al. 2010; Shweta et al. 2010; Kusari et al. 2009; Ran et al. 2017) (Table 8.3).

8.6 Nanoformulations

Nanotechnology is a booming branch that extends its application to all fields. In recent years, green synthesis of nanoparticles has been adopted as a sustainable means of development. Microbe-mediated synthesis of nanoparticles is more advantageous because of its easier cultivation. The intrinsic capabilities of microbes to reduce metal ions make them more utilizable as nano factories. Microbial NPs have been greatly explored for their antimicrobial, antitumor activities, reproductive medicine, drug delivery, and others (Fariq et al. 2017). Microbial synthesis includes bacteria, fungi, viruses, actinomycetes, and yeast. Fungal-mediated synthesis of Nps referred to as Mycosynthesis/Myconanotechnology is highly preferred since fungi produce a wide array of bioactive metabolites. Among fungal spp., *Fusarium* is greatly preferred, which was evident from the literature. *Fusarium* spp. for nanoparticle synthesis has various advantages:

- Easy downstream processing
- Nps are secreted extracellularly
- NP synthesis also occurs extracellularly
- NPs are produced in colloidal form
- Easier conversion to powder form
- Can produce various types of NP

The metabolites produced by fungus also act as stabilizing agents; therefore, stable NPs can be produced (Rai et al. 2021). It has been proposed that the NP synthesis occurs via the electron shuttle. The extracellular synthesis mechanism involves the nitrate reductase enzyme present in the cell-free extract of *Fusarium* spp. The electron donation by NADPH and quinone is responsible for the reduction of $M+$ to MNPs. Intracellular synthesis is mediated by the attachment of metal ions to negatively charged cell walls. NP synthesis occurs by the reductase enzyme

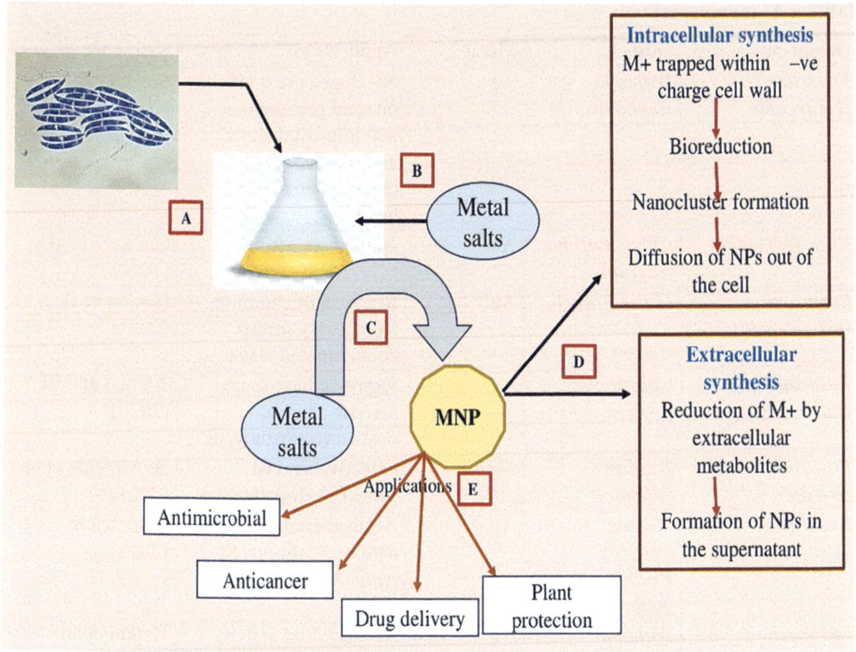

Fig. 8.2 Mechanism of synthesis of NP by *Fusarium* spp. (**a**) culturing of endophytic *Fusarium* spp., (**b**) addition of metal salts, (**c**) reduction of metal salts to metal nanoparticles, (**d**) mechanism involved in intracellular and extracellular synthesis, (**e**) applications of synthesized nanoparticles

Table 8.4 Nanoparticles synthesized from endophytic *Fusarium* strains

Organism	Host	Metal	Application	Reference
Fusarium sp.	*Withania somnifera*	Ag	Antibacterial activity against *S. typhi*, *S. aureus*, *E. coli*	Singh et al. (2015)
Fusarium solani	*Nothapodytes foetida*	Ag	Not explored	Musavi and Balakrishnan (2013), Sogra Fathima and Balakrishnan (2014)
Fusarium solani	*Withania somnifera*	Ag	Antibacterial activity against *K. pneumonia*, *P. aeruginosa*, *E. coli*, *S. typhi*, *S. aureus* and cytotoxicity studies	Vijayan et al. (2016)
Fusarium oxysporum	*Azadirachta indica*	Au	Anti-proliferative activity against specific cancer cell lines	Ahmad Siddiqui et al. (2016)
Fusarium acuminatum	*Zingiber officinale*	Ag	Antibacterial activity against human pathogens	Ingle et al. (2008)

(continued)

Table 8.4 (continued)

Organism	Host	Metal	Application	Reference
Fusarium semitectum	*Withania somnifera*	Ag	NPs had a great effect on seed germination and improved shoot, and root length in chickpea and mung bean	Singh et al. (2016)
Fusarium solani	*Chonemorpha fragrans*	Au	Anticancer activity	Clarance et al. (2020)
Fusarium verticillioides	*Maize grains*	Au	Removal of cadmium from heavy metal-contaminated water	Hassan et al. (2021)
Fusarium chlamydosporum	*Calotropis procera*	Ag	Expressed antifungal activity against postharvest pathogens	Ahmed and El (2020)
Fusarium concolor	*Paullinia cupana*	Ag	Effective against fungal phytopathogens	de Almeida et al. (2021)
Fusarium oxysporum	*Mentha arvensis Psidium guajava*	Ag	Antibacterial activity against *S. aureus*, *S. typhi*, *K. pneumoniae*, *E. coli*	Manoharan (2015)
Fusarium semitectum	*Withania somnifera*	Ag	Antibacterial activity against *Porphyromonas gingivalis*	Halkai et al. (2017)
Fusarium oxysporum	*Plant leaves*	Nitrogen-doped CO_3O_4	Photocatalytic degradation of methylene blue	Islam et al. (2021)
Fusarium oxysporum	*Panax ginseng*	Ag	Not explored	Bao et al. (2017)

bound to the cell membrane, leading to the formation of NPs (Rai et al. 2021). Fabrication of nanomaterials using bioactive metabolites would substantially enhance the functional capacity of the metabolites. The mechanism of synthesis of nanoparticles by *Fusarium* spp. is depicted in Fig. 8.2. Several *Fusarium* endophytic strains have been reported to synthesize nanoparticles, and their biological efficacy was analyzed, which are tabulated below (Table 8.4).

8.7 Insecticidal Activity

Several insects are known to cause damage to plants. They affect the plants in several ways by boring into the stem, feeding on fruits and vegetables, and cutting off roots and stems. Pest attacks cause significant economic losses. The utilization of synthetic insecticides and pesticides has controlled these pests to some extent, but in turn, they can have harmful effects on humans and the environment. Therefore, research on the utilization of natural insecticides is indispensable. Secondary metabolites of fungi have been greatly explored in many aspects. Little has been explored in terms of insecticides. Few studies highlight the importance of exploring *Fusarium* spp. as an insecticide.

Apicidin produced by *Fusarium* sp. has been reported to have insecticidal and nematicidal properties (Anke 2002). *F. sambucinum*, isolated from *Nicotiana tobacum*, expresses potent insecticidal activities. Indole alkaloids such as Sclerotiamide B secreted by the isolate were effective against *Helicoverpa armigera* larva-cotton bollworm (Zhang et al. 2019). Recent studies have explored the effect of *Fusarium* spp. on *Spodoptera frugiperda* smith and *Ceratitis capitata* Wiedemann. The organic extract of the fungi hindered the feeding capacity of *S. frugiperada* and altered the oviposition in *C. capitata.* The extract was mainly composed of sterols and fatty acids (Marcinkevicius et al. 2019). Further research into the bioactive metabolites of *Fusarium* can help ameliorate insect stress during harvest and also mitigate post-harvest loss.

8.8 Larvicidal Activity

In recent years, vector-borne diseases have been increasing in our environment due to varying climatic conditions. The emergence of various vector-borne diseases is a major health concern. New strategies to combat these vector-borne diseases are the need of the day (Chala and Hamde 2021). *Fusarium* isolated from *Vitex negundo* exhibited antagonistic activity against mosquitoes. The ethyl acetate extract of the fungi at 20 ppm concentration showed about 90% larvicidal activity against *Aedes aegypti.* The *Fusarium* isolates from *Azadirachta indica*, *Tagetes erecta*, *Mentha piperta*, *Andrographis paniculata*, and *Ocimum sanctum* exhibited significant larvicidal effects (Sathiyanathan and Umarajan 2019). A pigment produced by *F. moniliforme* was analyzed against *Ades aegypti* and *Anopheles stephensi* larva and pupa. The EtOAc extracts containing the pigment showed great potential to be used as mosquito larvicidal agents (Pradeep et al. 2015). *F. proliferatum* and *F. equiseti* expressed lethality against *Aedes aegypti* (Abrar et al. 2023). It has been proposed that the fungal metabolites penetrate the cuticle of the larva thereby deciphering larvicidal activities. Secondary metabolites of *F. oxysporum* were effective against the first and fourth instar stages of *Culex quinquefasciatus* and *Anopheles stephensi* (Prakash et al. 2010). Beauvericin produced by *Fusarium lateritium* at 20 µg/mL exhibited about 85% larvicidal activity against *Aedes aegypti* (Frederick Grove and Pople 1980). Mosquitoes are vectors for the spread of many diseases, namely chikungunya, West Nile virus, malaria, dengue, yellow fever, zika, and so on. Research to control these vectors from microbial sources would help develop eco-friendly larvicides.

8.9 Peptides

Peptides are short-chain amino acids that are naturally produced by plants and microorganisms. Their molecular weight is less than 10 kDa and they are endowed with greater biological activities, which makes them an excellent source of therapeutics (Wu et al. 2021). Microorganisms synthesize peptides to defend themselves by inhibiting the proliferation of other pathogenic microbes (Singh and Abraham 2014). This inherent nature of the peptides makes them an inevitable source for combating antimicrobial resistance. Fungi are well-known producers of cyclic peptides and depsipeptides such as cyclosporins, ergosterols, and so on, which have clinical relevance (Anke and Antelo 2009). The genus *Fusarium* has been found to harbor biosynthetic gene clusters comprising non-ribosomal peptide synthetases and polyketide synthases, which contribute to the synthesis of peptides (Nielsen et al. 2019). *Fusarium tricintum*, an endophyte of *Rhododendron tomentorum*, was identified to synthesize an intact peptide Trtesin. The isolated peptide was capable of inhibiting the growth of *Fusarium oxysporum Candida utilis*, *Staphylococcus aureus*, and *Candida albicans* (Tejesvi et al. 2013). Fusaristatin A and three other pentapeptides produced by *Fusarium decemcelluare* isolated from *Mahonia fortune* exhibited antibacterial activity against a wide range of gram-positive and gram-negative pathogens. These peptides also exerted an antagonistic effect against *Glomerella acutata* (Li et al. 2016). *Fusarium* spR5, an endophyte of *Muscari botryoides* was capable of synthesizing cyclic hexadepsipeptides, namely, Fusarihexin A and B. These peptides exhibited extensive antifungal activity against *Colletotrichum gloeosporioides*, *Colletotrichum musae*, and *Fusarium oxysporum* (Zhu et al. 2018). Fusaripeptide A produced by *Fusarium* sp. endophytic to *Mentha longifolia* exhibited greater anti-malarial activity against *Plasmodium falciparum* and also had a broad range of antifungal activity (Ibrahim et al. 2018). *Fusarium solani* isolated from *Narcissus tazetta* was capable of producing hexacyclopeptides which can act as communicating signals thereby promoting synergism in the niche (Wang et al. 2015). Secobeauveicin A, Beauvericin M-cyclic hexadepsipeptides were obtained from a *Fusarium* sp. endophytic to a tea plant (Chang et al. 2023). Several novel depsipeptides have been reported in *Fusarium* endophytic to tea plants (Lv et al. 2015). Application-oriented studies are essential to analyze the efficiency of the peptides secreted by endophytic *Fusarium* spp.

8.10 Exopolysaccharides

Microbial polysaccharides are a vibrant source of compounds that have both commercial and medicinal applications (Angelin and Kavitha 2020). Recent studies have highlighted the production of EPS by endophytic microorganisms, which are found to display structural diversity (Liu et al. 2017). *Fusarium* spp. are also known to secrete EPS. *Fusarium solani* an endophyte of *Alstonia scholaris* composed of, (1→2)-α-L-rhamnopyranosyl, (1→4,6)-β-D-galactopyranosyl,

Table 8.5 *Fusarium* spp. producing exopolysaccharides

S. no.	Organism	Host	Activity	Reference
1	*Fusarium* sp. A14	*Fritillaria unibracteata*	Antioxidant; antiproliferative effects	Pan et al. (2019)
2	*Fusarium solani* DO7	*Dendrobium officinale*	Immunomodulatory properties Antibacterial effects	Zeng et al. (2019a, b)
3	*Fusarium culmorum*	*Secale cereale* *Triticum aestivum*	Antioxidant	Jaroszuk-Ściseł et al. (2020)
4	*Fusarium equiseti* ANP2	Mangroves	Antioxidant	Prathyusha et al. (2018)
5	*Fusarium oxysporium* Dzf17	*Dioscorea zingiberensis*	Enhanced diosgenin production Antioxidant	Li et al. (2011), Li (2011)
6	*Fusarium oxysporum*	*Otoba gracilipes*	Antioxidant	Caicedo et al. (2019)
7	*Fusarium oxysporum* Y24-2	*Ipomoea pes-caprae*	Unexplored	Guo et al. (2013)

α-L-rhamnopyranosyl, (1→4)-β-D-galactopyranosyl moieties. This rhamnogalactan was found to have anti-allergic and anti-inflammatory properties and has no cytotoxicity (Mahapatra and Banerjee 2013a, b). Several other species of fusarium that are reported to secrete EPS are tabulated below (Table 8.5).

The exopolysaccharides have great potency for pharmaceuticals, cosmetics, functional food ingredients, and many more. Research must go beyond the structural elucidation and analyze their biological activities as the *Fusarium* strains could be excellent sources of EPS.

8.11 Plant Growth Promotors

Fungi grow without causing any symptoms of disease in the inner plant tissues; thus, they benefit from this interaction and also provide benefits to the host plant. Until now, many studies and research have mentioned the supportive aspects of endophytic fungi (Baron and Rigobelo 2022). In plant biology endophytes are an essential part of plants and serve as significant mechanisms for plant micro-ecosystems (Christian et al. 2016). After witnessing the positive result *of F. semitectum* on orchid seed germination, the concept of "Mycovitalism" gradually started being accepted (Jiang et al. 2011). Plants require endophytic microbes, according to the evidence provided by research data (Selosse et al. 2014). Bilal et al. (2018) reported that the endophytic fungal strain *Fusarium proliferatum* BRL1 isolated from *Osalis corniculata*'s root exhibited growth-promoting abilities. The isolate was assayed in mutant grain Waito-C for the capability as a growth promotor, and it showed increased length in root and shoot, higher production of biomass, and higher content of chlorophyll. These results indicate that the bioactive compounds synthesized by the isolated endophytic fungi play an important role as growth promotors.

Few other studies mention that their capability to sustain plant health and endophytic Fusarium plays an imperious role in preparing plants to counter abiotic stresses as well as in growth enhancement (Lata et al. 2018). For example, the presence of *F. culmorum* increased the capacity to tolerate stress and high salinity in *Leymus mollis* (Rodriguez et al. 2008). In 2018 Cui et al. (2018) stated that the germination of *C. songaricum* seeds was promoted by three endophytic isolates *F. nematophilum* KY379572, *F. redolens* KY379544 and *F. nematophilum* KY379764. This was supported by the finding of (Shah et al. 2018) that the growth and development of *Rhynchostylis retusa* were endorsed by the endophytic fungi *Fusarium* isolated from the root of *Dendrobium moniliforme*. Additionally, numerous *Fusarium* spp. Metabolites were well-reported as plant growth promotors. To mention a few, the antioxidant capacity, sprout growth, and flavonoid accumulation of tartary buckwheat were stimulated by polysaccharides obtained from *F. oxysporum* (Zhong et al. 2016). *F. oxysporum* Dzf17 induced diosgenin accumulation in the host plant *Dioscorea zingiberensis* (Li et al. 2011). The prospective to improve the defense system of a plant and the impact of endophytic *Fusarium* on its physiology have been investigated for numerous years.

Fusarium oxysporum GG22, which was isolated from *Rehmannia gutinosa* leaves, acts as a growth promotor of *R. glutinosa*. This strain was found to enhance plant leaf area, height, dry weight, and fresh weight plant tissue culture seedlings. Also, catapol and verbascoside are considerably improved. The SOD, POD, and CAT (antioxidant) of the plant were increased after the inoculation of the GG22 strain. So, these outcomes established that GG22 also boosts secondary metabolism and stimulates the growth of *R. glutinosa* (Zhu et al. 2022).

In a study, the endophytic fungi *Fusarium tricinctum* RSF-4L isolated from the *Solanum nigrum* leaves was analyzed for its capability as a plant growth promotor in Dongjin rice. All the growth features like chlorophyll content, biomass production, and root-shoot length were noticeably increased when treated with CFs of *F. tricinctum* (Khan et al. 2015). Similarly, an endophytic fungus *F. oxysporum* isolated from *Withania somnifera* leaves, synthesized ammonia and indole-3-acetic acid in significant amounts. Compared to the control plants, the CF strain considerably improved the growth of maize established by dry and fresh plant biomass and higher root and shoot length. This proves that *Fusarium oxysporum* can promote plant growth, and it was recommended as a biofertilizer to increase the yield in agriculture (Hussain and Khan 2018).

Another *F. oxysporum* strain GW improved wheat growth by inhibiting *A. fatua* a weed that competes with wheat and is responsible for the losses in yield. Dihyroxydiethoxyisoflavone, quercetagetin, calycosin, and isovitexin are the substances produced by the stains that suppressed the growth of weed without affecting the wheat seedlings (Asim et al. 2022). The given data altogether proved that fungal endophytes play a significant and valuable role in host plant physiology and improved root growth, providing nutrition and plant development even in higher stress tolerance. Different strains of endophytic *Fusarium* were commonly isolated from various host plants. It shows the competence of endophytic *Fusarium* as a most significant substance to the host plants. It establishes the ability to use endophytic *Fusarium* to increase the yield of the crop.

8.12 Biofuel Production

Biofuel production from endophytic fungi is of growing interest due to the production of diverse bioactive metabolites secreted by microorganisms, which can serve as precursors for the conversion of complex biomass (Bhagobaty 2015). Numerous research studies have shown that enzymes like lignases, cellulases, and hemicellulases, which are crucial for plant biomass decomposition, can be synthesized by endophytic fusarium. For instance, *Fusarium* fungi isolated from *Miscanthus sacchariflorus* root showed high levels of xylanase and cellulase activity. These are the key enzymes involved in the conversion of lignocellulosic biomass into biofuel such as ethanol (Zhang et al. 2017). Furthermore, it was found that bioactive metabolites synthesized by the *Fusarium* genus can help to improve biofuel production. The bioactive metabolites produced by endophytic *Fusarium* isolated from *Camellia sinensis* leaves have antifungal properties that are used during the fermentation process of biofuel production to prevent microbial contamination and improve productivity (Zhao et al. 2018).

In addition, different species from the genus of *Fusarium are* shown to play an important role in the ability to synthesize bioactive substances for the production of biofuel. A study established that diverse strains of endophytic *Fusarium* isolated from various plant species show evidence of differing levels of bioactive metabolites and enzyme production. This creates inquisitiveness for exploring the importance of endophytic *Fusarium* for finding new bioactive substances for biofuel production (Kusari et al. 2013).

F. sambucunum and two other *Fusarium* sp. exhibit a remarkable ability to synthesize laccase, lignin peroxidase, and manganese-dependent peroxidase which can act on the feedstock in the production of biofuel (Falade et al. 2021). Inulinase is a biocatalyst that hydrolyses Inulin into simple sugars, especially fructose. The same study has reported the ability of endophytic *Fusarium* to produce inulinase which could act on the inulin sugars and significantly impact fuel production (Falade et al. 2021).

A systems biology was applied to enhance biofuel production by *Fusarium strains. Fusarium oxyporum* produced more ethanol by expressing phosphoglucomutase in transaldolase and upper glycolysis in the Pentose Phosphate pathway and found that higher expression of two enzymes enhanced regeneration of NADPH, which leads to moving carbon flux towards ethanol production (Martien and Amador-Noguez 2017). Though the prospecting of *Fusarium* spp. for biofuel production seems to be more promising, there are several challenges that need to be unveiled. The presence of inhibitory substances in the substrate, specificity, and low yields need to be addressed to enhance the efficiency of biofuel production. Further understanding of the metabolic pathways involved in the production of biofuels would help in optimizing the production of biofuel.

8.13 Remediation Potential

The Endophytic *Fusarium* genus is commonly related to riparian plant species and is widely dispersed in numerous terrestrial plant species (Ruma et al. 2012). Research on the bioremediation potential of the said fungi is in emerging fronts albeit in a less explored area. They have the potential to degrade organic pollutants and promote growth.

Several experiments also have proved that *Fusarium* sp. A19 isolates broadly circulated in heavy metal-polluted land and has tolerance to several heavy metals, e.g., Zn, Cr, Co, Cu, Cd, and Pb (Pan et al. 2009). A study was conducted for *Fusarium* sp. isolated from *Salix variegata* Franch (Salicaceae) in China (three Gorges reservoir region). Inoculation experiments, molecular identification, and metal tolerance estimation are conducted to characterize Cd-resistant fungi (Nandy et al. 2020). In the same research, three different *Fusarium* spp. were isolated and they are identified as *F. nematophilium* and the other two were *F. solani* and all three strains are reported to have the ability to tolerate Cd (An et al. 2015).

Endophytic fungi *Fusarium* sp. CBRF44 isolated from *Brassica napus* can promote the remediation of Cd and Pb heavy metals (Sharma and Kumar 2021). *Fusarium* sp. isolated from the roots can enhance the growth of canola and improve the efficiency of phytoremediation (Deng and Cao 2017). The endophytic *Fusarium* genus has been found to remediate water contaminated by human and animal waste and soil contaminated by heavy metals. For instance, *Fusarium oxysporum* was used in phytoremediation of soil contaminated by heavy metal, and *Fusarium culmorum* strains were used for human and animal waste decomposition and decontamination (Ortega et al. 2020).

The shoot and root endophytes of *Aphelandra tetragona* were found to biotransform 2-hydroxy-1,4-benzoxazin-3-one, benzoxazinones, and 2-benzoxazolinone which were synthesized against pests by the plant defense system. An endophyte *Fusarium sambucinum* has been found to transform these substances into *N*-(2-hydroxyphenyl) acetamide and *N*-(2-hydroxyphenyl) acetamide analogues (Mishra and Venkateswara Sarma 2018). These findings suggest that the endophytic *Fusarium* can play a prominent role in remediating the soil thereby promoting the growth of the plants in a sustainable manner.

8.14 Enzymes

Enzymes are treasure troves for industries. Enzymes of microbial origin are being explored for their greater stability, ease of production, and performance criterion. Endophytic fungi secrete a broad range of enzymes that help the host plant defend themselves, and they also aid in nutrient enrichment for the microbes (Bhadra et al.

2022). Several endophytic fungi have been reported to produce cellulases, xylanases, lignases, and many others. *Fusarium* spp. has also been reported to produce pharmacologically and industrially significant enzymes. *Fusarium verticillioides* endophytic to *Zea mays* was capable of producing Endoglucanase, which aids in ethanol production (de Almeida et al. 2013). *Fusarium sambucinum* isolated from mangrove forests were found to produce lignin peroxidase and manganese peroxidase (Martinho et al. 2019). *Fusarium oxysporum* isolate PTM7, an endophyte of *Croton oblongifolius* Roxb, was capable of producing extracellular lipase (Tuangporn Panuthai 2012). *Fusarium sporotrichioides* isolated from *Ventilago madraspatana* secreted highly active proteases and other hydrolytic enzymes such as xylanase, cellulase, amylase, and pectinase. *Fusarium graminareum* isolated from the same host has quorum-sensing inhibitory potential, which might be attributed to the production of AHL-lactonases (Rajesh and Ravishankar Rai 2013). *Fusarium solani* from *Catharanthus roseus* was capable of producing amylase, cellulase and protease. These hydrolytic enzymes support penetration into plant cells, promoting symbiosis (Ayob and Simarani 2016). *Chrysanthemum indicum* harbors *Fusarium haematococcum*, which produces hydrolytic enzymes such as amylase, cellulase and protease that aid nutrient solubilization. This strain also helps to mitigate salinity stress through its antioxidant activity (Koshila Ravi et al. 2022). *Fusarium oxysporum* strains isolated from *Glycine max* were capable of extracellular production of carboxy methyl cellulase, amylase and xylanase (Montaser Farouk et al. 2020). L-Asparaginase is a pharmacologically significant enzyme that converts L-asparagine into ammonia and aspartic acid, making the cells starve and die through apoptosis, which has profound effects in cancer treatment. Endophytic fungus *Fusarium solani* from *Hedera helix* and *Fusarium proliferatum* from *Cymbopogon citratus* were capable of producing L-asparagine (Yap et al. 2021; Osama et al. 2023). Further optimization studies can improve the yield and efficiency of the enzymes.

8.15 Other Bioactivities

Fusarium isolated from *Calotropis gigantea* displayed a prominent α-glucosidase inhibitory activity. This suggests that *Fusarium* also has the potential to serve as a potential drug candidate for diabetes (Jasleen Kaur et al. 2018). *Fusarium* spp. endophytic to *Allium longicuspis* was found to inhibit pancreatic lipase, a major player in obesity and cardiovascular diseases (Gulyamova et al. 2022). *Fusarium* spp. capable of producing biosurfactant was isolated from *Eichhornia crassipes and* provided new avenues for the degradation of hydrocarbons (Lima et al. 2016).

The progressive degeneration of nerve cells caused by insufficiency in neurotransmitter secretion, abnormal protein aggregates, oxidative stress, and inflammation leads to neurodegenerative diseases such as Parkinson's disease, Alzheimer's disease, autism, and so on. Endophytic fungi are known to produce secondary metabolites capable of ameliorating neurodegenerative disorders (Semwal et al. 2022). *Camellia sinensis* harbors *Fusarium* spp., which is capable of disaggregating α-synuclein, thereby delaying or inhibiting the production of amyloid aggregates, which can have a profound effect on the treatment of Parkinson's disease (Vats and Saxena 2023). The crude extract of *Fusarium* sp. QQ-Fus-2-F isolated from *Euphorbia* exhibited anticholinergic activity (Omar S. Al-Qaralleh et al. 2021). 4,6′-Anhydrooxysporidinone produced by endophytic *F. lateritium* SSF2 displayed neuroprotective effects (Lee et al. 2020). *Fusarium solani* JS-0169 from *Morus alba* produces γ-pyrones, which exhibited neuroprotective effects by suppressing the synthesis of ubiquinone (Choi et al. 2020). Further studies on exploring the metabolites and improving the yield can provide valuable drugs to treat neuro disorders.

Several drugs have also been reported to be produced by endophytic *Fusarium* strains. Gingolide B, which is used in the treatment of cerebrovascular diseases, was produced by *Fusarium oxysporum* isolated from *Ginkgo biloba* (Cui et al. 2012). Vitexin, a compound with greater antioxidant and hepatoprotective effects, had been extracted from *Fusarium solani* G6 isolated from *Cajanus cajan* could enhance osteoblast proliferation, which could have a greater effect in the treatment of osteoporosis (Tang et al. 2021). *Fusarium* sp. ALAA-20 isolated *Aplysina fistularis* was capable of synthesizing Lovastatin—a drug used for lowering cholesterol levels. Recombination and optimization of the substrates improved the yield of Lovastatin. The recombinant lovastatin had improved antibacterial and antitumor activities (El-Bondkly et al. 2021). These studies highlight that using recombinant technology and media optimization can improve the yield and efficiency of the metabolites and help overcome the shortage of natural sources.

8.16 Conclusion

The *Fusarium* fungi are an untapped source of metabolites with novel biological activity and chemical diversity. But only a very few species have been well explored. There are a huge number of species to be explored, and the complete metabolome of the organism should be initiated, which could put forth the path for identifying novel therapeutics to satiate the needs of the expanding population. An improved understanding of the pathways involved in the synthesis of the bioactive metabolites could pave the way for improving production and efficiency through recombinant technology or genome shuffling. Studies have to be conducted to find possible ways to improve the production of bioactive metabolites by optimizing the media components, such as the utilization of low-cost media through the incorporation of agro-waste. This could help create cost efficiency, thereby having a direct impact on the economy. Further exploration can invoke insightful applications that could revolutionize pharma and other biotech industries and could also aid in mitigating various health problems and food insecurity-related issues.

References

Abrar A, Sarwar S, Abbas M, Chaudhry H, Ghani N, Fatima A, Tahir A (2023) Identification of locally isolated entomopathogenic Fusarium species from the soil of Changa Manga Forest, Pakistan and evaluation of their larvicidal efficacy against *Aedes aegypti*. Braz J Biol 83. https://doi.org/10.1590/1519-6984.246230

Adeleke BS, Babalola OO (2021) Pharmacological potential of fungal endophytes associated with medicinal plants: a review. J Fungi 7(2):1–16. https://doi.org/10.3390/jof7020147

Ahmad Siddiqui E, Ahmad A, Julius A, Syed A, Khan S, Kharat M, Pai K, Kadoo N, Gupta V (2016) Biosynthesis of anti-proliferative gold nanoparticles using endophytic *Fusarium oxysporum* strain isolated from neem (*A. indica*) leaves. Curr Top Med Chem 16(18):2036–2042. https://doi.org/10.2174/1568026616666160215160644

Ahmed M, El Y (2020) The efficiency of biosynthesized silver nanoparticles by endophytic *Fusarium chlamydosporum* F25 against plants postharvest fungal pathogen. Egypt Acad J Biol Sci G Microbiol 12(1):29–43. www.eajbs.eg.net

Ahmed AM, Mahmoud BK, Millán-Aguiñaga N, Abdelmohsen UR, Fouad MA (2023) The endophytic Fusarium strains: a treasure trove of natural products. RSC Adv 13(2):1339–1369. https://doi.org/10.1039/d2ra04126j

Al-Qaralleh OS, Al-Zereini WA, Al-Mustafa AH (2021) Antibacterial, antioxidant and neuroprotective activities of crude extract from the endophytic fungus *Fusarium* sp. isolate OQ-Fus-2-F from Euphorbia sp. plant. J Pharm Pharmacogn Res 9(6):755–765

Al-Rabia MW, Mohamed GA, Ibrahim SRM, Asfour HZ (2021) Anti-inflammatory ergosterol derivatives from the endophytic fungus *Fusarium chlamydosporum*. Nat Prod Res 35(23):5011–5020. https://doi.org/10.1080/14786419.2020.1762185

An H, Liu Y, Zhao X, Huang Q, Yuan S, Yang X, Dong J (2015) Characterization of cadmium-resistant endophytic fungi from *Salix variegata* Franch. in Three Gorges Reservoir Region, China. Microbiol Res 176:29–37. https://doi.org/10.1016/j.micres.2015.03.013

Angelin J, Kavitha M (2020) Exopolysaccharides from probiotic bacteria and their health potential. Int J Biol Macromol 162:853–865. https://doi.org/10.1016/j.ijbiomac.2020.06.190

Anke H (2002) Chapter 7: Insecticidal and nematicidal metabolites from fungi

Anke H, Antelo L (2009) Chapter 3: Cyclic peptides and depsipeptides from fungi. In: Physiology and genetics. Springer. https://doi.org/10.1007/978-3-642-00286-1_13

Asim S, Hussain A, Murad W, Hamayun M, Iqbal A, Rehman H, Tawab A, Irshad M, Alataway A, Dewidar AZ, Elansary HO, Lee I-J (2022) Endophytic *Fusarium oxysporum* GW controlling weed and an effective biostimulant for wheat growth. Front Plant Sci 13. https://doi.org/10.3389/fpls.2022.922343

Ateba JET, Toghueo RMK, Awantu AF, Mba'ning BM, Gohlke S, Sahal D, Rodrigues-Filho E, Tsamo E, Boyom FF, Sewald N, Lenta BN (2018) Antiplasmodial properties and cytotoxicity of endophytic fungi from *symphonia globulifera* (Clusiaceae). J Fungi 4(2). https://doi.org/10.3390/jof4020070

Ayob FW, Simarani K (2016) Endophytic filamentous fungi from a *Catharanthus roseus*: identification and its hydrolytic enzymes. Saudi Pharm J 24(3):273–278. https://doi.org/10.1016/j.jsps.2016.04.019

Bao J, Bi B, Yu H, Gao W, Wang N, Yang S (2017) Green biosynthesis of silver nanoparticles by the endophytic fungus *Fusarium oxysporum* from ginseng. Nanosci Nanotechnol Lett 9(7):1033–1037. https://doi.org/10.1166/NNL.2017.2433

Baron NC, Rigobelo EC (2022) Endophytic fungi: a tool for plant growth promotion and sustainable agriculture. Mycology 13(1):39–55. https://doi.org/10.1080/21501203.2021.1945699

Bhadra F, Gupta A, Vasundhara M, Reddy MS (2022) Endophytic fungi: a potential source of industrial enzyme producers. 3 Biotech 12(4). https://doi.org/10.1007/s13205-022-03145-y

Bhagobaty RK (2015) Endophytic fungi: prospects in biofuel production. Proc Natl Acad Sci India Sect B Biol Sci 85(1):21–25. https://doi.org/10.1007/s40011-013-0294-3

Bilal L, Asaf S, Hamayun M, Gul H, Iqbal A, Ullah I, Lee I-J, Hussain A (2018) Plant growth promoting endophytic fungi *Aspergillus fumigatus* TS1 and *Fusarium proliferatum* BRL1 produce gibberellins and regulates plant endogenous hormones. Symbiosis 76(2):117–127. https://doi.org/10.1007/s13199-018-0545-4

Brady SF, Clardy J (2000) CR377, a new pentaketide antifungal agent isolated from an endophytic fungus. J Nat Prod 63(10):1447–1448. https://doi.org/10.1021/np990568p

Bray F, Laversanne M, Sung H, Ferlay J, Siegel RL, Soerjomataram I, Jemal A (2024) Global cancer statistics 2022: GLOBOCAN estimates of incidence and mortality worldwide for 36 cancers in 185 countries. CA Cancer J Clin 74(3):229–263

Bungtongdee N, Sopalun K, Laosripaiboon W, Iamtham S (2019) The chemical composition, antifungal, antioxidant and antimutagenicity properties of bioactive compounds from fungal endophytes associated with Thai orchids. J Phytopathol 167(1):56–64. https://doi.org/10.1111/jph.12773

Caicedo NH, Davalos AF, Puente PA, Rodríguez AY, Caicedo PA (2019) Antioxidant activity of exo-metabolites produced by *Fusarium oxysporum*: an endophytic fungus isolated from leaves of Otoba gracilipes. MicrobiologyOpen 8(10). https://doi.org/10.1002/mbo3.903

Chakravarthi BVSK, Das P, Surendranath K, Karande AA, Jayabaskaran C (2008) Production of paclitaxel by *Fusarium solani* isolated from *Taxus celebica*. J Biosci 33(2):259–267. https://doi.org/10.1007/s12038-008-0043-6

Chala B, Hamde F (2021) Emerging and re-emerging vector-borne infectious diseases and the challenges for control: a review. Front Public Health 9. https://doi.org/10.3389/fpubh.2021.715759

Chang S, Yan B, Chen Y, Zhao W, Gao R, Li Y, Yu L, Xie Y, Si S, Chen M (2023) Cytotoxic hexadepsipeptides and anti-coronaviral 4-hydroxy-2-pyridones from an endophytic *Fusarium* sp. Front Chem 10. https://doi.org/10.3389/fchem.2022.1106869

Chen K, Sun W, Bie Q, Liu X, Chen C, Liu J, Xue Y, Wang J, Luo Z, Zhu H, Zhang Y (2018a) Fusopoltide A and fusosterede A, A polyketide with a pentaleno[1,2-c]pyran ring system and A degraded steride, from the fungus *Fusarium solani*. Tetrahedron Lett 59(27):2679–2682. https://doi.org/10.1016/j.tetlet.2018.05.082

Chen L, Deng H, Cui H, Fang J, Zuo Z, Deng J, Li Y, Wang X, Zhao L (2018b) Inflammatory responses and inflammation-associated diseases in organs. Oncotarget 9(6):7204. www.impactjournals.com/oncotarget/

Chen Y, Wang G, Yuan Y, Zou G, Yang W, Tan Q, Kang W, She Z (2022) Metabolites with cytotoxic activities from the mangrove endophytic fungus *Fusarium* sp. 2ST2. Front Chem 10. https://doi.org/10.3389/fchem.2022.842405

Choi HG, Song JH, Park M, Kim S, Kim C-E, Kang KS, Shim SH (2020) Neuroprotective γ-pyrones from *Fusarium solani* JS-0169: cell-based identification of active compounds and an informatics approach to predict the mechanism of action. Biomolecules 10(1):91. https://doi.org/10.3390/biom10010091

Chowdhary K, Arora H, Sharma S (2022) CRISPR/Cas9-based genome editing as a way ahead for inducing production of bioactive metabolites in endophytes. Natl Acad Sci Lett 45(3):275–280. https://doi.org/10.1007/s40009-022-01107-9

Christian N, Sullivan C, Visser ND, Clay K (2016) Plant host and geographic location drive endophyte community composition in the face of perturbation. Microb Ecol 72(3):621–632. https://doi.org/10.1007/s00248-016-0804-y

Clarance P, Luvankar B, Sales J, Khusro A, Agastian P, Tack JC, Al Khulaifi MM, AL-Shwaiman HA, Elgorban AM, Syed A, Kim HJ (2020) Green synthesis and characterization of gold nanoparticles using endophytic fungi *Fusarium solani* and its in-vitro anticancer and biomedical applications. Saudi J Biol Sci 27(2):706–712. https://doi.org/10.1016/j.sjbs.2019.12.026

Cui Y, Yi D, Bai X, Sun B, Zhao Y, Zhang Y (2012) Ginkgolide B produced endophytic fungus (*Fusarium oxysporum*) isolated from *Ginkgo biloba*. Fitoterapia 83(5):913–920. https://doi.org/10.1016/j.fitote.2012.04.009

Cui J-L, Vijayakumar V, Zhang G (2018) Partitioning of fungal endophyte assemblages in root-parasitic plant *Cynomorium songaricum* and its host *Nitraria tangutorum*. Front Microbiol 9. https://doi.org/10.3389/fmicb.2018.00666

Dame ZT, Silima B, Gryzenhout M, van Ree T (2016) Bioactive compounds from the endophytic fungus *Fusarium proliferatum*. Nat Prod Res 30(11):1301–1304. https://doi.org/10.1080/14786419.2015.1053089

de Almeida MN, Guimarães VM, Falkoski DL, Visser EM, Siqueira GA, Milagres AMF, de Rezende ST (2013) Direct ethanol production from glucose, xylose and sugarcane bagasse by the corn endophytic fungi *Fusarium verticillioides* and *Acremonium zeae*. J Biotechnol 168(1):71–77. https://doi.org/10.1016/j.jbiotec.2013.07.032

de Almeida ASF, Corrêa Junior A, da Silva Bentes JL (2021) Synthesis of silver nanoparticles (Agnps) by *Fusarium concolor* and inhibition of plant pathogens. Summa Phytopathol 47(1):9–15. https://doi.org/10.1590/0100-5405/235097

Deng Z, Cao L (2017) Fungal endophytes and their interactions with plants in phytoremediation: a review. Chemosphere 168:1100–1106. https://doi.org/10.1016/j.chemosphere.2016.10.097

Deng BW, Liu KH, Chen WQ, Ding XW, Xie XC (2009) *Fusarium solani*, Tax-3, a new endophytic taxol-producing fungus from *Taxus chinensis*. World J Microbiol Biotechnol 25(1):139–143. https://doi.org/10.1007/s11274-008-9876-2

Do Nascimento AM, Conti R, Turatti ICC, Cavalcanti BC, Costa-Lotufo LV, Pessoa C, De Moraes MO, Manfrim V, Toledo JS, Cruz AK, Pupo MT (2012) Bioactive extracts and chemical constituents of two endophytic strains of *Fusarium oxysporum*. Braz J Pharmacogn 22(6):1276–1281. https://doi.org/10.1590/S0102

Elavarasi A, Rathna GS, Kalaiselvam M (2012) Taxol producing mangrove endophytic fungi *Fusarium oxysporum* from *Rhizophora annamalayana*. Asian Pac J Trop Biomed 2(2):S1081–S1085. https://doi.org/10.1016/S2221-1691(12)60365-7

El-Bondkly AAM, El-Gendy MMAA, El-Bondkly AMA (2021) Construction of efficient recombinant strain through genome shuffling in marine endophytic *Fusarium* sp. ALAA-20 for improvement lovastatin production using agro-industrial wastes. Arab J Sci Eng 46(1):175–190. https://doi.org/10.1007/s13369-020-04925-5

Falade AO, Adewole KE, Ekundayo TC (2021) Aptitude of endophytic microbes for production of novel biocontrol agents and industrial enzymes towards agro-industrial sustainability. Beni-Suef Univ J Basic Appl Sci 10(1):61. https://doi.org/10.1186/s43088-021-00146-3

Fariq A, Khan T, Yasmin A (2017) Microbial synthesis of nanoparticles and their potential applications in biomedicine. J Appl Biomed 15(4):241–248. https://doi.org/10.1016/j.jab.2017.03.004

Fleming A (1929) On the antibacterial action of cultures of a Penicillium, with special reference to their use in the isolation of B. influenzae? Br J Exp Pathol 10(3):226–236

Fuska J, Proksa B, Fusková A (1975) New potential cytotoxic and antitumor substances I. In vitro effect of bikaverin and its derivatives on cells of certain tumors. Neoplasma 22(3):335–338

Gakuubi MM, Munusamy M, Liang ZX, Ng SB (2021) Fungal endophytes: a promising frontier for discovery of novel bioactive compounds. J Fungi 7(10). https://doi.org/10.3390/jof7100786

Garyali S (2013) Taxol production by an endophytic fungus, *Fusarium redolens*, isolated from Himalayan yew. J Microbiol Biotechnol 23(10):1372–1380. https://doi.org/10.4014/jmb.1305.05070

Gouda S, Das G, Sen SK, Shin HS, Patra JK (2016) Endophytes: a treasure house of bioactive compounds of medicinal importance. Front Microbiol 7:1538. https://doi.org/10.3389/fmicb.2016.01538

Grove JF, Pople M (1980) The insecticidal activity of beauvericin and the enniatin complex. Mycopathologia 70:103

Gurudatt PS, Priti V, Shweta S, Ramesha BT, Ravikanth G, Vasudeva R, Amna T, Deepika S, Ganeshaiah KN, Uma Shaanker R et al (2010) Attenuation of camptothecin production and negative relation between hyphal biomass and camptothecin content in endophytic fungal strains isolated from Nothapodytes nimmoniana Grahm (Icacinaceae). Curr Sci 98:1006–1009

Gulyamova TG et al (2022) Screening of pancreatic lipase inhibitors of endophytic fungi of medicinal plants in Uzbekistan. Biosci Biotechnol Res Asia 19(4):1037–1044. https://doi.org/10.13005/bbra/3053

Guo S, Mao W, Li Y, Tian J, Xu J (2013) Structural elucidation of the exopolysaccharide produced by fungus *Fusarium oxysporum* Y24-2. Carbohydr Res 365:9–13. https://doi.org/10.1016/j.carres.2012.09.026

Halkai KR, Mudda JA, Shivanna V, Rathod V, Halkai RS (2017) Biosynthesis, characterization and antibacterial efficacy of silver nanoparticles derived from endophytic fungi against *P. gingivalis*. J Clin Diagn Res 11(9):ZC92–ZC96. https://doi.org/10.7860/JCDR/2017/29434.10681

Hamzah TNT, Lee SY, Hidayat A, Terhem R, Faridah-Hanum I, Mohamed R (2018) Diversity and characterization of endophytic fungi isolated from the tropical mangrove species, *Rhizophora mucronata*, and identification of potential antagonists against the soil-borne fungus, Fusarium solani. Front Microbiol 9. https://doi.org/10.3389/fmicb.2018.01707

Handayani D, Mulia P, Andayani R, Wahyuni FS, Ariantari NP (2021) Secondary metabolite from mangrove endophytic fungus *Fusarium proliferatum* AED3. Rasayan J Chem 2021(Special Issue):150–155. https://doi.org/10.31788/RJC.2021.1456447

Hashem AH, Attia MS, Kandil EK, Fawzi MM, Abdelrahman AS, Khader MS, Khodaira MA, Emam AE, Goma MA, Abdelaziz AM (2023) Bioactive compounds and biomedical applications of endophytic fungi: a recent review. Microb Cell Factories 22(1). https://doi.org/10.1186/s12934-023-02118-x

Hassan EA, Nafady NA, Hassan SHA, Moustafa YS, Alamri SA, Hashem M (2021) Cadmium biosorption potential and kinetic behavior of endophytic *Fusarium verticillioides* and its green synthesized silver nanoparticles (Gsnps). Global NEST J 23(3):449–457. https://doi.org/10.30955/GNJ.003908

Hawas UW, Al-Farawati R, El-Kassem LTA, Turki AJ (2016) Different culture metabolites of the Red Sea fungus *Fusarium equiseti* optimize the inhibition of hepatitis C virus NS3/4A protease (HCV PR). Mar Drugs 14(10). https://doi.org/10.3390/md14100190

Hidayat I (2016) Three quinine and Cinchonidine producing *Fusarium* species from Indonesia. Curr Res Environ Appl Mycol 6(1):20–34. https://doi.org/10.5943/cream/6/1/3

Hoque N, Afroz F, Khatun F, Rony SR, Hasan CM, Rana MS, Sohrab MH (2022) Physicochemical, pharmacokinetic and cytotoxicity of the compounds isolated from an endophyte *Fusarium oxysporum*: in vitro and in silico approaches. Toxins 14(3):159. https://doi.org/10.3390/toxins14030159

Hussain A, Khan AW (2018) In vitro maize growth promotion by endophytic *Fusarium oxysporum* WLW. J Agric Biol Environ Stat. https://www.researchgate.net/publication/326347302

Ibrahim SRM, Elkhayat ES, Mohamed GAA, Fat'hi SM, Ross SA (2016a) Fusarithioamide A, a new antimicrobial and cytotoxic benzamide derivative from the endophytic fungus *Fusarium chlamydosporium*. Biochem Biophys Res Commun 479(2):211–216. https://doi.org/10.1016/j.bbrc.2016.09.041

Ibrahim SRM, Mohamed GA, Ross SA (2016b) Integracides F and G: new tetracyclic triterpenoids from the endophytic fungus *Fusarium* sp. Phytochem Lett 15:125–130. https://doi.org/10.1016/j.phytol.2015.12.010

Ibrahim SRM, Abdallah HM, Elkhayat ES, Al Musayeib NM, Asfour HZ, Zayed MF, Mohamed GA (2018) Fusaripeptide A: new antifungal and anti-malarial cyclodepsipeptide from the endophytic fungus *Fusarium* sp. J Asian Nat Prod Res 20(1):75–85. https://doi.org/10.1080/10286020.2017.1320989

Ingle A, Gade A, Pierrat S, Sonnichsen C, Rai M (2008) Mycosynthesis of silver nanoparticles using the fungus *Fusarium acuminatum* and its activity against some human pathogenic bacteria. Curr Nanosci 4(2):141–144. https://doi.org/10.2174/157341308784340804

Islam SKN, Naqvi SMA, Parveen S, Ahmad A (2021) Endophytic fungus-assisted biosynthesis, characterization and solar photocatalytic activity evaluation of nitrogen-doped Co3O4 nanoparticles. Appl Nanosci (Switzerland) 11(5):1651–1659. https://doi.org/10.1007/s13204-021-01824-5

Ivanova L, Skjerve E, Eriksen GS, Uhlig S (2006) Cytotoxicity of enniatins A, A1, B, B1, B2 and B3 from *Fusarium avenaceum*. Toxicon 47(8):868–876. https://doi.org/10.1016/j.toxicon.2006.02.012

Jaroszuk-Ściseł J, Nowak A, Komaniecka I, Choma A, Jarosz-Wilkołazka A, Osińska-Jaroszuk M, Tyśkiewicz R, Wiater A, Rogalski J (2020) Differences in production, composition, and antioxidant activities of exopolymeric substances (EPS) obtained from Cultures of endophytic *Fusarium culmorum* strains with different effects on cereals. Molecules 25(3). https://doi.org/10.3390/molecules25030616

Jiang W, Yang G, Chang Z, Chenxin F (2011) Species composition and molecular analysis of symbiotic fungi in roots of *Changnienia amoena* (Orchidaceae). Afr J Microbiol Res 5:222

Kaul S, Gupta S, Ahmed M, Dhar MK (2012) Endophytic fungi from medicinal plants: a treasure hunt for bioactive metabolites. Phytochem Rev 11(4):487–505. https://doi.org/10.1007/s11101-012-9260-6

Kaur J, Kaur R, Kaur A (2018) Evaluation of antidiabetic and antioxidant potential of endophytic fungi isolated from medicinal plants. Int J Green Pharm 12(1). https://doi.org/10.22377/ijgp.v12i01.1517

Kaushik NK, Murali TS, Sahal D, Suryanarayanan TS (2014) A search for antiplasmodial metabolites among fungal endophytes of terrestrial and marine plants of southern India. Acta Parasitol 59(4):745–757. https://doi.org/10.2478/s11686-014-0307-2

Khan AR, Ullah I, Waqas M, Shahzad R, Hong S-J, Park G-S, Jung BK, Lee I-J, Shin J-H (2015) Plant growth-promoting potential of endophytic fungi isolated from *Solanum nigrum* leaves. World J Microbiol Biotechnol 31(9):1461–1466. https://doi.org/10.1007/s11274-015-1888-0

Khan N, Afroz F, Begum MN, Roy Rony S, Sharmin S, Moni F, Mahmood Hasan C, Shaha K, Sohrab MH (2018) Endophytic *Fusarium solani*: a rich source of cytotoxic and antimicrobial napthaquinone and aza-anthraquinone derivatives. Toxicol Rep 5:970–976. https://doi.org/10.1016/j.toxrep.2018.08.016

Khayat MT, Ibrahim SRM, Mohamed GA, Abdallah HM (2019) Anti-inflammatory metabolites from endophytic fungus *Fusarium* sp. Phytochem Lett 29:104–109. https://doi.org/10.1016/j.phytol.2018.11.024

Koshila Ravi R, Prema Sundara Valli P, Muthukumar T (2022) Physiological characterization of root endophytic *Fusarium haematococcum* for hydrolytic enzyme production, nutrient solubilization and salinity tolerance. Biocatal Agric Biotechnol 43. https://doi.org/10.1016/j.bcab.2022.102392

Kour A, Shawl AS, Rehman S, Sultan P, Qazi PH, Suden P, Khajuria RK, Verma V (2008) Isolation and identification of an endophytic strain of *Fusarium oxysporum* producing podophyllotoxin from *Juniperus recurva*. World J Microbiol Biotechnol 24(7):1115–1121. https://doi.org/10.1007/s11274-007-9582-5

Kumar A, Abnave P, Ahmad A (2013a) Cultural, morphological and molecular characterization of vinca alkaloids producing endophytic fungus *Fusarium solani* isolated from *Catharanthus roseus*. Int J Bot Res 3(2): 1-12.

Kumar A, Patil D, Rajamohanan PR, Ahmad A (2013b) Isolation, purification and characterization of vinblastine and vincristine from endophytic fungus *Fusarium oxysporum* isolated from *Catharanthus roseus*. PLoS One 8(9):e71805. https://doi.org/10.1371/journal.pone.0071805

Kumar S, Ruma K, Sunil K, Prakash HS (2013c) Antioxidant, anti-inflammatory, antimicrobial and cytotoxic properties of fungal endophytes from Garcinia species. Int J Pharm Pharm Sci. https://www.researchgate.net/publication/267036276

Kumara PM, Soujanya KN, Ravikanth G, Vasudeva R, Ganeshaiah KN, Shaanker RU (2014) Rohitukine, a chromone alkaloid and a precursor of flavopiridol, is produced by endophytic fungi isolated from *Dysoxylum binectariferum* Hook.f and Amoora rohituka (Roxb).Wight & Arn. Phytomedicine 21(4):541–546. https://doi.org/10.1016/j.phymed.2013.09.019

Kusari S, Zühlke S, Spiteller M (2009) An endophytic fungus from *Camptotheca acuminata* that produces camptothecin and analogues. J Nat Prod 72(1):2–7. https://doi.org/10.1021/np800455b

Kusari P, Kusari S, Spiteller M, Kayser O (2013) Endophytic fungi harbored in Cannabis sativa L.: diversity and potential as biocontrol agents against host plant-specific phytopathogens. Fungal Divers 60(1):137–151. https://doi.org/10.1007/s13225-012-0216-3

Lara HH, Ayala-Núñez NV, del Turrent LCI, Padilla CR (2010) Bactericidal effect of silver nanoparticles against multidrug-resistant bacteria. World J Microbiol Biotechnol 26(4):615–621. https://doi.org/10.1007/s11274-009-0211-3

Lata R, Chowdhury S, Gond SK, White JF (2018) Induction of abiotic stress tolerance in plants by endophytic microbes. Lett Appl Microbiol 66(4):268–276. https://doi.org/10.1111/lam.12855

Lee D, Choi HG, Hwang JH, Shim SH, Kang KS (2020) Neuroprotective effect of tricyclic pyridine alkaloids from *Fusarium lateritium* SSF2, against glutamate-induced oxidative stress and apoptosis in the HT22 hippocampal neuronal cell line. Antioxidants 9(11):1115. https://doi.org/10.3390/antiox9111115

Li P (2011) In vitro antioxidant activities of polysaccharides from endophytic *fungus Fusarium oxysporum* Dzf17. Afr J Microbiol Res 5(32). https://doi.org/10.5897/ajmr11.1342

Li C-T, Li Y, Wang Q-J, Sung C-K (2008) Taxol production by *Fusarium Arthrosporioides* isolated from yew, Taxus Cuspidata. J Med Biochem 27(4):454–458. https://doi.org/10.2478/v10011-008-0022-3

Li Y-C, Tao W-Y, Cheng L (2009) Paclitaxel production using co-culture of Taxus suspension cells and paclitaxel-producing endophytic fungi in a co-bioreactor. Appl Microbiol Biotechnol 83(2):233–239. https://doi.org/10.1007/s00253-009-1856-4

Li P, Mou Y, Shan T, Xu J, Li Y, Lu S, Zhou L (2011) Effects of polysaccharide elicitors from endophytic *Fusarium oxysporium* Dzf17 on growth and diosgenin production in cell suspension culture of *Dioscorea zingiberensis*. Molecules 16(11):9003–9016. https://doi.org/10.3390/molecules16119003

Li G, Kusari S, Golz C, Strohmann C, Spiteller M (2016) Three cyclic pentapeptides and a cyclic lipopeptide produced by endophytic: *Fusarium decemcellulare* LG53. RSC Adv 6(59):54092–54098. https://doi.org/10.1039/c6ra10905e

Lima JMS, Pereira JO, Batista IH, de Queiroz Costa Neto P, dos Santos JC, de Araújo SP, Pantoja MC, da Mota AJ, de Azevedo JL (2016) Potential biosurfactant producing endophytic and epiphytic fungi, isolated from macrophytes in the Negro River in Manaus, Amazonas, Brazil. Afr J Biotechnol 15(24):1217–1223. https://doi.org/10.5897/AJB2015.15131

Liu J, Wang X, Pu H, Liu S, Kan J, Jin C (2017) Recent advances in endophytic exopolysaccharides: production, structural characterization, physiological role and biological activity. Carbohydr Polym 157:1113–1124. https://doi.org/10.1016/j.carbpol.2016.10.084

Lv F, Daletos G, Lin W, Proksch P (2015) Two new cyclic depsipeptides from the endophytic fungus Fusarium sp. Nat Prod Commun 10(10):1667

Mahapatra S, Banerjee D (2013a) Evaluation of in vitro antioxidant potency of exopolysaccharide from endophytic *Fusarium solani* SD5. Int J Biol Macromol 53:62–66. https://doi.org/10.1016/j.ijbiomac.2012.11.003

Mahapatra S, Banerjee D (2013b) Optimization of a bioactive exopolysaccharide production from endophytic *Fusarium solani* SD5. Carbohydr Polym 97(2):627–634. https://doi.org/10.1016/j.carbpol.2013.05.039

Maharjan S, Lee SB, Kim GJ, Cho SJ, Nam JW, Chin J, Choi H (2020) Isolation of unstable isomers of lucilactaene and evaluation of anti-inflammatory activity of secondary metabolites produced by the endophytic fungus *Fusarium* sp. QF001 from the roots of *Scutellaria baicalensis*. Molecules 25(4). https://doi.org/10.3390/molecules25040923

Manoharan S (2015) Green synthesis of silver nanoparticles and their antimicrobial property of endophytic fungi isolated from *Mentha arvensis* L. and *Psidium guajava* L. J Chem Pharm Res 7:340–346. https://www.researchgate.net/publication/282440342

Marcinkevicius K, Salvatore A, Murua G, Arena M, Vera N (2019) Phytochemical investigation and biological activities of *Fusarium* sp. an entomogenous fungus. Biocatal Agric Biotechnol 18. https://doi.org/10.1016/j.bcab.2019.101084

Martien JI, Amador-Noguez D (2017) Recent applications of metabolomics to advance microbial biofuel production. Curr Opin Biotechnol 43:118–126. https://doi.org/10.1016/j.copbio.2016.11.006

Martinho V, dos Santos Lima LM, Barros CA, Ferrari VB, Passarini MRZ, Santos LA, de Souza Sebastianes FL, Lacava PT, de Vasconcellos SP (2019) Enzymatic potential and biosurfactant production by endophytic fungi from mangrove forest in Southeastern Brazil. AMB Express 9(1). https://doi.org/10.1186/s13568-019-0850-1

Mishra R, Venkateswara Sarma V (2018) Current perspectives of endophytic fungi in sustainable development. In: Fungi and their role in sustainable development: current perspectives. Springer, Singapore, pp 553–584. https://doi.org/10.1007/978-981-13-0393-7_30

Montaser Farouk H, Zekry Attia E, El-Katatny MH (2020) Hydrolytic enzyme production of endophytic fungi isolated from soybean (Glycine max). J Mod Res. 2:1–7. www.jmr.journals.ekb.eg

Moron LS, Lim Y-W, Dela Cruz TEE (2018) Antimicrobial activities of crude culture extracts from mangrove fungal endophytes collected in Luzon Island, Philippines. Philippine Sci Lett 11:28

Musavi SF, Balakrishnan RM (2013) Biodiversity, antimicrobial potential, and phylogenetic placement of an endophytic *Fusarium oxysporum* NFX 06 isolated from *Nothapodytes foetida*. J Mycol 2013:1–10. https://doi.org/10.1155/2013/172056

Nadeem M (2012) *Fusarium solani*, P1, a new endophytic podophyllotoxin-producing fungus from roots of *Podophyllum hexandrum*. Afr J Microbiol Res 6(10). https://doi.org/10.5897/AJMR11.1596

Nandy S, Das T, Tudu CK, Pandey DK, Dey A, Ray P (2020) Fungal endophytes: futuristic tool in recent research area of phytoremediation. S Afr J Bot 134:285–295. https://doi.org/10.1016/j.sajb.2020.02.015

Nielsen MR, Sondergaard TE, Giese H, Sørensen JL (2019) Advances in linking polyketides and non-ribosomal peptides to their biosynthetic gene clusters in Fusarium. Curr Genet 65(6):1263–1280. https://doi.org/10.1007/s00294-019-00998-4

Nisa H, Kamili AN, Nawchoo IA, Shafi S, Shameem N, Bandh SA (2015) Fungal endophytes as prolific source of phytochemicals and other bioactive natural products: a review. Microb Pathogen 82:50–59. https://doi.org/10.1016/j.micpath.2015.04.001

Ortega HE, Torres-Mendoza D, Cubilla-Rios L (2020) Patents on endophytic fungi for agriculture and bio- and phytoremediation applications. Microorganisms 8(8):1237. https://doi.org/10.3390/microorganisms8081237

Osama S, El-Sherei MM, Al-Mahdy DA, Bishr M, Salama O, Raafat MM (2023) Optimization and characterization of antileukemic l-asparaginase produced by *Fusarium solani* endophyte. AMB Express 13(1). https://doi.org/10.1186/s13568-023-01602-2

Pan R, Cao L, Zhang R (2009) Combined effects of Cu, Cd, Pb, and Zn on the growth and uptake of consortium of Cu-resistant Penicillium sp. A1 and Cd-resistant *Fusarium* sp. A19. J Hazard Mater 171(1–3):761–766. https://doi.org/10.1016/j.jhazmat.2009.06.080

Pan F, Hou K, Li DD, Su TJ, Wu W (2019) Exopolysaccharides from the fungal endophytic Fusarium sp. A14 isolated from *Fritillaria unibracteata* Hsiao et KC Hsia and their antioxidant and antiproliferation effects. J Biosci Bioeng 127(2):231–240. https://doi.org/10.1016/j.jbiosc.2018.07.023

Panuthai T (2012) An extracellular lipase from the endophytic fungi *Fusarium oxysporum* isolated from the Thai medicinal plant, Croton oblongifolius Roxb. Afr J Microbiol Res 6(11). https://doi.org/10.5897/ajmr11.965

Pradeep FS, Palaniswamy M, Ravi S, Thangamani A, Pradeep BV (2015) Larvicidal activity of a novel isoquinoline type pigment from *Fusarium moniliforme* KUMBF1201 against Aedes aegypti and Anopheles stephensi. Process Biochem 50(9):1479–1486. https://doi.org/10.1016/j.procbio.2015.05.022

Prakash S, Singh G, Soni N, Sharma S (2010) Pathogenicity of *Fusarium oxysporum* against the larvae of *Culex quinquefasciatus* (Say) and *Anopheles stephensi* (Liston) in laboratory. Parasitol Res 107(3):651–655. https://doi.org/10.1007/s00436-010-1911-1

Prathyusha AMVN, Mohana Sheela G, Bramhachari PV (2018) Chemical characterization and antioxidant properties of exopolysaccharides from mangrove filamentous fungi *Fusarium equiseti* ANP2. Biotechnol Rep 19. https://doi.org/10.1016/j.btre.2018.e00277

Rai M, Bonde S, Golinska P, Trzcińska-Wencel J, Gade A, Abd-Elsalam K, Shende S, Gaikwad S, Ingle AP (2021) *Fusarium* as a novel fungus for the synthesis of nanoparticles: mechanism and applications. J Fungi 7(2):1–24. https://doi.org/10.3390/jof7020139

Rajamanikyam M, Vadlapudi V, Amanchy R, Upadhyayula SM (2017) Endophytic fungi as novel resources of natural therapeutics. Braz Arch Biol Technol 60. https://doi.org/10.1590/1678-4324-2017160542

Rajesh PS, Ravishankar Rai V (2013) Hydrolytic enzymes and quorum sensing inhibitors from endophytic fungi of *Ventilago madraspatana* Gaertn. Biocatal Agric Biotechnol 2(2):120–124. https://doi.org/10.1016/j.bcab.2013.01.002

Ran X, Zhang G, Li S, Wang J (2017) Characterization and antitumor activity of camptothecin from endophytic fungus *Fusarium solani* isolated from *Camptotheca acuminate*. Afr Health Sci 17(2):566. https://doi.org/10.4314/ahs.v17i2.34

Rodriguez RJ, Henson J, Van Volkenburgh E, Hoy M, Wright L, Beckwith F, Kim Y-O, Redman RS (2008) Stress tolerance in plants via habitat-adapted symbiosis. ISME J 2(4):404–416. https://doi.org/10.1038/ismej.2007.106

Ruma K, Shailasree S, Sampath KKK, Ramachandrappa Niranjana S, Sripathy Prakash H (2012) Endophytic fungal assemblages in *Calophyllum* and *Garcinia* spp. of Clusiaceae Family in Western Ghats, India. Curr Biotechnol 1(2):109–114. https://doi.org/10.2174/2211550111201020109

Sathiyanathan M, Umarajan KM (2019) Larvicidal activity of endophytic fungi isolated from selected medicinal plants on *Aedes aegypti*. J Pharmacogn Phytochem 8(2):247–253

Selosse M-A, Bessis A, Pozo MJ (2014) Microbial priming of plant and animal immunity: symbionts as developmental signals. Trends Microbiol 22(11):607–613. https://doi.org/10.1016/j.tim.2014.07.003

Semwal P, Painuli S, Anand J, Martins NC, Machado M, Sharma R, Batiha GE-S, Yaro CA, Lorenzo JM, Rahman MM (2022) The neuroprotective potential of endophytic fungi and proposed molecular mechanism: A current update. Evid Based Complement Alternat Med 2022:1–12. https://doi.org/10.1155/2022/6214264

Shah A, Rather MA, Hassan QP, Aga MA, Mushtaq S, Shah AM, Hussain A, Baba SA, Ahmad Z (2017) Discovery of anti-microbial and anti-tubercular molecules from *Fusarium solani*: an endophyte of Glycyrrhiza glabra. J Appl Microbiol 122(5):1168–1176. https://doi.org/10.1111/jam.13410

Shah S, Shrestha R, Maharjan S, Selosse M-A, Pant B (2018) Isolation and characterization of plant growth-promoting endophytic fungi from the roots of *Dendrobium moniliforme*. Plants 8(1):5. https://doi.org/10.3390/plants8010005

Sharma P, Kumar S (2021) Bioremediation of heavy metals from industrial effluents by endophytes and their metabolic activity: recent advances. Bioresour Technol 339:125589. https://doi.org/10.1016/j.biortech.2021.125589

Shiono Y, Shibuya F, Koseki T, Harizon, Supratman U, Uesugi S, Kimura K (2015) A new α-pyrone metabolite from a mangrove plant endophytic fungus, *Fusarium* sp. J Asian Nat Prod Res 17(4):403–408. https://doi.org/10.1080/10286020.2014.961919

Shu RG, Wang FW, Yang YM, Liu YX, Tan RX (2004) Antibacterial and xanthine oxidase inhibitory cerebrosides from *Fusarium* sp. IFB-121, and endophytic fungus in *Quercus variabilis*. Lipids 39(7):667–673. https://doi.org/10.1007/s11745-004-1280-9

Shweta S, Zuehlke S, Ramesha BT, Priti V, Mohana Kumar P, Ravikanth G, Spiteller M, Vasudeva R, Uma Shaanker R (2010) Endophytic fungal strains of *Fusarium solani*, from *Apodytes dimidiata* E. Mey. ex Arn (Icacinaceae) produce camptothecin, 10-hydroxycamptothecin and 9-methoxycamptothecin. Phytochemistry 71(1):117–122. https://doi.org/10.1016/j.phytochem.2009.09.030

Sim JH, Khoo CH, Lee LH, Cheah YK (2010) Molecular diversity of fungal endophytes isolated from *Garcinia mangostana* and *Garcinia parvifolia*. J Microbiol Biotechnol 20(4):651–658. https://doi.org/10.4014/jmb.0909.09030

Singh N, Abraham J (2014) Ribosomally synthesized peptides from natural sources. J Antibiot 67(4):277–289. https://doi.org/10.1038/ja.2013.138

Singh AK, Rathod V, Singh D, Ninganagouda S, Kulkarni P, Mathew J, ul Haq M (2015) Bioactive silver nanoparticles from endophytic fungus *Fusarium* sp. isolated from an ethanomedicinal plant *Withania somnifera* (Ashwagandha) and its antibacterial activity. Int J Nanomater Biostruct 5(1):15–19

Singh AK, Rathod V, Singh D, Mathew J, Kulkarni P (2016) Effect of silver nanoparticles (AgNps) produced by an endophytic fungus *Fusarium semitectum* isolated from a medicinal plant *Withania Somnifera* (Ashwagandha) on seed germination. Int J Res Stud Agric Sci 2(5):6–12

Singh A, Singh DK, Kharwar RN, White JF, Gond SK (2021) Fungal endophytes as efficient sources of plant-derived bioactive compounds and their prospective applications in natural product drug discovery: insights, avenues, and challenges. Microorganisms 9(1):1–42. https://doi.org/10.3390/microorganisms9010197

Sogra Fathima B, Balakrishnan RM (2014) Biosynthesis and optimization of silver nanoparticles by endophytic fungus *Fusarium solani*. Mater Lett 132:428–431. https://doi.org/10.1016/j.matlet.2014.06.143

Sonaimuthu V, Krishnamoorthy S, Johnpaul M (2010) Taxol producing endophytic fungus *Fusarium culmorum* SVJM072 from medicinal plant of *Tinospora cordifolia*—a first report. J Biotechnol 150:425–425. https://doi.org/10.1016/j.jbiotec.2010.09.588

Strobel G (2018) The emergence of endophytic microbes and their biological promise. J Fungi 4(2):57. https://doi.org/10.3390/jof4020057

Strobel G, Daisy B (2003) Bioprospecting for microbial endophytes and their natural products. Microbiol Mol Biol Rev 67(4):491–502. https://doi.org/10.1128/MMBR.67.4.491-502.2003

Strobel G, Hess W, Ford E, Sidhu R, Yang X (1996) Taxol from fungal endophytes and the issue of biodiversity. J Ind Microbiol Biotechnol 17(5–6):417–423. https://doi.org/10.1007/BF01574772

Tang P, Zhang Z, Niu L, Gu C, Zheng W, Cui H, Yuan X (2021) *Fusarium solani* G6, a novel vitexin-producing endophytic fungus: characterization, yield improvement and osteoblastic proliferation activity. Biotechnol Lett 43(7):1371–1383. https://doi.org/10.1007/s10529-021-03118-w

Tang N, Zhang R, Zheng Y, Wang J, Khatib M, Jiang X, Zhou C, Omar R, Saliba W, Wu W, Yuan M, Cui D, Haick H (2022) Highly efficient self-healing multifunctional dressing with antibacterial activity for sutureless wound closure and infected wound monitoring. Adv Mater 34(3). https://doi.org/10.1002/adma.202106842

Tayung K, Barik BP, Jha DK, Deka DC (2011) Identification and characterization of antimicrobial metabolite from an endophytic fungus, *Fusarium solani* isolated from bark of Himalayan yew. Mycosphere 2(3):203–213

Tejesvi MV, Segura DR, Schnorr KM, Sandvang D, Mattila S, Olsen PB, Neve S, Kruse T, Kristensen HH, Pirttilä AM (2013) An antimicrobial peptide from endophytic *Fusarium tricinctum* of *Rhododendron tomentosum* Harmaja. Fungal Divers 60(1):153–159. https://doi.org/10.1007/s13225-013-0227-8

Tiwari P, Bae H (2022) Endophytic fungi: key insights, emerging prospects, and challenges in natural product drug discovery. Microorganisms 10(2). https://doi.org/10.3390/microorganisms10020360

Tiwari S, Singh S, Pandey P, Saikia SK, Negi AS, Gupta SK, Pandey R, Banerjee S (2014) Isolation, structure determination, and antiaging effects of 2,3-pentanediol from endophytic fungus of *Curcuma amada* and docking studies. Protoplasma 251(5):1089–1098. https://doi.org/10.1007/s00709-014-0617-0

Toghueo RMK (2020) Bioprospecting endophytic fungi from Fusarium genus as sources of bioactive metabolites. Mycology 11(1):1–21. https://doi.org/10.1080/21501203.2019.1645053

Toghueo R, Zeuko E, Ines Mbekou Kanko M, Ngo Mback MN, Rmk T, Kanko MM, Marie Arc-en-Ce J, Mback NN, de Aldana VB, Boyom FF (2016) Antimicrobial and antiradical activities of ethyl acetate extracts from endophytic fungi isolated from Cameroonian medicinal plants. J Med Plants Stud 290(4). https://www.researchgate.net/publication/306039866

Toghueo RMK, Sahal D, Zabalgogeazcoa Í, Baker B, Boyom FF (2018) Conditioned media and organic elicitors underpin the production of potent antiplasmodial metabolites by endophytic fungi from Cameroonian medicinal plants. Parasitol Res 117(8):2473–2485. https://doi.org/10.1007/s00436-018-5936-1

Toghueo RMK, Kemgne EAM, Eke P, Kanko MIM, Dize D, Sahal D, Boyom FF (2019) Antiplasmodial potential and GC-MS fingerprint of endophytic fungal extracts derived from

Cameroonian Annona muricata. J Ethnopharmacol 235:111–121. https://doi.org/10.1016/j.jep.2019.02.010

Tung CY, Yang DB, Gou M (2002) A preliminary study on the condition of the culture and isolate of endophytic fungus producing Vincristine. J Chuxiong Normal Univ 6:39–41

Vasundhara M, Baranwal M, Kumar A (2016) Fusarium tricinctum, an endophytic fungus exhibits cell growth inhibition and antioxidant activity. Indian J Microbiol 56:433–438

Vats S, Saxena S (2023) Endophytic *Fusarium* species, a unique bioresource for disaggregator of misfolded alpha-synuclein. Arch Microbiol 205(6):224. https://doi.org/10.1007/s00203-023-03575-z

Venugopalan A, Potunuru UR, Dixit M, Srivastava S (2016) Reprint of: effect of fermentation parameters, elicitors and precursors on camptothecin production from the endophyte *Fusarium solani*. Bioresour Technol 213:311–318. https://doi.org/10.1016/j.biortech.2016.05.023

Vijayan S, Koilaparambil D, George TK, Manakulam Shaikmoideen J (2016) Antibacterial and cytotoxicity studies of silver nanoparticles synthesized by endophytic *Fusarium solani* isolated from *Withania somenifera* (L.). Autumn 2016L. J Water Environ Nanotechnol 1(2):91–103. https://doi.org/10.7508/jwent.2016.02.003

Wang Q, Xu L (2012) Beauvericin, a bioactive compound produced by fungi: a short review. Molecules 17(3):2367–2377. https://doi.org/10.3390/molecules17032367

Wang QX, Li SF, Zhao F, Dai HQ, Bao L, Ding R, Gao H, Zhang LX, Wen HA, Liu HW (2011) Chemical constituents from endophytic fungus *Fusarium oxysporum*. Fitoterapia 82(5):777–781. https://doi.org/10.1016/j.fitote.2011.04.002

Wang WX, Kusari S, Sezgin S, Lamshöft M, Kusari P, Kayser O, Spiteller M (2015) Hexacyclopeptides secreted by an endophytic fungus *Fusarium solani* N06 act as crosstalk molecules in *Narcissus tazetta*. Appl Microbiol Biotechnol 99(18):7651–7662. https://doi.org/10.1007/s00253-015-6653-7

Wang Y, Zhang Y, Cong H, Li C, Wu J, Li L, Jiang J, Cao X (2023) Cultivable endophyte resources in medicinal plants and effects on hosts. Life 13(8). https://doi.org/10.3390/life13081695

Wei J, Wu B (2020) Chemistry and bioactivities of secondary metabolites from the genus *Fusarium*. In: Fitoterapia, vol 146. Elsevier B.V. https://doi.org/10.1016/j.fitote.2020.104638

Weller MG (2012) A unifying review of bioassay-guided fractionation, effect-directed analysis and related techniques. Sensors (Switzerland) 12(7):9181–9209. https://doi.org/10.3390/s120709181

World Health Organization (2018) World malaria report 2018. World Health Organization

Wu Z, Li Y, Zhang L, Ding Z, Shi G (2021) Microbial production of small peptide: pathway engineering and synthetic biology. Microb Biotechnol 14(6):2257–2278. https://doi.org/10.1111/1751-7915.13743

Xiong Z-Q, Yang Y-Y, Zhao N, Wang Y (2013) Diversity of endophytic fungi and screening of fungal paclitaxel producer from Anglojap yew, Taxus x media. BMC Microbiol 13(1):71. https://doi.org/10.1186/1471-2180-13-71

Xu F, Tao W, Cheng L, Guo L (2006) Strain improvement and optimization of the media of taxol-producing fungus *Fusarium maire*. Biochem Eng J 31(1):67–73. https://doi.org/10.1016/j.bej.2006.05.024

Yan C, Liu W, Li J, Deng Y, Chen S, Liu H (2018) Bioactive terpenoids from: *Santalum album* derived endophytic fungus Fusarium sp. YD-2. *RSC*. Advances 8(27):14823–14828. https://doi.org/10.1039/c8ra02430h

Yap LS, Lee WL, Ting ASY (2021) Optimization of L-asparaginase production from endophytic *Fusarium proliferatum* using OFAT and RSM and its cytotoxic evaluation. J Microbiol Methods 191. https://doi.org/10.1016/j.mimet.2021.106358

Zaher AM, Makboul MA, Moharram AM, Tekwani BL, Calderón AI (2015) A new enniatin antibiotic from the endophyte *Fusarium tricinctum* Corda. J Antibiot 68(3):197–200. https://doi.org/10.1038/ja.2014.129

Zeng YJ, Yang HR, Wang HF, Zong MH, Lou WY (2019a) Immune enhancement activity of a novel polysaccharide produced by Dendrobium officinale endophytic fungus *Fusarium solani* DO7. J Funct Foods 53:266–275. https://doi.org/10.1016/j.jff.2018.12.038

Zeng YJ, Yang HR, Zong MH, Yang JG, Lou WY (2019b) Novel antibacterial polysaccharides produced by endophyte *Fusarium solani* DO7. Bioresour Technol 288. https://doi.org/10.1016/j.biortech.2019.121596

Zhan J, Burns AM, Liu MX, Faeth SH, Gunatilaka AAL (2007) Search for cell motility and angiogenesis inhibitors with potential anticancer activity: beauvericin and other constituents of two endophytic strains of *Fusarium oxysporum*. J Nat Prod 70(2):227–232. https://doi.org/10.1021/np060394t

Zhang L, Guo B, Li H et al (2000) Preliminary study on the isolation of endophytic fungus of Catharanthus roseus and its fermentation to produce products of therapeutic value Zhong cao yao = Chinese Traditional and Herbal Drugs. 2000;31(11):805–807

Zhang J, Liu D, Wang H, Liu T, Xin Z (2015) Fusartricin, a sesquiterpenoid ether produced by an endophytic fungus *Fusarium tricinctum* Salicorn 19. Eur Food Res Technol 240(4):805–814. https://doi.org/10.1007/s00217-014-2386-6

Zhang R et al (2017) Isolation of cellulolytic endophytes from *Miscanthus sacchariflorus* and their application in the production of lignocellulolytic enzymes. Biotechnol Lett 39(10): 1381–1388

Zhang P, Yuan XL, Du YM, Zhang HB, Shen GM, Zhang ZF, Liang YJ, Zhao DL, Xu K (2019) Angularly prenylated indole alkaloids with antimicrobial and insecticidal activities from an endophytic fungus *Fusarium sambucinum* TE-6L. J Agric Food Chem. https://doi.org/10.1021/acs.jafc.9b05827

Zhao J, Fu Y, Luo M, Zu Y, Wang W, Zhao C, Gu C (2012) Endophytic fungi from pigeon pea [Cajanus cajan (L.) Millsp.] produce antioxidant Cajaninstilbene acid. J Agric Food Chem 60(17):4314–4319. https://doi.org/10.1021/jf205097y

Zhao S et al (2018) Endophytic fungi from *Camellia sinensis* and their bioactive metabolites: a comprehensive review. Phytochem Rev17(4): 781–798

Zhong L, Niu B, Tang L, Chen F, Zhao G, Zhao J (2016) Effects of polysaccharide elicitors from endophytic *Fusarium oxysporum* Fat9 on the growth, flavonoid accumulation and antioxidant property of *Fagopyrum tataricum* sprout cultures. Molecules 21(12):1590. https://doi.org/10.3390/molecules21121590

Zhu X, Zhong Y, Xie Z, Wu M, Hu Z, Ding W, Li C (2018) Fusarihexins A and B: novel cyclic hexadepsipeptides from the mangrove endophytic fungus *Fusarium* spR5 with antifungal activities. Planta Med 84(18):1355–1362. https://doi.org/10.1055/a-0647-7048

Zhu Y, Shao Y, Li L, Zhao L, Zhang M, Dong C (2022) The plant growth-promoting endophytic *Fusarium oxysporum* GG22 enhances *Rehmannia glutinosa* secondary metabolites accumulation. Ind Crop Prod 182:114881. https://doi.org/10.1016/j.indcrop.2022.114881

9 Volatile Organic Compounds from Endophytic Fungi and Their Role in Plant Growth and Disease Management

B. Shankar Naik, Rekha. D, and Surekha R. Deshpande

Abstract

Endophytic fungi are microbes that colonize the internal tissues of plants. They produce a wide spectrum of gaseous volatile organic compounds with diverse physicochemical, biotechnological, antimicrobial, and plant growth-promoting properties. These compounds are useful in pharmacological, agricultural, and industrial sectors and include terpenes, alkaloids, quinones, cyclohexanes, and hydrocarbons. Many exhibit antimicrobial and antioxidant properties that promote plant growth. This review discusses volatile organic compounds from endophytic fungi and their potential role in plant growth promotion and biological control.

Keywords

Endophytic fungi · VOCs · Plant growth · Antimicrobial · Disease management

B. S. Naik (✉)
Department of P.G. Studies and Research in Applied Botany, Bio Science Complex, Kuvempu University, Jnana Sahyadri, Shimoga, Karnataka, India

Government PU College for Girls, Basavanahalli, Chikkamagaluru, Karnataka, India

R. D.
Department of P.G. Studies and Research in Applied Botany, Bio Science Complex, Kuvempu University, Jnana Sahyadri, Shimoga, Karnataka, India

S. R. Deshpande
Basaveshwara Science College, Bagalkote, Karnataka, India

K. A. Abd-Elsalam, S. F. AbuQamar (eds.), *Fungal Endophytes Volume I*,
https://doi.org/10.1007/978-981-97-7312-1_9

9.1 Introduction

The ever-increasing human population is a major hindrance to meeting global food security (Verma et al. 2022). During the last three decades, agricultural produce has been reduced to 5% (Newbery et al. 2016). According to a report the global temperature may rise from 1.7 to 4.8 °C with odd climatic changes by the twenty-first century. Various abiotic stress conditions, such as high carbon dioxide (CO_2), drought, high temperature, and salinity, hamper food production (Keswani et al. 2019). Generation of ROS during biotic and abiotic stress conditions such as singlet oxygen ($^1O_2$), hydrogen peroxide (H_2O_2), superoxide radical (O_2), hydroxyl radical (OH^-), etc., decreases crop yield (Keswani et al. 2019). Inadequate and uncontrolled application of pesticides and chemical fertilizers harms the environment, human health, and non-target organisms present in aquatic and soil environments (Baron and Rigobelo 2021). Excess use of chemicals also results in the accumulation of heavy metals, natural aging of rivers and lakes, soil acidification, contamination of aquifers and water reservoirs, and release of greenhouse gases (Baron and Rigobelo 2021). Sustainable agriculture is one of the major alternatives to reduce the ill effects of pesticides and chemical fertilizers. Hence, knowledge about the symbiotic relationship between plants and soil microbiota and their mode of action in stressful conditions is essential to produce more agricultural yield with balanced ecosystem.

Endophytic fungi are asymptomatic, ecological, nonpathogenic, diverse polyphyletic groups of fungi colonizing host plant tissues and offer several beneficial impacts on host plants (Gupta et al. 2020). Endophytes enhance root and shoot growth, help in nutrient uptake and seed germination, and confer resistance to host plants against various biotic and abiotic stresses (drought, salinity, high temp, CO_2 and metal toxicity (Grabka et al. 2022). Many endophytes have been reported as biocontrol agents and biofertilizers as chemical alternatives (Adeleke et al. 2022). Endophytic fungi have also been reported to enhance bioremediation, bio assimilation and mineralization (Gavrila et al. 2022). They have been reported as nanosensors in detecting contaminants (Rana et al. 2020). They can also be used as bio-stimulants to produce biologically active compounds, phytohormones, phosphate solubilization factors (Rana et al. 2020), and siderophores (Srinivas et al. 2020), which play a major role in solubilization of various metals like cadmium, molybdenum, manganese, zinc, nickel and copper, etc. (Rana et al. 2020). Endophytes enhance Phyto immobilization by degrading biomass and help host plants to uptake nutrients like nitrogen, phosphate, zinc, etc. (Yung et al. 2021). Endophytes have also been reported to increase crop production by solubilizing and mobilizing phosphorus, potassium and zinc salts (Yung et al. 2021). They also enhance plant growth through Phytostimulation by lowering ethylene by 1-aminoc yclopropanLe-1-carboxylate deaminase (ACC) (Singh et al. 2015). Endophytic fungi are known to neutralize toxic soil pollutants through Phytotransformation. Several studies reported that fungi like *Piriformospora indica*, *Exophiala* sp., *Neotyphodium* sp., etc., are associated with various agricultural crops that reduce phytotoxicity by metal pollutants and help in soil bioremediation (Ningxiao et al. 2016) (Table 9.1, Fig. 9.1). Endophytic fungi are known to produce a mixture of

Table 9.1 Volatile organic compounds produced from endophytic fungi and their biological activity

Endophytic fungi	Crop	VOC	Target organism/ disease	Reference
Aureobasidium pullulans, *Sarcocladium summerbelleii*	Olive tree	z-3-Hexen-1-ol, benzyl alcohol nonanol	Antifungal	Sdiri et al. (2022)
Daldinia cf. *concentrica*	*Olea europaea* L.	Alcohols, dienes, ketones, aldehydes, and sesquiterpenes	Antifungal	Liarzi et al. (2016)
Diaporthe sp. CEL 3	*Chloranthus elatior* Sw.	Imidazole compounds-(2-aminoethyl)-1Himidazol-2-carbaldehyde, pyrazole4,5-imidazole, 1-formyl 3 ethyl. Phenyl compounds-phenol, 4-[2-(methylamino) ethyl]-. 6-Nitro-3-chlorophenol, Phenol,2,4,6-tritert-butyl	Antifungal	Santra and Banerjee (2023)
Fusarium oxysporum	*Gossypum herbaceum*	Farnesol, B-caryophellene, limonene, hexanoic acid, octanoic acid	Antifungal	Zhang et al. (2014)
Fusarium proliferatum	*Tetrastigram harmandii*	Volatile organic compounds	Banana and mango anthracnose	Cagadas et al. (2021)
Gliocladium sp.	*Eucryphia cordifolia*	1-Butanol, 3-methyl-phenylethyl alcohol, acetic acid, 2-phenyl ethyl ester	Antifungal	Stinson et al. (2003)
Hypoxylon anthochroum		Eucalyptol, 2-mthyl-butanol, 3-methyl-butanol, ocinene, Terpinolene	Antifungal	Macias-Rubalcava et al. (2018)
Hypoxylon anthochroum	*Bursera lancifolia*	Phenylethyl alcohol, eucalyptol	Antifungal and phytotoxic	Ulloa-Benítez et al. (2016)
Hypoxylon sp.	*Persea indica*	1,8-Cineole, 1-methyl-1,4-cyclohexadiene	Antifungal	Tomsheck et al. (2010)

(continued)

Table 9.1 (continued)

Endophytic fungi	Crop	VOC	Target organism/ disease	Reference
M. albus	*Cinnamomum zeylanicum*	3-Methyl-acetate, alcohol, 2-methyl-1-butanol, and isobutyric acid	Antifungal	Strobel (2006)
M. albus	*Grevilea pterifolia*, *Kennedia nigriscans*, *Terminalia prostrata*	Naphthalene, 2-methyl propionic acid	Antifungal	Ezra and Strobel (2003)
M. albus		2-Methyl propionic acid, 3-methyl-1-butanol, ethyl alcohol	Antifungal	Corcuff et al. (2011)
M. albus		*N*-methyl-*N*-nitrosolsobutyramide	Antifungal	Hutchings et al. (2017)
M. brasiliensiss	*Schinus terebinthifolius*	Pogostol, 2-phenyl-ethyl acetate, phenyl ethyl alcohol	Antifungal	Pena et al. (2019)
M. musae	*Musa acuminata*	Isobutyric acid, 3-methyl-1-butanol, ethyl-1-2-methylbutyrate		Suwannarach et al. (2013)
Muscodor cinnamomi	*C. bejalghota*	Isobutyric acid, 3-methyl-1-butanol, ethyl-2-methyl-butyrate	Antifungal	Suwannarach et al. (2013)
Muscodor darjeelingensis	*C. camphora*	Isobutyric acid, 3-methyl-1-butanol, ethyl-2-methyl-butyrate	Antifungal	Sanjai et al. (2014)
Muscodor equiseti	*Equisetum debile*	Isobutyric acid, 3-methyl-1-butanol, ethyl-2-methyl-butyrate	Antifungal	Suwannarach et al. (2013)
Muscodor kashyyum	*Aegle marmelos*	Cyclohex-3-en-1-ol, b-bisabolol	Antifungal	Meshram et al. (2013)
Muscodor oryzae	*Oryza rufipogon*	Isobutyric acid, 3-methyl-1-butanol, ethyl-2-methyl-butyrate	Antifungal	Suwannarach et al. (2013)
Muscodor suthepensis	*C. bejalghota*	Isobutyric acid, 3-methyl-1-butanol, ethyl-2-methyl-butyrate	Antifungal	Suwannarach et al. (2013)
Myrothecium inundatum	*Acalypha indica*	3-Octanone, 3-octanol, 7-octen-4-ol	Antifungal	Banerjee et al. (2010)
Neurospora dictyophora 3zf-02	*Torreya grandis*	Benzene and naphthalene compounds oleic acid, methyl abietate, terpinene-4-ol	Antifungal and antioxidant	Zhang et al. (2022)

(continued)

Table 9.1 (continued)

Endophytic fungi	Crop	VOC	Target organism/ disease	Reference
Nodulisporium sp.	*Thelypteris angustifolia*	1-Methyl-1, 4-cyclohexadiene, 2-methyl-1-pentanol, 1-heptanol, 1-octanol	Antifungal	Riyaz-Ul-Hassan et al. (2013)
Nodulisporium sp. CMU-UPE34	*L. loudoni* Teijsm&Binn	Acids, alcohols, esters and monoterpenes, eucalyptol(1,8-cineole)	Antifungal	Suwannarach et al. (2013)
Oxyporus latemarginatus EF069	*Capsicum annum*	2-Furanmethanol	Antifungal	Lee et al. (2009)
Oxysporus latemarginatus	*Acalypha indica*	5-Pentyl-2-furaldehyde	Antifungal	
Penicillium commune, *P. canescence*	*Olea europaea*	3-Methyl-1-butanol and phenylethyl alcohol	Antimicrobial	Malhadas et al. (2017)
Phomopsis sp.	*Odontoglossum* sp.	Benzene, ethanol, 2-propanone, sabinene	Antifungal	Lee et al. (2009)
Trichoderma asperellum T1	*Lactuca sativa*	VOCs	Antifungal, plant growth	Wonglom et al. (2020)
Trichoderma harzianum	*Solanum lycopersicum*	Diterpene	Antifungal	Malmierca et al. (2015)
T. longibrachiatum	*Rice*	Caryophyllene, longofolenee	Antifungal	Sridharan et al. (2020)
T. viridae	*Capsicum frutescence*	6-Pentyl-2H-pyran-2-one	Antifungal	Schalchli et al. (2016)
T. koningiopsis	*Panax notoginseng*	Cycloocta-2, 4-dien-1-ol, cis-1-butyl-2-methylcyclo propane	Antifungal	Chen et al. (2016)
T. asperelloides PSU-P1		Chitinase (CHI) and Β-1,3-glucanase (GLU)	Plant growth promotion, antifungal	Phoka et al. (2020)
T. gamsii	*Panax notoginseng*	Dimethyl disulphide, dibenzofuran, methane thiol	Antimicrobial	Chen et al. (2016)
T. harzianum TE5	*Cucumis sativus*	Diterpene	Antifungal	Zhang et al. (2014)
T. saturisporum, *T. harzianum*, *T. reesei*	*Capsicum frutescence*	Ethylene, hydrogen cyanide	Antifungal	Ajith and Lakshmidevi (2010)
Urnula sp.	*Dicksonia antarctica*	4-Decene, 2-dodecene	Antifungal	Strobel et al. (2017)
Xylaria sp.	*Haematoxylon brasiletto*	2-Methyl-1-propanoland 2-methyl-1-butanol	Antifungal, phytotoxic	Sanchez-Ortiz et al. (2016)

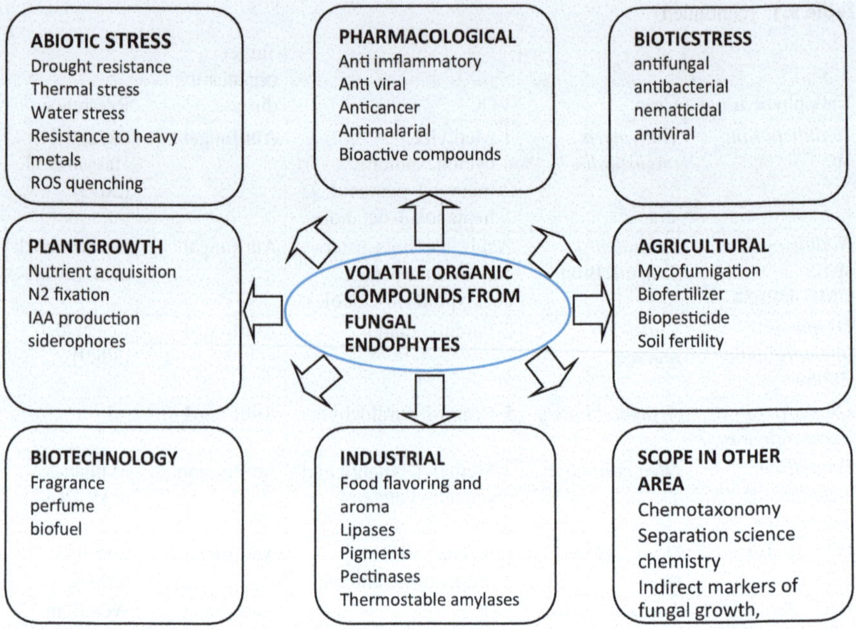

Fig. 9.1 Fungal volatile organic compounds and their applications

VOC molecules that interact with host plants and allow plants to communicate with the immediate environment. These emissions provide resistance to host plants against herbivores, pathogens and communications with different organs of plants and attract pollinators, seed dispersers and microbes (Kaddes et al. 2019). It has been reported that more than 300 secondary metabolites produced by endophytic fungi are VOCs (Lugtenberg et al. 2016; Kaddes et al. 2019; Strobel et al. 2001). This review focuses on recent data on volatile organic compounds produced by endophytic fungi and their potential application in biological control and plant growth development for sustainable agriculture.

9.2 Volatile Compounds from Endophytic Fungi

Volatile organic compounds (VOCs) are hydrophobic organic molecules with a low molecular weight (<300 Da) and high vapor pressure (>0.01 kPa at 20 °C) (Pagans et al. 2006). Due to their physiochemical properties, these molecules can diffuse plant cell membranes and thus play an important role in the functioning of soil ecosystems (Pichersky et al. 2006). VOCs are anthropogenic or natural compounds belonging to five major chemical groups such as terpenoids, fatty acid derivatives, benzenoid compounds, phenyl propanoids and amino acid derivatives, which are able to inhibit/kill several plant pathogens (Delory et al. 2016). The four major biosynthetic pathways, namely, Shikimate/phenylalanine, the mevalonic acid (MVA), the methylerythritol phosphate (MEP) and Lipoxygenase (LOX), needed four precursors, namely, PEP, E4P, E4P pyruvate, and acetyl CoA, which requires the availability of carbon, nitrogen sulfur, and energy released by host metabolism. These pathways lead to production of benzenoids/phenyl propanoids, sesquiterpenoids, monoterpenes, hemiterpenes diterpenes, volatile carotenoid derivatives and methyl jasmonate/green leaf volatiles (Dudareva et al. 2006) (Fig. 9.2).

Terpenes are derived from C5 isoprene units and form the largest group of natural products obtained from plant and microbial sources (Galindo-Solis and Fernandez 2022). Terpenes mediate host fungal interactions and also between fungi-micro biome (Farh and Jeon 2020). So far, about 80,000 terpenoids have been characterized by plant and microbial sources (Galindo-Solis and Fernandez 2022). These compounds have been shown to have a wide range of biological activities such as antifungal, antibacterial, anti-inflammatory, anticancer, antioxidant, antidepressant, and antimalarial (Chen et al. 2016). B caryophyllene is a volatile sesquiterpenoid extracted from Arabidopsis thaliana flowers, which have been shown to be defensive against pathogens (Gershenzon and Dudareva 2007).

Endophytic fungi can produce alcohols such as ethanol, acting as signaling molecules or antimicrobial properties. A *Phomopsis* sp. isolated from *Odontoglossum* sp. produced several VOCs, which are composed of sabinene, 1-butanolpheethyl alcohol, 1-propanol, and acetone, which are shown to inhibit a broad spectrum of fungi belonging to Deuteromycetes, Ascomycetes, and Oomycetes (Kaddes et al. 2019). Similarly *Muscodor* CZ-620 and strain MFC2 produced a mixture of volatiles composed of isobutyl alcohol, 2-methyl-1-butanol, and iso-butyric acid

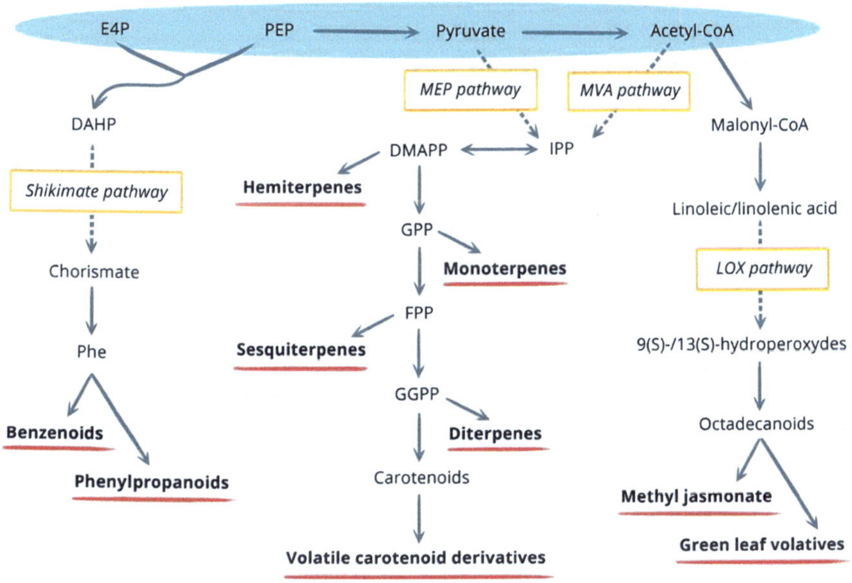

Fig. 9.2 Biosynthetic pathways in the production of volatile compounds in host plants (Kaddes et al. 2019)

(Mercier et al. 2007). Suwannarach et al. (2013) reported VOC production from *Nodulisporium* sp. CMU-UPE 34 consisted mainly of alcohols, acids, esters, and monoterpenes. An endophytic *Gliocladium roseum* (NRRL 50072 isolated from *Eucryphia cordifolia* was shown to produce a series of acetic acid esters of straight-chain compounds, including those of hexyl, heptyl, octyl, sec octyl and decyl alcohols (Strobel et al. 2008). Likewise an endophytic *Nodulisporium* sp. (*Hypoxylon* sp.) obtained from *Thelypteris anguistifolia* produced a wide range of VOCs, predominantly 1-butanol, 2-methyl and phenyl ethyl alcohol (Riyaz-Ul-Hassan et al. 2013). Recently, an endophytic *Cladosporium* sp. isolated from leaves of black poplar (*Populus nigra*) produced aliphatic alcohols, ketones and esters (Walther et al. 2021). Fungal endophytes produce aldehydes and ketones, which can act as antimicrobial compounds or have antimicrobial properties (Kaddes et al. 2019).

Esters are natural compounds with insecticidal and antifungal properties known to be produced from plants and microbes. An endophytic *Fusarium* sp. isolated from mangrove *Kandelia candel* known to produce a new isoflavone (5-*O*-methyl-2-methoxy-3-methyl alpinum isoflavone) and known compounds (6-methoxy-5,7,4-trihydroxy isoflavone, 6-methoxy-7,4-dihydroxy isoflavone, (+)-marmesin and 4-methyl benzoic acid carboxy methyl ester) (Huang et al. 2010). An enophytic fungi *Paecilomyces* sp. (JN227071.1) isolated from *Moringa olifera* leaves has shown to produce 9, cis-13-octa decanoic acid, methyl ester: 1-heptacosanol, 1-nonadecene, cyclotetra cosine: 1,2-benzene dicarboxylic acid: butyl 2-methyl propyl ester: di-sec-butylphthalate: 1-nonadecene, *n*-nonadecanol-1, behenic alcohol, *n*-heptadecanol-1, 1-hexadecanol: *n*-pentadecanol: dodecanoic acid: 2,3-dihydroxypropyl ester, oleic acid, 9-octa decanol (Hawar et al. 2023).

9.3 Endophytic Fungal VOCs as Antifungal Agents

Volatiles play an important role in the functioning of both above-ground and below-ground ecosystems (Morath et al. 2012). The below-ground interactions result in soil fungi stasis. In the failure of fungal propagules to germinate or the inhibition of fungal hyphal growth under favorable temperature and moisture conditions (Watson and Ford 1972). It has been hypothesized that soil fungi stasis might be due to competition for nutrients or the release of inhibitory compounds in the soil (Watson and Ford 1972). Strobel et al. (2001) introduced the concept of mycofumigation through VOCs produced by endophytic fungi to control postharvest diseases of fruits and vegetables. The volatiles produced by *Muscodor* species included esters, alcohols, acids, lipids and ketones. They reported about 28 VOCs from Xylariaceous endophytic fungi, *M. albus* isolated from *Cinnomomum zeylanicum.* These VOCs belonged to five classes comprising alcohols, acids, esters, ketones and lipids. 1-butanol, 3-methyl-acetate, were active against several plant pathogens such as *Pythium ultimum, Phytophthora cinnamomi, Rhizoctonia solani, Ustilago hordei, Stagnospora nodorum, Sclerotinia sclerotium, Aspergillus fumigatus, Verticillium dahlia, Tapesiayallundae, Candida albicans, E. coli, Staphylococcus aureus, Micrococcus luteus, Bacillus subtilis*, etc. In 2003, Stinson and his colleagues reported the healthy seedling establishment of sugar beets mycofumigated with *M. albus*. Lee et al. (2009) reported the antifungal nature of VOCs (5-pentyl-2-furaldehyde) produced by *Oxyporus latemarginatus* inhibited several pathogens, including *Alternaria alaternata*, *Colletotrichum gleosporioides* and *Fusarium oxysporum* sp. *Lycoperscici*. The genera *Nodulisporium*, *Muscodor* and *Trichoderma* have been known to produce a wide range of VOCs in host plants and play a major role in inhibiting the fungal pathogens (Garnica-Vergara et al. 2016). *Muscodor* species have been reported to produce VOCs that are lethal to a wide range of pathogenic fungi. *Muscodor* CZ-620 and strain MFC2 controlled the fungal rot of fruits (Mercier et al. 2007). Suwannarach et al. (2013) reported the fumigating activity of *M. suthepensis* against pathogenic fungi *P. digitatum*. Mycofumigation with *M. suthepensis* completely inhibited the mandarin fruit rot caused by *P. digitatum*. It has been also reported that fumigation with *M. albus* controlled bluemold caused by *Penicillium expansum* and greymold caused by *B. cinerea* of apples. Similarly fumigation with *M. albus* controlled the brown rot caused by *Monilinia fruticola*. In a report, VOCs produced by *Muscodortigerii* completely inhibited the growth of *Alternaria alternata* and *Cercospora beticola* (Saxena et al. 2015). A thermolabile molecule *N*-methyl-*N*-nitrosoisobutyramide (MNIBA) was identified by Hutchings et al. (2017) from *M. albus* known to cause DNA damage by methylation. Suwannarach et al. (2013) reported that 31 VOCs were analyzed by GC-MS from *Nodulisporium* sp. CMU-UPE34, which include mainly alcohols, acids esters and monoterpenes. One of the main molecules was eucalyptol (Synonym: 1,8-cineole). *Nodulisporium* sp. CMU-UPE34 controlled the growth of many pathogenic fungi in citrus fruits, including *Penicillium digitatum*, *P. expansum*, *Aspergillus fumigatus* and *Rhizoctonia solani* (Schalchli et al. 2016). The VOCs known to be detected by GC are isobutyl alcohol, 2-methyl-1-butanol and isobutyric acid (Mercier et al. 2007). *Muscodor cinnamomi* CMU-cib461 controlled the growth of *Rhizoctonia solani*, which can induce leaf blight, leaf spot, damping off and leaf rot (Suwannarach et al. 2016). Similarly, *O. latemarginatus* EF069 is known to inhibit the growth of *Botrytis cineria* and *Rhizoctonia solani*, which can cause postharvest apple decay and moth orchids' root rot, respectively. In vitro studies show that *Trichoderma arundinaceum* controls *B. cinerea* (Malmierca et al. 2012). It has been suggested that VOCs inhibit *B. cineria* by inducing the expression of defense genes with the production of salicylic acid and jasmonic acid with hydrolyte enzymes (Malmierca et al. 2012). In another experiment, VOCs produced by *Trichoderma* species such as *T. harzianum*, *T. saturnisporum* and *T. resei* effectively inhibited the growth of *Colletotrichum capscici* (Ajith and Lakshmidevi 2010). Similarly, *F. oxysporum* inhibited the growth of *B. cineria* in addition to *Rhizoctonia solani*, *Penicillium digitatum* and *Aspergillus niger* by inhibiting pectin methyl esterase, cellulose and polyphenols oxidase secretions (Zhang et al. 2014). *Phomopsis* sp. isolated from *Odontoglossum* sp. (Orchidaceae) from the forest of northern Ecuador is known to produce alcohols and carboxylic acids, especially sabinene, 1-butanol, phenethyl alcohol, 1-propanol and acetone. These molecules have been shown to inhibit a wide range of fungi from Deuteromycetes, Ascomycetes, and Oomycetes (Kaddes et al. 2019). Fumigation of jasmine rice (*Oryzaindica*) known with *Nodulisporium* sp. CMU-UPE34 culture and fumigation of wheat/rice bran with *Nodulisporium* sp. CF016 controlled cabbage green mold and grey and blue mold lesions (Park et al. 2010). However, several studies reported the relationships between VOC and phytotoxicity (Arimura et al. 2000). It is apparent that there are species and isolate specific differences in the production of volatiles. When the mycelia of basidiomycetes fungi (wood decayfungi) *Hyplonoma fascicularis* and *Resinicium bicolor* interacted physically, a new set of sesquiterpenes such as alpha-murolene and gammaamorphene were produced (Hynes et al. 2006). The antibiotic activity of *Trichoderma* VOCs has been documented early in 1971 (Dennis and Webster). Several species (*T. harzianum*, *T. viridae*, *T. lignorum*, *T. hamatum* and *T. reesei*) are able to inhibit the growth of *Fusarium solani*, a phytopathogen (Chakraborty and Chatterjee 2008).

A *Gliocladium* sp., an endophytic fungus of *Eucryphiala cordifolia*, was reported to produce annulene, a potent antimicrobial VOC (Stinson et al. 2003). The endophytic fungus *Daldinia concentrica* isolated from an olive tree in Israel produced at least 28 VOCs promising postharvest control. These volatiles protect dried fruits of apricot, plum, and raisin from rotting. Moreover, these protect the pea nuts against *Aspergillus niger*, oranges and tomato paste against *Penicillium digitatum* and grapes against *Botrytis cinerea* (Liarzi et al. 2016). Volatiles of *Fusarium oxysporum* and its bacterial consortium inhibited the fungal growth of a pathogenic strain, *F. oxysporum* (Minerdi et al. 2011). The VOCs of *M. albus*is have the ability to control soil-borne diseases. It effectively inhibited the growth of *Rhizoctonia solani*, which is responsible for damping off broccoli and also *Phytophthora capsici* responsible for root rot of bell pepper (Mercier and Manker 2005).

Stimulation or enhancement of soil-borne biocontrol agents is another property of certain VOCs; for example, the volatiles of *Trichoderma atroviridae* stimulated the expression of biocontrol gene in *Pseudomonas florescence* (Lutz et al. 2004). The fungal VOC 1-octen-3-ol (mushroom alcohol) effectively upregulated the defense genes in *Arabidopsis thaliana* and protected *Botrytis cinerea* (Kishimoto et al. 2017). The endophyte such as *M. yucatanensis*, *M. albus*, and *M. fengyangensis* produced VOCs that inhibit and kill several plant pathogenic bacteria and fungi (Strobel et al. 2001). The VOCs produced by *M. crispans* produced a mixture of VOCs that inhibited a wide range of plant pathogens, including *Mycosphaerella fiziensis* (the black sigatoka pathogen of bananas) and bacterial pathogen of citrus *Xanthomonas axonopodis* pv *citri* (Mitchell et al. 2010). Sabinene (a monoterpene with peppery odor) is produced from *Phomopsis* sp., which was earlier known from higher plants along with 1-butanol, 3-methyl; benzene ethanol, 1-propanol, 2 methyl and 2-propanone (Singh et al. 2011). The VOC profile of *Phoma* sp. isolated from *Creósote* bush consists of transcaryophyllene, sesquiterpenoids alcohols and naphthalene derivatives. These VOCs inhibited the growth of a wide range of plant pathogens, including *Verticillium*, *Ceratocystis*, *Cercospora*, and *Sclerotinia* (Strobel et al. 2011). In addition, fungal VOCs may enable their host plants to outcompete neighboring plants. The VOCs produced by *M. yucatanensis* were reported as toxic to root and seed germination and have an inhibitory effect on root growth in amaranth, tomato and barnyard grass (Macias-Rubalcava et al. 2010).

Plaszko et al. (2020) examined VOCs produced by fungal endophytes isolated from horseradish (*Armoracia rusticana*) using headspace-gas chromatography-mass spectroscopy headspace-GC-MS). Forty-three isolates produced several typical VOCs acetone, methyl acetate, methyl formate, ethyl acetate, methyl butanol isomers, styrene, and betaphyllandrene with disulfide and dimethyl sulfide. The fungi like *Setophoma*, *paraphoma*, *Plectosphaerella*, *Pyrenochaeta*, *Volutella*, *Cadophora*, *Notophoma*, and *Curvularia* were reported for the first time. Santra and Banerjee (2023) investigated biofumigation activity of VOCs produced from *Diaporthe* sp. CEL3 isolated from leaves of *Chloranthus elatior* Sw. The volatiles of this fungi exhibited inhibitory effects and minimized the infections of cherry fruit rot pathogen *Monilinia fruticola* along with *Rhizoctonia solani*, *Botrytis cineria*, *Pythium ultimum*. Wonglom et al. (2020) reported the antifungal activity of VOCs obtained from *Trichoderma asperellum* in lettuce. Lettuces showed increased concentrations of cell wall degrading enzymes chitinase and B-1,3-glucanase. The VOCs emitted by *T. asperellum* significantly increased the numbers of leaves and roots, plant biomass and total chlorophyll content in lettuce. Zhang et al. (2022) reported the VOCs produced from endophyte *Neurospora dictyophora* 3ZF-02 isolated from arils of *Torreya grandis*. The main components were esters, acids, alcohols, and ketones, along with benzene and naphthalene compounds. The extracts showed antioxidant activity withIC50 of 1.44 and 0.95 g/L against diphenyl picryl hydrazinyl and hydroxyl radicals, respectively. The VOCs also exhibited strong antifungal activities against *Candidaalbicans* with MIC 1.56 g/L. Malhadas et al. (2017) examined the antimicrobial potential of VOCs from *Alternaria alternata* and *Penicillium commune*. These VOCs belonged to six volatiles involving 3-methyl-1-butanol and phenylethyl alcohol.

Endophytic *Sarocladium brachiariae* HND5 produced VOCs, of which three (2-methoxy-4-vinylphenol, 3,4-dimethoxystyrol and caryophyllene) exhibited antifungal activity against *Fusarium oxysporum* f. sp. cubens (FOC) which causes vascular wilt disease in banana plants. TEM and double florescence staining revealed that 2-methoxy-4-vinyl phenol and 3,4-dimethoxystyrol damaged the plasma membranes, leading to cell death. 3,4-Dimethoxy styrol also promoted the synthesis of chitin synthase genes and altered the cell walls of FOC hyphae. B-caryophyllene induced the accumulation of ROS in FOC hyphae (Yang et al. 2015). Recently, endophytic fungi from wild Rubiaceae plants have shown to be effective in controlling the American leaf spot of coffee caused by mycena citricolor. One of the isolate *Daldinia eschschaltzii* (ylariale) inhibited the growth of *M. citricolor*, and the VOC 1,8-dimethoxy naphthalene induced the damage in pathogens' reproductive structures (Levya et al. 2023). Endophytic *Aspergillus* AFEG-2017 isolated from *Eleacharis geniculata* was reported to have produced 25 organic and bioactive compounds. Of them, 2-(2-hydroxy-3-isobutoxypropyl)pent-4-enoic acid, hydrazide was the most abundant VOC. The VOCs of AFEG-2017 enhanced the seedling growth of *Trigonella foenum graecum*, *Solanum lycopersicum*, *Portulaca oleracea* and *Lepidium sativum* and also inhibited the pathogenic fungi *Fusarium oxysporum* (Fakhry et al. 2022). In a recent study endophytic *Sarocladium brachiariae* HND5 produced several VOCs of which three volatiles (2-methoxy-4-vinyl phenol,

3,4-dimethoxystyrol and caryophyllene) exhibited antifungal activity against *Fusarium oxysporum* f. sp. *cubense* (FOC), which causes vascular wilt disease in banana plants. TEM and double florescence staining revealed that 2-methoxy-4-vinyl phenol and 3,4-dimethoxystyrol damaged the plasma membrane leading to cell death. 3,4-Methoxy styrol also promoted the synthesis of chitin synthases gene and altered the cell wall of FOC hyphae. B-caryophyllene induced the accumulation of ROS in FOC hyphae (Yang et al. 2015).

9.4 Endophytic Fungal Volatiles as Antibacterial Agents

VOCs produced by fungal endophytes have shown antibacterial activity against several pathogens. *Musodor* spp. has been reported as an antibacterial agent. The volatiles of *M. albus* Cz620 isolated from inhibited *Escherichia coli* and *Bacillus subtilis* and other isolates such as KN-26, KN-27, GP-100, Gp-115, TP-21 inhibited *E. coli* (Ezra et al. 2004). *Muscodor albus* isolated from unidentified vine from Indonesia produced several VOCs, including tetrohydrofuran, 2-methyl furan; 2-butanone; aciphyllene and a large quantity of azulene derivatives. The isolate I-41.3s was effective against *B. subtilis*, *E. coli* and *Salmonella typhi* (Atmosukarto et al. 2005). An endophytic *Muscodor crispans* isolated from *Ananas ananaoides* from Bolivian Amazon Basin produced wide spectrum of volatiles comprised propanoic acid, 2-methyl-, methyl ester; propanoic acid, 2-methyl-, 1-butanol, 3-methyl-, 1-butanol, 3-methyl-, acetate; propanoic acid, 2-methyl-, 2-methylethyl ester and ethanol. Exhibited inhibitory activity against *Mycobacterium marianum*, *S. aureus*, *Salmonella cholereasus* and *Yersinia pestis* (Mitchell et al. 2010). Similarly, *M. fengyangensis* isolated from *Cinnamomum zeylanicum* from South east China exhibited strong antimicrobial activity against *E. coli* (Zhang et al. 2010). *M. kashayum* an endophyte isolated from leaves of *Agle marmelos* from Western Ghats of India reported to produce 23 different volatile compounds predominantly 3-cyclohexen-1-ol, 1-(1,5-dimethyl-4-hexenyl)-4-methyl; 1,6-dioxacyclododecane-7,12-dione; 2,6-bis(1,1-dimethylethyl)-4-(1-oxopropyl) phenol; 2,4-di-tert-butylthiophenol and 4-octadecylmorpholine. These volatile exhibited a broad spectrum antimicrobial activity against human bacterial pathogens such as *E. coli* and *P. aeruginosa* (Meshram et al. 2013). Suwannarach et al. (2013) isolated four novel species of *Muscodor* vi *M. musae*, *M. oryzae*, *M. suthepensis*, and *M. equiseti* from *Musa acuminata*, *Oryza rufipogon*, *Cinnamomum bejolghota* and *Euisetum debile* respectively from Northern Thailand. 2-Methyl propionic acid was the main compound extracted and exhibited significant antibacterial activity against *Enterococcus faecalis*, *E. coli*, *Proteus mirabilis*, *S. aureus* and *Pseudomonas pneumoniae*. Recently, an endophytic fungus *Curvularia eragrostidis* HelS1 isolated from *Helecteris isora* exhibited antibacterial activity effectively against six Gram +ve and seven Gram −ve pathogens with a MLC value ranging from 12.5 to 400 μg/mL. Thevolatiles were found to be 1-H-indene 1 methanol acetate, tetroquinone, *N,N*-diphenyl-2-2-nitro-thio-benzamide, trans-1,2-diethyl-trans-2-decolinol, naphthalene and azulene (Santra and Banerjee 2022). Malhadas et al. (2017) reported the antibacterial activity of VOCs produced by endophytic fungi from olive trees. *Alternaria alternata* and *Penicillium commune* were effective against several bacterial and fungal pathogens. The main composition of VOCs was 3-methyl-1-butanol and phenylethyl alcohol.

9.5 Endophytic Fungal VOCs in Plant Growth and Development

Over-application of pesticides can lead to the development of pesticide resistance, rendering them less effective. The production and application of inorganic nitrogen fertilizer has resulted in case of human alteration to the global nitrogen cycle. Continuous application of inorganic nitrogen leads to enhanced production of microbial nitrous oxide production (N_2O) (Kanchiswamy et al. 2015). The genetically modified crops are able to exhibit enhanced disease resistance and can reduce depending on chemical fertilizers inputs, and the regulation frameworks required to commercialize the crops are lengthy (Kanchiswamy et al. 2015). Hence there is a need to explore alternate control strategies to develop sustainable agriculture and the application of beneficial mutualistic endophytes due to their ability to antagonize soil pathogens and enrich the plant biomass. These organisms produce a wide spectrum of secondary metabolites, which are systematically associated with the host plants (Shankar Naik et al. 2006).

Fungal VOCs, which are termed mycofumigaters, have been used for plant growth promotion and prevention of postharvest fungal growth. VOCs are also sources of fuels termed as mycodiesel. In agriculture, the potential use of microbial VOCs includes bio-stimulation of crops (Wonglom et al. 2020). The microbial volatile profile is dependent on multiple parameters such as geography, time and environmental conditions, including nutrient content, microbial community composition, temperature, humidity, and pH (Inseam and Seewald 2010). These factors lead to drastically different emission profiles, concentrations and quantities observed in nature (Wonglom et al. 2020). Microbial VOCs also have the potential to enhance plant growth, enabling them to potentially exploit a new category of fertilizer previously described as gaseous fertilizer (Shankar Naik 2018). Similarly, 2-pentyfuran produced from the culture of *Bacillus megaterium* showed dose-dependent growth promotion in *A. thaliana* (Zou et al. 2010). A range of VOCs including dimethyl sulphide from *Burkholderia nambifuran* increased *A. thaliana* biomass (Groenhagen et al. 2013). VOCs, especially 6-pentyl-2H-pyran-2-one from *Trichoderma* sp., are known to influence plant growth and reduce the disease symptoms caused by fungal pathogens (Kottb et al. 2015). Garnica-Vergara et al. (2016) reported increased plant biomass and root branching in *A. thaliana* exposed to 6-pentyl-2H-pyran-2-2one. Similarly, 2 heptanol, 3-octanol and 2-heptanone produced from *Trichoderma atroviridae* increased plant growth in *A. thalian* (Estrada-Rivera et al. 2019). A secondary metabolite called b-caryophyllene produced from *Fusarium oxysporum* increased lettuce biomass (Minerdi et al. 2011). B-caryophyllene is also known to enhance biomass of *Brassica campestris* in addition to suppression of fungal pathogen *Colletotrichum hignicianum* (Yamagiwa et al. 2011). Several studies indicate VOCs may promote plant growth through modulating plant growth hormone responses. A *Trichoderma* VOC 1-decene stimulated differential expression of auxin-related genes (Lee et al. 2019). Dimethyldisulphide has been commercialized and used as a soil fumigant and a pesticide in the name of PALADIN (De Boer et al. 2019).

9.6 Conclusions

Endophytic fungi represent a relatively unexplored pool of a variety of metabolites with potential applications. The study of these volatiles promises the discovery of new products of human exploitation in plant growth, biocontrol, fuel, pharmacognosy and biotechnology. There is a need to discover efficient endophytes with advanced approaches such as profiling and analyzing VOCs, genome sequencing and genomic tools to understand physiological, molecular, and cellular changes in plant microbial systems for sustainable agriculture.

References

Adeleke BS, Ayilara MS, Akinola SA et al (2022) Biocontrol mechanisms of endophytic fungi. Egypt J Biol Pest Control 32:46. https://doi.org/10.1186/s41938-022-00547-1

Ajith PS, Lakshmidevi N (2010) Effect of volatile and non-volatile compounds from *Trichoderma* spp. against *Colletotrichum capsici* incitant of anthracnose on bell peppers. Nat Sci 8:265–269

Arimura GI et al (2000) Herbivory induced volatiles elicit defense gene in lima bean leaves. Nature 406:512–515

Atmosukarto I, Castillo U, Hess W, Sears J, Strobel G (2005) Isolation and characterization of *Muscodor albus*; a volatile antibiotic producing fungus. Plant Sci 169:854–861

Banerjee D, Strobel G, Geary B, Sears J, Ezra D, Liarji O et al (2010) *Muscodor albus* strain GBA an endophytic fungus of *Ginkgo biloba* from United States of America produces volatile antimicrobials. Mycology 1:179–186

Baron NC, Rigobelo EC (2021) Endophytic fungi a tool for plant growth promotion and sustainable agriculture. Mycology 13:39–55

Cagadas RB, Ceballo FA, Cumagun CJR, Pangga IB (2021) Mycofumigation with the endophytic fungi *Fusarium proliferatum* (Matsushima) *Nirenberg* and *Diaporthe* sp, for the control of banana and mango anthracnose. Ann Trop Res 43(2):13–24

Chakraborty MR, Chatterjee NC (2008) Control of *Fusarium* wilt of *Solanumm elongena* by *Trichoderma* spp. Biol Plant 52:582–586

Chen JL, Sun SZ, Miao CP, Wu K, Chen YW, Xu LH et al (2016) Endophytic *Trichodermagamsii* YIMPH30019: a promising biocontrol agent with hyperosmolar, mycoparasitism and antagonistic activities of induced volatile organic compounds on root rot pathogenic fungi of *Panaxnotoginseng*. J Ginseng Res 40:3015–3324

Corcuff R, Mercier J, Twedell R, Arul J (2011) Effect of water activity on the production of volatile organic compounds by *Muscodoralbus* and their effect on three pathogens in stored potato. Fungal Biol 115:220–227

De Boer W, Li X, Meisner A, Garbeva P (2019) Pathogen suppression by microbial volatile compounds in soils. FEMS Microbiol 95:105

Delory BM, Delaplace P, Du Jardin P, Fauconnier ML (2016) Barley (*Hordeum distichon* L.) roots synthesise volatile aldehydes with a strong age dependent pattern and release (e)-non-2-enal and (e, z)-nona-2, 6-dienal after mechanical injury. Plant Physiol Biochem 104:134–145

Dudareva N, Negre F, Nagegowda DA, Orlova I (2006) Plant volatiles: recent advances and future perspectives. Crit Rev Plant Sci 25:417–440

Estrada-Rivera M, Rebolledo-Prudencio OG, Perez-Robels DA, Rocha-Medina MD, Gonzalez-Lopez MDC, Casas-Flores S (2019) *Trichoderma* histone deacetylase HDA-2 modulates multiple responses in *Arabidopsis*. Plant Physiol 179:1343–1361

Ezra D, Strobel GA (2003) Effect of substrate on the bioactivity of volatile antimicrobials produced by *Muscodor albus*. Plant Sci 165:1229–1238

Ezra D, Jasper J, Rogers T, Knighton B et al (2004) Proton-transfer reaction-mass spectroscopy as technique to measure volatile emissions of Muscodor albus. Plant Sci 166:1471–1477

Fakhry F, Abdel-Motaal, Kamel NM, El-Zayat SA (2022) Plant seedling growth stimulation and antifungal activities of volatile organic compounds emitted by *Aspergillus flavus* endophyte. J Appl Biotechnol Rep. https://doi.org/10.30491/JABR.2022.335430.1514

Farh ME, Jeon J (2020) Roles OF fungal volatiles from perspective of distinct life styles in filamentous fungi. Plant Pathol J 36(3):193–203

Galindo-Solis JM, Fernandez FJ (2022) Endophytic fungal terpenoids: natural role and bioactivities. Microorganisms 10(2):339

Garnica-Vergara A, Barrera-Ortiz S, Muñoz-Parra E, Raya-González J, Méndez-Bravo A, Macías-Rodríguez L et al (2016) The volatile 6-pentyl-2H-pyran-2-one from *Trichoderma atroviridae* regulates Arabidopsisthaliana root morphogenesis via auxin signalling and ethylene insensitive 2 functioning. New Phytol 209:1496–1512

Gavrila SS, Ursachi CS, Perta-Crisan S, Munteanu FD (2022) Recent trends in biosensors for environmental quality monitoring. Sensors 22:1513. https://doi.org/10.3390/s22041513

Gershenzon J, Dudareva N (2007) The function of terpene natural products in the natural world. Nat Chem Biol 3:408–414

Grabka R, d'Entremont TW, Adams SJ, Walker AK, Tanney JB, Abbasi PA et al (2022) Fungal endophytes and their role in agricultural plant protection against pests and pathogens. Plants 11:384. https://doi.org/10.3390/plants11030384

Groenhagen U, Baumgartner R, Bailly A, Gardiner A, Eberl L, Schulz S, Weisskopf L (2013) Production of bioactive volatiles by different *Burkholderia ambifaria* strains. J Chem Ecol 39:892–906

Gupta S, Chaturveedi P, Kulkarni MG, Staden JV (2020) A critical review on exploiting the pharmaceutical potential of plant endophytic fungi. Biotechnol Adv 39:107462. https://doi.org/10.1016/j.biotechadv.2019.107462

Hawar SN, Taha ZK, Hameid AF, Al-Shmgani HS, Sulaiman GM, Elsilk SE (2023) Antifungal activity of bioactive compounds produced by endophytic fungus *Paecilomyces* sp (JN227071.1) against Rhizoctonia solani. Int J Biomath 2023. https://doi.org/10.1155/2023/2411555

Huang Z, Yang J, She Z, Lin Y (2010) Isoflavons from the mangrove endophytic fungus *Fusarium* sp. (ZZF41). Nat Prod Commun 5:1771–1773

Hutchings ML, Hiller DA, Berro J, Strobel SA (2017) Mycofumigation through production of the volatile DNA methylating agent n-methyl-n-nitroso iso butyramide by fungi in the genus *Muscodor*. J Biol Chem 292:7358–7371

Hynes J, Muller CT et al (2006) Changes in volatile production during the course of fungal mycelia interactions between *Hypholoma fasciculare* and *Resinicium bicolor*. J Chem Ecol 33:43–57

Inseam H, Seewald SA (2010) Volatile organic compounds (VOCs) in soils. Biol Fertil Soil 46:199–213

Kaddes A, Fauconnierm L, Sassi K, Nasraoui B, Ji Jakli MH (2019) Endophytic fungal volatile compounds as a solution for sustainable agriculture. Molecules 3(24):16

Kanchiswamy CN, Malnoy M, Maffei ME (2015) Chemical diversity of microbial volatiles and their potential for plant growth and productivity. Front Plant Sci 6. https://doi.org/10.3389/fpls.2015.00151

Keswani C, Dilnashin H, Birla H, Singh S (2019) Unravelling efficient applications of agriculturally important microorganisms for alleviation of induced inter cellular oxidative stress in crops. Acta Agric Slovanica 114:121. https://doi.org/10.14720/aas.2019.114.1.14

Kishimoto K, Matsui K, Ozawa R, Takabayashi J (2017) Volatile 1-octen-3-ol induces a defensive response in Arabidopsis thaliana. J Gen Plant Pathol 73:35–37

Kottb M, Gogolashvili T, Grobkinsky DK, Piechulla B (2015) *Trichoderma* volatiles effecting *Arabidopsis*: from inhibition to protection against phytopathogenic fungi. Front Microbiol 6:995

Lee SO, Kim HY, Choi GJ, Lee HB, Jang KS, Choi YH, Kim JC (2009) Mycofumigation with *Oxyporuslatemarginatus* EF069 for control of postharvest apple decay and *Rhizoctonia* root rot on moth orchid. J Appl Microbiol 106:1213–1219

Lee S, Behringer G, Hung R, Bennet J (2019) Effects of fungal volatile organic compounds on Arabidopsis thaliana growth and gene expression. Fun Ecol 37:1–9

Levya EE, Montero MMG, Ortiz CO, valvarde EA, Picado EA, JMC F, Wolfe LA, Chaverri P (2023) The endophytobiome of wild Rubiaceae as a source of antagonistic fungi against the American leaf spot of coffee (Mycena citricolor). J Appl Microbiol 134:1–12

Liarzi O, Bar E, Lewinsohn E, Ezra D (2016) Use of endophytic fungus Daldinia cf concentrica and its volatiles as biocontrol agents. PLoS One 11(12):e0168242

Lugtenberg BJJ, Caradus JR, Johnson LJ (2016) Fungal endophytes for sustainable crop production. FEMS Microbiol Ecol 92(12):17

Lutz MP, Wenger S, Maurhofer M et al (2004) Signalling between bacterial and fungal biocontrol agents in a strain mixture. FEMS Microbiol Ecol 48:447–455

Macias-Rubalcava ML, Hernadez-Bautista BE, Oropeza F, Duarte G, Gonzalez MC, Glenn AE, Hanlin R, Anaya AL (2010) Allelochemical effects of volatile compounds and organic extracts from *Muscodor yucatanensis*, a tropical endophytic fungus from *Bursera simaruta*. J Chem Ecol 36:1122–1131

Macias-Rubalcava ML, Sanchez-Fernandez RE, Roque-Flore G, Lappe-Oliveras M-RYM (2018) Volatile organic compounds from *Hypoxylon anthechroum* endophytic strains as postharvest mycofumigation alternative for cherry tomatoes. Food Microbiol 76:363–373

Malhadas C, Malheiro R, Pereira JA, de Pinho G, Baptista P (2017) Antimicrobial activity of endophytic fungi from Olive tree leaves. World J Microbiol Biotechnol 33:1–12

Malmierca M, Cardoza R, Alexander N, McCormick S, Hermosa R, Monte E, Gutiérrez S (2012) Involvement of *Trichoderma trichothecenes* in the biocontrol activity and in the induction of plant defence related genes. Appl Environ Microbiol 78:4856–4868

Malmierca MG, Mc Cormick SP, Cardoza RE, Alexander NJ, Monte E, Gutierrez S (2015) Production of trichodiene by *Trichoderma harzianum* alters the perception of this biocontrol strain by plants and antagonized fungi. Environ Microbiol 17:2628–2646

Mercier J, Manker D (2005) Biocontrol of soil borne diseases and plant growth enhancement in greenhouse soilless mix by the volatile producing fungus *Muscodor albus*. Crop Prot 24:355–362

Mercier J, Jiménez-Santamaría JI, Tamez-Guerra P (2007) Development of the volatile-producing fungus *Muscodor albus* as a novel antimicrobial biofumigant. Rev Mex Fitopatol 25:173–179

Meshram V, Kapoor N, Saxena S (2013) *Muscodor kashayum* sp. Nov.—a new volatile antimicrobial producing endophytic fungus. Mycology 4:196–204

Minerdi D, Bossi S, Maffei ME et al (2011) Fusarium oxysporum and its bacterial consortium promote lettuce growth and expansin A5 gene expression through microbial volatile organic compound (BVOC) emission. FEMS Microbiol Ecol 76:342–351

Mitchell AM, Strobel GA, Moore E, Robison R, Sears J (2010) Volatile antimicrobials from *Muscodorcrispens*, a novel endophytic fungus. Microbiology 156:270–277

Morath SU, Hung R, Bennet JW (2012) Fungal volatile organic compounds: a review with emphasis on their biotechnological potential. Fungal Biol Rev 26:73–83

Newbery F, Qi A, Fitt BDL (2016) Modelling impacts of climate change on arable crop diseases, progress, challenges and applications. Curr Opin Plant Biol 32:101–109. https://doi.org/10.1016/j.pbi.2016.07.002

Ningxiao LI, Alfiky A, Vaughan MM, Kang S (2016) Stop and smell the fungi: fungal volatile metabolites are overlooked signals involved in fungal interaction with plants. Fungal Biol Rev 30:134–144

Pagans E, Font X, Sánchez A (2006) Emission of volatile organic compounds from composting of different solid wastes: abatement by biofiltration. J Hazard Mater 131:179–186

Park MS, Ahn J, Choi GJ, Choi YH, Jang KS, Kim JC (2010) Potential of the volatile-producing fungus Nodulisporium sp. CF016 for the control of postharvest diseases of apple. Plant Pathol J 26:253–259

Pena LC, Jungklaus GH, Savi DC, Ferreira-Maba L, Servienski A, Maia BH et al (2019) *Muscodor brasiliensis* sp. Nov. produces volatile organic compunds with activity against *Penicillium digitatum*. Microbiol Res 221:28–35

Phoka N, Suwannarach N, Lumyong S, Ito S-I, Matsui K, Arikit S, Sunpapao A (2020) Role of volatiles from the endophytic fungus *Trichoderma asperelloides* PSU-P1 in the biocontrol potential and in promoting the plant growth of *Arabidopsis thaliana*. J Fungi 6:341

Pichersky E, Noel JP, Dudareva N (2006) Biosynthesis of plant volatiles: Nature's diversity and ingenuity. Science 311:808–811

Plaszko T, Szucs Z, Kallai Z, Csoma H, Vasas G, Gonda S (2020) Volatile organic compounds (VOCs) of endophytic fungi growing on extracts of the host horseradish (*Aromacia rusticana*). Metabolites 10:451

Rana KL, Kour D, Kaur T, Devi R, Yadav AN et al (2020) Endophytic microbes: biodiversity plant growth promoting mechanisms and potential applications for agricultural sustainability. Anton Van Leeuwenhoek (Int J Gen Mol Microbiol) 113:1075–1107

Riyaz-Ul-Hassan S, Strobel G, Geary B, Sears J (2013) An endophytic *Nodulisporium* sp. from central America producing volatile organic compounds with both biological and fuel potential. J Micobiol Biotechnol 23(1):29–35

Sanchez-Ortiz BL, Sanchez-Fernandez RE, Duarte G, Lappe Oliveras P, Macias Rubalcava ML (2016) Antifungal, anti-oomycete and phytotoxic effects of volatile organic compounds from the endophytic fungus *Xylaria* sp strain PB 3f3 isolated from *Haematoxylon brasiletto*. J Appl Microbiol 120:1313–1325

Sanjai S, Vineet M, Neha K (2014) *Muscodor darjeelingensis*, a new endophytic fungus of *Cinnamomum camphora* collected from northeastern Himalaya. Sydowia 66:55–67

Santra HK, Banerjee D (2022) Broad spectrum antimicrobial action of cell free culture extracts and volatile organic compounds produced by endophytic fungi Curvularia eragrostidis. Front Microbiol 13:1–19

Santra HK, Banerjee D (2023) Broad antimicrobial action of cell free culture extracts and volatile organic compounds produced by endophytic fungi *Curvularia eragrostidis*. Fornt Microbiol 13:1–19

Saxena S, Meshram V, Kapoor N (2015) *Muscodortigerii* sp. Nov.-volatile antibiotic producing endophytic fungus from the northeastern Himalayas. Ann Microbiol 65:47–57

Schalchli H, Tortella G, Rubilar O, Parra L, Hormazabal E, Quiroz A (2016) Fungal volatiles: an environmentally friendly tool to control pathogenic microorganisms in plants. Crit Rev Biotechnol 36:144–152

Sdiri Y, Lopes T, Rodrigues N, Silva K, Rodrigues I, Pereira JA, Baptista P (2022) Biocontrol ability and production of volatile organic compounds as a potential mechanism of action of olive endophytes against *Colletotrichum acutatum*. Microrganism 10:571

Shankar Naik B (2018) Volatile hydrocarbons from endophytic fungi and their efficacy in fuel production and disease control. Egypt J Biol Pest Control 28:69

Shankar Naik B, Shashikala J, Krishnamurthy YL (2006) Study on the diversity of endophytic communities from rice (*Oryzasativa* L.) and their antagonistic activities *in vitro*. Microbiol Res 3:290–296

Singh SK, Strobel GA, Knighton B, Geary B, Sears J, Ezra D (2011) An endophytic *Phomopsis* sp. possessing bioactivity and fuel potential with its volatile organic compounds. Microb Ecol 61:729–739

Singh RP, Shelke GM, Kumar A, Jha PN (2015) Biochemistry and genetics of ACC deaminase: a weapon to stress ethylene produced in plants. Front Microbiol 6:937. https://doi.org/10.3389/fmicb.2015.00937

Sridharan AP, Hankappan S, Karthikeyan G, Uthandi S (2020) Comprehensive profiling of the VOCs of *Trichoderma longibrachiatum* EF5 while interacting with *Sclerotium rolfsii* and *Macrophomina phaseolina*. Microb Res 236:126436

Srinivas C, Shubha S, Narasimha Murthy K, Arakere U, Sudisha J (2020) Detection and characterization of antibacterial siderophores secreted by endophytic fungi from *Cymbidium aloifolium*. Biomolecules 10:1412. https://doi.org/10.3390/biom10101412

Stinson M, Ezra D, Hess WM, Sears J, Strobel G (2003) An endophytic *Gliocladium* sp. of *Eucryphia cordifolia* producing selective volatile antimicrobial compounds. Plant Sci 165:913–922

Strobel G (2006) *Muscodor albus* and its biological promise. J Indus Microbiol Biotechnol 33:514–522

Strobel GA, Dirkse E, Sears J, Markworth C (2001) Volatile antimicrobials from *Muscodor albus*, a novel endophytic fungus. Microbiology 147:2943–2950

Strobel GA, Knighton B, Kluck K, Ren Y, Livinghouse T, Griffen M, Spakowicz D, Sears J (2008) The production of myco-diesel hydrocarbons and their derivatives by the endophytic fungus *Gliocladium roseum* (NRRL 50072). Microbiology 154:3319–3328

Strobel G, Singh SK, Riyaz-Ul-Hassan S, Mitchell AM, Geary B, Sears J (2011) An endophytic pathogenic *Phoma* sp. from creosote bush producing biologically active volatile compounds having fuel potential. FEMS Microbiol Lett 320(2):87–94

Strobel G, Ericksen A, Sears J, Xie J, Geary B, Blatt B (2017) *Urnula* sp. an endophyte of Dicksonia Antarctica making a fragrant mixture of biologically active volatile organic compounds. Microb Ecol 74:312–321

Suwannarach N, Kumlaa J, Bussabana B et al (2013) Biofumigation with the endophytic fungus *Nodulisporium* spp. CMU-UPE34 to control postharvest decay of citrus fruit. Crop Prot 45:63–70

Suwannarach N, Bussaban B, Nuangmek W, Pithakpol W, Jirawattanakul B, Matsui K, Lumyong S (2016) Evaluation of *Muscodor suthepensis* strain cmu-cib 462 as a postharvest biofumigant for tangerine fruit rot caused by *Penicillium digitatum*. J Sci Food Agric 96:339–345

Tomsheck AR, Strobel GA, Booth E, Geary B, Spakowicz D, Knighton B, Floerchinger C, Sears J, Liarji O, Ezra D (2010) Hypoxylon sp., an endophyte of *Persia indica* producing 1,8-cineole and other bioactive volatiles with fuel potential. Microb Ecol 60(4):903–914

Ulloa-Benítez Á, Medina-Romero YM, Sánchez-Fernández RE, Lappe-Oliveras P, Roque-Flores G, Duarte Lisci G et al (2016) Phytotoxic and antimicrobial activity of volatile and semi-volatile organic compounds from the endophyte Hypoxylon anthochroum strain Blaci isolated from *Bursera lancifolia* (Burseraceae). J Appl Microbiol 121(2):380–400

Verma A, Shameen N, Jatav HS, Satyanarayana E, Parray JA, Poczai P, Sayyed RZ (2022) Fungal endophytes to combat biotic and abiotic stresses for climate-smart and sustainable agriculture. Front Plant Sci 13:953836. https://doi.org/10.3389/fpls.2022.953836

Walther C, Baumann P, Luck K, Rothe B, Biedermann HWP, Gershenzon J, Kollner TG, Unsicker SB (2021) Volatile emission and biosynthesis in endophytic fungi colonizing black poplar leaves. Beilstein J Org Chem 17:1698–1711

Watson AG, Ford EJ (1972) Soil fungistasis—a reappraisal. Annu Rev Phytopathol 10:327

Wonglom P, Ito S, Sunpapao A (2020) Volatile organic compounds emitted from endophytic fungus Trichoderma asperellum T 1 mediate antifungal activity defense response and promote plant growth in lettuce (*Lactuca sativa*). Fungal Ecol 43:100867. https://doi.org/10.1016/j.funeco.2019.100867

Yamagiwa Y, Inagaki Y, Ichinose Y, Toyoda K, Hyakumachi M, Shiraishi T (2011) Talaromyceswortmanii FS2 emits b-caryophyllene which promotes plant growth and induces resistance. J Gen Plant Pathol 77:336–341

Yang Y, Chen Y, Yu D, Wang J, Shao X, Xu F, Wang H (2015) Antifungal modes of action of tea tree oil and its two characteristic components against Botrytis cinerea. J Appl Microbiol 119:1253–1262

Yung L, Sieguey C, Azou-Barre A, Blaudez D (2021) Natural fungal endophytes from Noccaeacaerulescens mediate neutral to positive effects on plant biomass, mineral nutrition and Zn phytoextraction. Front Microbiol 12:689367. https://doi.org/10.3389/fmicb.2021.689367

Zhang CL, Wang GP, Mao LJ, Komon-Zelazowska M et al (2010) Muscodor fengyangensis sp. nov. from southeast China: morphology, physiology and production of volatile compounds. Fungal Biol 114:797–808

Zhang F, Yang X, Ran W, Shen Q (2014) *Fusarium oxysporum* induces the production of proteins and volatile organic compounds by *Trichoderma harzianum* t-e5. FEMS Microbiol Lett 359:116–123

Zhang H, Xu L, Kong X, Ji X, Yang S (2022) Changes and bioactivities on volatile organic compounds of endophytic fungi *Neurospora dictyophora* 3ZF-02 in different ages. J Basic Microbiol 63(1):104–115

Zou C, Li Z, Yu D (2010) Bacillus megaterium Strain XTBG34 promotes plant growth by producing 2-pentylfuran. J Microbiol 48(4):460–466

Endophytic Fungal Terpenoids: Biosynthesis and Biological Activity

10

Juan M. Galindo-Solís and Francisco J. Fernández

Abstract

Endophytic fungi, a diverse group with phylogenetic variations, intricately colonize plant tissues, establishing complex interactions with host plants and microbiomes. At the beginning of the chapter, the multifaceted chemical interkingdom dynamics (where endophytic fungi modulate host defense mechanisms, fungal virulence factors, and microbiome interactions) is explored. The delicate balance achieved contributes to plant health, even under stress conditions. Endophytic fungi produce a plethora of secondary metabolites, particularly terpenoids, which play pivotal roles in ecological functions and trophic interactions. Terpenoids, the largest and most diverse bioactive secondary metabolites, exhibit various functions, including defense, signaling, and growth regulation. Their biosynthesis involves intricate steps, resulting in structurally diverse compounds with potential therapeutic applications. This chapter delves into the role of terpenoids in mediating endophytic fungus-host plant and fungus-microbiome interactions. It also discusses the classification, biosynthesis, and biological activities of terpenoids from endophytic fungi and emphasizes their potential as therapeutic agents. Despite significant strides, challenges remain in understanding the molecular mechanisms and utilizing endophytic fungi for industrial production. Future research directions incorporate multidisciplinary approaches, such as genomics, metabolomics, and synthetic biology, to unlock the full potential of terpenoids from endophytic fungi.

J. M. Galindo-Solís · F. J. Fernández (✉)
Biotechnology Department, Universidad Autónoma Metropolitana (Unidad Iztapalapa), Avenida Ferrocarril San Rafael Atlixco, 186, Colonia Leyes de Reforma, 1A Sección, Alcaldía Iztapalapa, Mexico City, Mexico
e-mail: fjfp@xanum.uam.mx

K. A. Abd-Elsalam, S. F. AbuQamar (eds.), *Fungal Endophytes Volume I*,
https://doi.org/10.1007/978-981-97-7312-1_10

Keywords

Biotechnological applications · Endophytic fungi · Plant–microbe interactions · Secondary metabolites · Terpenoids

10.1 Introduction

Endophytic fungi are a phylogenetically diverse group of fungi that asymptomatically and intermittently colonize plant tissues and drive saprophytic, commensal, or mutualistic interactions with their host plant (Schouten 2019; Stone et al. 2004). The endophytic fungal association is a complex chemical interkingdom interaction that involves the modulation of host plant defense mechanisms and regulation of fungal virulence factors, as well as antagonistic and cooperative interactions with other members of the microbiome (Chagas et al. 2018). A delicate balance between host plants with endophytic fungi and the rest of the microbiome results in healthy plants (Schulz et al. 2015); moreover, endophytic fungi help to maintain plant health even under biotic and abiotic stress conditions (Eid et al. 2019; Terhonen et al. 2019). As a result of these interactions, endophytic fungi produce a variety of secondary metabolites that can be exploited biotechnologically (Eid et al. 2019).

Some secondary metabolites have known functions as competitive chemicals that act against other organisms, such as metal transporters, growth stimulators, hormones, or differentiation effectors (Demain and Fang 2000). Endophytic fungi are a reservoir of a variety of secondary metabolites with unknown natural functions but interesting bioactivities (Hacquard et al. 2016). In terms of trophic interactions, comparative genomics and transcriptomics studies support that secondary metabolites may play a role in establishing beneficial endophytic interactions with the host plant (Hacquard et al. 2016; Zhou et al. 2017, 2018).

Terpenoids are the largest and most diverse group of bioactive secondary metabolites. They perform important ecological functions as defense and signaling molecules (Dewick 2009). Endophytes biosynthesize a structurally and functionally diverse portfolio of terpenoids that act as long-distance communication between fungi and bacteria, insect attractors, antagonistic chemicals, plant growth promoters, or signaling molecules during symbiotic interactions (Kramer and Abraham 2011). The natural roles of terpenoids include antagonistic and beneficial interactions among organisms (Gershenzon and Dudareva 2007). In general, volatile terpenoids are remarkable molecules that interact among different organisms. In particular, fungal volatile terpenoids may play a key role in mediating endophytic fungus-host plant and endophytic fungus-microbiome interactions (Farh and Jeon 2020; Kramer and Abraham 2011).

Terpenoid biosynthesis occurs in four general steps. First, universal C5 precursors are formed; second, linear prenyl precursors are formed by a family of proteins commonly referred to as isoprenyl diphosphate synthases. The third step is a central part of terpenoid biosynthesis, which is carried out by terpenoid cyclases in a complex cyclization cascade. In the final stage, tailoring enzymes decorate the terpenoid scaffold skeleton, enhancing the chemical diversity and bioactivities of terpenoids

(Christianson 2017). Later in this chapter, terpenoid biosynthesis will be discussed in more detail.

Terpenoids are used as therapeutic drugs in human health. Remarkable drugs such as paclitaxel, camptothecin, vinca alkaloids, and fusidic acid are natural terpenoid products (Uzma et al. 2018). Hundreds of new bioactive terpenoids have been isolated and characterized from endophytic fungi from sources such as plants used in traditional medicine, crops, algae, and mangroves (De Souza et al. 2011). The bioactivities of terpenoids include anticancer, anti-inflammatory, antibacterial, antiviral, antimalarial, and hypoglycemic effects. Recent studies suggest their potential use as insect repellents, immune modulators, antioxidants, anti-aging agents and neuroprotectants (Yang et al. 2020a, b). Approximately one-third of all terpenoids listed in the Dictionary of Natural Products have antineoplastic-related activities (Guillaume et al. 2019). Even though many terpenoids exhibit a variety of important activities, the molecular mechanisms of most terpenoids remain unclear; therefore, further studies on the mechanisms of activity are essential for a better understanding and application of these structurally diverse molecules (Huang et al. 2012).

10.2 Terpenoids Modulate Endophytic Interactions

Endophytism is a highly complex process of interactions between host plants and endophytes driven by evolutionary selective pressures for mutual benefit (Mattoo and Nonzom 2021). The theory of "balanced antagonism," proposed decades ago, explains how the association of endophytic fungi involves a complex interaction involving averting/suppressing host plant defense mechanisms, regulating fungal virulence factors, and mediating coexistence with other microorganisms, leading to multiple balanced antagonisms in which secondary metabolites play an important role as defense and signaling molecules. In this scenario, environmental factors influence the interactions of all players in some way (Galindo-Solís and Fernández 2022; Schulz et al. 2015) (Fig. 10.1). The imbalance of these interactions could affect the health of the host plant, leading to disease symptoms and the elimination of the fungus by host defenses (Kogel et al. 2006). In support of this theory, some studies have provided evidence that microbial-plant interactions trigger the production of secondary metabolites (Marmann et al. 2014). For example, transcriptomic analysis during co-culture of *Bupleurum scorzonerifolium* suspension cells with an endophytic fungus showed how co-culture promoted the accumulation of the bioactive compound saikosaponin D by regulating the mevalonic acid (MVA) pathway, the calcium signaling pathway, and the AMPK signaling pathway (Cheng et al. 2024).

Terpenoids are widely distributed natural products produced by animals, fungi, bacteria, and archaea. Ecologically, they serve important functions as antagonistic and beneficial interactions between phylogenetically distant organisms (Kramer and Abraham 2011). Plants, animals, and microorganisms use terpenoids as weapons to defend themselves against competitors, pathogens, and predators, but also to transmit specific messages regarding symbiotic or pathogenic interactions, the presence of food and mates. However, the natural role of most terpenoid molecules is

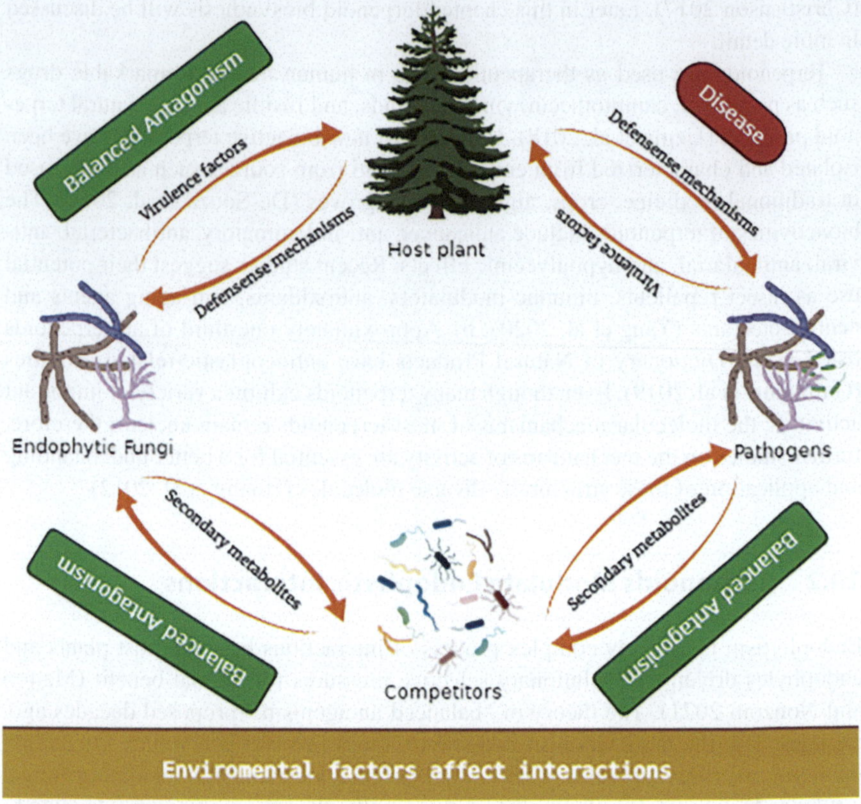

Fig. 10.1 Multiple balanced antagonisms in endophytic fungal interactions (Galindo-Solís and Fernández 2022)

unknown (Gershenzon and Dudareva 2007). Patterns of fungal volatile sesquiterpenoid emissions are useful for predicting fungal lifestyle in ecosystems (ectomycorrhizal, pathogenic and saprophytes) (Farh and Jeon 2020). Especially in endophytic fungi, some volatile sesquiterpenoids may play a key role in mediating endophytic fungus-host plant and endophytic fungus-microbiome interactions (Farh and Jeon 2020; Kramer and Abraham 2011). Terpenoids may play a key role in endophytism by contributing to signaling communication with the host plant. For example, transcriptome analysis and quantitative PCR during colonization of the root endophytic fungus *Phomopsis liquidambari* showed that amino acid metabolism, carbohydrate metabolism, fatty acid biosynthesis, secondary metabolism, and enzymes of sesquiterpenoid and triterpenoid biosynthesis are upregulated (Zhou et al. 2017, 2018). In addition, *Lacaria bicolor* produces the active sesquiterpenoid (−)-thujopsene while interacting with its host plant and stimulates lateral root production, an important step in this early mutualistic interaction. The authors suggest that this stimulation, facilitated by this sesquiterpenoid signal, creates a win-win situation by increasing root surface area for plant nutrient uptake and improving fungal access to plant-derived carbon via root exudates (Ditengou et al. 2015; Hung et al. 2013). An additional metatranscriptomic study revealed that terpenoid biosynthesis is highly expressed in symbiotic tissues, suggesting that it is directly related to the fungal response to its environment. Nevertheless, further studies are needed to determine the role of terpene-associated genes in this interaction (Liao et al. 2014).

Despite the above examples, most research correlating fungal lifestyles, effects, and classes of compounds produced has some limitations because it does not accurately represent compounds produced under natural conditions and may assume that antagonisms govern endophytic interactions. For example, most reported laboratory experiments are not close to natural conditions (biotic and abiotic factors), and we have limited chemical and biological data in this area (Farh and Jeon 2020). In addition, balanced antagonism (Fig. 10.1) works with other mechanisms, such as cooperation, to maintain endophytism, i.e., it may be one of the ways used by endophytes to survive within the host plant (Mattoo and Nonzom 2021). In summary, secondary metabolites play a central role in mediating biochemical communication between endophytic fungi and host plants, directing the complex interactions among endophytic fungi, host plants, and pathogens in their community, and thereby determining the host plant range of an endophytic fungal and endophytic microbiome.

10.3 Biosynthesis of Terpenoids

Terpenoid natural products are a large group of chemical compounds composed of activated isoprene (C5) units joined head-to-tail or tail-to-tail. According to TeroKit, a database-driven web server for terpenome research, more than 18,000 terpenoids are known, representing the most chemically and structurally diverse group of natural products (Zeng et al. 2020). The unique hydrocarbon backbone of terpenoids is modified by cyclization reactions and the addition of functional groups, increasing their chemical diversity and biological functions. Based on the number of C5 units, terpenoids are classified as monoterpenoids (C10), sesquiterpenoids (C15), diterpenoids (C20), sesterterpenoids (C25), triterpenoids (C30), and tetraterpenoids (C40) (Dewick 2009). Due to their hydrocarbon backbone and functional groups, their lipophilic nature makes them soluble in non-polar organic solvents, contributing to their volatility and characteristic aromas. The enormous structural diversity of terpenoids is due to the evolutionary malleability of the components of the biosynthetic pathways, including terpene scaffold-forming terpene synthases and scaffold-decorating enzymes monooxygenases, glycosyltransferases, and acyltransferases (Sandeep and Ghosh 2020) (Fig. 10.2).

It is commonly said that the isoprene moiety is the building block of terpenoids, but the diphosphate ester of isoprene, isopentenyl diphosphate (IPP), and its isomer, dimethylallyl pyrophosphate (DMAPP), are the chemically activated building blocks that can be readily incorporated into various terpenoid metabolites (Ro 2011). Therefore, the formation of the biologically relevant precursors DMAPP and IPP is the first step in terpenoid biosynthesis. In fungi, these key precursors are biosynthesized via the mevalonate (MVA) and methylerythritol phosphate (MEP) pathways (Figs. 10.3 and 10.4).

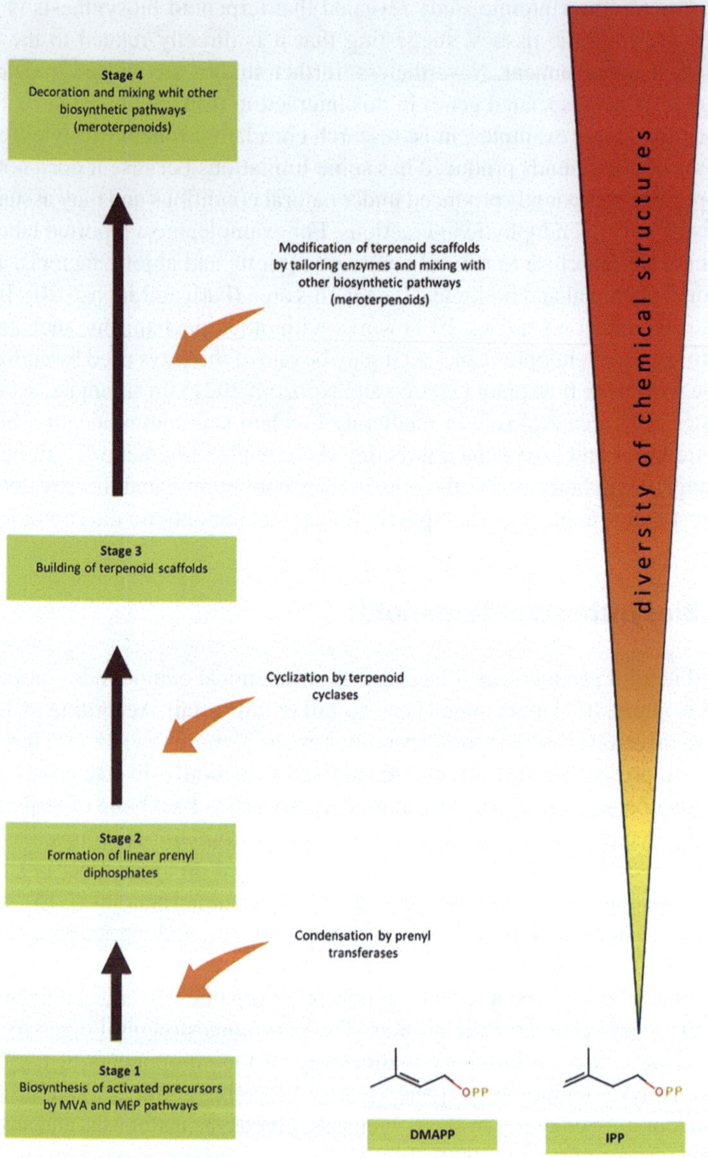

Fig. 10.2 General view of terpenoid biosynthesis in fungi. The biosynthesis of terpenoids involves four general stages that contribute to the vast structural diversity of terpenoids

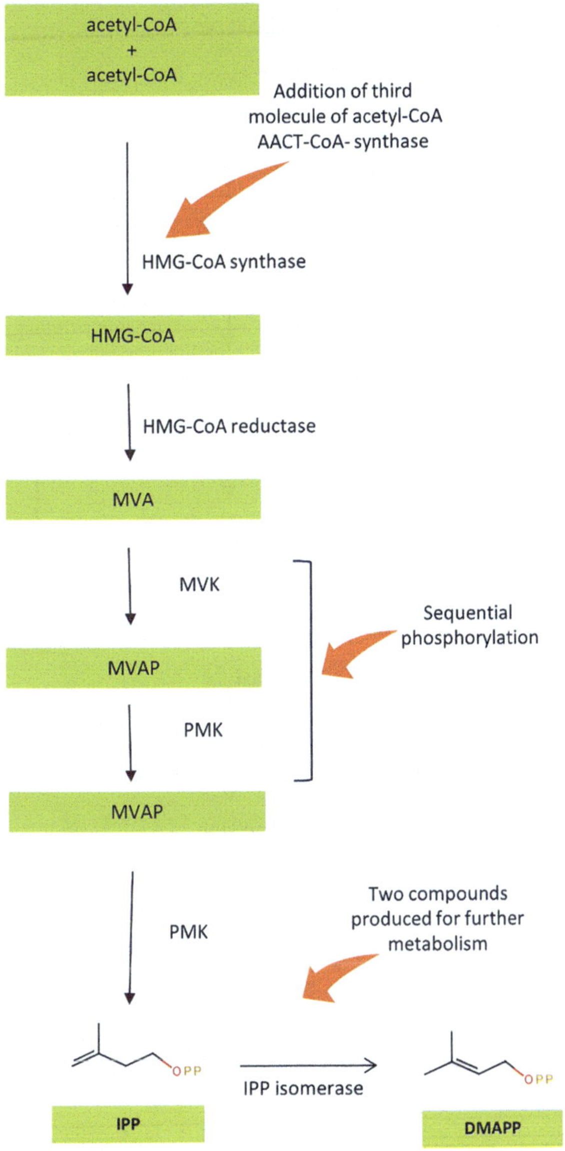

Fig. 10.3 MVA pathway. The MVA pathway produces the biologically active precursors IPP and DMAPP from acetyl-CoA

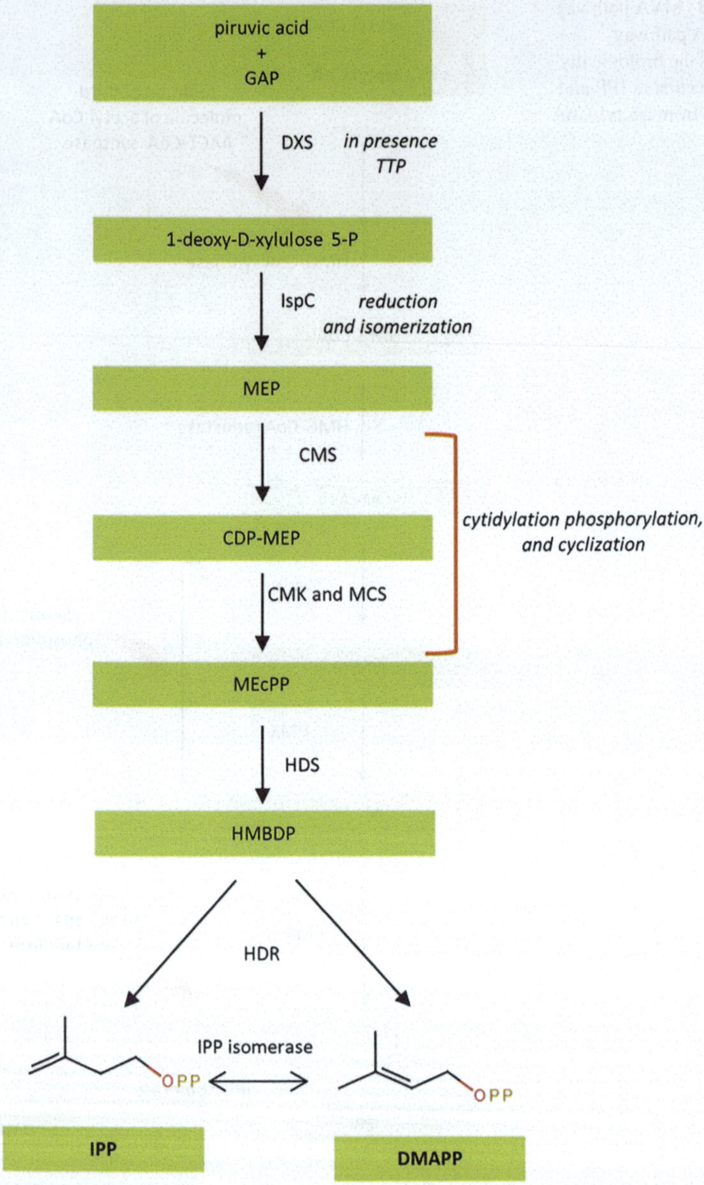

Fig. 10.4 MEP pathway, which generates the biologically active precursors IPP and DMAPP from pyruvic acid and GAP

10.3.1 Mevalonate Pathway

The mevalonate pathway starts with the activated form of acetic acid, acetyl-CoA, to form mevalonic acid and finally IPP and DMAPP. The pathway is found in most eukaryotes, archaea, and bacteria. Although some variations of MVA have been reported in archaea, the pathway is essentially the same in all eukaryotes (Hoshino and Villanueva 2023). Figure 10.3 provides a general overview of the MVA pathway. In the first step, two molecules of acetyl-CoA are condensed by Claisen reaction, then a third molecule of acetyl-CoA is added by stereospecific aldol reaction and hydrolysis to give 3-hydroxy-3-methylglutaril-CoA (HMG-CoA); these reactions are catalyzed by acetoacetyl-CoA synthase and 3-hydroxy-3-methylglutaril-CoA synthase (HMG-CoA synthase). In the second step, 3-hydroxy-3-methylglutaril-CoA reductase (HMG-CoA reductase) catalyzes the reduction of thioester to aldehyde via hemithioacetal and the reduction of aldehyde to alcohol to mevalonic acid (MVA); reactions of HMG-CoA reductase use two equivalents of NADPH. The third step involves sequential phosphorylation, resulting in the production of mevalonic acid phosphate (MVAP) and mevalonic acid diphosphate (MVAP). Subsequently, decarboxylation-elimination generates IPP. Mevalonate kinase (MVK) and phosphomevalonate kinase (PMK) catalyze phosphorylation in the presence of two molecules of ATP, while mevalonate 5-diphosphate decarboxylase (MVD) catalyzes the decarboxylation-elimination reaction. Finally, IPP undergoes stereospecific allylic isomerization to DMAPP by IPP isomerase (Dewick 2009; Miziorko 2011; Brock and Dickschat 2013).

10.3.2 MEP Pathway

In 1988, Gérard Flesch and Michel Rohmer proposed that terpenoids could be synthesized by a no-mevalonate pathway while studying hopanoid biogenesis in the purple non-sulfur and bacteria *Rhodopseudomonas palustris* and *Rhodopseudomonas acidophila* and the facultative methylotroph *Methylobacterium organophilum* (Flesch and Rohmer 1988). Later, based on experiments with ^{13}C and/or ^{2}H-labeled precursors, Rohmer formally proposed the existence of a completely new biosynthetic pathway for the biosynthesis of triterpenoids in bacteria (Rohmer 1993; Rohmer et al. 1993). However, detailed studies in bacteria, parasites and plants were necessary to elucidate the MEP pathway (Eisenreich et al. 2004).

The MEP pathway (Fig. 10.4) converts pyruvic acid and glyceraldehyde 3-phosphate (GAP) into methylerythritol phosphate (MEP), which is then transformed into IPP and DMAPP. The first step involves the enzyme deoxyxylulose phosphate synthase (DXS), which, in the presence of thiamine diphosphate (TPP), converts pyruvic acid and GAP into 1-deoxyxylulose 5-phosphate. This compound is then transformed into MEP by 1-deoxy-D-xylulose-5-phosphate reductoisomerase (DXR). MEP is transformed into the cyclic intermediate methylerythritol 2,4-cyclodiphosphate (MEcDP) through three subsequent reactions of cytidylation phosphorylation and cyclization. These reactions are catalyzed by CDP-ME synthase,

4-(cytidine-5′-diphospho)-2-C-methyl-D-erythritol kinase, and 2-C-methyl-D-erythritol-2,4-cyclodiphosphate synthase, respectively. MEcDP is then converted into hydroxymethylbutenyl diphosphate (Dewick 2009; Frank and Groll 2017).

Contrary to the conventional understanding, there is increasing evidence that the MEP pathway is active in fungi, despite the initial belief that they lack this metabolic pathway (Frank and Groll 2017). Studies of taxol production by endophytic fungi have revealed the presence of a putative ortholog of 2-C-methyl-D-erythritol 2,4-cyclodiphosphate synthase (DXS), the enzyme immediately preceding DXR in the plant MEP pathway (Soliman et al. 2011). In addition, the enzymatic activity of 1-deoxy-D-xylulose 5-phosphate reductoisomerase (DXR) was detected in the taxan-producing endophytic fungus *Cryptosporiopsis tarraconensis* (Mohammadi Ballakuti and Ghanati 2023). In the taxol-producing endophytic fungus Aspergillus aculeatinus, both MVA and MEP pathways were found through transcriptome analysis (Qiao et al. 2020). A recent study using stable isotope and chemical inhibition analysis in *Yarrowia lipolytica* suggests that the MEP pathway may also function in yeast (Dissook et al. 2021).

10.3.3 Biosynthesis of Terpenoid Backbones

After the formation of the key intermediates IPP and DMAPP via the MVA and MEP pathways (Figs. 10.3 and 10.4), the second step is the synthesis of linear isoprenoid diphosphate via a condensation reaction catalyzed by isoprenoid diphosphate synthases (Fig. 10.2). The reaction starts with the formation of a reactive DMAPP carbocation. Then, IPP condensates with the DMAPP carbocation by nucleophilic reaction to form a C10 carbocation. Finally, stereospecific deprotonation yields GPP. Similarly, condensation of two or three IPP units on a DMAPP yields C15 FPP and C20 GGPP (Ro 2011).

In the third step of terpenoid biosynthesis, terpenoid cyclases are responsible for the complex cyclization reaction of linear isoprenoid diphosphates (GPP, FPP, GGPP, and GFPP), which form the vast variety of terpenoid hydrocarbon skeletons (Fig. 10.2). During the cyclization cascade, more than half of the carbons in linear isoprenoids undergo bonding, hybridization, and stereochemical changes in a single enzymatic multistep cyclization reaction (Brock and Dickschat 2013; Christianson 2017). Terpenoid cyclases are classified into two classes based on the chemical strategy used to generate the initial carbocation that initiates the cyclization cascade. Class I terpenoid cyclases (TCIs) are metal-dependent enzymes with a trinuclear cluster that initiates an ionization-dependent cyclization of linear prenyl diphosphates to generate an allylic cation. In contrast, class II terpenoid cyclases (TCII) use a general acid to generate a protonation-dependent cascade and yield a tertiary carbocation. UbiA-type terpenoid cyclases are a third class of terpenoid cyclases found in fungi and bacteria. To date, 22 UbiA-type terpenoid cyclases have been described in fungi, producing 10 different carbon skeletons, mainly sesquiterpenes and diterpenes (Luo et al. 2022, 2024).

In the final stage of terpenoid biosynthesis, several enzymes decorate terpenoid scaffolds, such as cytochrome P450 enzymes (P450s), flavin-dependent monooxygenases (FMOs), acyltransferases, short-chain dehydrogenase/reductases (SDRs), and glycosyltransferases (Luo et al. 2024; Quin et al. 2014) (Fig. 10.2). Currently, research on terpenoid tailoring enzymes in fungi is focused on those clustered with terpenoid cyclases, including P450 enzymes, oxidoreductase, FMO, acyltransferase, glycosyltransferase, and SDR (Luo et al. 2024).

10.3.4 Classification of Fungal Terpenoids

A classification based on related terpenoid cyclases is shown in Table 10.1 (Luo et al. 2024). Because monoterpenoids are poorly described in fungi, we focused only on sesquiterpenoids, diterpenoids, triterpenoids, and sesterterpenoids. Meroterpenoids were not included in this table due to their mixed biosynthetic origin. Only one monoterpenoid cyclase has been described in *Hypoxylon* sp. Hyp3 is a monoterpenoid cyclase that catalyzes the production of 1,8-cineole, a component of eucalyptus oil, by the same mechanism used by an unrelated plant homologue (Shaw et al. 2015).

10.3.4.1 Sesquiterpenoids

Sesquiterpenoids are a diverse group of terpenoids with a cyclic C-15 skeleton derived from the linear precursor FPP (Table 10.1). The cyclization of FPP is mainly catalyzed by TCI, followed by UbiA-type terpenoid cyclases and less frequently by TCII (Luo et al. 2024). Cyclization by fungal class I sesquiterpenoid cyclases (TCI)

Table 10.1 General classification of terpenoids and associated terpenoid cyclases

Linear prenyl precursor	Common name	Terpenoid cyclases involved	The most common cyclization mechanisms known in fungi
GPP (C10)	Monoterpenoid	Hyp3	Via formation of α-terpinyl cation
FPP (C15)	Sesquiterpenoid	Class I, Class II and UbiA type	C1–C10 via formation of *E,E*-germacradienyl cation C1–C11 via formation of *trans*-humulyl cation C1–C6 via formation of bisabolyl cation C1–C7 via formation of cycloheptenyl cation C1–C10 via formation of *Z,E*-germacradienyl cation C1–C11 via formation of *Z,E*-humulyl cation
GGPP (C20)	Diterpenoid	Class I, Class II and UbiA type	Type-a, Type-b, Type-c and Type-d
GFPP (C25)	Sesterterpenoid	Class I	Type-a and Type-b
FPP + FPP	Triterpenoid	Class I and Class II	Vía formation of squalene or 2,3(*S*)-oxidosqualene

can proceed via C1–C6, C1–C7, C1–C10, and C1–C11 cyclization reactions. The cyclization mechanism can proceed via two pathways, depending on the allylic cation formed: (1) formation of a trans-allylic cation intermediate, yielding *E,E*-germacradienyl (C1–C10 bond formation) and trans-humulyl (C1–C11 bond formation) carbocation intermediates; (2) formation of a cis-allylic cation intermediate by nerolidyl diphosphate, yielding the bisabolyl (C1–C6 bond formation), Z,E-germacradienyl (C1–C10 bond formation), and Z,E-humulyl (C1–C11 bond formation) carbocation intermediates, respectively (Quin et al. 2014). The generated carbocation intermediates can undergo further cyclization reactions, such as ring closure, methyl migration, hydride transfer, etc., before the completion of the cyclization cascade to ultimately generate a variety of sesquiterpenoid carbon skeletons (Christianson 2017).

Currently, only two TCIIs have been described in fungi. The first one, DrtB, is involved in the biosynthesis of calidoustene A and calidoustene B, drimane-type sesquiterpene esters isolated from *Aspergillus calidoustus* (Huang et al. 2021a, b). The second, AstC, is involved in the biosynthesis of astelloides, a group of esterified drimane-type sesquiterpene lactones produced by some *Aspergillus* species (Dai et al. 2023; Shinohara et al. 2016).

10.3.4.2 Diterpenoids

The first committed step in diterpenoid biosynthesis is the cyclization reaction of the linear prenyl diphosphate, GGPP (Table 10.1). The reaction is catalyzed by diterpenoid cyclase to yield the C20 diterpene scaffold (Quin et al. 2014). In fungi, cyclization is performed by class I terpenoid cyclases, although some class II and UbiA-type terpenoid cyclases have been reported. Depending on the differences in the cyclization cascade, class I diterpenoid cyclases can be divided into four types. Type a and type b cyclization mechanisms imply deprotonation-protonation of carbocation, while type c involves ring expansion and type d a ring expansion followed by ring contraction. In addition, the most common class I fungal diterpene-cyclases are chimeric enzymes with C-terminal prenyltransferase (PT) and N-terminal terpenoid cyclase (TC) domains and relatively conserved cyclization patterns (Luo et al. 2024). Fusicoccadiene synthase (PaFS) was the first class I chimeric terpenoid cyclase described in the plant-pathogenic fungus *Phomopsis amygdali*. PaFS catalyzes the formation of fusicocca-2,10 (14)-diene directly from isoprene units (Toyomasu et al. 2007). It is interesting to note that the chimeric strategy provides some advantages, such as increasing the yield of products and reducing the natural promiscuity of the PT domain, which ultimately optimizes the use of resources (Mitsuhashi and Abe 2018).

Class II diterpene cyclases from fungi are bifunctional, with class I and class II domains yielding labdane-related terpenoids that are rarely found in fungi. Nevertheless, genome mining of some plant-associated fungi predicts the production of labdane-related terpenoids, which may play a role in plant-fungus interactions (Xu et al. 2017). In the case of UbiA-type terpenoid cyclases, only three diterpene cyclases have been described in medicinal basidiomycetes through genomic, biochemical, and structural analyses (Li et al. 2019a, b, c; Yang et al. 2017).

10.3.4.3 Sesterterpenoids

Sesterterpenoids are a small group of terpenoids widely distributed in fungi, bacteria, lichens, higher plants, insects, and marine sponges. The linear isoprenoid diphosphate precursor of sesterterpenoids is GFPP (Table 10.1) (Li and Gustafson 2021). All known sesterterpenoid cyclases from fungi are chimeric class I terpenoid cyclases with PT and TC domains, like the diterpenoid synthases described above. Ophiobolin F synthase (AcOs) was the first fungal sesterterpenoid cyclase from *Aspergillus clavatus* (Chiba et al. 2013). Later, various sesterterpenoid cyclases were discovered in fungi through genome mining and heterologous expression approaches, such as EvQs from *Emericella variecolor*, which yields (+)-quiannulatene (Okada et al. 2016). Further investigations in *Aspergillus ustus* 094102 revealed that ophiobolin biosynthesis involves five gene clusters, including the sesquiterpenoids drimane, isochromane, sterols, and dipeptides (Chai et al. 2016). Systematic mining of 477 fungal genomes looking for chimeric terpene synthases (di and sesterterpenoid cyclases) led to the identification of 34 new active enzymes and two new sesterterpenoids (sesterevisene and sesterorbiculene). Most of the new enzymes yielded sesterterpenoid products. It is important to note that prior to this study, only 20 chimeric di terpenoid and sesterterpenoid synthases had been functionally characterized, suggesting the diversity of sesterterpenoids and sesterterpenoid synthases waiting to be discovered in fungi (Chen et al. 2021).

10.3.4.4 Triterpenoids

Triterpenoids are C30 scaffolds ranging from acyclic to hexacyclic with thousands of structural variations (Table 10.1). The classical biosynthetic pathway of triterpenoids in fungi starts with the dimerization of FPP by squalene synthase to squalene, then monofunctional class II terpene synthases utilize the C30 prenyl chain to yield (S)-2,3-oxidosqualene and subsequently lanosterol, the precursor of triterpenoids. Examples of triterpenoids include clavaric acid from *Hypheloma sublaterium*, ganoderic acid from *Ganoderma lucidum*, or helvolic acid from *Aspergillus fumigatus* (Quin et al. 2014). A recent report by Tao and coworkers revealed a novel pathway for terpenoid biosynthesis in fungi that does not involve the formation of squalene. Talaropentaene synthase (TvTS) from *Talaromyces verruculosus* and macrophomene synthase (MpMS) from *Macrophomina phaseolina* are chimeric class I terpene synthases that catalyze both the synthesis of hexaprenyl diphosphate (PT domain) and its cyclization (cyclase domain) to yield the triterpenoids talaropentaene, macrophomene, and colleterpenol (Tao et al. 2022). This finding demonstrates the fundamental biosynthetic versatility of triterpenoids waiting to be discovered.

10.3.4.5 Meroterpenoids

Meroterpenoids are widely distributed natural products derived in part from terpenoid pathways. They represent a great structural and biosynthetic diversity resulting from mixed biosynthetic origin and successive chemical modification. Therefore, some important secondary metabolites can be considered as meroterpenoids, such as vitamin E, ubiquinone-10, vinblastine, and tetrahydrocannabinol (Geris and

Table 10.2 General classification of fungal meroterpenoids

Ketides and polyketides	Indoles	Shikimates	Others
• Triketides (acyl chain type, pyrone type and pyrone derivatives type) • Tetraketides (orsellinic acid/ orcinaldehyde type, orcinaldehyde derivative type and methyl orsellinic acid type) • Pentaketide • Hexaketide • Other polyketides	• Indole diterpenoids (alkaloids) • Indole sesquiterpenoids	• Tricycloalternarene (TCA)-type • Diisoprenyl-cyclohexene/ane-type • Vibralactone/ fomannoxin type • *Ganoderma* meroterpenoids • Others	• Derived from the putative sugar-terpenoid pathway • Derived from amino acid-shikimic acid-mevalonic acid-pathway • Derived from shikimic acid-polyketide–terpenoid pathway • Derived from other mixed complex terpenoid pathways

Simpson 2009; Matsuda and Abe 2015). Despite several classifications of meroterpenoids reported in the past decades, analysis of available chemical and biosynthetic information suggests that the non-terpenoid moiety of meroterpenoids determines their structural diversity and complexity (Zhao et al. 2021). Table 10.2 summarizes the general classification of fungal meroterpenoids according to their non-terpenoid starting scaffold (Jiang et al. 2021). The most studied meroterpenoids are ketide and polyketide terpenoids, in which the polyketide moiety is predominantly assembled by the multidomain enzymes polyketide synthase (PKS) (Han et al. 2021).

Meroterpenoid cyclases are fundamentally different from all the terpenoid cyclases mentioned above. The first meroterpenoid cyclase, named Pyr4, was discovered while studying the biosynthesis of the meroterpenoid pyripyropene in the fungus *Aspergillus fumigatus*. Since then, a family of membrane-bound meroterpenoid cyclases catalyzing polyketide terpenoids and indole terpenoids has been referred to as Pyr4-like cyclases (Itoh et al. 2010). To illustrate, the biosynthesis of the polyketide meroterpenoid pyripyropene proceeds in a modular fashion with four general steps as follows: the first and second steps consist of the assembly of the terpenoid and non-terpenoid building blocks by a tandem configuration of type I polyketide synthase, CoA ligase, malonyl-CoA and UbiA-type prenyltransferase; in the third step, the cyclization cascade is initiated by Pyr4 working in tandem with a flavin-dependent monooxygenase; the fourth step involves tailoring of the generated core scaffold by two P450 monooxygenases and two acetyltransferases (Barra and Abe 2021).

10.4 Biological Activities of Terpenoids from Endophytic Fungi

The biological activities of terpenoids from endophytic fungi include anticancer, anti-inflammatory, antibacterial, antiviral, antimalarial and hypoglycemic. Meanwhile, growing evidence suggests their application as insect resistance, immunoregulation, antioxidant, antiaging, and neuroprotective agents (Yang et al. 2020a, b). Although many terpenoids show several important activities, the molecular mechanisms of action of most terpenoids remain unclear; therefore, further studies on the mechanisms of action are essential for a better understanding and application of these molecules (Huang et al. 2012).

According to Amirzakariya and Shakeri's update of the 516 terpenoids isolated from endophytic fungi from 2011 to 2020, the most reported terpenoids were sesquiterpenoids, followed by meroterpenoids, diterpenoids, triterpenoids, sesterterpenoids, and monoterpenoids (Amirzakariya and Shakeri 2022). Similarly, Zhao and colleagues summarized bioactive terpenoids from endophytic fungi isolated by Chinese researchers between 2017 and 2019 (Zhao et al. 2020). However, they report a higher number of sesterterpenoids, all of which were produced by endophytic fungi isolated from plants in China. The most chemically diverse sesquiterpenoids were bisabolane, cadinane, cyclonerane, guaiane, hirsutane, tremulane, protoilludane, eudesmane, illudalane, eremophilane, drimane, sativene, trichothecene, and other different types (Zhao et al. 2020).

Below, we present examples of endophytic fungal terpenoids that exhibited strong or moderate activities related to human health, emphasizing test methods and those terpenoids with mechanism of action studies.

10.4.1 Anticancer

The discovery of novel natural products from any source is a multidisciplinary and complex work. The extensive process requires many investigations into the biological activity and mechanisms of action of the new terpenoid. Activity-guided screening tools to discover terpenoids for cancer therapy include organoids, animal models, cell cultures, and biochemical assays. Subsequently, the mechanism of action is investigated using chemical and structural biology, proteomics, and bioinformatics methods (Huang et al. 2021a, b). The screening tools predict the antitumor activity of molecules. The key is to select a simple and effective method to assess anticancer activity using in vitro assays. Cytotoxic assays are the most popular in the field to quickly assess antitumor activity in the early stage of investigation (Setiawati et al. 2022), since cell viability and proliferation are indicators for understanding how exposure to toxic agents affect genes, proteins, and pathways on cell survival or death. The selection of the optimal assay depends on the specific cell type, culture conditions, and research questions (Adan et al. 2016).

The 3-(4,5-dimethylthiazol-2-yl-)2,5-diphenyltetrazolium bromide (MTT) assay is a commonly used colorimetric assay in the early stages of drug discovery

(Setiawati et al. 2022). Table 10.3 provides examples of cytotoxic terpenoids that were initially screened using the MTT assay. For example, *Paraconiothyrium brasiliense* produces a tricyclic sesquiterpenoid called brasilamide E, which exhibits selective cytotoxicity against breast (MCF-7) and gastric (MGC) cancer cell lines. Further experiments showed that brasilamide E at 50 μM had no detectable cytotoxicity against the non-cancerous HaCaT (human keratinocyte) and NIH-3 T3 (mouse embryo fibroblast) cells. Mechanistic studies showed that brasilamide E inhibited the expression of hexokinase II (HK2) in MCF-7 cells, resulting in glucose metabolism dysfunction, ATP depletion, and ultimately inhibition of breast cancer cell proliferation (Liu et al. 2015). The meroterpenoid isopenicin A is an antimitotic terpenoid that was initially identified as a potent inhibitor of the Wnt/β--catenin signaling pathway (Tang et al. 2019), which is highly activated in human cancers and is normally involved in many physiological processes, including development, tissue homeostasis, and tissue regeneration (Jung and Park 2020). Studies on the mechanism of action of this meroterpenoid using cytotoxicity assay, tubulin polymerization assay, cell cycle analysis, and apoptosis by flow cytometry showed that it is an inhibitor of tubulin polymerization in vitro, which induces G2/M cycle arrest and subsequent apoptosis (Chen et al. 2020a, b).

Apoptosis evasion is a unique characteristic of cancer that is not specific to the type or cause of cancer. For this reason, apoptosis is targeted in anticancer therapy. Therefore, the study of apoptosis in cancer cells is crucial for the evaluation of cytotoxic activity (Pfeffer and Singh 2018). A group of researchers investigated the induction of apoptosis using a novel sesquiterpenoid isolated from a liquid culture of the endophytic fungus *Penicillium* sp. The terpenoid was found to be cytotoxic against the human lingual cancer cell line (Tca8113). Oral administration of the terpenoid to nude mice resulted in potent antitumor activity against human osteosarcoma. Remarkably, administration of 30 mg/kg of the compound produced results like those of the chemotherapeutic agent cisplatin at 5 mg/kg. The DNA fragmentation assay showed the induction of apoptosis in MG-63 cells (Zheng et al. 2014).

10.4.2 Anti-inflammatory

Inflammation is a complex physiological process that can lead to various diseases. The production of inflammatory mediators via the arachidonic acid pathway (AAP) contributes to chronic inflammatory diseases such as osteoarthritis, rheumatoid arthritis, asthma, cancer, and neurological disorders. AAP is mediated by cyclooxygenases (COX1, COX2), which produce prostaglandins. These prostaglandins can either inhibit or promote platelet aggregation and cause vasoconstriction or vasodilation. Inflammatory mediators are also produced via 5-lipoxygenase, which leads to the production of leukotrienes. Leukotrienes can cause bronchospasm and increased vascular permeability. The enzymes COX1, COX2 and 5-lipoxygenase are key in the first phase of the inflammatory response (Rubbo and Wood 2020).

Nonsteroidal anti-inflammatory drugs (NSAIDs) are commonly used to treat inflammation by inhibiting the activity of oxygenases. However, NSAIDs have

Table 10.3 Anticancer terpenoids isolated from endophytic fungi

Class of terpenoid	Name	Endophytic fungi/host plant	Observations	Reference
Diterpenoid	3,4-seco-sonderianol	Endophytic fungus J37/*Ceriops tagal*	Cytotoxicity against K562, SGC-7901, and BEL-7402 cell lines with IC50 values of 9.2 ± 0.1, 15.7 ± 0.1, and 25.4 ± 0.3 μg/mL, respectively	Zeng et al. (2015)
Sesquiterpenoid	Zopfellin C	*Zopfella* sp./*Actinidia chinensis*	Cytotoxicity against MCF-7 and A549 cell lines with IC50 values of 18.3 and 21.7 μM, respectively	Sun et al. (2020)
Sesquiterpenoid	Proversilins C and E	*Aspergillus versicolor* F210/*Lycoris radiate*	Cytotoxicity against HL-60 cells with IC50 values of 7.3 and 9.9 μM, respectively	Li et al. (2020a, b, c, d)
Sesquiterpenoid	Aspergillusone D	*Aspergillus clavatus*/*Tripterygium hypoglaucum* (Levl.) Hutch	Cytotoxicity against A549 and MCF-7 cell lines with IC50 values of 0.2 and 5.9 μM, respectively	Wang et al. (2015)
Sesquiterpenoid	3-epi-Steperoxide A, steperoxide A, and merulin C	*Pseudolagarobasidium acaciicola*/*Bruguiera gymnorrhiza*	Cytotoxicity against MOLT-3, HuCCA-1, A549, HepG2, HL-60, MDA-MB-231, T47D, and HeLa cell lines with IC50 ranging from 0.19 to 5.25 μM	Wibowo et al. (2014)

(continued)

Table 10.3 (continued)

Class of terpenoid	Name	Endophytic fungi/host plant	Observations	Reference
Sesquiterpenoid	Brasilamide E	Paraconiothyrium brasiliense/Acer truncatum	Cytotoxicity against MCF-7and MGC cell lines with IC50 values of 8.4 Bo et al. (2024)14.7 μM, respectively	Liu et al. (2015)
Sesquiterpenoid	Nectrianolins A–C	*Nectria pseudotrichia*/*Gliricidia sepium*	Cytotoxicity against HL60 and HeLa cell lines with IC50 values between 1.5 to 52.1 μM	Ariefta et al. (2017)
Triterpenoids	Kadanguslactone A	*Penicillium ochrochloron*/*Kadsura angustifolia*	Cytotoxicity against HepG2 cell line with IC50 values of 10.2 and 17.3 μM, respectively	Qin et al. (2020)
Meroterpenoid	Emeridones B, D and F	*Emericella* sp. TJ29/*Hypericum perforatum*	Selective cytotoxicity against A-549, SMMC-7721 and SW-480 in the range of 8.19–18.8 μM	Li et al. (2019a, b, c)
Meroterpenoid	biscognienyne M	*Aspergillus* QG1a/*Kandelia candel*	Cytotoxicity against the A2780 cell line with IC50 value of 6.8 μM	Bo et al. (2024)
Sesterterpenoid	Griffithiiene and scalaradial	*Hypomontagnella monticulosa* /*Zingiber griffithii* Baker	Cytotoxicity against PANC-1, NBT-T2, and HCT116 cell lines with IC50 values ranging from 0.05 and 0.75 ppm	Lutfia et al. (2021)

Sesquiterpenoid	Perenniporin A, decahydro-2,2,5,8-tetramethyl-2H-Naphtho[1,8-bc]genfuran-5-ol and albicanol (4)	*Perenniporia tephropora* Z41/*Taxus chinensis* var. Mairei	Cytotoxicity against HeLa, SMMC-7721, and PANC-1 cell lines with IC50 values ranging from 6 to 58 μg/mL	Wu et al. (2013)
Sesquiterpenoid	Rhinomilisins A and G and gliocladic acid (monoterpenoid)	*Rhinocladiella similis*/*Acrostichum aureum*	Cytotoxicity against the L5178Y cell line with IC50 values of 5.0, 8.7, and 24.4 μM, respectively	Liu et al. (2019a, b, c)
Sesquiterpenoid	Fumagillenes A and B	Aspergillus fumigatus/Ligusticum wallichii	Cytotoxicity against MDA-ME-231 and MV4–11 cell lines with IC50 values ranging from 8.4 to 17.3 μg/mL	Li et al. (2020a, b, c, d)
Meroterpenoid	Isopenicin A	*Penicillium* sp. sh18/*Isodon eriocalyx*	G2/M cell cycle arrest followed by cell apoptosis	Chen et al. (2020a, b)
Triterpenoid	Integracides F, G, H and J	*Fusarium sp.*/*Mentha longifolia L*	Cytotoxicity against BT-549, SKOV-3, and KB cell lines with IC50 values ranging from 0.8 to 3 μM	Ibrahim et al. (2016a, b)
Triterpenoid	3β,4α-dihydroxy-26-methoxyergosta-7,24(28)-dien-6-one	*Aspergillus flavus*/*Tylophora ovata*	Selective cytotoxicity against MCF-7 cell line with IC50 value of 2.6 μM	Liu et al. (2019a, b, c)
Meroterpenoid	Periconone E	*Periconia sp. F-31*/*Annona muricata*	Cytotoxicity against MCF-7 cell line with IC50 value of 4.2 μM	Liu et al. (2017a, b)

(continued)

Table 10.3 (continued)

Class of terpenoid	Name	Endophytic fungi/host plant	Observations	Reference
Sesquiterpenoid	Diaporol R	*Diaporthe* sp./*Rhizophora stylosa*	Cytotoxicity against SW480 cell line with IC50 value of 8.72 μM	Chen et al. (2015)
Sesquiterpenoid	15-hydroxy-6α,12-epoxy-7β,10αH,11βH-spiroax-4-ene-12-one	*Penicillium* sp. FJ-1/*Avicennia marina*	Cytotoxicity against the Tca8113 cell line and antitumor activity against human osteosarcoma	Zheng et al. (2014)
Sesquiterpenoid	Penisarin B	*Penicillium* sp. *KMU18029* /*Aconitum brachypodum* Diels	Cytotoxicity against HL-60 and SMMC-7721cell lines with IC50 of 3.6 and 3.7 μM, respectively	Li et al. (2020a, b, c, d)
Sesquiterpenoid	Talaperoxides A and D	*Talaromyces flavus*/*Sonneratia apetala*	Cytotoxicity against MCF- 7, MDA-MB-435, HepG2, HeLa, and PC-3 with IC50 values between 0.70 and 2.78 μg/mL	Li et al. (2011)

adverse effects, including gastrointestinal disorders and renal failure, which limit their use in some cases. Therefore, there is a need to develop new anti-inflammatory agents that can target both the COX and 5-lipoxygenase pathways to provide both efficacy and safety (Mukhopadhyay et al. 2023).

During the inflammatory response to infection, the bacterial endotoxin lipopolysaccharide (LPS) induces proinflammatory cytokines and subsequently induces the expression of inducible nitric oxide synthase (iNOS) in macrophages, neutrophils, and many other cells (Sharma et al. 2007). For this reason, bacterial-derived LPS is the stimulatory model for triggering an inflammatory cascade in macrophages through Toll-like receptor 4 (TLR4). This response is characterized by the production of TNF-α (tumor necrosis factor-alpha), nitric oxide (NO), and other inflammatory mediators (Reis et al. 2011). NO is a paracrine mediator released by endothelial cells and certain neurons that only affects cells close to the cell that produces it. NO has several physiological roles in the cardiovascular system and inflammation. Under normal physiological conditions, NO has an anti-inflammatory effect. However, under certain circumstances, NO is a pro-inflammatory mediator that causes inflammation due to overproduction (Sharma et al. 2007).

Table 10.4 summarizes some examples of terpenoids with important anti-inflammatory activity as measured by anti-NO production in LPS-activated macrophage cells. Some noteworthy examples are then discussed. Amestolkolides A-D are 3,5-dimethylorsellinic acid (DMOA)-derived meroterpenoids isolated from the endophytic fungus *Talaromyces amestolkiae* YX1 obtained from the Asian mangrove *Kandelia obovate*. Only amestolkolides A and B showed anti-inflammatory activities. Amestolkolide B showed strong inhibitory effects on NO production with IC50 values of 1.6 μM, significantly lower than the positive control indomethacin (IC50 = 26.3 μM). Evaluation of cytotoxicity (up to 100 μM) showed that the anti-inflammatory activity was not related to the cytotoxic effects of amestolkolides (Chen et al. 2018a, b). Studies by Liu and coworkers led to the identification of 12 ophiobolin-type tetracyclic sesterterpenoids (bipolaricins and ophiobolins) from the endophytic fungus *Bipolaris* sp. TJ403-B1 isolated from the leaves of wheat (Liu et al. 2019a, b, c). Bipolaricins B, C, 3-anhydro-6-epi-ophiobolin A (6-epi-OP-A), 8-epi-ophiobolin (8-epi-OP-J) J and ophiobolin J (OP-J) showed significant anti-NO production. Cytotoxicity assay showed that the anti-inflammatory potency against NO production was not associated with general cytotoxic activity. Further experiments suggested that 6-epi-OP A suppressed LPS-induced NO production and pro-inflammatory cytokine release (IL-1β, RANTES, MIP-1β, and TNF-α) in macrophages through inhibition of heme oxygenase-1 (oxidative injury protection) induction as well as the NF-κB pathway (Liu et al. 2019a, b, c). Another prospective study led to the identification of two steroids with a tetracyclic triterpene skeleton, phomosterols A and B, with anti-NO production activity. Additional molecular docking studies revealed an interesting binding energy of phomosterols with iNOS, which may involve a mechanism of NO inhibition (Chen et al. 2020a, b). Han and colleagues used the TNF-α regulation assay in dendritic cells through TLR 3 regulatory effect to screen anti-inflammatory activities. They isolated nine meroterpenoids, guignardones, from *Guignardia mangiferae* found in the leaves of *Ilex*

Table 10.4 Anti-inflammatory terpenoids isolated from endophytic fungi

Class of terpenoid	Name	Endophytic fungi/host plant	Observations	Reference
Sesquiterpenoid	Gxsespene A	*Aspergillus* sp. GXNU-MA1/*Acanthus ilicifolius* L.	Anti-NO production: IC50 = 16.15 μM	Zhou et al. (2022)
Sesquiterpenoid	Trichoderiols A-B	*Trichoderma atroviride* S361/*Cephalotaxus fortune*	Anti-NO production: IC50 = 15.3 and 9.1 μM, respectively	Zheng et al. (2011)
Sesquiterpenoid	Pestaloporinate B	*Pestalotiopsis* sp./*Melia azedarach* Linn.	Anti-NO production: IC50 = 19.0 μM	Liu et al. (2016)
Diterpenoid	Botrysphins G and H	*Botrysphaeria laricina*/*Rhodobryum umgiganteum*	Anti-NO production: IC50 = 13.9 μM and 41.9 μM, respectively	Yang et al. (2020a, b)
Meroterpenoid	Amestolkolides A and B	*Talaromyces amestolkiae* YX1/*Kandelia obovata*	Anti-NO production: IC50 = 30 and 1.6 respectively	Chen et al. (2018a, b)
Sesterterpenoid	Bipolaricins B, C, 3-anhydro-6-*epi*-ophiobolin A, 8-*epi*-ophiobolin J and ophiobolin J	*Bipolaris* sp. TJ403-B1/*Triticum*	Anti-NO production: IC50 = 19, 19, 5.1, 20 and 6.7 μM, respectively	Liu et al. (2019a, b, c)
Triterpenoids	Chlamydosterols A and B	*Fusarium chlamydosporum*/*Anvillea garcinii*	5-LOX inhibitory activity: IC50 of 3.06 and 7.23 μM	Al-Rabia et al. (2021)
Triterpenoids	Integracides G, H, J, L, B and F	*Fusarium sp.*/*Mentha longifolia*	5-LOX inhibitory activity: IC50 ranging from 1.46 to 3.97 μM	Khayat et al. (2022)
Triterpenoids	Phomosterols A and B	*Phoma* sp. SYSU-SK-7/*Kandelia candel*	Anti-NO production: IC50 = 13.5, 25.0 and 12.5 μM	Chen et al. (2020a, b)
Meroterpenoid	Guignardones J-L, M-Ñ and P	*Guignardia mangiferae*/*Ilex cornuta*	TNF-α regulation at 10 μM	Han et al. (2015)
Sesquiterpenoid	Annulohpoxylotol A and B	Annulohypoxylon truncatum/Zizania caduciflora	NF-κB activation: IC50 = 7.11 and 19.24	Li et al. (2017)
Sesquiterpenoid	Chondroterpenes (A–H) and hirsutanols (A, C and E)	*Chondrostereum* sp. NTOU4196/*Pterocladiella capillacea*	Anti-NO production: 4–14 μM of NO in LPS-Induced NO Production in Microglial BV-2 Cells at 20 μM	Hsiao et al. (2017)

Sesquiterpenoid	Monocyclic sesquiterpenediol acorane sesquiterpenes	*Trichoderma* sp./Xylocarpus granatum	Inhibitory activity of NO production: inhibition rates of 75.0% and 39.2% at 0.1 μM, respectively	Zhang et al. (2017)
Sesquiterpenoid	Periconianone A and periconianone B	*Periconia* sp./*Annonsa muricata*	Anti-NO production: IC50 = 0.15 and 0.38 μM	Zhang et al. (2014)
Meroterpenoid	Fusariumin A and asperterpenoid A	*Fusarium* sp. YD-2/*Santalum album*	Anti-NO production: IC50 = 50 and 1.6 μM, respectively	Yan et al. (2018)
Sesterterpenoid	Asperophiobolins H-J	*Aspergillus* sp. ZJ-68/*Kandelia candel*	Anti-NO production: IC50 = 21, 24, 16 μM, respectively	Cai et al. (2019)

cornuta (Han et al. 2015). TLRs are important inflammatory mediators in the gut; they play an important role in mediating immune responses to a variety of pathogen-derived ligands, linking adaptive and innate immunity (Sameer and Nissar 2021). In addition, they play a role in tumor immunity, and consequently, their activation is a target for cancer treatment (Cen et al. 2018).

The transcription factor NF-κB regulates several innate and adaptive immune processes and serves as an essential mediator of inflammatory responses, inducing the expression of numerous pro-inflammatory genes. Thus, deregulated NF-κB activation contributes to the pathogenic processes of various inflammatory diseases (Liu et al. 2017a, b). Annulohpoxylotols obtained from a foliar endophytic fungus can inhibit NF-κB activation in a dose-dependent manner (Li et al. 2017).

Neuronal inflammation caused by activated microglia (release of NO) is a feature of certain neurodegenerative diseases, including Alzheimer's disease, Parkinson's disease, multiple system atrophy, amyotrophic lateral sclerosis, frontotemporal dementia, progressive supranuclear palsy, corticobasal degeneration, dementia with Lewy bodies, and Huntington's disease (Azargoonjahromi 2023; Gao et al. 2023; Hunot et al. 1996; Ransohoff 2016). In addition, acetylcholinesterase (AChE) is involved in the pathogenesis of neurodegenerative diseases by influencing the inflammatory response. Therefore, NO production in activated microglia and AChE inhibition assays are used to discover novel terpenoids for treating neurodegenerative diseases (Walczak-Nowicka and Herbet 2021). Currently, AChE inhibitors are the most used drugs for the treatment of Alzheimer's disease. However, they have significant side effects. AChE inhibitors produced by plants and other organisms have fewer side effects (Ahmed et al. 2021). Therefore, fungal terpenoids that inhibit AChE may provide an alternative to currently approved drugs for the treatment of neurodegenerative diseases.

The sesquiterpenoids colletotrichine A and asperterpenols, together with the meroterpenoids spiroterreusnoids, are AChE inhibitors in vitro (Table 10.4). Colletotrichine A is produced by *Colletotrichum gloeosporioides* GT-7, an endophytic fungus of the Chinese medicinal plant *Uncaria rhynchophylla*. This sesterterpenoid showed AChE inhibitory activity with an IC50 value of 28 μg/mL (Chen et al. 2018a, b). Asperterpenol A and B are tetracyclic sesterterpenoids isolated from the mangrove endophyte *Aspergillus* sp. 085242. They inhibited AChE with IC50 values of 2.3 and 3 μg/mL, respectively. Butyrylcholinesterase assay suggested that asperterpenol B is selective for AChE (Xiao et al. 2013). *Aspergillus terreus* Thom, isolated from *Tripterygium wilfordii* Hook. f., produced six dimethylorsellinic acid meroterpenoids called spiroterreusnoids, which showed moderate AChE inhibition. However, they showed interesting inhibitory effects on the important protein that initiates the production of the β-amyloid peptide involved in Alzheimer's disease, beta-site amyloid precursor protein cleaving enzyme 1 (BACE1) (Qi et al. 2019). In addition, a series of rotoilludane, illudalane, and botryane sesquiterpenoids from *Phomopsis* sp. TJ507A showed potent inhibition of BACE1, ranging from 19.4% to 43.8% at 40 μM, with no apparent hepatotoxicity (Xie et al. 2018).

Knowledge of the biology and biochemistry of inflammatory processes is used to identify potential therapeutic targets for the treatment of diseases caused by inflammation. The most commonly used screening methods for the discovery of new terpenoids with potential anti-inflammatory activity include (1) enzymatic assays to detect the inhibitory activity of COX1, COX2, 5-LOX or iNOS; (2) genetic expression, e.g., expression of toll-like receptors (TLR) or nuclear factor-kappa B (NF-κB) related to mediators of inflammatory responses (Liu et al. 2017a, b; Sameer and Nissar 2021); and (3) quantification of TNF-α or NO production in specific activated cells. The most common screening test used in the examples presented in Table 10.4 is the quantification of overproduction of NO in LPS-induced macrophages and microglial cells.

10.4.3 Antimicrobial

Natural product research has made remarkable contributions to the field of antimicrobial discovery. For example, penicillin from *Penicillium notatum*, streptomycin or avermectins from actinomycetes, and artemisins from *Artemisia annua* were discovered as potent antimicrobials. The researchers who discovered these antimicrobials were awarded the Nobel Prize (Shen 2015). However, due to the rise of antimicrobial resistance, there is a need for new antimicrobial compounds with innovative chemical structures and modes of action. Unfortunately, current research and funding for the development of new classes of antimicrobials is insufficient (Miethke et al. 2021).

Terpenoids typically exhibit bacteriostatic and bactericidal activity against human pathogenic bacteria. They work by disrupting the cell membrane, modulating efflux pumps, blocking oxygen uptake, altering oxidative phosphorylation, reducing the ability of cells to adhere, or suppressing biofilm development (Mahizan et al. 2019). The general process of antimicrobial drug development is initiated by an in vitro screening phase, followed by in vivo profiling and validation (preclinical studies), and finally clinical trials and approval (Miethke et al. 2021). During this process, the most used antibacterial and antifungal assays for the screening stage are broth microdilution and agar dilution, which determine the minimum inhibitory concentration (MIC) of the terpenoid required to inhibit the growth of a specific pathogen strain (Balouiri et al. 2016).

10.4.4 Antibacterial and Antifungal

Antibacterial and antifungal activities are found together in some terpenoids. Trichocadinins B-G are cadinane sesquiterpenes with broad antibacterial and antifungal activities, including human, plant, and animal pathogens (Shi et al. 2019). A similar pattern was observed for meroterpenoids chermesins E-H, from the marine red algae endophytic *Penicillium chermesinum* EN-480 (Hu et al. 2022). Nevertheless, some terpenoids show selective antimicrobial activities against a group of pathogens. To illustrate, *Leptosphaeria* sp. XL026 produces sesquiterpenoids, leptosphins A and B, and diterpenoids, leptosphin C and conidiogenones.

Some of those diterpenoids display weak antimicrobial activities against human pathogenic bacteria but are stronger against a set of microbial plant pathogens (Chen et al. 2019). Fusartricin is a sesquiterpenoid produced by *Fusarium tricinctum* Salicorn 19, an endophyte of *Salicornia bigelovii*. Sesquiterpenoids exhibit antibacterial and antifungal activities against *Enterobacter aerogenes*, *Micrococcus tetragenus* and *Candida albicans* (Zhang et al. 2015).

Some terpenoids show interesting activities against relevant human pathogenic bacteria, such as methicillin-resistant *Staphylococcus aureus* (MRSA). Sesquiterpenoid, peniterester from *Penicillium* sp. T2-M20 showed antibacterial activity against Bacillus *subtilis*, *Escherichia coli* and *Staphylococcus aureus* with MICs of 8, 8 and 4 μg/mL, respectively, comparable to the positive control (Duan et al. 2020). Unfortunately, the authors do not report data on the origin and biochemical characteristics of the stains used in the antimicrobial assay. In contrast, Li and colleagues reported stronger antibacterial activities against clinical isolates, which is relevant because clinical isolates often have acquired resistance (Zhu et al. 2022). The compound is a tetracyclic triterpenoid, 19-nor-lanosta-5(10),6,8,24-tetraene-1α,3β,12β,22S-tetraol, isolated from *Diaporthe* sp. LG23, which is active against Gram-positive and Gram-negative clinical isolates of the human pathogens *Streptococcus pyogenes*, *Pseudomonas aeruginosa*, *S. aureus*, and *E. coli*, with MICs comparable to the positive controls streptomycin and gentamicin (Li et al. 2015). In addition, harzianol, a diterpenoid from *Trichoderma atroviride* (Li et al. 2020a, b, c, d), (S)-18,19-dihydroxyneogrifolin, a meroterpenoid produced by *Penicillium* sp. (Duan et al. 2016), and eutyscoparin G, a sesquiterpenoid from *Eutypella scoparia* (Zhang et al. 2021), showed similar MICs against MRSA ranging from 4 to 6.25 μg/mL.

One of the main mechanisms leading to multidrug resistance (MDR) is the overexpression of efflux pumps. As inhibitors of numerous efflux pumps in MDR bacterial strains, terpenoids can act synergistically with conventional antibiotics. Terpenoids appear to be models for the development of new antibiotics with innovative structures and pharmacological properties, although specific molecular mechanisms are not fully understood (Dias et al. 2022). Despite the evidence of the importance of terpenoids as antimicrobials and the growing number of new terpenoid structures from endophytic fungi, there are no studies investigating the mechanisms of action of endophytic fungal terpenoids.

Fungal terpenoids active against *Mycobacterium tuberculosis* are poorly reported. Some mangrove entophytic *Aspergillus sp.* produce antimycobacterial terpenoids. Asperophiobolins (Cai et al. 2019) and asperterpenoid A (Huang et al. 2013) are sesterterpenoid inhibitors of the protein tyrosine phosphate B (MptpB), a virulence factor secreted by *M. tuberculosis* to attenuate the human host immune response. *Aplosporella javeedii*, an endophyte of *Orychophragmus violaceus*, produces fusaproliferin and terpestacin, sesterterpenoids that inhibit the growth of *M. tuberculosis* in broth microdilution (Gao et al. 2020).

10.4.5 Antiparasitic

Some studies on terpenoids derived from endophytic fungi have shown efficacy in combating parasitic tropical diseases such as malaria, trypanosomiasis, and leishmaniasis. For example, phomoarcherin B (Hemtasin et al. 2011) and eremoxylarin C (Isaka et al. 2014) are sesterterpenoids with antimalarial activity against *Plasmodium falciparum* with IC50 values of 0.79 and 3.1 μg/mL, respectively. About trypanosomiasis, *Aspergillus calidoustus,* the endophyte of the medicinal plant *Acanthospermum austral*, produced the antimalarial sesterterpenoids ophiobolin K and 6-epi-ophiobolin K against *Trypanosoma cruzi* (IC50 values of 13.0 and 9.62 mM, respectively). Nevertheless, complementary experiments revealed that these compounds showed low selective activity against the parasite and were cytotoxic to *T. cruzi* fibroblast host cells (De Carvalho et al. 2016). Furthermore, the meroterpenoid austin accumulated in co-cultures of *Talaromyces purpurogenus* H4 and *Phanerochaete* sp. H2, endophytes of *Handroanthus impetiginosus*, showed anti-*T. cruzi* activity with an IC50 value of 36.6 μg/mL (do Nascimento et al. 2020). In the field of leishmaniasis, *Fusarium* sp. biosynthesizes a group of tetracyclic triterpenoids called integracides with multiple bioactivities, integracides F, G, H, and J exhibit significant antileishmanial activity against *Leishmania donovani* with IC50 values of 3.74, 2.53, 4.75 and 3.29 μM, respectively (Ibrahim et al. 2016a, b).

In the field of antimalarial compounds, terpenoids are emphasized because the sesquiterpenoid artemisin is an important drug to treat patients infected with *Plasmodium vivax* and *Plasmodium falciparum* (Shen 2015). The challenge of drug resistance is to produce new drugs with novel action modes and identify new targets. The challenge is enormous. New drug candidates should have high potency, low propensity for resistance, specific pharmacokinetics for infrequent dosing, low cost, safety in women and children, and ideally the ability to block *Plasmodium* transmission to *Anopheles* mosquito vectors (Siqueira-Neto et al. 2023). In preclinical studies, only a few plant monoterpenoids and sesquiterpenoids have shown promising results (Gabriel et al. 2018). However, there is a lack of exhaustive studies on antimalarial, anti-trypanosomal or anti-leishmanial terpenoids from fungi.

10.4.6 Antiviral

Viral infections are the most common diseases affecting humans, and new viruses are constantly emerging, while there are a limited number of vaccines and antivirals to treat viral diseases. Endophytes, marine fungi, and edible/medicinal mushrooms are sources of new antiviral molecules (Linnakoski et al. 2018). However, we have found only a few examples of antiviral terpenoids from endophytic fungi. The meroterpenoids talaromyolides (Cao et al. 2020), aperterpenes (Liu et al. 2019a, b, c) and periconones (Liu et al. 2017a, b) showed antiviral activities against pseudorabies virus (PRV), influenza (H1N1) virus, and human immunodeficiency virus 1 (HIV-I), respectively. *Talaromyces purpureogenus*, isolated from the seaweed *Grateloupia filicina* C. Ag (Wulf.) produced polycyclic meroterpenoids talaromyolides E-K actives against the Herpesviridae member, PRV. Talaromyolides I and K showed the most potent inhibition effect rates (48.69% and 60.11%), blocking effect rates of 45.71% and 53.24% and killing effect rates of 48.69% and 62.16% at a concentration of 50 μg/mL, respectively. These figures are relevant because talaromyolide K has been shown to have potent virucidal activity with low host cell toxicity (Cao et al. 2020). Aperterpenes N and O are accumulated in solid cultures of the marine algae-derived *Aspergillus terreus* EN-539. Aperterpene N showed inhibitory activity against influenza virus neuraminidase, with an IC50 value of 18.0 nM (Liu et al. 2019a, b, c). Polyketide-terpenoid hybrid periconones B-E are produced by *Periconia* sp., an endophyte of *Annona muricata*. Periconone B showed anti-HIV activity with an IC50 value of 18.0 μmol/L. Bergamotane sesquiterpenoids, brasilamides A-D and 14-nordrimane-type sesquiterpenoid, and phomanolide are other examples of anti-HIV-1 and anti-H1N1 influenza virus activities. Surprisingly, mechanism of action studies were not performed in any of the examples mentioned, perhaps because the most common methods for evaluating antiviral activity only provide information about the antiviral potential of compounds (Linnakoski et al. 2018).

10.4.7 Others

Finally, additional activities of terpenoids from endophytic fungi included antidiabetic and lipid metabolism. The sesquiterpenoids paraconiothins C and I inhibit the liver receptors LXRα and LXRβ at 50 μM (Nakashima et al. 2019). LXRs are nuclear receptors involved in the transcriptional control of phospholipid metabolism, which means they regulate membrane composition and function. Therefore, LXR inhibitors are new targets for diseases associated with dysregulated lipid metabolism, including atherosclerosis, diabetes mellitus, and cancer (Wang and Tontonoz 2018). Another therapeutic target is pancreatic lipase, which plays a key role in the digestion and absorption of dietary fats (Zhu et al. 2021). The endophytic *Penicillium purpurogenum* produces a series of sesquiterpene lactones inhibitors of pancreatic lipases, with IC50 values ranging from 2.8 to 6.63 μM, comparable to the positive control kaempferol (Wang et al. 2018). Related to antidiabetic terpenoids, α-glucosidase inhibition is a therapeutic potential against type 2 diabetes mellitus. A phenolic bisabolane sesquiterpenoid from the mangrove endophytic *Aspergillus flavus* QQSG-3 (Wu et al. 2018) and the sesterterpenoid YW3548 (Bilal et al. 2018) from the endophytic *Paecilomyces formosus* LHL10 showed α-glucosidase inhibitory activity.

10.5 Future Directions and Challenges

Fungi have long been exploited to produce valuable compounds. However, we still have limited knowledge and understanding of their fundamental cellular and developmental processes, which directly affect the implementation of industrial production processes (Meyer et al. 2016). Endophytic fungi produce a repertoire of secondary metabolites with innovative chemical structures, bioactivities and applications in human health, food production and agriculture. However, the complexity of specific biosynthetic pathways, low production yields, and yield loss with repeated subculturing make them unattractive for industrial production (Tiwari and Bae 2022). Even when modest yields are achieved in stable and sustainable fermentations, some important challenges are still faced, such as the ignorance of biosynthetic pathways and their regulation (rate-limiting genes/enzymes), which is the case of the important chemotherapeutic terpenoids such as taxol or camptothecin (Pu et al. 2013; Subban and Kempken 2023; Tiwari and Bae 2022). In general, these difficulties stem from the limited knowledge of fungal biology and endophytism. Additional challenges are related to the study of bioactive terpenoids. For example, most studies describing new bioactive endophytic fungal terpenoids lack specific experiments that would shed light on the mechanisms of action, thereby limiting their real potential as therapeutic agents.

It has been observed that there is a sustained interest in the study of endophytic fungal terpenoids. This interest can be attributed to two facts, although the mechanisms behind both are not fully understood: (1) secondary metabolites shape the phylogenesis of the symbiosis between endophytic fungi and their host plants; (2) endophytic fungi produce an enormous chemical diversity of terpenoids with interesting bioactivities. Therefore, it is important to understand the molecular and biochemical mechanisms of communication between host plants, fungi, and the whole microbiome to fully exploit the potential of endophytic fungi.

The "one strain many compounds" (OSMAC) approach is a traditional method in natural product drug discovery applied to endophytic fungi. Most of the compounds presented in the Sect. 10.4 were discovered using this approach. However, due to the complex scenario of endophytic fungi, it is necessary to incorporate high-throughput strategies that integrate genome mining methods, whole genome sequencing, deep learning approaches, synthetic biology, and metabolomics along with traditional bioactivity-guided isolation and culture-based methods (Aghdam and Brown 2021; Tiwari and Bae 2022).

Moreover, structural and biochemical studies on terpenoid cyclases have revealed a huge untapped potential in terpenoid chemo diversity that can be explored through protein engineering (Christianson 2017). Nevertheless, an in-depth understanding of the function and catalytic mechanisms of terpenoid cyclases and related tailoring enzymes is required to harness the potential of these biocatalysts and thereby discover second-generation terpenoids through pathway engineering (Luo et al. 2024; Yin and Dickschat 2023). In this regard, recent work has focused on discovering unknown biosynthetic pathways, reconstructing biosynthetic pathways, and creating high-yielding production strains (Yin and Dickschat 2023).

The unique nature of endophytism, together with the complex biosynthesis of terpenoids and the limited knowledge of fungal biology and biochemistry, poses important challenges for the use of fungal terpenoids as therapeutic agents. To overcome these challenges, future research directions aim to utilize multidisciplinary and innovative methods, such as multi-omics approaches, to reveal the true potential of terpenoids from endophytic fungi.

10.6 Conclusion

Endophytic fungi are a diverse group of fungi that inhabit almost all plants in all habitats studied to date. They can communicate with host plants and other inhabitants by producing secondary metabolites, including diterpenoids, triterpenoids, sesterterpenoids, and meroterpenoids. These metabolites have the potential to discover new terpenoids with unique chemical structures and bioactivities. Terpenoids isolated from endophytic fungi have shown unusual chemical arrays and bioactivities such as anti-inflammatory, anticancer, antimicrobial, enzyme inhibitor, and antidiabetic properties.

The central step of terpenoid biosynthesis is the cyclization of linear prenyl diphosphate. This process is catalyzed by a diverse group of proteins called terpenoid cyclases, and research is focused on understanding their genetics, reaction mechanisms, and structural characterization.

In summary, applications of terpenoid natural products in drug discovery and other fields require multidisciplinary approaches to gain a deep understanding of the mechanisms driving endophytism and the biosynthetic pathways of terpenoids and their regulation.

References

Adan A, Kiraz Y, Baran Y (2016) Cell proliferation and cytotoxicity assays. Curr Pharm Biotechnol 17(14):1213–1221. https://doi.org/10.2174/1389201017666160808160513

Aghdam SA, Brown AMV (2021) Deep learning approaches for natural product discovery from plant endophytic microbiomes. Environ Microb 16(1):1–20. https://doi.org/10.1186/s40793-021-00375-0

Ahmed S, Khan ST, Zargaham MK, Khan AU, Khan S, Hussain A, Uddin J, Khan A, Al-Harrasi A (2021) Potential therapeutic natural products against Alzheimer's disease with reference of acetylcholinesterase. Biomed Pharmacother 139:111609. https://doi.org/10.1016/j.biopha.2021.111609

Al-Rabia MW, Mohamed GA, Ibrahim SRM, Asfour HZ (2021) Anti-inflammatory ergosterol derivatives from the endophytic fungus *Fusarium chlamydosporum*. Nat Prod Res 35(23):5011–5020. https://doi.org/10.1080/14786419.2020.1762185

Amirzakariya BZ, Shakeri A (2022) Bioactive terpenoids derived from plant endophytic fungi: an updated review (2011–2020). Phytochemistry 197:113130. https://doi.org/10.1016/j.phytochem.2022.113130

Ariefta NR, Kristiana P, Nurjanto HH, Momma H, Kwon E, Ashitani T, Tawaraya K, Murayama T, Koseki T, Furuno H, Usukhbayar N, Ichi KK, Shiono Y (2017) Nectrianolins A, B, and C, new metabolites produced by endophytic *fungus Nectria pseudotrichia* 120-1NP. Tetrahedron Lett 58(43):4082–4086. https://doi.org/10.1016/j.tetlet.2017.09.032

Azargoonjahromi A (2023) Dual role of nitric oxide in Alzheimer's disease. Nitric Oxide 1(134–135):23–37. https://doi.org/10.1016/j.niox.2023.03.003

Balouiri M, Sadiki M, Ibnsouda SK (2016) Methods for in vitro evaluating antimicrobial activity: a review. J Pharmaceut Anal 6(2):71. https://doi.org/10.1016/j.jpha.2015.11.005

Barra L, Abe I (2021) Chemistry of fungal meroterpenoid cyclases. Nat Prod Rep 38(3):566–585. https://doi.org/10.1039/d0np00056f

Bilal S, Ali L, Khan AL, Shahzad R, Asaf S, Imran M, Kang SM, Kim SK, Lee IJ (2018) Endophytic fungus *Paecilomyces formosus* LHL10 produces sester-terpenoid YW3548 and cyclic peptide that inhibit urease and α-glucosidase enzyme activities. Arch Microbiol 200(10):1493–1502. https://doi.org/10.1007/S00203-018-1562-7

Bo X, Zhao Y, Qin F, Wu F, Tan M, Ju S, Song Z, Li W, He F, Wei Y, Xu H, Huang X (2024) Cytotoxic metabolites from the mangrove endophytic fungus *Aspergillus sp.* GXNU QG1a. Nat Prod Res 16:1–8. https://doi.org/10.1080/14786419.2024.2303601

Brock NL, Dickschat JS (2013) Biosynthesis of terpenoids. In: Natural products: phytochemistry, botany and metabolism of alkaloids, phenolics and terpenes, pp 2693–2732. https://doi.org/10.1007/978-3-642-22144-6_121

Cai R, Jiang H, Mo Y, Guo H, Li C, Long Y, Zang Z, She Z (2019) Ophiobolin-type sesterterpenoids from the mangrove endophytic fungus *Aspergillus* sp. ZJ-68. J Nat Prod 82(8):2268–2278. https://doi.org/10.1021/acs.jnatprod.9b00462

Cao X, Shi Y, Wu S, Wu X, Wang K, Sun H, He S, Dickschat JS, Wu B (2020) Polycyclic meroterpenoids, talaromyolides E – K for antiviral activity against pseudorabies virus from the endophytic fungus *Talaromyces purpureogenus*. Tetrahedron 76(30):131349. https://doi.org/10.1016/j.tet.2020.131349

Cen X, Liu S, Cheng K (2018) The role of Toll-like receptor in inflammation and tumor immunity. Front Pharmacol 9. https://doi.org/10.3389/fphar.2018.00878

Chagas FO, Pessotti RDC, Caraballo-Rodríguez AM, Pupo MT (2018) Chemical signaling involved in plant-microbe interactions. Chem Soc Rev 47(5):1652–1704. https://doi.org/10.1039/c7cs00343a

Chai H, Yin R, Liu Y, Meng H, Zhou X, Zhou G, Bi X, Yang X, Zhu T, Zhu W, Deng Z, Hong K (2016) Sesterterpene ophiobolin biosynthesis involving multiple gene clusters in *Aspergillus ustus*. Sci Rep 6(1):1–11. https://doi.org/10.1038/srep27181

Chen CJ, Liu XX, Zhang WJ, Zang LY, Wang G, Ng SW, Tan RX, Ge HM (2015) Sesquiterpenoids isolated from an endophyte fungus *Diaporthe* sp. RSC Adv 5(23):17559–17565. https://doi.org/10.1039/c4ra13136c

Chen S, Ding M, Liu W, Huang X, Liu Z, Lu Y, Liu H, She Z (2018a) Anti-inflammatory meroterpenoids from the mangrove endophytic fungus *Talaromyces amestolkiae* YX1. Phytochemistry 146:8–15. https://doi.org/10.1016/j.phytochem.2017.11.011

Chen XW, Yang ZD, Sun JH, Song TT, Zhu BY, Zhao JW (2018b) Colletotrichine A, a new sesquiterpenoid from *Colletotrichum gloeosporioides* GT-7, a fungal endophyte of *Uncaria rhynchophylla*. Nat Prod Res 32(8):880–884. https://doi.org/10.1080/14786419.2017.1365071

Chen HY, Liu TK, Shi Q, Yang XL (2019) Sesquiterpenoids and diterpenes with antimicrobial activity from *Leptosphaeria* sp. XL026, an endophytic fungus in *Panax notoginseng*. Fitoterapia 137:104243. https://doi.org/10.1016/j.fitote.2019.104243

Chen L, Fan DM, Tang JW, An T, Li X, Kong LM, Puno PT, Li Y (2020a) Discovery of isopenicin A, a meroterpenoid as a novel inhibitor of tubulin polymerization. Biochem Biophys Res Commun 525(2):303–307. https://doi.org/10.1016/j.bbrc.2020.02.026

Chen Y, Yang W, Zou G, Yan Z, Qiu P, Long Y, She Z (2020b) Metabolites with anti-inflammatory and α-glucosidase inhibitory activities from the mangrove endophytic fungus *Phoma* sp. SYSU-SK-7. Tetrahedron Lett 61(48):152578. https://doi.org/10.1016/j.tetlet.2020.152578

Chen R, Jia Q, Mu X, Hu B, Sun X, Deng Z, Chen F, Bian G, Liu T (2021) Systematic mining of fungal chimeric terpene synthases using an efficient precursor-providing yeast chassis. Proc Natl Acad Sci U S A 118(29):e2023247118. https://doi.org/10.1073/pnas.2023247118

Cheng Y, Liu Y, Li X, Liu G, Li Z, Liu B, Gao N (2024) Transcriptome analysis of the mechanism of endophytic fungus CHS3 promoting saikosaponin d synthesis in *Bupleurum scorzonerifolium* Willd. suspension cells. Fitoterapia 173:105778. https://doi.org/10.1016/j.fitote.2023.105778

Chiba R, Minami A, Gomi K, Oikawa H (2013) Identification of ophiobolin F synthase by a genome mining approach: a sesterterpene synthase from *Aspergillus clavatus*. Org Lett 15(3):594–597. https://doi.org/10.1021/ol303408a

Christianson DW (2017) Structural and chemical biology of terpenoid cyclases. Chem Rev 117(17):11570–11648. https://doi.org/10.1021/acs.chemrev.7b00287

Dai G, Sun J, Peng X, Shen Q, Wu C, Sun Z, Sui H, Ren X, Zhang Y, Bian X (2023) Astellolides R-W, drimane-type sesquiterpenoids from an *Aspergillus parasiticus* strain associated with an isopod. J Nat Prod 86(7):1746–1753. https://doi.org/10.1021/acs.jnatprod.3c00215

De Carvalho CR, De Lourdes Almeida Vieira M, Cantrell CL, Wedge DE, Alves TMA, Zani CL, Pimenta RS, Sales PA, Murta SMF, Romanha AJ, Rosa CA, Rosa LH (2016) Biological activities of ophiobolin K and 6-epi-ophiobolin K produced by the endophytic fungus *Aspergillus calidoustus*. Nat Prod Res 30(4):478–481. https://doi.org/10.1080/14786419.2015.1022777

De Souza JJ, Vieira IJC, Rodrigues-Filho E, Braz-Filho R (2011) Terpenoids from endophytic fungi. Molecules 16(12):10604–10618. https://doi.org/10.3390/molecules161210604

Demain AL, Fang A (2000) The natural functions of secondary metabolites. Adv Biochem Eng Biotechnol 69:1–39. https://doi.org/10.1007/3-540-44964-7_1

Dewick PM (2009) The mevalonate and methylerythritol phosphate pathways: terpenoids and steroids. Med Nat Prod:187–310. https://doi.org/10.1002/9780470742761.ch5

Dias KJSDO, Miranda GM, Bessa JR, De Araújo ACJ, Freitas PR, De Almeida RS, Paulo CLR, Neto JBDA, Coutinho HDM, Ribeiro-Filho J (2022) Terpenes as bacterial efflux pump inhibitors: a systematic review. Front Pharmacol 13:953982. https://doi.org/10.3389/fphar.2022.953982

Dissook S, Kuzuyama T, Nishimoto Y, Kitani S, Putri S, Fukusaki E (2021) Stable isotope and chemical inhibition analyses suggested the existence of a non-mevalonate-like pathway in the yeast *Yarrowia lipolytica*. Sci Rep 11(1):1–8. https://doi.org/10.1038/s41598-021-85170-0

Ditengou FA, Müller A, Rosenkranz M, Felten J, Lasok H, Van Doorn MM, Legué V, Palme K, Schnitzler JP, Polle A (2015) Volatile signalling by sesquiterpenes from ectomycorrhizal fungi reprogrammes root architecture. Nat Commun 6(1):1–9. https://doi.org/10.1038/ncomms7279

do Nascimento JS, Silva FM, Magallanes-Noguera CA, Kurina-Sanz M, dos Santos EG, Caldas IS, Luiz JHH, de Silva EO (2020) Natural trypanocidal product produced by endophytic fungi through co-culturing. Folia Microbiol 65(2):323–328. https://doi.org/10.1007/s12223-019-00727-x

Duan R, Zhou H, Yang Y, Li H, Dong J, Li X, Chen G, Zhao L, Ding Z (2016) Antimicrobial meroterpenoids from the endophytic fungus *Penicillium* sp. T2-8 associated with *Gastrodia elata*. Phytochem Lett 18:197–201. https://doi.org/10.1016/j.phytol.2016.10.013

Duan RT, Yang RN, Li HT, Tang LH, Liu T, Yang YB, Zhou H, Ding ZT (2020) Peniterester, a carotane-type antibacterial sesquiterpene from an artificial mutant *Penicillium* sp. T2-M20. Fitoterapia 140:104422. https://doi.org/10.1016/j.fitote.2019.104422

Eid AM, Salim SS, Hassan SED, Ismail MA, Fouda A (2019) Role of endophytes in plant health and abiotic stress management. In: Microbiome in plant health and disease: challenges and opportunities, pp 119–144. https://doi.org/10.1007/978-981-13-8495-0_6

Eisenreich W, Bacher A, Arigoni D, Rohdich F (2004) Biosynthesis of isoprenoids via the non-mevalonate pathway. Cell Mol Life Sci 61(12):1401–1426. https://doi.org/10.1007/s00018-004-3381-z

Farh MEA, Jeon J (2020) Roles of fungal volatiles from perspective of distinct lifestyles in filamentous fungi. Plant Pathol J 36(3):193–203. https://doi.org/10.5423/ppj.rw.02.2020.0025

Flesch G, Rohmer M (1988) Prokaryotic hopanoids: the biosynthesis of the bacteriohopane skeleton: formation of isoprenic units from two distinct acetate pools and a novel type of carbon/carbon linkage between a triterpene and d-ribose. Eur J Biochem 175(2):405–411. https://doi.org/10.1111/J.1432-1033.1988.tb14210.x

Frank A, Groll M (2017) The methylerythritol phosphate pathway to isoprenoids. Chem Rev 117(8):5675–5703. https://doi.org/10.1021/acs.chemrev.6b00537

Gabriel HB, Sussmann RA, Kimura EA, Rodriguez AAM, Verdaguer IB, Leite GCF, Katzin AM, Gabriel HB, Sussmann RA, Kimura EA, Rodriguez AAM, Verdaguer IB, Leite GCF, Katzin

AM (2018) Terpenes as potential antimalarial drugs. Terpenes Terpenoids 1:39–57. https://doi.org/10.5772/intechopen.75108

Galindo-Solís JM, Fernández FJ (2022) Endophytic fungal terpenoids: natural role and bioactivities. Microorganisms 10(2):339. https://doi.org/10.3390/microorganisms10020339

Gao Y, Stuhldreier F, Schmitt L, Wesselborg S, Wang L, Müller WEG, Kalscheuer R, Guo Z, Zou K, Liu Z, Proksch P (2020) Sesterterpenes and macrolide derivatives from the endophytic fungus *Aplosporella javeedii*. Fitoterapia 146:104652. https://doi.org/10.1016/j.fitote.2020.104652

Gao C, Jiang J, Tan Y, Chen S (2023) Microglia in neurodegenerative diseases: mechanism and potential therapeutic targets. Signal Transduct Target Ther 8(359):2–37. https://doi.org/10.1038/s41392-023-01588-0

Geris R, Simpson TJ (2009) Meroterpenoids produced by fungi. Nat Prod Rep 26(8):1063–1094. https://doi.org/10.1039/b820413f

Gershenzon J, Dudareva N (2007) The function of terpene natural products in the natural world. Nat Chem Biol 3(7):408–414. https://doi.org/10.1038/nchembio.2007.5

Guillaume FC, Gilles C, Bruno H, Marti DG, Chassagne F, Cabanac ÁG, Hubert ÁG, Marti ÁG, David B (2019) The landscape of natural product diversity and their pharmacological relevance from a focus on the Dictionary of Natural Products®. Phytochem Rev 18(3):601–622. https://doi.org/10.1007/S11101-019-09606-2

Hacquard S, Kracher B, Hiruma K, Münch PC, Garrido-Oter R, Thon MR, Weimann A, Damm U, Dallery JF, Hainaut M, Henrissat B, Lespinet O, Sacristán S, Ver Loren Van Themaat E, Kemen E, McHardy AC, Schulze-Lefert P, O'Connell RJ (2016) Survival trade-offs in plant roots during colonization by closely related beneficial and pathogenic fungi. Nat Commun 7(1):1–13. https://doi.org/10.1038/ncomms11362

Han WB, Dou H, Yuan WH, Gong W, Hou YY, Ng SW, Tan RX (2015) Meroterpenes with toll-like receptor 3 regulating activity from the endophytic fungus *Guignardia mangiferae*. Planta Med 81(2):145–151. https://doi.org/10.1055/s-0034-1383392

Han J, Jiang L, Zhang L, Quinn RJ, Liu X, Feng Y (2021) Peculiarities of meroterpenoids and their bioproduction. Appl Microbiol Biotechnol 105(10):3987–4003. https://doi.org/10.1007/s00253-021-11312-z

Hemtasin C, Kanokmedhakul S, Kanokmedhakul K, Hahnvajanawong C, Soytong K, Prabpai S, Kongsaeree P (2011) Cytotoxic pentacyclic and tetracyclic aromatic sesquiterpenes from *Phomopsis archeri*. J Nat Prod 74(4):609–613. https://doi.org/10.1021/np100632g

Hoshino Y, Villanueva L (2023) Four billion years of microbial terpenome evolution. FEMS Microbiol Rev 47(2). https://doi.org/10.1093/femsre/fuad008

Hsiao G, Chi WC, Pang KL, Chen JJ, Kuo YH, Wang YK, Cha HJ, Chou SC, Lee TH (2017) Hirsutane-type sesquiterpenes with inhibitory activity of microglial nitric oxide production from the red alga-derived fungus *Chondrostereum* sp. NTOU4196. J Nat Prod 80(5):1615–1622. https://doi.org/10.1021/acs.jnatprod.7b00196

Hu XY, Li XM, Liu H, Wang BG, Meng LH (2022) Mining new meroterpenoids from the marine red alga-derived endophytic *Penicillium chermesinum* EN-480 by comparative transcriptome analysis. Bioorg Chem 128:106021. https://doi.org/10.1016/j.bioorg.2022.106021

Huang M, Lu JJ, Huang MQ, Bao JL, Chen XP, Wang YT (2012) Terpenoids: natural products for cancer therapy. Expert Opin Investig Drugs 21(12):1801–1818. https://doi.org/10.1517/13543784.2012.727395

Huang X, Huang H, Li H, Sun X, Huang H, Lu Y, Lin Y, Long Y, She Z (2013) Asperterpenoid A, a new sesterterpenoid as an inhibitor of *Mycobacterium tuberculosis* protein tyrosine phosphatase B from the culture of *Aspergillus* sp. 16-5c. Org Lett 15(4):721–723. https://doi.org/10.1021/ol303549c

Huang Y, Hoefgen S, Valiante V (2021a) Biosynthesis of fungal drimane-type sesquiterpene esters. Angew Chem Int Ed 60(44):23763–23770. https://doi.org/10.1002/anie.202108970

Huang M, Lu JJ, Ding J (2021b) Natural products in cancer therapy: past, present and future. Nat Prod Bioprospect 11(1):5–13. https://doi.org/10.1007/S13659-020-00293-7

Hung R, Lee S, Bennett JW (2013) *Arabidopsis thaliana* as a model system for testing the effect of *Trichoderma volatile* organic compounds. Fungal Ecol 6(1):19–26. https://doi.org/10.1016/j.funeco.2012.09.005

Hunot S, Boissière F, Faucheux B, Brugg B, Mouatt-Prigent A, Agid Y, Hirsch EC (1996) Nitric oxide synthase and neuronal vulnerability in Parkinson's disease. Neuroscience 72(2):355–363. https://doi.org/10.1016/0306-4522(95)00578-1

Ibrahim SRM, Abdallah HM, Mohamed GA, Ross SA (2016a) Integracides H-J: New tetracyclic triterpenoids from the endophytic fungus *Fusarium* sp. Fitoterapia 112:161–167. https://doi.org/10.1016/j.fitote.2016.06.002

Ibrahim SRM, Mohamed GA, Ross SA (2016b) Integracides F and G: new tetracyclic triterpenoids from the endophytic fungus *Fusarium* sp. Phytochem Lett 15:125–130. https://doi.org/10.1016/j.phytol.2015.12.010

Isaka M, Yangchum A, Supothina S, Chanthaket R, Srikitikulchai P (2014) Isopimaranes and eremophilanes from the wood-decay fungus *Xylaria allantoidea* BCC 23163. Phytochem Lett 8(1):59–64. https://doi.org/10.1016/j.phytol.2014.01.011

Itoh T, Tokunaga K, Matsuda Y, Fujii I, Abe I, Ebizuka Y, Kushiro T (2010) Reconstitution of a fungal meroterpenoid biosynthesis reveals the involvement of a novel family of terpene cyclases. Nat Chem 2(10):858–864. https://doi.org/10.1038/nchem.764

Jiang M, Wu Z, Liu L, Chen S (2021) The chemistry and biology of fungal meroterpenoids (2009-2019). Org Biomol Chem 19(8):1644–1704. https://doi.org/10.1039/d0ob02162h

Jung YS, Park JIL (2020) Wnt signaling in cancer: therapeutic targeting of Wnt signaling beyond β-catenin and the destruction complex. Exp Mol 52(2):183–191. https://doi.org/10.1038/s12276-020-0380-6

Khayat MT, Mohammad KA, Mohamed GA, Safo MK, Ibrahim SRM (2022) Integracides: tetracyclic triterpenoids from *Fusarium* sp.—their 5-lipoxygenase inhibitory potential and structure–activity relation using *in vitro* and molecular docking studies. Life 12(12). https://doi.org/10.3390/life12122095

Kogel KH, Franken P, Hückelhoven R (2006) Endophyte or parasite—what decides? Curr Opin Plant Biol 9(4):358–363. https://doi.org/10.1016/j.pbi.2006.05.001

Kramer R, Abraham WR (2011) Volatile sesquiterpenes from fungi: what are they good for? Phytochem Rev 11(1):15–37. https://doi.org/10.1007/S11101-011-9216-2

Li K, Gustafson KR (2021) Sesterterpenoids: chemistry, biology, and biosynthesis. Nat Prod Rep 38(7):1251–1281. https://doi.org/10.1039/d0np00070a

Li H, Huang H, Shao C, Huang H, Jiang J, Zhu X, Liu Y, Liu L, Lu Y, Li M, Lin Y, She Z (2011) Cytotoxic norsesquiterpene peroxides from the endophytic fungus *Talaromyces flavus* isolated from the mangrove plant *Sonneratia apetala*. J Nat Prod 74(5):1230–1235. https://doi.org/10.1021/np200164k

Li G, Kusari S, Kusari P, Kayser O, Spiteller M (2015) Endophytic *Diaporthe* sp. LG23 produces a potent antibacterial tetracyclic triterpenoid. J Nat Prod 78(8):2128–2132. https://doi.org/10.1021/acs.jnatprod.5b00170

Li W, Sun YN, Lee C, Bang SH, Kim S, Ma JY, Kim YH, Shim SH (2017) Annulohpoxylotol A and B, new sesquiterpenoids from the endophytic fungus *Annulohypoxylon truncatum*, are natural NF-κB inhibitors. Arch Pharm Res 40(2):152–158. https://doi.org/10.1007/s12272-016-0865-y

Li Q, Chen C, Cheng L, Wei M, Dai C, He Y, Gong J, Zhu R, Li XN, Liu J, Wang J, Zhu H, Zhang Y (2019a) Emeridones A-F, a series of 3,5-demethylorsellinic acid-based meroterpenoids with rearranged skeletons from an endophytic fungus *Emericella* sp. TJ29. J Org Chem 84(3):1534–1541. https://doi.org/10.1021/acs.joc.8b02830

Li HL, Li XM, Li X, Yang SQ, Wang BG (2019b) Structure, absolute configuration and biological evaluation of polyoxygenated meroterpenoids from the marine algal-derived *Aspergillus terreus* EN-539. Phytochem Lett 32:138–142. https://doi.org/10.1016/j.phytol.2019.05.017

Li XL, Xu YX, Li Y, Zhang R, Hu TY, Su P, Zhou M, Tang T, Zeng Y, Yang YL, Gao W (2019c) Rapid discovery and functional characterization of diterpene synthases from basidiomycete fungi by genome mining. Fungal Genet Biol 128:36–42. https://doi.org/10.1016/j.fgb.2019.03.007

Li S, Chen JF, Qin LL, Li XH, Cao ZX, Gu YC, Le Guo D, Deng Y (2020a) Two new sesquiterpenes produced by the endophytic fungus *Aspergillus fumigatus* from *Ligusticum wallichii*. J Asian Nat Prod Res 22(2):138–143. https://doi.org/10.1080/10286020.2018.1540606

Li WY, Liu Y, Lin YT, Liu YC, Guo K, Li XN, Luo SH, Li SH (2020b) Antibacterial harziane diterpenoids from a fungal symbiont *Trichoderma atroviride* isolated from *Colquhounia coccinea* var. mollis. Phytochemistry 170:112198. https://doi.org/10.1016/j.phytochem.2019.112198

Li W, Shao YT, Yin TP, Yan H, Shen BC, Li YY, Xie HD, Sun ZW, Ma YL (2020c) Penisarins A and B, sesquiterpene coumarins isolated from an endophytic *Penicillium* sp. J Nat Prod 83(11):3471–3475. https://doi.org/10.1021/acs.jnatprod.0c00393

Li H, Zhang R, Cao F, Wang J, Hu Z, Zhang Y (2020d) Proversilins A-E, drimane-type sesquiterpenoids from the endophytic *Aspergillus versicolor*. J Nat Prod 83(7):2200–2206. https://doi.org/10.1021/acs.jnatprod.0c00298

Liao HL, Chen Y, Bruns TD, Peay KG, Taylor JW, Branco S, Talbot JM, Vilgalys R (2014) Metatranscriptomic analysis of ectomycorrhizal roots reveals genes associated with Piloderma-Pinus symbiosis: improved methodologies for assessing gene expression *in situ*. Environ Microbiol 16(12):3730–3742. https://doi.org/10.1111/1462-2920.12619

Linnakoski R, Reshamwala D, Veteli P, Cortina-Escribano M, Vanhanen H, Marjomäki V (2018) Antiviral agents from fungi: diversity, mechanisms and potential applications. Front Microbiol 9. https://doi.org/10.3389/fmicb.2018.02325

Liu L, Chen X, Li D, Zhang Y, Li L, Guo L, Cao Y, Che Y (2015) Bisabolane sesquiterpenoids from the plant endophytic fungus *Paraconiothyrium brasiliense*. J Nat Prod 78(4):746–753. https://doi.org/10.1021/np5009569

Liu Y, Yang MH, Wang XB, Li TX, Kong LY (2016) Caryophyllene sesquiterpenoids from the endophytic fungus, *Pestalotiopsis* sp. Fitoterapia 109:119–124. https://doi.org/10.1016/j.fitote.2015.12.003

Liu T, Zhang L, Joo D, Sun SC (2017a) NF-κB signaling in inflammation. Signal Transduct Target Ther 2(1):1–9. https://doi.org/10.1038/sigtrans.2017.23

Liu JM, Zhang DW, Zhang M, Chen RD, Yan Z, Zhao JY, Zhao JL, Wang N, Dai JG (2017b) Periconones B–E, new meroterpenoids from endophytic fungus *Periconia* sp. Chin Chem Lett 28(2):248–252. https://doi.org/10.1016/j.cclet.2016.07.031

Liu M, Sun W, Shen L, Hao X, Al Anbari WH, Lin S, Li H, Gao W, Wang J, Hu Z, Zhang Y (2019a) Bipolaricins A-I, ophiobolin-type tetracyclic sesterterpenes from a phytopathogenic *Bipolaris* sp. fungus. J Nat Prod 82(10):2897–2906. https://doi.org/10.1021/ACS.JNATPROD.9B00744

Liu S, Zhao Y, Heering C, Janiak C, Müller WEG, Akoné SH, Liu Z, Proksch P (2019b) Sesquiterpenoids from the endophytic fungus *Rhinocladiella similis*. J Nat Prod 82(5):1055–1062. https://doi.org/10.1021/acs.jnatprod.8b00938

Liu Z, Zhao JY, Sun SF, Li Y, Qu J, Liu HT, Liu YB (2019c) Sesquiterpenes from an endophytic *Aspergillus flavus*. J Nat Prod 82(5):1063–1071. https://doi.org/10.1021/acs.jnatprod.8b01084

Luo P, Lv JM, Xie YF, Xiao LY, Qin SY, Chen GD, Luo XZ, Hu D, Gao H (2022) Discovery and characterization of a novel sub-group of UbiA-type terpene cyclases with a distinct motif I. Org Chem Front:3057–3060. https://doi.org/10.1039/d2qo00408a

Luo P, Huang JH, Lv JM, Wang GQ, Hu D, Gao H (2024) Biosynthesis of fungal terpenoids. Nat Prod Rep. https://doi.org/10.1039/d3np00052d

Lutfia A, Munir E, Yurnaliza Y, Basyuni M (2021) Chemical analysis and anticancer activity of sesterterpenoid from an endophytic fungus *Hypomontagnella monticulosa* Zg15SU and its host *Zingiber griffithii* Baker. Heliyon 7(2). https://doi.org/10.1016/j.heliyon.2021.e06292

Mahizan NA, Yang SK, Moo CL, Song AAL, Chong CM, Chong CW, Abushelaibi A, Erin Lim SH, Lai KS (2019) Terpene derivatives as a potential agent against antimicrobial resistance (AMR) pathogens. Molecules 24(14). https://doi.org/10.3390/molecules24142631

Marmann A, Aly AH, Lin W, Wang B, Proksch P (2014) Co-cultivation—a powerful emerging tool for enhancing the chemical diversity of microorganisms. Mar Drugs 12(2):1043–1065. https://doi.org/10.3390/md12021043

Matsuda Y, Abe I (2015) Biosynthesis of fungal meroterpenoids. Nat Prod Rep 33(1):26–53. https://doi.org/10.1039/c5np00090d

Mattoo AJ, Nonzom S (2021) Endophytic fungi: understanding complex cross-talks. Symbiosis 83(3):237–264. https://doi.org/10.1007/S13199-020-00744-2

Meyer V, Andersen MR, Brakhage AA, Braus GH, Caddick MX, Cairns TC, de Vries RP, Haarmann T, Hansen K, Hertz-fowler C, Krappmann S, Mortensen UH, Peñalva MA, Ram AFJ, Head RM (2016) Current challenges of research on filamentous fungi in relation to human welfare and a sustainable bio-economy: a white paper. Fungal Biol Biotechnol 3(1):1–17. https://doi.org/10.1186/s40694-016-0024-8

Miethke M, Pieroni M, Weber T, Brönstrup M, Hammann P, Halby L, Arimondo PB, Glaser P, Aigle B, Bode HB, Moreira R, Li Y, Luzhetskyy A, Medema MH, Pernodet JL, Stadler M, Tormo JR, Genilloud O, Truman AW et al (2021) Towards the sustainable discovery and development of new antibiotics. Nat Rev Chem 5(10):726–749. https://doi.org/10.1038/s41570-021-00313-1

Mitsuhashi T, Abe I (2018) Chimeric terpene synthases possessing both terpene cyclization and prenyltransfer activities. Chembiochem 19(11):1106–1114. https://doi.org/10.1002/cbic.201800120

Miziorko HM (2011) Enzymes of the mevalonate pathway of isoprenoid biosynthesis. Arch Biochem Biophys 505(2):131–143. https://doi.org/10.1016/j.abb.2010.09.028

Mohammadi Ballakuti N, Ghanati F (2023) Developed network between taxoid and phenylpropanoid pathways in *Cryptosporiopsis tarraconensis*, taxan-producing endophytic fungus by Debiased Sparse Partial Correlation (DSPC) algorithm. PLoS One 18(2):e0282010. https://doi.org/10.1371/journal.pone.0282010

Mukhopadhyay N, Shukla A, Makhal PN, Kaki VR (2023) Natural product-driven dual COX-LOX inhibitors: overview of recent studies on the development of novel anti-inflammatory agents. Heliyon 9(3):e14569. https://doi.org/10.1016/j.heliyon.2023.e14569

Nakashima KI, Tomida J, Hirai T, Kawamura Y, Inoue M (2019) Paraconiothins A-J: sesquiterpenoids from the endophytic fungus *Paraconiothyrium brasiliense* ECN258. J Nat Prod 82(12):3347–3356. https://doi.org/10.1021/acs.jnatprod.9b00638

Okada M, Matsuda Y, Mitsuhashi T, Hoshino S, Mori T, Nakagawa K, Quan Z, Qin B, Zhang H, Hayashi F, Kawaide H, Abe I (2016) Genome-based discovery of an unprecedented cyclization mode in fungal sesterterpenoid biosynthesis. J Am Chem Soc 138(31):10011–10018. https://doi.org/10.1021/jacs.6b05799

Pfeffer CM, Singh ATK (2018) Apoptosis: a target for anticancer therapy. Int J Mol Sci 19(2). https://doi.org/10.3390/ijms19020448

Pu X, Qu X, Chen F, Bao J, Zhang G, Luo Y (2013) Camptothecin-producing endophytic fungus *Trichoderma atroviride* LY357: isolation, identification, and fermentation conditions optimization for camptothecin production. Appl Microbiol Biotechnol 97(21):9365–9375. https://doi.org/10.1007/s00253-013-5163-8

Qi C, Zhou Q, Gao W, Liu M, Chen C, Li XN, Lai Y, Zhou Y, Li D, Hu Z, Zhu H, Zhang Y (2019) Anti-BACE1 and anti-AchE activities of undescribed spiro-dioxolane-containing meroterpenoids from the endophytic fungus *Aspergillus terreus* Thom. Phytochemistry 165:112041. https://doi.org/10.1016/j.phytochem.2019.05.014

Qiao W, Tang T, Ling F (2020) Comparative transcriptome analysis of a taxol-producing endophytic fungus, *Aspergillus aculeatinus* Tax-6, and its mutant strain. Sci Rep 10(1):1–11. https://doi.org/10.1038/s41598-020-67614-1

Qin D, Shen W, Gao T, Zuo S, Song H, Xu J, Yu B, Peng Y, Guo J, Tang W, Dong J (2020) Kadanguslactones A-E, further oxygenated terpenoids from *Kadsura angustifolia* fermented by a symbiotic endophytic fungus, *Penicillium ochrochloron* SWUKD4.1850. Phytochemistry 174:112335. https://doi.org/10.1016/j.phytochem.2020.112335

Quin MB, Flynn CM, Schmidt-Dannert C (2014) Traversing the fungal terpenome. Nat Prod Rep 31(10):1449. https://doi.org/10.1039/C4NP00075G

Ransohoff RM (2016) How neuroinflammation contributes to neurodegeneration. Science 353(6301):777–783. https://doi.org/10.1126/science.aag2590

Reis J, Guan XQ, Kisselev AF, Papasian CJ, Qureshi AA, Morrison DC, van Way CW, Vogel SN, Qureshi N (2011) LPS-Induced Formation of Immunoproteasomes: TNF-α and nitric oxide production are regulated by altered composition of proteasome-active sites. Cell Biochem Biophys 60(1–2):77–88. https://doi.org/10.1007/S12013-011-9182-8

Ro DK (2011) Terpenoid biosynthesis. Plant Metab Biotechnol:217–240. https://doi.org/10.1002/9781119991311.ch8

Rohmer M (1993) The biosynthesis of triterpenoids of the hopane series in the Eubacteria: a mine of new enzyme reactions. Pure Appl Chem 65(6):1293–1298. https://doi.org/10.1351/pac199365061293

Rohmer M, M'hamed Knani T, Simonin P, Sutter B, Sahmt H (1993) Isoprenoid biosynthesis in bacteria: a novel pathway for the early steps leading to isopentenyl diphosphate. Biochem J 295:517. https://doi.org/10.1042/bj2950517

Rubbo H, Wood I (2020) 5-LOX inhibition by natural products. Nat Chem Biol 16(7):714–715. https://doi.org/10.1038/s41589-020-0568-z

Sameer AS, Nissar S (2021) Toll-like receptors (TLRs): structure, functions, signaling, and role of their polymorphisms in colorectal cancer susceptibility. Biomed Res Int 2021:1157023. https://doi.org/10.1155/2021/1157023

Sandeep, Ghosh S (2020) Triterpenoids: structural diversity, biosynthetic pathway, and bioactivity. Stud Nat Prod Chem 67:411–461. https://doi.org/10.1016/B978-0-12-819483-6.00012-6

Schouten A (2019) Endophytic fungi: definitions, diversity, distribution and their significance in plant life. In: Endophyte biotechnology: potential for agriculture and pharmacology. https://doi.org/10.1079/9781786399427.0006

Schulz B, Haas S, Junker C, Andrée N, Schobert M (2015) Fungal endophytes are involved in multiple balanced antagonisms. Curr Sci 109(1):39–45

Setiawati A, Candrasari D, Setyajati FE, Prasetyo V, Setyaningsih D, Hartini Y (2022) Anticancer drug screening of natural products: in vitro cytotoxicity assays, techniques, and challenges. Asian Pac J Trop Biomed 12(7):279. https://doi.org/10.4103/2221-1691.350176

Sharma JN, Al-Omran A, Parvathy SS (2007) Role of nitric oxide in inflammatory diseases. Inflammopharmacology 15(6):252–259. https://doi.org/10.1007/s10787-007-0013-x

Shaw JJ, Berbasova T, Sasaki T, Jefferson-George K, Spakowicz DJ, Dunican BF, Portero CE, Narváez-Trujillo A, Strobel SA (2015) Identification of a fungal 1,8-cineole synthase from *Hypoxylon* sp. with specificity determinants in common with the plant synthases. J Biol Chem 290(13):8511–8526. https://doi.org/10.1074/jbc.m114.636159

Shen B (2015) A new golden age of natural products drug discovery. Cell 163(6):1297–1300. https://doi.org/10.1016/j.cell.2015.11.031

Shi XS, Meng LH, Li XM, Li X, Wang DJ, Li HL, Zhou XW, Wang BG (2019) Trichocadinins B-G: antimicrobial cadinane sesquiterpenes from *Trichoderma virens* QA-8, an endophytic fungus obtained from the medicinal plant *Artemisia argyi*. J Nat Prod 82(9):2470–2476. https://doi.org/10.1021/acs.jnatprod.9b00139

Shinohara Y, Takahashi S, Osada H, Koyama Y (2016) Identification of a novel sesquiterpene biosynthetic machinery involved in astellolide biosynthesis. Sci Rep 6(1):1–11. https://doi.org/10.1038/srep32865

Siqueira-Neto JL, Wicht KJ, Chibale K, Burrows JN, Fidock DA, Winzeler EA (2023) Antimalarial drug discovery: progress and approaches. Nat Rev Drug Discov 22(10):807–826. https://doi.org/10.1038/s41573-023-00772-9

Soliman SSM, Tsao R, Raizada MN (2011) Chemical inhibitors suggest endophytic fungal paclitaxel is derived from both mevalonate and non-mevalonate-like pathways. J Nat Prod 74(12):2497–2504. https://doi.org/10.1021/np200303v

Stone J, Polishook J, White JF (2004) Endophytic fungi. In: Biodiversity of fungi. Elsevier Academic Press, pp 241–270. https://doi.org/10.13140/rg.2.1.2497.0726

Subban K, Kempken F (2023) Insights into Taxol® biosynthesis by endophytic fungi. Appl Microbiol Biotechnol 107(20):6151–6162. https://doi.org/10.1007/s00253-023-12713-y

Sun LT, Chen Y, Yang HX, Li ZH, Liu JK, Wang GK, Feng T (2020) Bisabolane sesquiterpenes and α-pyrone derivative from endophytic fungus *Zopfiella* sp. Phytochem Lett 37:29–32. https://doi.org/10.1016/j.phytol.2020.03.008

Tang JW, Kong LM, Zu WY, Hu K, Li XN, Yan BC, Wang WG, Sun HD, Li Y, Puno PT (2019) Isopenicins A–C: two types of antitumor meroterpenoids from the plant endophytic fungus *Penicillium* sp. sh18. Org Lett 21(3):771–775. https://doi.org/10.1021/acs.orglett.8b04020

Tao H, Lauterbach L, Bian G, Chen R, Hou A, Mori T, Cheng S, Hu B, Lu L, Mu X, Li M, Adachi N, Kawasaki M, Moriya T, Senda T, Wang X, Deng Z, Abe I, Dickschat JS, Liu T (2022) Discovery of non-squalene triterpenes. Nature 606(7913):414–419. https://doi.org/10.1038/S41586-022-04773-3

Terhonen E, Blumenstein K, Kovalchuk A, Asiegbu FO (2019) Forest tree microbiomes and associated fungal endophytes: functional roles and impact on forest health. Forests 10(1):42. https://doi.org/10.3390/f10010042

Tiwari P, Bae H (2022) Endophytic fungi: key insights, emerging prospects, and challenges in natural product drug discovery. Microorganisms 10(2):360. https://doi.org/10.3390/microorganisms10020360

Toyomasu T, Tsukahara M, Kaneko A, Niida R, Mitsuhashi W, Dairi T, Kato N, Sassa T (2007) Fusicoccins are biosynthesized by an unusual chimera diterpene synthase in fungi. Proc Natl Acad Sci U S A 104(9):3084–3088. https://doi.org/10.1073/pnas.0608426104

Uzma F, Mohan CD, Hashem A, Konappa NM, Rangappa S, Kamath PV, Singh BP, Mudili V, Gupta VK, Siddaiah CN, Chowdappa S, Alqarawi AA, Abd-Allah EF (2018) Endophytic fungi-alternative sources of cytotoxic compounds: a review. Front Pharmacol 9:328484. https://doi.org/10.3389/fphar.2018.00309

Walczak-Nowicka ŁJ, Herbet M (2021) Acetylcholinesterase inhibitors in the treatment of neurodegenerative diseases and the role of acetylcholinesterase in their pathogenesis. Int J Mol Sci 22(17):9290. https://doi.org/10.3390/ijms22179290

Wang B, Tontonoz P (2018) Liver X receptors in lipid signalling and membrane homeostasis. Nat Rev Endocrinol 14(8):452–463. https://doi.org/10.1038/s41574-018-0037-x

Wang J, Bai G, Liu Y, Wang H, Li Y, Yin W, Wang Y, Lu F (2015) Cytotoxic metabolites produced by the endophytic fungus *Aspergillus clavatus*. Chem Lett 44(8):1148–1149. https://doi.org/10.1246/cl.150417

Wang YN, Xia GY, Wang LY, Ge GB, Zhang HW, Zhang JF, Wu YZ, Lin S (2018) Purpurolide A, 5/5/5 spirocyclic sesquiterpene lactone in nature from the endophytic fungus *Penicillium purpurogenum*. Org Lett 20(22):7341–7344. https://doi.org/10.1021/acs.orglett.8b03323

Wibowo M, Prachyawarakorn V, Aree T, Wiyakrutta S, Mahidol C, Ruchirawat S, Kittakoop P (2014) Tricyclic and spirobicyclic norsesquiterpenes from the endophytic fungus *Pseudolagarobasidium acaciicola*. Eur J Org Chem 2014(19):3976–3980. https://doi.org/10.1002/ejoc.201402262

Wu LS, Hu CL, Han T, Zheng CJ, Ma XQ, Rahman K, Qin LP (2013) Cytotoxic metabolites from *Perenniporia tephropora*, an endophytic fungus from *Taxus chinensis* var. *mairei*. Appl Microbiol Biotechnol 97(1):305–315. https://doi.org/10.1007/s00253-012-4189-7

Wu Y, Chen Y, Huang X, Pan Y, Liu Z, Yan T, Cao W, She Z (2018) α-Glucosidase inhibitors: diphenyl ethers and phenolic bisabolane sesquiterpenoids from the mangrove endophytic fungus *Aspergillus flavus* QQSG-3. Mar Drugs 16(9):307. https://doi.org/10.3390/md16090307

Xiao Z, Huang H, Shao C, Xia X, Ma L, Huang X, Lu Y, Lin Y, Long Y, She Z (2013) Asperterpenols A and B, new sesterterpenoids isolated from a mangrove endophytic fungus *Aspergillus* sp. 085242. Org Lett 15(10):2522–2525. https://doi.org/10.1021/ol401005j

Xie S, Wu Y, Qiao Y, Guo Y, Wang J, Hu Z, Zhang Q, Li X, Huang J, Zhou Q, Luo Z, Liu J, Zhu H, Xue Y, Zhang Y (2018) protoilludane, illudalane, and botryane sesquiterpenoids from the endophytic fungus *Phomopsis* sp. TJ507A. J Nat Prod 81(6):1311–1320. https://doi.org/10.1021/acs.jnatprod.7b00889

Xu M, Hillwig ML, Tiernan MS, Peters RJ (2017) Probing labdane-related diterpenoid biosynthesis in the fungal genus Aspergillus. J Nat Prod 80(2):328–333. https://doi.org/10.1021/acs.jnatprod.6b00764

Yan C, Liu W, Li J, Deng Y, Chen S, Liu H (2018) Bioactive terpenoids from *Santalum album* derived endophytic fungus *Fusarium* sp. YD-2. RSC Adv 8(27):14823–14828. https://doi.org/10.1039/c8ra02430h

Yang Y, Zhang S, Ma K, Xu Y, Tao Q, Chen Y, Chen J, Guo S, Ren J, Wang W, Tao Y, Yin W, Liu H (2017) Discovery and characterization of a new family of diterpene cyclases in bacteria and fungi. Angew Chem 129(17):4827–4830. https://doi.org/10.1002/ange.201700565

Yang W, Chen X, Li Y, Guo S, Wang Z, Yu X (2020a) Advances in pharmacological activities of terpenoids. Nat Prod Commun 15(3). https://doi.org/10.1177/1934578x20903555

Yang H, Liu XY, Zhang PL, Gao HM, Zhang LT, Shen T, Ren DM, Lou HX, Wang XN (2020b) New terpenoids and triketides from culture of the fungus *Botrysphaeria laricina*. Fitoterapia 147:104758. https://doi.org/10.1016/j.fitote.2020.104758

Yin Z, Dickschat JS (2023) Engineering fungal terpene biosynthesis. Nat Prod Rep 40(1):28–45. https://doi.org/10.1039/d2np00020b

Zeng YB, Gu HG, Zuo WJ, Zhang LL, Bai HJ, Guo ZK, Proksch P, Mei WL, Dai HF (2015) Two new sesquiterpenoids from endophytic fungus J3 isolated from mangrove plant *Ceriops tagal*. Arch Pharm Res 38(5):673–676. https://doi.org/10.1007/S12272-014-0448-8

Zeng T, Liu Z, Zhuang J, Jiang Y, He W, Diao H, Lv N, Jian Y, Liang D, Qiu Y, Zhang R, Zhang F, Tang X, Wu R (2020) Tero Kit: a database-driven web server for terpenome research. J Chem Inf Model 60(4):2082–2090. https://doi.org/10.1021/acs.jcim.0c00141

Zhang D, Ge H, Zou JH, Tao X, Chen R, Dai J (2014) Periconianone A, a new 6/6/6 carbocyclic sesquiterpenoid from endophytic fungus *Periconia* sp. with neural anti-inflammatory activity. Org Lett 16(5):1410–1413. https://doi.org/10.1021/ol500197x

Zhang J, Liu D, Wang H, Liu T, Xin Z (2015) Fusartricin, a sesquiterpenoid ether produced by an endophytic fungus *Fusarium tricinctum* Salicorn 19. Eur Food Res Technol 240(4):805–814. https://doi.org/10.1007/s00217-014-2386-6

Zhang M, Zhao JL, Liu JM, Chen RD, Xie KB, Chen DW, Feng KP, Zhang D, Dai JG (2017) Neural anti-inflammatory sesquiterpenoids from the endophytic fungus *Trichoderma* sp. Xy24. J Asian Nat Prod Res 19(7):651–658. https://doi.org/10.1080/10286020.2016.1251908

Zhang W, Lu X, Huo L, Zhang S, Chen Y, Zou Z, Tan H (2021) Sesquiterpenes and steroids from an endophytic *Eutypella scoparia*. J Nat Prod 84(6):1715–1724. https://doi.org/10.1021/acs.jnatprod.0c01167

Zhao Y, Cui J, Liu M, Zhao L (2020) Progress on terpenoids with biological activities produced by plant endophytic fungi in China between 2017 and 2019. Nat Prod Commun 15(7). https://doi.org/10.1177/1934578x20937204

Zhao M, Tang Y, Xie J, Zhao Z, Cui H (2021) Meroterpenoids produced by fungi: occurrence, structural diversity, biological activities, and their molecular targets. Eur J Med Chem 209:112860. https://doi.org/10.1016/j.ejmech.2020.112860

Zheng CJ, Sun PX, Jin GL, Qin LP (2011) Sesquiterpenoids from *Trichoderma atroviride*, an endophytic fungus in *Cephalotaxus fortunei*. Fitoterapia 82(7):1035–1038. https://doi.org/10.1016/j.fitote.2011.06.010

Zheng C, Chen Y, Jiang LL, Shi XM (2014) Antiproliferative metabolites from the endophytic fungus *Penicillium* sp. FJ-1 isolated from a mangrove *Avicennia marina*. Phytochem Lett 10:272–275. https://doi.org/10.1016/j.phytol.2014.10.011

Zhou J, Li X, Chen Y, Dai CC (2017) *De novo* transcriptome assembly of *Phomopsis liquidambari* provides insights into genes associated with different lifestyles in rice (*Oryza sativa* L.). Front Plant Sci 8:244752. https://doi.org/10.3389/fpls.2017.00121

Zhou J, Li X, Huang PW, Dai CC (2018) Endophytism or saprophytism: decoding the lifestyle transition of the generalist fungus *Phomopsis liquidambari*. Microbiol Res 206:99–112. https://doi.org/10.1016/j.micres.2017.10.005

Zhou D, Zhang W, Hao L, Qin X, Yang R, Li J, Huang X (2022) A new sesquiterpene from mangrove endophytic fungus *Aspergillus* sp. GXNU-MA1. Nat Prod Res 36(7):1857–1863. https://doi.org/10.1080/14786419.2020.1824225

Zhu G, Fang Q, Zhu F, Huang D, Yang C (2021) Structure and function of pancreatic lipase-related protein 2 and its relationship with pathological states. Front Genet 12(July):1–7. https://doi.org/10.3389/fgene.2021.693538

Zhu Y, Huang WE, Yang Q (2022) Clinical perspective of antimicrobial resistance in bacteria. Infect Drug Resist 15:735. https://doi.org/10.2147/idr.s345574

Hidden Biocontrol Agents: The World of Insect-Pathogenic Fungi

11

Kamel A. Abd-Elsalam, Rawan K. Hassan, Toka E. Abdelkhalek, Hassan Almoammar, and Ali A. A. El-Sayed

Abstract

In this chapter, the intricate relationships between insect-pathogenic fungus and their arthropod hosts are explored. These fungi utilize cunning strategies to infect insects despite their defenses, causing behavioral and physiological alterations in their hosts. It delves into the mechanisms of invasion, manipulation of hosts, and the evolutionary competition between fungi and insects. It also addresses issues like environmental conditions and ethical considerations while talking about the ecological and agricultural roles of these fungi as biocontrol agents against pests. The diversity of insect-pathogenic fungi is demonstrated through case studies and examples, emphasizing their evolutionary dynamics, ecological significance, and useful applications in agriculture and natural habitats.

Keywords

Insect-pathogenic fungi · Microbial pathogenic · Metarhizium · Hidden bioagents

K. A. Abd-Elsalam (✉)
Plant Pathology Research Institute, Agricultural Research Center, Giza, Egypt

R. K. Hassan · T. E. Abdelkhalek
Biotechnology English Program, Faculty of Agriculture, Cairo University, Giza, Egypt

H. Almoammar
Wellness & Preventive Medicine Institute, Health Sector, King Abdulaziz City for Science and Technology (KACST), Riyadh, Saudi Arabia

A. A. A. El-Sayed
Plant Protection Research Institute, Agricultural Research Center, Dokki, Giza, Egypt

K. A. Abd-Elsalam, S. F. AbuQamar (eds.), *Fungal Endophytes Volume I*,
https://doi.org/10.1007/978-981-97-7312-1_11

11.1 Introduction

"Exploring the Enigmatic Realm: In-depth Insights into Insect-Pathogenic Fungi" offers a thorough examination of endophytic insect-pathogenic fungi, including their classification, traits, and ecological functions (Boomsma et al. 2014). The introduction portion of the research by Islam et al. (2021) provides a comprehensive explanation of the life cycle, reproduction, and study methodology of the subject. The book explores the evolutionary adaptations that allow these fungi to flourish in different habitats and discusses their infection methods and pathology (Boucias et al. 2018). The article by Lovett et al. (2017) presents in-depth information on toxic metabolites, enzymes, and host immune responses. It also explores the impact of environmental factors on the spread and progression of infections. The study by Araújo et al. (2016) thoroughly examines the many aspects of endophytic insect-pathogenic fungi, including their variety, interactions, uses, and ecological consequences. This resource is extremely important for researchers, students, and enthusiasts who are eager to study the complex world of these hidden entomopathogenic organisms. The current chapter serves as a comprehensive reference that helps readers comprehend the importance of these organisms and their possible uses in several sectors.

11.2 Introduction to Endophytic Insect-Pathogenic Fungi

Endophytic insect-pathogenic fungi constitute a captivating convergence of ecology, entomology, and mycology by inhabiting plant internal tissues while retaining the capacity to parasitize and regulate insect hosts. This distinctive ecological niche sets them apart from conventional insect-pathogenic fungi, providing fresh perspectives on the intricacies of symbiotic relationships within ecosystems (Parveen and Rashtrapal 2024). The introductory overview of endophytic insect-pathogenic fungi encompasses their varied forms, life cycles, and mechanisms of interaction with both plants and insects.

11.2.1 What Are Endophytic Insect-Pathogenic Fungi?

Endophytic entomopathogenic fungi, belonging to the phylum Ascomycota, colonize plants without causing harm but instead infect and exploit insects as a nutritional resource. Their complex association with both plants and insects may be observed in many environments throughout the globe, with species such as Beauveria acting as prominent illustrations (Branine et al. 2019). These fungi have the ability to produce specific structures and substances, such as tiny spores, that initiate infection on the outer layer of insects' bodies (Moonjely et al. 2016; Behie et al. 2015). Their life cycle includes the colonization of plants by spores, proliferation inside plant tissues, and ultimately invading insects upon eating. Studying these fungi is difficult because of their intricate life cycles and interconnections. This requires using several approaches like microscopy, molecular techniques, laboratory studies, and field investigations (Barelli et al. 2016; Hu and Bidochka 2021). They have a crucial function in naturally controlling pests, demonstrating potential as agents for biocontrol in integrated pest management, and in creating biopesticides and medicines. Furthermore, they have significant economic potential for the synthesis of enzymes and biofuels. It is crucial to comprehend the methods and effects of pests to achieve sustainable pest control and to optimize their use in agriculture, medicine, and industry (Sasan and Bidochka 2012; Boomsma et al. 2014).

11.2.2 Importance of Endophytic Insect-Pathogenic Fungi

Endophytic insect-pathogenic fungi are a one-of-a-kind team with the ability to penetrate and get rid of bugs from within. They create equally helpful partnerships with plants, dwelling within their cells without triggering damage. Nevertheless, when bugs come into call with contaminated plants, these fungi can go against the bug's body, causing a deadly infection (Pathan et al. 2021). This vibrant interaction amongst fungi, plants, and bugs holds considerable eco-friendly and farming value.

11.2.2.1 Natural Pest Control

Endophytic insect pathogenic fungi are also an eco-friendly approach against devastating insect pests of various crops (Ahmad et al. 2020). The fungi attack and kill myriad insect pests such as aphids, nematode worms, caterpillars, beetles, locusts, and flies. They grow rapidly within the host's body and secrete lethal metabolites upon host death, thus curbing insect pests. They help in suppressing chemical pesticide pollution and in lowering human risks. When combined with natural resources, endophytic insect pathogenic fungi can pave the way for efficient biological control of insect pests while maintaining efficient agro-ecosystems.

11.2.2.2 Biodiversity and Ecosystem Functioning

In conclusion, these endophytes insect-pathogenic fungi play an imperative role in the conservation of biodiversity and environment homeostasis through insect population control, complete elimination of disruptive insect pests, as well as supporting essential ecological features like pollination and decomposition (Garrido-Jurado et al. 2023; Samal et al. 2023). Because they are selective in that way, they are ecological resilience enhancers: their presence helps keep pests' predators under control and keeps the diversity of organisms. Through the regulation of insect populations, they discourage a single species from dominating and help keep the ecological framework in balance. Moreover, these fungi are also involved in nutrient cycling by degrading dead or injured insects through which cell contents are dispersed on the soil surface for plant production and promote ecosystem functioning (Francis et al. 2022).

11.2.2.3 Conservation of Endophytic Insect-Pathogenic Fungi

The conservation of endophytic insect-pathogenic fungi is essential for their ecological and agricultural benefits. Unfortunately, these fungi are at risk of habitat destruction and chemical pesticide application. Conservation efforts should focus on preserving the natural habitats of these fungi and encouraging more sustainable farming practices (Bamisile et al. 2023). The loss of their natural hosts can result in a decrease in the population of these fungi, as they are dependent on specific plant hosts. Thus, the conservation plans must prioritize preserving natural habitats, restoring degraded habitats, and promoting organic farming methods. This will help us to keep these fungal species alive (Behie et al. 2015).

11.2.2.4 Potential for Biotechnological Applications

Endophytic fungi, which infect and kill insects, could be a game changer that goes far beyond ecology. They are highly selective against insect pests and can produce toxic compounds. When such fungi are used, some biocontrol agents can be produced that are invaluable in producing eco-friendly and safe biopesticides for human use (Russell 1998). The utility of biopesticides cannot be overstated. Fungi, a key component of biodiversity, can produce specific biopesticides suitable for controlling specific pests while being safe for non-target and beneficial pest control agents. The insecticidal compounds produced by endophytic fungi offer great potential for pharmacological, agricultural, and industrial applications. They can provide a wide range of novel drugs and bioactive compounds (Malavasi et al. 2020).

11.2.2.5 Understanding Microbial Interactions

The study of endophytic insect-pathogenic fungi offers valuable insights into intricate interactions among microorganisms, plants, and insects (Weiland-Bräuer 2021). These fungi have evolved sophisticated mechanisms to infect hosts, evade immune systems, and manipulate behaviors, inspiring innovative strategies for pest control and biotechnology (Hassani et al. 2018). Understanding these interactions deepens comprehension of ecosystems, informing conservation, agriculture, and sustainable pest management (Braga et al. 2016).

11.2.3 Life Cycle of Endophytic Insect-Pathogenic Fungi

Endophytic insect-pathogenic fungi play a crucial role in regulating insect populations and serve as potent biocontrol agents. Understanding their life cycle is essential for effective pest management strategies. Typically, their life cycle consists of four main stages: spore dispersal, infection, colonization, and sporulation (Behie and Bidochka 2014). Variations may exist among species, but these stages collectively contribute to the fungi's efficacy in controlling insect pests.

1. Spore Dispersal: The life cycle of endophytic insect-pathogenic fungi begins with spore dispersal, where small, lightweight spores equipped with specialized structures, such as appendages or adhesive substances, spread to new hosts

through mechanisms like wind, water, insects, and other animals (Boomsma et al. 2014).
2. Infection: Endophytic insect-pathogenic fungi initiate infection upon contacting a suitable host. They produce enzymes and toxins to breach the insect's protective barriers like the cuticle. Germinating spores form specialized structures like appressoria or infection pegs, facilitating penetration into the host's tissues. Once inside, the fungus establishes itself and begins colonization (Barelli et al. 2016).
3. Colonization: During the colonization phase, endophytic insect-pathogenic fungi grow and spread within the insect's body, utilizing nutrients from the insect for growth and reproduction. Hyphae, thread-like structures, extend throughout the insect's tissues, absorbing nutrients and breaking down host tissues. Specialized structures like mycelial fans or cords facilitate the fungus's spread within the host (Branine et al. 2019).
4. Sporulation: In the final phase of the life cycle, called sporulation, endophytic insect-pathogenic fungi produce spores. Under favorable conditions, the fungus forms reproductive structures like conidiophores or fruiting bodies bearing spores. These spores are released into the environment, infecting new hosts and continuing the life cycle. Spore production timing and conditions vary among species, influenced by factors such as humidity, temperature, and nutrient availability (Dash et al. 2018).

Environmental factors, such as temperature, humidity, and the availability of hosts, play a crucial role in determining the success or timing of endophytic insect-pathogenic fungi's life cycle. Different species may possess unique adaptations to cater to different hosts or ecological niches (Bamisile et al. 2023). Identifying their life cycle is essential for developing sustainable practices and creating effective pest management products, such as biocontrol agents and biopesticides. By examining the stages, it is possible to identify potential weaknesses and triggers for intervention, leading to the creation of tailored pest control measures.

11.2.4 Methods of Studying Endophytic Insect-Pathogenic Fungi

Studying endophytic insect-pathogenic fungi is crucial for understanding their biology, ecology, and potential applications in pest control. These organisms have developed unique strategies for infecting and eliminating insects (Yerukala et al. 2022). Researchers employ various methods and techniques to investigate these **Pathogenic**, aiming to uncover their secrets and potential applications.

11.2.4.1 Isolation and Cultivation

Isolation and growth of endophytic insect-pathogenic fungi are among the most important approaches in this field. Samples are obtained from different habitats and then placed in selective media for isolation. The utilization of dilution and direct plating techniques allows the production of pure culture (Bamisile et al. 2023). 2023 is the year in which it occurred. Specific media is used to cultivate cultures to replicate natural conditions, which can be further characterized by their morphological and molecular identifications, growth patterns, and reproductive strategies. The cultivation of fungi facilitates the production of sufficient bacteria for experimental purposes like bioassays and biochemical analyses (Sasan and Bidochka 2012; Sayed et al. 2020).

11.2.4.2 Morphological and Molecular Identification

The identification of endophytic insect-pathogenic fungi is crucial for comprehending their taxonomy, diversity, and ecological roles. Molecular identification involves the examination of physical traits, such as growth patterns, reproductive structures, and morphology and mycelial growth (Moonjely et al. 2016). Light and electron microscopy are commonly used for detailed examination. PCR and DNA sequencing techniques have transformed the fungal taxonomy field (EL-Lebody et al. 2021). By analyzing specific parts of the fungal genome and comparing them with existing databases, these techniques can be employed to identify closely related species or cryptic taxa more precisely (Alsohaili and Bani-Hasan 2018).

11.2.4.3 Bioassays and Pathogenicity Testing

Researchers assess the pathogenicity and virulence of endophytic insect-pathogenic fungi through bioassays and pathogenicity testing (Landa et al. 1994). Bioassays involve exposing insects to fungi under controlled conditions and monitoring survival, behavior, and physiological responses to determine susceptibility (Butt and Goettel 2000). Pathogenicity testing quantifies virulence by measuring parameters like mortality percentages, time to death, spore production, and fungal biomass in infected insects (Meyer et al. 1990). Examining the virulence of *Beauveria bassiana* from the spotted lanternfly *Lycorma delicatula* in eastern Pennsylvania, the study emphasizes the value of comparative evaluations in determining the efficacy of insect pathogens (Fig. 11.1).

11.2.4.4 Genomics and Transcriptomics

Advances in genomics and transcriptomics have changed the study of endophytic insect-pathogenic fungi at the molecular level (Martino et al. 2018). Whole-genome sequencing provides complete genetic information for, and reveals insights into, evolutionary history, genetic diversity, and ability to produce bioactive compounds. Transcriptomics focuses on gene expression in response to specific conditions or disease bacterial stages. Genes and pathways involved in pathogen recognition, infection, and production of metabolites are identified (Bhadauria et al. 2001; Baldrian and López-Mondéjar 2014). Understanding these molecular mechanisms allows researchers to develop targeted strategies to control pests and develop biocides.

11.2.4.5 Ecological Studies

Biology researchers conduct field studies to understand the ecological roles of endophytic insect pathogenic fungi while investigating their distribution, abundance and interactions in natural ecosystems (Tóth and Barta 2010). These studies collect

Fig. 11.1 Spotted lanternfly specimens with a non-target containing Beauveria fungal outgrowth. (**a**) Adult spotted lanternfly with abundant conidia (infectious spores) and Beauveria outgrowth. Spores of white powder are evident on the surrounding garbage and ground. (**b**) Adult Spotted Lanternfly remains adhered to tree bark after being destroyed by Beauveria. (**c**) *Beauveria brongniartii* growth on a non-target yellowjacket wasp. (Reprinted from Clifton et al. (2023))

samples from different locations and examine fungal presence and diversity using molecular methods (Sharma and Parveen 2011). Researchers monitor insect populations over time, comparing sites with and without fungi, and study the effects of fungal diseases on insect behavior, reproduction, and population density considering environmental influences (Warcup 1951). Such studies provide insight into the biological control of these fungi and their role in ecological balance.

11.2.4.6 Experimental Modifications

Experimental manipulations are important in the ecology and ecology of endophytic insect-pathogenic fungi (Murphy et al. 2015). Laboratory studies investigate fungal biology, including environmental factors related to growth and infection, as well as interactions with insect hosts (Aguilar-Trigueros and Rillig 2016). Researchers control for changes in conditions such as temperature or nutrient availability to assess their effects on fungal growth and pathology and genetic variability to study specific genes involved in pathogenicity -can use methods (Arendt et al. 2016). Methods for studying these fungi include isolation, cultivation, genomics and ecological studies, which allow comprehensive assessment of their potential for pest control for permanent edges and other applications.

11.3 Diversity of Endophytic Insect-Pathogenic Fungi

Endophytic insect pathogenic fungi establish symbiotic relationships with plants, increasing the resistance of insect pests. They demonstrate the ability to infect and repel a variety of pests, making them promising agents for pest control in agriculture. This group includes fungal species with unique characteristics, life cycles, and modes of action (Trizelia et al. 2023).

11.3.1 Taxonomy and Classification of Endophytic Insect-Pathogenic Fungi

Endophytic insect-pathogenic fungi belong to the kingdom Fungi, within the phylum Ascomycota and the class Sordariomycetes (Anwar et al. 2017). Sordariomycetes is recognized for its diverse range of fungi renowned for their insect-pathogenic properties (Table 11.1).

11.3.1.1 Order Hypocreales

Endophytic insect-pathogenic fungi primarily belong to the order Hypocreales within the class Sordariomycetes (Lovett and St Leger 2017). This order is known for its diversity and encompasses various families such as Clavicipitaceae, Cordycipitaceae, and Ophiocordycipitaceae, all of which contain numerous species with insect-pathogenic attributes.

11.3.1.1.1 Family Clavicipitaceae

The family Clavicipitaceae is notable for harboring endophytic insect-pathogenic fungi, including the genera Metarhizium and Beauveria (Torres and White 2009; Rehner 2009). Species like Metarhizium anisopliae and Metarhizium robertsii are found in soil and infect a wide range of insect hosts. Similarly, Beauveria bassiana and Beauveria brongniartii are effective in infecting and eliminating various insect pests.

11.3.1.1.2 Family Cordycipitaceae

The family Cordycipitaceae contains significant endophytic insect-pathogenic fungi, with the genus Cordyceps being prominent (Wang et al. 2020). Species like Cordyceps militaris and Cordyceps sinensis follow a unique life cycle involving infection and parasitization of insects, leading to the formation of characteristic

Table 11.1 Taxonomy and classification of endophytic, insect-pathogenic fungi (EIPF)

Category	Description	Examples
Kingdom	**Fungi**	–
Phylum	**Ascomycota**	–
Class	**Sordariomycetes**	–
Order	**Hypocreales**	**Most EIPF belong here**
Family		
*	**Cordycipitaceae**	Beauveria
*	**Clavicipitaceae**	Metarhizium
*	**Ophiocordycipitaceae**	Ophiocordyceps
*	**Others**	Acremonium, Aschersonia, etc.
Genus	Further classification based on specific characteristics	See examples in the Family section
Species	Unique identifier within a genus	Beauveria bassiana, Metarhizium robertsii, etc.

fruiting bodies. Cordyceps fungi have been traditionally used in medicine due to perceived health benefits (Kepler et al. 2017).

11.3.1.1.3 Family Ophiocordycipitaceae

The family Ophiocordycipitaceae is closely related to Cordycipitaceae and includes fungi such as Ophiocordyceps unilateralis and Ophiocordyceps sinensis (Xiao et al. 2023). These fungi are known for manipulating the behavior of their insect hosts, leading to phenomena like "zombie ants." Ophiocordyceps fungi have garnered attention due to their intriguing life cycle and potential applications across various fields (Quandt et al. 2014).

11.3.1.2 Molecular Techniques for Taxonomic Identification

Traditionally, taxonomic identification of endophytic insect-pathogenic fungi relied on morphological characteristics, but molecular techniques have become pivotal (Singh 2012). DNA sequencing, particularly of regions like the internal transcribed spacer (ITS), allows for precise identification and understanding of genetic relatedness among species (Das et al. 2014; Purty and Chatterjee 2016). Polymerase chain reaction (PCR) and DNA barcoding are also used to amplify and analyze specific DNA regions for identification (Curran 1990). These molecular methods enhance taxonomy, revealing new species and evolutionary relationships.

11.3.1.3 Challenges in Taxonomy and Classification

Despite advancements, taxonomy and classification of endophytic insect-pathogenic fungi pose challenges due to morphological and genetic variation (Godfray 2002). Subtle morphological differences and cryptic species complicate accurate identification (Rouhan and Gaudeul 2014). Cryptic species, though morphologically similar, may have distinct ecological roles. To address this, researchers employ integrative approaches, combining morphological and molecular methods along with ecological data for comprehensive taxonomy (Khanday and Sofi 2021; Sleeman et al. 2022).

11.3.2 Common Genera of Endophytic Insect-Pathogenic Fungi

Endophytic insect-pathogenic fungi are diverse, with distinct genera. They infect and eliminate insects internally, each with unique characteristics and infection methods (Bamisile et al. 2023).

11.3.2.1 Beauveria

Beauveria is a well-studied family of endophytic insect pathogenic fungi, including species such as *Beauveria bassiana* and *Beauveria brongniarti.* These beetles have a wide variety of hosts transmit a variety of diseases to insects such as aphids, whiteflies, beetles, and caterpillars (Moonjely et al. 2016). They use conidia to attach to the insect's cuticle, germinate and penetrate, and eventually give toxic metabolites responsible for insect decomposition (Yerukala et al. 2022). Fungal

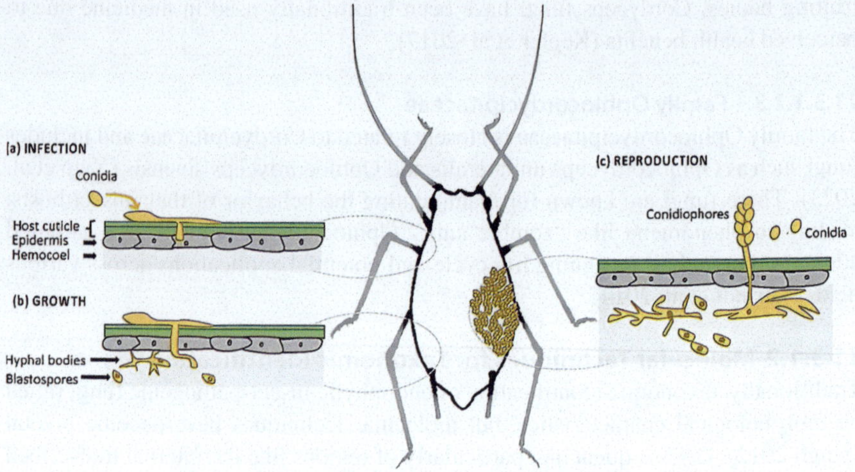

Fig. 11.2 An example of the *Beauveria bassiana* infection process in an arthropod host, in this case, aphids. (**a**) Infection: shown by conidial attachment, germination, appressorial structure development, and host penetration. (**b**) Growth: shown by the hyphal bodies and blastospores produced by the fungus, together with the hemocoel's generation of secondary metabolites. (**c**) Reproduction: brought about by the fungal expansion, which passively releases the asexual conidia and starts a fresh infection cycle. (Reprinted from Francis et al. (2022))

proliferation attaches to cuticles due to their water repellency and specific recognition systems. Under appropriate conditions, it germinates, producing extracellular fluid enzymes. The appressoria are placed on the surface of the cuticle, and the infected hairs penetrate the hemocoel. Tumors grow, proliferate, and invade, and immune responses are suppressed by toxins (Fig. 11.2).

11.3.2.2 Metarhizium

Metarhizium encompasses species like *Metarhizium anisopliae* and *Metarhizium robertsii*, which are commonly found in soil. They employ an "appressorium-mediated penetration" strategy, producing appressoria to breach the insect's cuticle (Zimmermann 2007; Roberts and St Leger 2004). Once inside, Metarhizium fungi release enzymes and toxic metabolites, leading to the insect's demise by degrading its tissues and suppressing its immune response.

11.3.2.3 Cordyceps

Cordyceps fungi, including species like *Cordyceps militaris* and Cordyceps sinensis, infect various insect hosts and manipulate their behavior. They penetrate the insect's cuticle, grow within its body, and control its nervous system, compelling it to climb to an elevated location where the fungus emerges, forming a fruiting body to release spores strategically (Ng and Wang 2005).

11.3.2.4 Isaria

Isaria, with species like *Isaria fumosorosea* and *Isaria javanica,* infects diverse insect pests using blastospores to attach to the insect's cuticle (Kang et al. 2018). Once attached, blastospores germinate, penetrate the exoskeleton, and proliferate within the insect, producing toxic metabolites that cause its demise (Zhao et al. 2021).

11.3.2.5 Hirsutella

Hirsutella fungi, including species like *Hirsutella thompsonii* and *Hirsutella citriformis,* infect aphids, mites, and whiteflies using an "infection cushion" strategy (Minter and Brady 1980; Reddy et al. 2020). They form cushion-like structures on the insect's cuticle, releasing conidia to penetrate the exoskeleton (Seifert and Boulay 2004). Inside the insect, Hirsutella fungi produce toxic metabolites, disrupting physiological processes and causing its eventual demise (Table 11.2).

11.3.3 Ecological Roles of Endophytic Insect-Pathogenic Fungi

Endophytic insect-pathogenic fungi are crucial for ecosystem balance, contributing to natural pest control, nutrient cycling, and influencing interactions among organisms, underscoring their pivotal ecological roles (Parveen and Rashtrapal 2024).

11.3.3.1 Natural Pest Control

Endophytic insect-pathogenic fungi contribute significantly to natural pest control by employing sophisticated mechanisms to infect and eliminate specific hosts, thus

Table 11.2 Some common genera of endophytic, insect-pathogenic fungi (EIPF)

Genus	Family	Key characteristics	Examples
Beauveria	Cordycipitaceae	Broad host range, white to green conidia, infects via cuticle	*Beauveria bassiana*
Metarhizium	Clavicipitaceae	Broad host range, green to brown conidia, infects via cuticle	*Metarhizium anisopliae*
Ophiocordyceps	Ophiocordycipitaceae	Primarily infects ants and other social insects, mummifies host	*Ophiocordyceps unilateralis*
Lecanicillium	Clavicipitaceae	Primarily infects scale insects and mealybugs, white to cream conidia	*Lecanicillium muscatellum*
Paecilomyces	Cordycipitaceae	Diverse ecological roles, including insect pathogenicity and endophytism	*Paecilomyces varioti*
Acremonium	Clavicipitaceae	Diverse ecological roles, some species endophytic and insect-pathogenic	*Acremonium neomonospora*
Aschersonia	Aschersoniaceae	Primarily infects scale insects and other soft-bodied insects	*Aschersonia aleyrodis*

regulating pest populations in agricultural and forestry systems without reliance on chemical pesticides (Ahmad et al. 2020; Perumal et al. 2024). This targeted approach offers an environmentally friendly and sustainable alternative, mitigating economic losses while avoiding the negative impacts of chemical interventions.

11.3.3.2 Nutrient Cycling

By breaking down diseased insect corpses and releasing necessary nutrients back into the environment, endophytic insect-pathogenic fungi improve the recycling of organic matter and supply essential elements like nitrogen and phosphorus for the growth of plants and other organisms (Hu and Bidochka 2021). This technique helps overcome restrictions on ecosystem production in nutrient-poor areas, where it is especially important (Bamisile et al. 2023).

11.3.3.3 Interactions with Other Organisms

Endophytic insect-pathogenic fungi engage in both mutualistic and potentially negative interactions within ecosystems. Some insects benefit from mutualistic relationships with these fungi, using them for defense against insects or parasites (Nicoletti and Becchimanzi 2022). However, caution is needed as these fungi, although host-specific, can inadvertently harm non-target organisms like beneficial pollinators. Striking a balance to minimize unintended harm to non-target organisms is crucial (Francis et al. 2022).

11.3.3.4 Effects on Insect Populations and Communities

Endophytic insect-pathogenic fungi significantly impact insect populations and communities by regulating susceptible insect populations, influencing resource availability, and altering competitive dynamics within ecosystems (Yang et al. 2021). Changes in herbivorous insect populations can affect plant abundance, subsequently impacting resources for other herbivores. Additionally, these fungi can reshape insect community composition, impacting overall ecosystem structure (Fenoglio et al. 2021).

11.3.3.5 Implications for Agriculture and Forestry

Endophytic insect-pathogenic fungi offer eco-friendly alternatives to chemical pesticides in agriculture, reducing reliance on harmful compounds (Mohan et al. 2021). In forestry, they aid in managing destructive insect pests and protecting trees and forests. However, successful implementation requires considering factors like host specificity and environmental impacts (McCarl 2010). Ongoing research aims to optimize their application for effective and responsible pest management in diverse ecosystems.

11.3.4 Distribution and Habitat of Endophytic Insect-Pathogenic Fungi

Endophytic insect-pathogenic fungi are widely distributed across diverse habitats globally, including forests, grasslands, wetlands, and agricultural fields (Leger 2024). Their distribution spans temperate to tropical regions and is influenced by climate, host availability, and environmental conditions.

11.3.4.1 Distribution Patterns

The distribution of endophytic insect-pathogenic fungi varies depending on species and host insects: Some, like Beauveria, are widely distributed across continents, while others have more limited ranges (Delmas et al. 2024). Dispersal mechanisms, including wind or insect-carried spores, influence colonization abilities, shaping distribution patterns (Dietzel et al. 2019).

11.3.4.2 Habitat Preferences

Endophytic insect pathogenic fungi exhibit versatility in preferred habitats, thriving in natural and modified humans. They have been isolated from various substrates such as soil, plant roots, decomposing organic matter, insect carcasses, and micro-habitats such as leaf surfaces, insect cuticles and spines (Bidochka et al. 2002). In basic structures, they control insect populations, contribute to biodegradation in forests, and protect plants from herbivores in grasslands (Chaverri and Samuels 2013). These fungi also provide natural pests in agriculture and contribute to urban ecosystems such as gardens, parks, and green spaces (Bidochka et al. 2001).

11.3.4.3 Environmental Factors

Various environmental factors, including climate, moisture availability, soil type, nutrient availability, and pH, significantly influence the distribution and abundance of endophytic insect-pathogenic fungi (de Araujo et al. 2017; Bernardo-Cravo et al. 2020). Climate conditions, such as temperature and humidity, shape their growth and reproduction requirements, while moisture availability affects their adaptation to different environments. Soil characteristics, including type and nutrient content, also play a role, with some species exhibiting specific soil preferences.

11.3.4.4 Host Specificity

The distribution of endophytic insect-pathogenic fungi is closely linked to the availability of suitable host insects, with differing host preferences among fungal species contributing to their specific habitats (Li et al. 2020). While some fungi have broad host ranges, infecting various insect species across different orders like Metarhizium, others exhibit specialization, targeting specific insect groups such as ants or caterpillars (van der Does and Rep 2007). Habitat diversity and environmental conditions directly influence the availability of suitable host insects, with forests supporting a higher diversity of these fungi due to diverse insect hosts, while more homogenous insect populations in agricultural fields influence their distribution (Borah et al. 2018).

11.4 Mechanisms of Infection by Endophytic Insect-Pathogenic Fungi

Understanding the infection mechanisms of endophytic insect-pathogenic fungi is crucial for developing targeted interventions in pest management (Moonjely et al. 2016; Branine et al. 2019). Environmental factors like temperature, humidity, and host availability play pivotal roles in influencing these mechanisms (Table 11.3).

Table 11.3 Mechanisms of infection by endophytic insect-pathogenic fungi (EIPF)

Stage	Description	Details
Spore adhesion	Initial contact with insect	– EIPF spores (usually conidia) adhere to the insect's cuticle through hydrophobic interactions and specific recognition systems (lectins) – Spore surface features like hydrophobins and lectin-binding sites play a role
Spore germination and penetration	Fungal growth initiation and entry	– Spores germinate on the cuticle under favorable conditions (moisture, temperature) – Germ tube grows and secretes enzymes to degrade the insect cuticle – Mechanical pressure from the germ tube can also aid in penetration
Nutrient acquisition and growth	Fungal development within the insect	– Once inside, the fungus differentiates into yeast-like blastospores within the insect's hemolymph (blood) – Blastospores absorb nutrients from the insect's body fluids – Fungal mycelium may also grow throughout the insect's body cavity
Toxin production and insect mortality	Fungal attack and insect death	– Blastospores produce insecticidal toxins like beauvericin and destruxins – These toxins disrupt insect physiology, leading to paralysis, organ failure, and death
Nutrient recycling and dispersal	Fungal benefit and continuation of cycle	– After insect death, the fungus may further colonize the body and utilize its nutrients for growth and reproduction – New spores are produced on or within the dead insect, ready to disperse and infect new hosts
Endophytic colonization (concurrent with insect infection)	Fungal presence within plant tissues	– EIPF can establish a symbiotic relationship with plants by colonizing their internal tissues without causing harm – The fungus may receive nutrients from the plant while offering potential benefits like insect pest control – The exact mechanisms of plant colonization and the benefits to the plant are still being researched

11.4.1 Adaptations for Recognition of Insect Hosts

Endophytic insect-pathogenic fungi have evolved diverse adaptations for precise host recognition and infection, ensuring their viability and reproductive success (Barelli et al. 2016; Leger et al. 1997). This section explores the mechanisms and strategies employed by these fungi in host recognition.

11.4.1.1 Chemical Signaling

Endophytic insect-pathogenic fungi employ chemical signals, known as semiochemicals, as a pivotal adaptation for insect host recognition. These volatile compounds act as attractants, guiding insects towards fungal spores or mycelium, with varying specificity targeting species or offering a broader appeal. Furthermore, the fungi release semiochemicals that influence insect behavior, facilitating infection by manipulating feeding, mating, or oviposition behaviors (Ahmad et al. 2020).

11.4.1.2 Morphological Adaptations

Endophytic insect-pathogenic fungi employ morphological adaptations to enhance host recognition, utilizing specialized structures or appendages for physical attachment to the insect cuticle. Adhesive pads or hooks secure the fungus to the insect's body, increasing the likelihood of successful infection. Additionally, unique infection structures like appressoria or infection cushions enable these fungi to breach the insect's cuticle and initiate infection upon exposure to specific cues or the presence of the insect host (Butt et al. 2016; Lovett and St Leger 2017).

11.4.1.3 Enzymatic Adaptations

Endophytic insect-pathogenic fungi utilize diverse enzymatic adaptations to facilitate host recognition and infection. Key enzymes such as chitinases play a pivotal role in breaking down chitin in the insect cuticle, while proteases and lipases aid in degrading proteins and lipids within the insect's body. These enzymatic activities support nutrient acquisition and fungal growth, contributing to successful infection (Anwar et al. 2017; Sharma et al. 2020).

11.4.1.4 Immune Evasion

Endophytic insect-pathogenic fungi possess mechanisms to evade or suppress the insect's immune response, which is crucial for successful infection. They produce immunosuppressive compounds that inhibit the insect's immune system, preventing immune cell activation or antimicrobial peptide production. Additionally, some fungi modify their cell surface molecules to evade recognition, disguising themselves to hinder the immune response (Hu and Bidochka 2021; Boucias et al. 2018).

11.4.1.5 Genetic Adaptations

Genetic adaptations play a crucial role in host recognition by endophytic insect-pathogenic fungi. Genes and gene clusters, regulated by environmental cues or insect host presence, control the production of virulence factors and host recognition molecules. Genetic diversity within fungal populations leads to variations in host range and virulence, facilitating adaptation to diverse ecological niches and the exploitation of various insect hosts (Moonjely et al. 2016; Branine et al. 2019).

11.4.2 Infection Strategies of Endophytic Insect-Pathogenic Fungi

Genetic adaptations in endophytic insect-pathogenic fungi govern the production of virulence factors and host recognition molecules (Moonjely et al. 2016; Branine et al. 2019). These adaptations, influenced by environmental cues and host presence, contribute to variations in host range and virulence, enabling adaptation to diverse ecological niches and insect hosts.

11.4.2.1 Cuticle Penetration

Endophytic insect-pathogenic fungi penetrate insect cuticles using specialized structures like appressoria, overcoming this protective barrier to gain access to the insect's internal tissues (Hu and Bidochka 2021; Mora et al. 2018).

11.4.2.2 Infection Structure Formation

Upon cuticle penetration, fungi create infection structures like hyphal bodies or blastospores within the insect host, facilitating colonization and establishing a foothold for further growth (Ahmad et al. 2020).

11.4.2.3 Nutrient Acquisition

These fungi employ enzymatic mechanisms, including chitinases and other enzymes, to extract essential nutrients from insect tissues, supporting their growth and development (Jaber et al. 2018).

11.4.2.4 Immune Evasion

Despite the insect's immune defenses, fungi produce compounds that interfere with the immune system, inhibiting immune proteases and suppressing antimicrobial peptide production to establish successful infections (Zhang et al. 2021; Boucias et al. 2018).

11.4.2.5 Manipulation of Host Behavior

Certain fungi manipulate host behavior through bioactive compounds, inducing infected insects to exhibit behaviors that aid in the fungus's spread, such as climbing to elevated positions for increased fungus spread (Shang et al. 2015).

11.4.2.6 Horizontal Transmission

Fungi employ horizontal transmission strategies by producing dispersible spores with specialized structures that facilitate transfer to new hosts, ensuring their survival and spread within insect populations (Francis et al. 2022; Parveen and Rashtrapal 2024).

11.4.2.7 Covert Infection

Some fungi establish covert infections within hosts without causing visible symptoms, allowing them to persist undetected and potentially infect new hosts when conditions are favorable (Hu and Bidochka 2021; Francis et al. 2022).

11.4.3 Toxic Metabolites Produced by Endophytic Insect-Pathogenic Fungi

Endophytic insect-pathogenic fungi utilize various strategies, with a key mechanism being the production of toxic metabolites (Table 11.4). These compounds are crucial for host recognition, circumventing the insect immune system and ultimately causing host mortality (Samal et al. 2023).

11.4.3.1 Types of Toxic Metabolites

Secondary Metabolites: Endophytic insect-pathogenic fungi produce secondary metabolites like alkaloids, terpenoids, polyketides, and peptides, which play crucial roles in interactions with other organisms, such as acting as neurotoxins or cytotoxic agents (Madariaga-Mazón et al. 2019).

Alkaloids: Alkaloids are widely produced toxic compounds by these fungi, with diverse structures and activities, including neurotoxic and cytotoxic effects, leading to insect paralysis or cell damage (Matsuura and Fett-Neto 2015).

Table 11.4 Toxic metabolites produced by endophytic insect-pathogenic fungi (EIPF)

Metabolite class	Examples	Target	Mode of action	Effect on insect
Cyclic peptides	Beauvericin, Destruxins	Insect cell membranes	Disrupts membrane integrity and leads to leakage of cellular contents	Paralysis, disruption of organ function, death
Depsipeptides	Tenellins, Bassianins	Chitin synthesis	Inhibits the formation of insect cuticle, essential for molting and growth	Impaired molting, developmental failure, death
Polyketides	Oosporein, Spiculisporin	Not fully understood	May disrupt cell signaling pathways or have multiple effects	Reduced growth, developmental abnormalities, death
Enzymes	Chitinases, Proteases	Insect cuticle, proteins	Degrade insect cell wall and internal tissues, facilitate fungal invasion	Increased susceptibility to fungal colonization
Secondary metabolites	Beauverolides, Lolitrem B	Varied	May have insecticidal, deterrent, or immunosuppressive effects	Disrupts feeding behavior, weakens immune system, death

Terpenoids: Derived from isoprene units, terpenoids serve various biological functions, including acting as feeding deterrents and antimicrobials, aiding in insect control and competition inhibition (McGarvey and Croteau 1995).

Polyketides: Synthesized by polyketide synthase enzyme complexes, polyketides have diverse activities, including antifungal and insecticidal effects, contributing to fungal pathogenicity (Cano et al. 2018).

Peptides: Fungal peptides, synthesized by ribosomes, exhibit insecticidal, antimicrobial, and immunomodulatory properties, disrupting insect cell membranes or inhibiting essential proteases for digestion and development (Palma 2011).

Enzymes: Endophytic insect-pathogenic fungi produce enzymes like chitinases, proteases, and lipases, which degrade insect structural components, aiding in fungal penetration and nutrient acquisition within the host (Veliz et al. 2017).

Chitinases: Chitinases produced by endophytic insect-pathogenic fungi hydrolyze chitin, a key component of the insect exoskeleton, facilitating fungal penetration and colonization of host tissues (Veliz et al. 2017).

Endophytic insect-pathogenic fungi produce toxic metabolites such as secondary metabolites, peptides, and enzymes, which aid in efficient host recognition, immune system evasion, and successful infection.

11.4.3.2 Mode of Action of Toxic Metabolites

Endophytic insect-pathogenic fungi produce toxic metabolites that disrupt insect physiology, leading to mortality. These metabolites affect processes like the nervous system, cellular functions, and the insect's immune response, enhancing the fungi's pathogenicity (Wink and Schimmer 2010; Paschapur et al. 2021). Understanding their mode of action is vital for pest control and biotechnological applications.

Neurotoxic Effects: Toxic metabolites synthesized by endophytic insect-pathogenic fungi often act as neurotoxins, disrupting the insect's nervous system by interfering with neurotransmission, ion channels, or key enzymes. This disruption leads to outcomes like paralysis, convulsions, or death, contributing to the fungus's pathogenicity (Anamika et al. 2018).

Disruption of Cellular Processes: Certain toxic metabolites target specific cellular processes within the insect host, such as protein synthesis, DNA replication, or energy metabolism. By disrupting these essential cellular functions, the metabolites induce cellular dysfunction, ultimately leading to the insect's demise (Aliferis and Jabaji 2011).

Immune Suppression: Toxic metabolites produced by endophytic insect-pathogenic fungi can also suppress the immune response of the insect host (Banerjee 1999). These compounds inhibit immune cell activity or interfere with the production of immune-related molecules, allowing the fungus to evade the host's defense mechanisms and establish a successful infection.

Coevolutionary Arms Race: The synthesis of toxic metabolites reflects the ongoing coevolutionary arms race between endophytic insect-pathogenic fungi and their insect hosts. In response to insects' evolving defense mechanisms, fungi have developed diverse toxic compounds to overcome these defenses, leading to the diversification of toxic metabolites among fungal species (Boomsma et al. 2014).

This arms race influences the adaptation and pathogenicity of these fungi, shaping their interactions with insect hosts over time.

11.4.4 Interaction with Insect Immune System

Endophytic insect-pathogenic fungi employ sophisticated mechanisms to overcome the immune defenses of their insect hosts. Insects exhibit a complex immune system encompassing cellular and humoral responses to combat microbial infections (Hu and Bidochka 2021). Nevertheless, these fungi have evolved diverse strategies to elude or suppress these immune responses, ensuring successful infections and eventual host mortality (Nicoletti and Becchimanzi 2022).

11.4.4.1 Recognition and Avoidance of Immune Responses

Endophytic insect-pathogenic fungi employ mechanisms to evade detection by the insect immune system, such as producing enzymes that degrade crucial components like antimicrobial peptides (Bamisile et al. 2023). By degrading these molecules, fungi can remain undetected within the insect, avoiding triggering an immune response and ensuring successful infection (Schmid-Hempel 2005).

11.4.4.2 Suppression of Cellular Immune Responses

Endophytic insect-pathogenic fungi counter cellular immune responses by producing toxins targeting and killing hemocytes, hindering their ability to eliminate fungal infections. Some fungi manipulate hemocyte behavior, reducing their efficiency in recognizing and engulfing fungal spores (Bamisile et al. 2023; Hu and Bidochka 2021).

11.4.4.3 Inhibition of Humoral Immune Responses

Endophytic insect-pathogenic fungi inhibit humoral immune responses by synthesizing or activating antimicrobial peptides and other immune molecules. They produce enzymes that degrade antimicrobial peptides, making them ineffective against fungal infections (Dietzel et al. 2019). In addition, some fungi interfere with signaling pathways that regulate the immune system, effectively suppressing humoral immune responses (Vilcinskas et al. 1999). Analysis of interactions between *Metarhizium* spp. Our understanding of the relationship between fungal infection and resistance has increased for bees. Metarhizium hyphae that have penetrated the intestinal mucosa transform into yeast-like hyphal bodies (HBs) to form the hemocoel of the host. Furthermore, fungal cell walls regenerate to evade antimicrobial resistance. In addition, released HB enzymes and metabolites suppress the host immune system or interfere with host nutrition. These mechanisms help the fungus form the haemocoel and settle well. Enzymes produced by the hyphal bodies in the hemolymph inhibit proteins associated with the immune system and hydrolyze disaccharides to provide nutrients (Fig. 11.3).

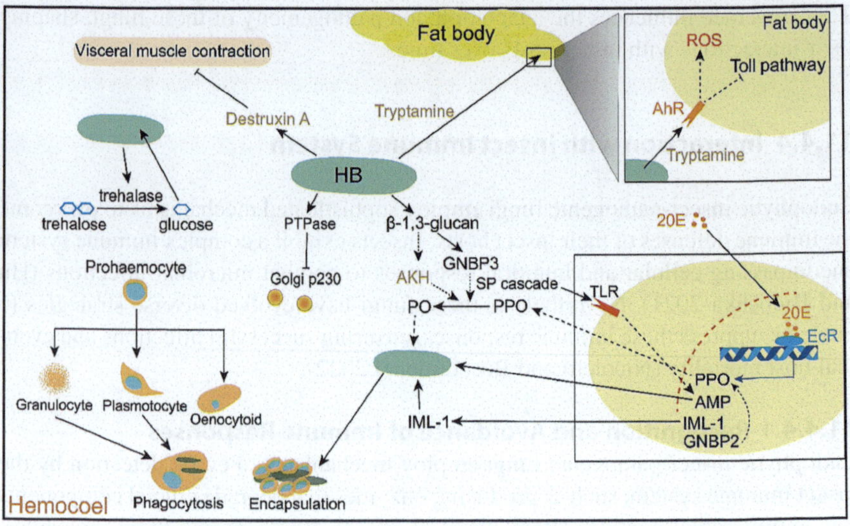

Fig. 11.3 Late-phase interactions between Metarhizium species and locusts. The formation of yeast-like hyphal bodies (HB) in locust hemolymph depends on food consumption. Metarhizium secretes proteins and metabolites that suppress humoral and cellular immunity. Ecdysone and Toll pathways inhibit HB development. The dotted arrows, or T, reflect indirect activities that promote or inhibit another component. Solid arrows or T indicate direct promotion or restriction of another component. *20E* 20-Hydroxyecdysone, *EcR* ecdysone receptor, *TLR* Toll-like receptor, *IML-1* immunlectin-1, *GNBP3* Gram-negative bacteria binding protein 3, *AhR* aryl hydrocarbon receptor, *SP* serine protease, *Serpin* serine protease inhibitor, *ROS* reactive oxygen species, *PPO* prophenoloxidase, *PO* phenoloxidase, *AMP* antimicrobial peptides, *PTPase* protein tyrosine phosphatase (Reprinted from Li and Xia 2022)

11.4.4.4 Immune Evasion Through Surface Modifications

Endophytic insect-pathogenic fungi employ surface modifications to evade recognition and attack by the insect immune system. By producing surface molecules mimicking those found in insect cells, they effectively camouflage themselves (Brivio and Mastore 2020; Orapiriyakul et al. 2018). This disguising mechanism hinders immune cells from distinguishing between self and non-self, facilitating the establishment of a successful infection.

11.4.4.5 Manipulation of Host Immune Signaling

Endophytic insect-pathogenic fungi have the ability to manipulate the host's immune signaling pathways to their advantage. They produce molecules that mimic or interfere with the signaling molecules involved in immune responses, effectively modulating the host's immune system (Dietzel et al. 2019; Schmid-Hempel 2005). Through this manipulation, these fungi create an environment conducive to their growth and reproduction while concurrently inhibiting the host's immune defenses.

11.4.4.6 Co-evolutionary Arms Race

The interaction between endophytic insect-pathogenic fungi and the insect immune system represents a co-evolutionary arms race. As fungi evolve strategies to overcome host immune defenses, insects, in turn, develop new mechanisms to recognize and eliminate fungal infections (Joop and Vilcinskas 2016). This dynamic relationship has led to the diversification and specialization of both the insect immune system and the virulence factors of the fungi, shaping their ongoing evolution (Cummins and Huitema 2017).

11.4.5 Host Specificity of Endophytic Insect-Pathogenic Fungi

Host specificity is vital in the interaction between endophytic insect-pathogenic fungi and their hosts, indicating their ability to infect specific insect species (Branine et al. 2019). Understanding this aids in effective biological control and ecological understanding (Moonjely et al. 2016).

11.4.5.1 Factors Influencing Host Specificity

Host specificity in endophytic insect-pathogenic fungi results from genetic, physiological, and ecological factors influencing their ability to recognize and infect specific insect hosts. It includes:

11.4.5.1.1 Genetic Factors

Genetic factors strongly impact host specificity in endophytic insect-pathogenic fungi, shaping their ability to produce enzymes and toxins essential for infecting insect hosts. Unique genetic traits evolve, enabling fungi to target specific insect species (Li et al. 2020).

11.4.5.1.2 Physiological Factors

Physiological factors in both fungi and insect hosts influence host specificity. Fungi must adapt to insect hosts' physiological conditions like temperature and pH for infection success. Similarly, insect traits such as cuticle composition and immune response affect susceptibility to fungal infection (Bernays and Lee 1988).

11.4.5.1.3 Ecological Factors

Ecological factors, including habitat, seasonality, and the presence of other organisms, impact host specificity by influencing the availability and suitability of potential hosts for fungi. Certain fungi may specialize in infecting insects in specific ecological niches or with particular behaviors.

11.4.5.2 Mechanisms of Host Recognition

Endophytic insect-pathogenic fungi utilize various mechanisms to recognize and infect specific insect hosts, relying on cues and molecules to facilitate infection. Comprehending these mechanisms is crucial for developing efficient biological control strategies (Branine et al. 2019).

11.4.5.2.1 Chemical Cues

Chemical cues are integral for host recognition in endophytic insect-pathogenic fungi, eliciting responses that enable fungi to locate and infect preferred hosts (Milutinović and Schmitt 2022). These cues, derived from the insect or environment, often induce the production of specific enzymes and toxins facilitating infection.

11.4.5.2.2 Cuticle Penetration

Endophytic insect-pathogenic fungi employ various mechanisms to breach the insect cuticle, a vital step for successful infection. These include enzyme production to degrade the cuticle and secretion of adhesive substances to aid attachment and penetration, ensuring effective colonization (Ortiz-Urquiza and Keyhani 2013).

11.4.5.2.3 Immune Evasion

To establish successful infection within the insect host, endophytic insect-pathogenic fungi employ strategies to evade or suppress the immune response. These include producing immunosuppressive molecules and manipulating the host's immune signaling pathways, enabling fungi to persist and proliferate (Bowie and Unterholzner 2008).

11.4.5.3 Host Range and Specificity

Endophytic insect-pathogenic fungi exhibit varied host ranges, ranging from generalists infecting diverse hosts to specialists targeting specific species (Moonjely et al. 2016). This diversity impacts biological control and ecological dynamics.

11.4.5.3.1 Generalist Fungi

Generalist endophytic insect-pathogenic fungi have broad host ranges, utilizing diverse enzymes and toxins to target multiple insect species (Branine et al. 2019). They offer the potential for biological control against various pests, but their effectiveness may vary.

11.4.5.3.2 Specialist Fungi

Specialist endophytic insect-pathogenic fungi are highly specific to certain insect hosts, employing intricate mechanisms for host recognition. While effective against specific pest species, they may have minimal impact on non-target insects (Hu and Bidochka 2021).

11.4.5.4 Implications for Biological Control

Understanding host specificity in endophytic insect-pathogenic fungi is crucial for effective biological control. Selecting fungi with suitable host ranges minimizes non-target impacts, while insights into host recognition mechanisms aid in formulating effective applications (Jaber et al. 2018). Ongoing research is necessary to address potential shifts in host specificity and ensure sustained efficacy in biological control strategies.

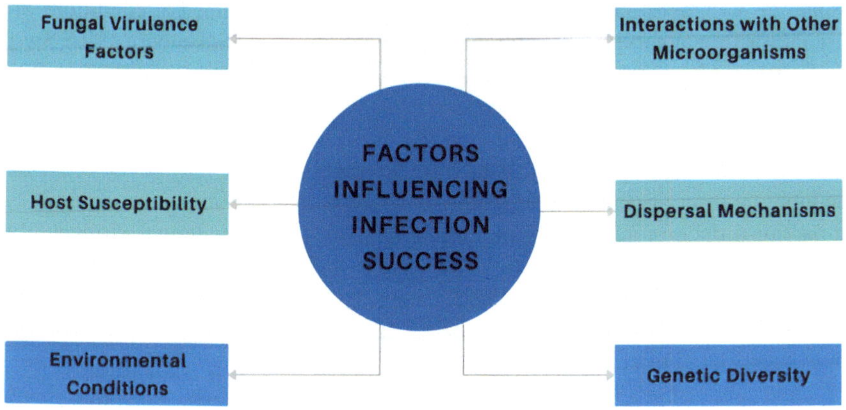

Fig. 11.4 Different factors are responsible for the success of entomopathogenic fungi infections in insects

11.4.6 Factors Affecting the Success of Infection

Understanding factors influencing the success of infections by endophytic insect-pathogenic fungi is vital for pest management. These factors encompass intrinsic aspects related to the fungus itself and extrinsic factors associated with the environment and host insects (Fig. 11.4). Grasping these elements is essential for leveraging fungal applications effectively in pest management strategies (Hu and Bidochka 2021; Bamisile et al. 2023).

11.4.6.1 Virulence Factors of the Fungus

Endophytic insect-pathogenic fungi possess diverse virulence factors facilitating insect infection, including toxic metabolites, cuticle-degrading enzymes, and immune evasion mechanisms (Sahab and Sabbour 2011). Variability in virulence among fungal species influences infection success, with some producing potent insecticidal secondary metabolites like destruxins and beauvericin, enhancing virulence and infection outcomes (Valero-Jiménez et al. 2016).

11.4.6.2 Susceptibility of the Host

The success of fungal infection depends on the susceptibility of the host insect, influenced by factors like natural defenses, immune response efficiency, and physiological characteristics. Some insect species exhibit high susceptibility to specific fungal pathogens, while others show resistance or tolerance (Graham et al. 2004). Environmental conditions such as temperature and humidity also impact host susceptibility by affecting fungal growth and activity.

11.4.6.3 Environmental Conditions

Environmental conditions significantly influence the success of endophytic insect-pathogenic fungal infections. Variations in temperature, humidity, and light impact fungal growth, sporulation, and infectivity. Suitable environmental conditions are crucial for fungal survival, proliferation, and host infection (Stromberg 1997; Swift et al. 2017). Changes driven by climate or habitat alteration can profoundly affect fungal distribution and infection success.

11.4.6.4 Interactions with Other Microorganisms

Endophytic insect-pathogenic fungi interact within diverse microbial communities, impacting infection success. Some microorganisms enhance fungal virulence by producing metabolites synergistic with fungal toxins, while others compete for resources, constraining fungal growth (Casadevall and Pirofski 2000). Understanding these interactions is crucial for optimizing the use of these fungi in pest management strategies (Fuller and Freter 1992).

11.4.6.5 Dispersal Mechanisms

Endophytic insect-pathogenic fungi rely on diverse dispersal mechanisms for infection success. Aerial conidia, lightweight spores, travel through air currents, enabling long-distance dispersal (Fitt et al. 2006; Matthysen and Clobert 2012). Spore-bearing structures release spores near infected hosts, enhancing the chances of infection. Insect-mediated transmission involves infected insects acting as vectors, spreading fungal spores to healthy individuals. The efficiency of these mechanisms is critical for infection success.

11.4.6.6 Genetic Diversity

Genetic diversity within populations of endophytic insect-pathogenic fungi significantly impacts infection success by influencing virulence, adaptability, and host range (Turner et al. 2011). High genetic diversity enhances the likelihood of encountering favorable traits, aiding in overcoming host defenses and environmental challenges. Additionally, genetic diversity allows for adaptive evolution in response to changing conditions or host populations (Ekroth et al. 2021).

11.5 Ecological Impacts of Endophytic Insect-Pathogenic Fungi

Endophytic insect-pathogenic fungi exert multifaceted ecological effects beyond direct interactions with insect hosts, impacting broader ecosystem dynamics (Barelli et al. 2016). Understanding these effects is crucial for assessing their ecological roles and potential applications.

11.5.1 Role in Natural Pest Control

Endophytic insect-pathogenic fungi serve as natural adversaries of insect pests, employing unique mechanisms to infect and eliminate their hosts. Effective biocontrol agents contribute to regulating pest populations and maintaining ecological balance by selectively targeting specific insect species (Branine et al. 2019).

11.5.1.1 Infection Dynamics

Endophytic insect-pathogenic fungi are effective in natural pest control due to their ability to infect and eliminate insect pests. They employ sophisticated strategies, including specialized structures like spores or hyphae that penetrate the insect's cuticle or access natural openings (May et al. 2001; Sasmal et al. 2016). Once inside the host, fungal proliferation leads to the insect's demise.

11.5.1.2 Host Specificity

Endophytic insect-pathogenic fungi exhibit host specificity, targeting specific insect hosts for efficient pest control while minimizing harm to non-target organisms (Brodeur 2012; Van Klinken 1999). This characteristic ensures precise pest targeting, reducing pest populations without adversely affecting beneficial insects or the ecosystem.

11.5.1.3 Impact on Pest Populations

Endophytic insect-pathogenic fungi can regulate pest populations by initiating outbreaks under conducive conditions, leading to declines in pest numbers and reducing reliance on chemical pesticides (Naylor and McShane 1997).

11.5.1.4 Interactions with Other Organisms

Endophytic insect-pathogenic fungi contribute to natural pest control by forming symbiotic relationships with plants, indirectly protecting them from insect pests (Wilby 2022). Additionally, these fungi interact with other natural enemies of pests, such as predators and parasitoids, enhancing overall pest control efforts (Naylor and McShane 1997; Francis et al. 2022).

11.5.1.5 Implications for Agriculture and Forestry

Endophytic insect-pathogenic fungi offer a sustainable alternative to chemical pesticides in agriculture and forestry, mitigating environmental and health risks associated with pesticide use. Integrating these fungi into pest management strategies promotes the conservation of beneficial organisms and fosters sustainable agricultural practices (Furlan and Kreutzweiser 2015).

11.5.1.6 Future Directions

Research on endophytic insect-pathogenic fungi should delve into factors affecting their biocontrol effectiveness, such as environmental conditions and interactions (Herz et al. 2019; Chidawanyika et al. 2012). Genetic studies may reveal mechanisms driving pathogenicity and host specificity, enhancing targeted pest management. Understanding host specificity, shaped by genetic, physiological, and ecological elements, is crucial for efficient biological control applications (Jeger 1999).

11.5.2 Interactions with Other Organisms in Ecosystems

Endophytic insect-pathogenic fungi interact extensively with organisms beyond their insect hosts, including plants, animals, and microorganisms, shaping ecosystem dynamics significantly (Nicoletti and Becchimanzi 2022). These interactions have profound implications for ecosystem functioning. This section delves into the diverse interactions of these fungi and their consequences, emphasizing their role in maintaining ecological balance (Branine et al. 2019).

11.5.2.1 Interactions with Plants

Endophytic insect-pathogenic fungi form nuanced relationships with plants, offering both benefits and drawbacks. While some fungi provide protection against herbivores through deterrent metabolites, others induce diseases in plants, affecting their health and productivity. Understanding these interactions is crucial for effective plant disease management and ecological balance (Hu and Bidochka 2021; Bamisile et al. 2023).

11.5.2.2 Interactions with Animals

Endophytic insect-pathogenic fungi influence various animal species directly and indirectly, impacting behavior, population dynamics, and community structures. These fungi alter pathogen-prey dynamics among insects and indirectly affect vertebrate animals consuming infected insects. The presence of these fungi can shape the behavior and distribution of vertebrates by influencing insect prey availability (Moonjely et al. 2016; Barelli et al. 2016).

11.5.2.3 Interactions with Microorganisms

Endophytic insect-pathogenic fungi engage in intricate relationships with microorganisms, involving both competitive and cooperative interactions (Braga et al. 2016). They compete with other microorganisms for resources while forming cooperative associations with beneficial bacteria or fungi, enhancing their ability to colonize and infect insect hosts.

11.5.2.4 Implications for Ecosystem Functioning

Interactions between endophytic insect-pathogenic fungi and other organisms significantly influence ecosystem functioning. These fungi regulate insect populations, indirectly affecting plant communities and higher trophic levels (Hooper et al. 2005). Moreover, their presence influences nutrient cycling and carbon sequestration processes, impacting ecosystem dynamics and stability. Understanding these interactions is crucial for sustainable agriculture, conservation, and ecosystem management (Goudard and Loreau 2008; Lundberg and Moberg 2003).

11.5.3 Effects on Insect Populations and Communities

Endophytic insect-pathogenic fungi play a crucial role in shaping insect populations and communities, significantly impacting population dynamics and community structure (Agrawal and Maron 2022).

11.5.3.1 Population Regulation

Endophytic insect-pathogenic fungi significantly regulate insect populations by causing considerable mortality, influenced by factors such as fungal virulence, host susceptibility, and environmental conditions (Berryman et al. 1987; Harcourt and Leroux 1967). The impact varies across insect species and life stages, inducing changes in population dynamics.

11.5.3.2 Indirect Effects on Insect Communities

Endophytic insect-pathogenic fungi's influence extends to indirect impacts on insect communities, disrupting interactions and dynamics due to substantial reductions in specific insect species (van Veen et al. 2005; Filazzola et al. 2021). This disruption may lead to shifts in community composition and structure, potentially affecting other ecosystem organisms (Strauss 1987).

11.5.3.3 Trophic Interactions

These fungi significantly affect trophic interactions within insect communities by providing infected insects as food sources for predators, parasitoids, and scavengers (Srivastava 2006). Parasitoid wasps, for instance, utilize infected insects as hosts, influencing dynamics between insects and parasitoids. Additionally, the fungi's presence can alter the foraging behavior of predators and parasitoids, impacting their distribution and abundance (Tscharntke et al. 2005).

11.5.3.4 Ecological Succession

Endophytic insect-pathogenic fungi contribute to ecological succession within insect populations and communities by creating ecological niches through the decline of heavily impacted insect populations (Coombs and Woodroffe 1963; Schowalter 1981). This succession process, characterized by changes in species composition and abundance, influences ecosystem functioning and stability.

11.5.4 Implications for Agriculture and Forestry

Endophytic insect-pathogenic fungi have significant implications for agriculture and forestry, including effective insect pest management, promotion of sustainable practices, and contributions to ecosystem health (Stamps and Linit 1997; McCarl 2010). This section explores their diverse implications in agricultural and forestry contexts (Table 11.5).

Table 11.5 Applications of endophytic insect-pathogenic fungi (EIPF) in agriculture and forestry

Application	Description	Benefits	Challenges
Biological pest control	EIPF spores are formulated into biopesticides and applied to crops or forestry plantations	– Reduced reliance on chemical insecticides – Environmentally friendly approach to pest management – Broad host range for some EIPF species – Potential for long-term pest suppression through endophytic colonization of plants	– Lower efficacy compared to some chemical insecticides in certain situations – Environmental factors like moisture and UV light can affect spore viability and performance – Specificity of some EIPF species may limit their application against certain pests – Regulatory hurdles for commercialization of biopesticides
Plant disease control	Some EIPFs have been shown to induce plant defense mechanisms or directly antagonize plant pathogens	– Reduced reliance on fungicides for disease control – Potential for integrated pest management (IPM) strategies	– Mechanisms of plant disease control by EIPF are not fully understood for all species – Identifying and selecting effective EIPF strains for specific diseases requires further research
Plant growth promotion	Certain EIPFs may enhance plant growth through mechanisms like nutrient mobilization or hormone production	– Increased crop yields and improved plant health – Reduced need for chemical fertilizers in some cases	– The mechanisms of plant growth promotion by EIPF are still being elucidated – Consistency of growth promotion effects across different plant species and environments needs further investigation
Forestry protection	EIPF applications can target defoliating insects and other pests in forest ecosystems	– Reduced insect damage to trees, leading to healthier forests – Potential for sustainable forest management practices	– Large-scale application of EIPF in forests can be challenging due to logistical and economic factors – Long-term monitoring of EIPF impact on forest ecosystems is necessary

11.5.4.1 Biological Control of Insect Pests

Endophytic insect-pathogenic fungi offer a promising avenue for biological control of insect pests in agriculture and forestry (Mantzoukas and Eliopoulos 2020). These fungi, adept at infecting and eliminating a wide range of pests, provide an environmentally friendly alternative to chemical pesticides (Stamps and Linit 1997; Garnas et al. 2012). By establishing symbiotic relationships with host plants, they infect pests when consumed, minimizing impacts on beneficial insects and non-target organisms.

11.5.4.2 Enhanced Plant Health and Growth

Endophytic insect-pathogenic fungi contribute to improved plant health and growth by enhancing nutrient uptake, water absorption, and stress tolerance (Gullino et al. 2022). Through mutualistic relationships with host plants, these fungi provide various benefits, including enzymatic breakdown of organic compounds, solubilization of minerals, and production of growth-promoting substances (Verma et al. 2019; Pautasso et al. 2010).

11.5.4.3 Reduced Environmental Impact

Utilizing endophytic insect-pathogenic fungi in agriculture and forestry reduces environmental impacts compared to conventional pest control methods (García-González et al. 2020). By minimizing reliance on chemical pesticides, these fungi help preserve biodiversity, safeguard populations of beneficial insects, and mitigate risks to human health and the environment.

11.5.4.4 Economic Benefits

Adopting endophytic insect-pathogenic fungi translates into economic benefits by reducing costs associated with chemical pesticides, increasing crop yields, and improving product quality. Additionally, the potential for developing biopesticides and biocontrol products offers opportunities for innovation and sustainable pest management solutions (Knickel and Maréchal 2018).

11.6 Conservation and Management of Endophytic Insect-Pathogenic Fungi

The conservation and effective management of endophytic insect-pathogenic fungi are critical for preserving their ecological roles, ensuring sustainable pest control, and harnessing their potential for various applications (Moonjely et al. 2016). In this section, we explore strategies and considerations for the conservation and management of these fungi.

11.6.1 Strategies for the Conservation of Endophytic Insect-Pathogenic Fungi

Endophytic insect-pathogenic fungi constitute a captivating group of microorganisms pivotal for natural insect pest control. These fungi exhibit a unique ability to reside within plant tissues without causing harm to their host (Sharma et al. 2020). However, upon contact with an insect or insect fed on host plants, the fungi swiftly infect and eliminate the pest, offering a natural and sustainable form of pest control (Pathan et al. 2021).

Conserving these fungi is imperative to secure their long-term survival and efficacy in pest management. The strategies outlined below aim to safeguard and promote their habitats, integrate them into agricultural and forestry practices, and reduce reliance on chemical pesticides (Mishra 2023).

11.6.1.1 Preservation of Natural Habitats

Conservation of the natural habitats of endophytic insect pathogenic fungi is important to maintain their biodiversity and ensure their long-term survival (Segoli et al. 2020; Rotchés-Ribalta et al. 2021). Efforts should focus on protecting natural vegetation, reducing habitat degradation caused by human activities, and maintaining plant diversity to support species-specific host preferences of fungi. These approaches and conservation measures are necessary to maintain the biodiversity and biological roles of endophytic insect pathogens in the ecosystem (Kurniawan et al. 2021).

11.6.1.2 Integration into Agricultural and Forestry Systems

The use of endophytic fungi in agriculture and forestry requires the adoption of practices that support their growth and reduce the use of synthetic pesticides (Clauser et al. 2021). Integrated pest management (IPM) incorporating these fungi provides effective pest control through environmental reduction (Dalla Corte et al. 2020). Agricultural practices such as crop rotation, intercropping, and cover cropping increase plant diversity, which produces endophytic fungi (Aguiar et al. 2020). Reduced tillage and organic farming promote good soil health and microbial diversity, which helps these worms. In forests, natural ecosystem conservation and sustainable logging practices maintain endophytic fungal diversity by providing suitable habitats, such as deadwood preservation and the growth of tree species.

11.6.1.3 Ethical Considerations

Ethical considerations are paramount in the utilization of endophytic insect-pathogenic fungi for pest control, aiming to safeguard non-target organisms (Jacobsen et al. 2012). Rigorous monitoring and assessment are vital to mitigate unintended harm to beneficial insects, birds, and mammals (Toner et al. 2021). Additionally, adherence to scientific research and evidence-based practices is crucial to prevent indiscriminate use and comprehend ecological interactions and associated risks.

11.6.1.4 Research and Conservation Priorities

Research priorities include understanding biology, ecology, interactions, taxonomy, genetics, and application techniques for effective conservation and management (Heilmann-Clausen et al. 2015). Research efforts should prioritize:

1. **Taxonomy and Identification:** Enhancing taxonomy aids in identifying new species and understanding the ecological roles of endophytic insect-pathogenic fungi (Hibbett et al. 2016).
2. **Host Specificity:** Investigating host specificity informs biocontrol potential and impact on non-target organisms for endophytic insect-pathogenic fungi (Zhou and Hyde 2001).
3. **Ecological Interactions:** Studying ecological interactions elucidates endophytic fungi's role in pest control and biodiversity impact (Perotto et al. 2013).
4. **Conservation Genetics:** Assessing genetic diversity informs conservation strategies for endophytic fungi (Douhan et al. 2011).
5. **Application Techniques:** Efficient application techniques boost endophytic fungi's biocontrol efficacy (Aslam et al. 2017). Prioritizing research and conservation ensures their sustainable use (Anderson and Cairney 2004).

11.7 Challenges and Future Directions

Endophytic insect pathogenic fungi offer a promising solution for pest control in agriculture and forestry, but challenges remain. The development of effective implementation strategies is important for their practical implementation (Khandel and Shahi 2018; Pathak et al. 2021). Understanding the interactions between fungi, host plants and insects is essential for adaptation (Baggs 2011). Educating stakeholders on the benefits of these fungi and appropriate management techniques is essential for widespread adoption (Deng et al. 2017). Overall, further research and practical applications are needed to fully realize the potential of these fungi in agriculture and forestry (Baggs 2011).

11.8 Conclusion

Endogenous insect pathogens are an important component of natural parasites, contributing to ecosystem conservation and ecosystem functioning, and their unique characteristics of global distribution and transformation strategies species, including sustainable alternatives to pesticides, emphasize the importance of effective conservation management. These fungi exhibit sophisticated strategies for recognizing insect hosts and pathogens, allowing them to efficiently exploit hosts through various strategies. Their research provides valuable insights into microbial interactions that lead to improvements in another pest control method. In the family Ascomycota and the Sordariomycetes group, factors such as climate and the availability of hosts determine their distribution pattern and habitat preferences. The inclusion of these fungi in pest control methods reduces their reliance on pesticides, their antimicrobial effects and chemical synthetic chemicals for environmental promotion stewardship in agriculture and forestry development, including extensive biotechnological applications. Overall, the integration of endophytic insect-pathogenic fungi into pest management practices holds promise for environmental protection, enhancement of biological resistance, and promoting sustainable agroforestry practices.

References

Agrawal AA, Maron JL (2022) Long-term impacts of insect herbivores on plant populations and communities. J Ecol 110(12):2800–2811

Aguiar AS, Dos Santos FN, Cunha JB, Sobreira H, Sousa AJ (2020) Localization and mapping for robots in agriculture and forestry: a survey. Robotics 9(4):97

Aguilar-Trigueros CA, Rillig MC (2016) Effect of different root endophytic fungi on plant community structure in experimental microcosms. Ecol Evol 6(22):8149–8158

Ahmad I, Jiménez-Gasco MDM, Barbercheck ME (2020) The role of endophytic insect-pathogenic fungi in biotic stress management. Plant Stress Biol: Strategies Trends:379–400

Aliferis KA, Jabaji S (2011) Metabolomics—a robust bioanalytical approach for the discovery of the modes-of-action of pesticides: a review. Pestic Biochem Physiol 100(2):105–117

Alsohaili SA, Bani-Hasan BM (2018) Morphological and molecular identification of fungi isolated from different environmental sources in the Northern Eastern desert of Jordan. Jordan J Biol Sci 11(3)

Anamika JS, Sahgal M, Sahu S, Prakash A (2018) Fungal endophytes and their secondary metabolites: role in sustainable agriculture. In: Fungi and their role in sustainable development: current perspectives, pp 121–146

Anderson IC, Cairney JW (2004) Diversity and ecology of soil fungal communities: increased understanding through the application of molecular techniques. Environ Microbiol 6(8):769–779

Anwar W, Shahid AA, Haider MS (2017) Entomopathogenic fungi: introduction, history, classification, infection mechanism, enzymes, and toxins. In: Biopesticides and bioagents. Apple Academic Press, pp 101–151

Araújo JP, Hughes DP (2016) Diversity of entomopathogenic fungi: which groups conquered the insect body? Adv Genet 94:1–39

Arendt KR, Hockett KL, Araldi-Brondolo SJ, Baltrus DA, Arnold AE (2016) Isolation of endohyphal bacteria from foliar Ascomycota and in vitro establishment of their symbiotic associations. Appl Environ Microbiol 82(10):2943–2949

Aslam S, Tahir A, Aslam MF, Alam MW, Shedayi AA, Sadia S (2017) Recent advances in molecular techniques for the identification of phytopathogenic fungi–a mini review. J Plant Interact 12(1):493–504

Baggs EM (2011) Soil microbial sources of nitrous oxide: recent advances in knowledge, emerging challenges and future direction. Curr Opin Environ Sustain 3(5):321–327

Baldrian P, López-Mondéjar R (2014) Microbial genomics, transcriptomics and proteomics: new discoveries in decomposition research using complementary methods. Appl Microbiol Biotechnol 98:1531–1537

Bamisile BS, Afolabi OG, Siddiqui JA, Xu Y (2023) Endophytic insect pathogenic fungi-host plant-herbivore mutualism: elucidating the mechanisms involved in the tripartite interactions. World J Microbiol Biotechnol 39(12):326

Banerjee BD (1999) The influence of various factors on immune toxicity assessment of pesticide chemicals. Toxicol Lett 107(1–3):21–31

Barelli L, Moonjely S, Behie SW, Bidochka MJ (2016) Fungi with multifunctional lifestyles: endophytic insect pathogenic fungi. Plant Mol Biol 90:657–664
Behie SW, Bidochka MJ (2014) Ubiquity of insect-derived nitrogen transfer to plants by endophytic insect-pathogenic fungi: an additional branch of the soil nitrogen cycle. Appl Environ Microbiol 80(5):1553–1560
Behie SW, Jones SJ, Bidochka MJ (2015) Plant tissue localization of the endophytic insect pathogenic fungi Metarhizium and Beauveria. Fungal Ecol 13:112–119
Bernardo-Cravo AP, Schmeller DS, Chatzinotas A, Vredenburg VT, Loyau A (2020) Environmental factors and host microbiomes shape host–pathogen dynamics. Trends Parasitol 36(7):616–633
Bernays EA, Lee JC (1988) Food aversion learning in the polyphagous grasshopper Schistocerca americana. Physiol Entomol 13(2):131–137
Berryman AA, Stenseth NC, Isaev AS (1987) Natural regulation of herbivorous forest insect populations. Oecologia 71:174–184
Bhadauria V, Popescu L, Zhao WS, Peng YL (2001) Fungal transcriptomics. Microbiol Res 162(4):285–298
Bidochka MJ, Kamp AM, Lavender TM, Dekoning J, De Croos JA (2001) Habitat association in two genetic groups of the insect-pathogenic fungus Metarhizium anisopliae: uncovering cryptic species? Appl Environ Microbiol 67(3):1335–1342
Bidochka MJ, Menzies FV, Kamp AM (2002) Genetic groups of the insect-pathogenic fungus Beauveria bassiana are associated with habitat and thermal growth preferences. Arch Microbiol 178:531–537
Boomsma JJ, Jensen AB, Meyling NV, Eilenberg J (2014) Evolutionary interaction networks of insect pathogenic fungi. Annu Rev Entomol 59:467–485
Borah N, Albarouki E, Schirawski J (2018) Comparative methods for molecular determination of host-specificity factors in plant-pathogenic fungi. Int J Mol Sci 19(3):863
Boucias DG, Zhou Y, Huang S, Keyhani NO (2018) Microbiota in insect fungal pathology. Appl Microbiol Biotechnol 102:5873–5888
Bowie AG, Unterholzner L (2008) Viral evasion and subversion of pattern-recognition receptor signalling. Nat Rev Immunol 8(12):911–922
Braga RM, Dourado MN, Araújo WL (2016) Microbial interactions: ecology in a molecular perspective. Braz J Microbiol 47:86–98
Branine M, Bazzicalupo A, Branco S (2019) Biology and applications of endophytic insect-pathogenic fungi. PLoS Pathog 15(7):e1007831
Brivio MF, Mastore M (2020) When appearance misleads: The role of the entomopathogen surface in the relationship with its host. Insects 11(6):387
Brodeur J (2012) Host specificity in biological control: insights from opportunistic pathogens. Evol Appl 5(5):470–480
Butt TM, Goettel MS (2000) Bioassays of entomogenous fungi. CABI
Butt TM, Coates CJ, Dubovskiy IM, Ratcliffe NA (2016) Entomopathogenic fungi: new insights into host–pathogen interactions. Adv Genet 94:307–364
Cano PM, Puel O, Oswald IP (2018) Mycotoxins: fungal secondary metabolites with toxic properties. In: Fungi. CRC Press, pp 318–371
Casadevall A, Pirofski LA (2000) Host-pathogen interactions: basic concepts of microbial commensalism, colonization, infection, and disease. Infect Immun 68(12):6511–6518
Chaverri P, Samuels GJ (2013) Evolution of habitat preference and nutrition mode in a cosmopolitan fungal genus with evidence of interkingdom host jumps and major shifts in ecology. Evolution 67(10):2823–2837
Chidawanyika F, Mudavanhu P, Nyamukondiwa C (2012) Biologically based methods for pest management in agriculture under changing climates: challenges and future directions. Insects 3(4):1171–1189
Clauser NM, Felissia FE, Area MC, Vallejos ME (2021) A framework for the design and analysis of integrated multi-product biorefineries from agricultural and forestry wastes. Renew Sust Energ Rev 139:110687

Clifton EH, Castrillo LA, Jaronski ST, Hajek AE (2023) Cryptic diversity and virulence of Beauveria bassiana recovered from Lycorma delicatula (spotted lanternfly) in eastern Pennsylvania. Front Insect Sci 3:1127682
Coombs CW, Woodroffe GE (1963) An experimental demonstration of ecological succession in an insect population breeding in stored wheat. J Anim Ecol 32(2):271–279
Cummins M, Huitema E (2017) Effector–Decoy pairs: another countermeasure emerging during host–microbe co-evolutionary arms races? Mol Plant 10(5):662–664
Curran J (1990) Molecular techniques in taxonomy. Entomopathogenic Nematodes Biol Control:63–74
Dalla Corte AP, Rex FE, Almeida DRAD, Sanquetta CR, Silva CA, Moura MM et al (2020) Measuring individual tree diameter and height using GatorEye High-Density UAV-Lidar in an integrated crop-livestock-forest system. Remote Sens 12(5):863
Das S, Dash HR, Mangwani N, Chakraborty J, Kumari S (2014) Understanding molecular identification and polyphasic taxonomic approaches for genetic relatedness and phylogenetic relationships of microorganisms. J Microbiol Methods 103:80–100
Dash CK, Bamisile BS, Keppanan R, Qasim M, Lin Y, Islam SU et al (2018) Endophytic entomopathogenic fungi enhance the growth of Phaseolus vulgaris L.(Fabaceae) and negatively affect the development and reproduction of Tetranychus urticae Koch (Acari: Tetranychidae). Microb Pathog 125:385–392
de Araujo GR, Souza WD, Frases S (2017) The hidden pathogenic potential of environmental fungi. Future Microbiol 12(16):1533–1540
Delmas CE, Bancal MO, Leyronas C, Robin MH, Vidal T, Launay M (2024) Monitoring the phenology of plant pathogenic fungi: why and how? Biol Rev 99(3):1075–1084
Deng H, Gao R, Liao X, Cai Y (2017) CRISPR system in filamentous fungi: current achievements and future directions. Gene 627:212–221
Dietzel K, Valle D, Fierer N, U'Ren JM, Barberán A (2019) Geographical distribution of fungal plant pathogens in dust across the United States. Front Ecol Evol 7:304
Douhan GW, Vincenot L, Gryta H, Selosse MA (2011) Population genetics of ectomycorrhizal fungi: from current knowledge to emerging directions. Fungal Biol 115(7):569–597
Ekroth AK, Gerth M, Stevens EJ, Ford SA, King KC (2021) Host genotype and genetic diversity shape the evolution of a novel bacterial infection. ISME J 15(7):2146–2157
EL-Lebody K, Soliman MS, EL-Metwally EM, Abd-Elaziz M, Moustafa HZ (2021) Isolation and pathogenicity of endophytic fungi associated with some maize hybrids against certain Lepidoptera pests. Egypt J Agric Res 99(1):49–60
Fenoglio MS, Calviño A, González E, Salvo A, Videla M (2021) Urbanisation drivers and underlying mechanisms of terrestrial insect diversity loss in cities. Ecol Entomol 46(4):757–771
Filazzola A, Matter SF, MacIvor JS (2021) The direct and indirect effects of extreme climate events on insects. Sci Total Environ 769:145161
Fitt BD, McCartney HA, West JS (2006) Dispersal of foliar plant pathogens: mechanisms, gradients and spatial patterns. Epidemiol Plant Dis:159–192
Francis F, Fingu-Mabola JC, Ben Fekih I (2022) Direct and endophytic effects of fungal entomopathogens for sustainable aphid control: a review. Agriculture 12(12):2081
Fuller R, Freter R (1992) Factors affecting the microecology of the gut. In: Fuller R (ed) Probiotics: the scientific basis. Chapman & Hall, London, UK, pp 111–144
Furlan L, Kreutzweiser D (2015) Alternatives to neonicotinoid insecticides for pest control: case studies in agriculture and forestry. Environ Sci Pollut Res 22(1):135–147
García-González Á, Achón M, Carretero Krug A, Varela-Moreiras G, Alonso-Aperte E (2020) Food sustainability knowledge and attitudes in the Spanish adult population: a cross-sectional study. Nutrients 12(10):3154
Garnas JR, Hurley BP, Slippers B, Wingfield MJ (2012) Biological control of forest plantation pests in an interconnected world requires greater international focus. Int J Pest Manag 58(3):211–223
Garrido-Jurado I, González-Mas N, Yousef-Yousef M (2023) Ecosystem services of entomopathogenic ascomycetes. J Invertebrate Pathol 108015

Godfray HCJ (2002) Challenges for taxonomy. Nature 417(6884):17–19
Goudard A, Loreau M (2008) Nontrophic interactions, biodiversity, and ecosystem functioning: an interaction web model. Am Nat 171(1):91–106
Graham JH, Gottwald TR, Cubero J, Achor DS (2004) Xanthomonas axonopodis pv. citri: factors affecting successful eradication of citrus canker. Mol Plant Pathol 5(1):1–15
Gullino ML, Albajes R, Al-Jboory I, Angelotti F, Chakraborty S, Garrett KA et al (2022) Climate change and pathways used by pests as challenges to plant health in agriculture and forestry. Sustain For 14(19):12421
Harcourt DG, Leroux EJ (1967) Population regulation in insects and man. Am Sci 55(4):400–415
Hassani MA, Durán P, Hacquard S (2018) Microbial interactions within the plant holobiont. Microbiome 6:1–17
Heilmann-Clausen J, Barron ES, Boddy L, Dahlberg A, Griffith GW, Nordén J et al (2015) A fungal perspective on conservation biology. Conserv Biol 29(1):61–68
Herz A, Cahenzli F, Penvern S, Pfiffner L, Tasin M, Sigsgaard L (2019) Managing floral resources in apple orchards for pest control: Ideas, experiences and future directions. Insects 10(8):247
Hibbett D, Abarenkov K, Kõljalg U, Öpik M, Chai B, Cole J et al (2016) Sequence-based classification and identification of Fungi. Mycologia 108(6):1049–1068
Hooper DU, Chapin FS III, Ewel JJ, Hector A, Inchausti P, Lavorel S et al (2005) Effects of biodiversity on ecosystem functioning: a consensus of current knowledge. Ecol Monogr 75(1):3–35
Hu S, Bidochka MJ (2021) Root colonization by endophytic insect-pathogenic fungi. J Appl Microbiol 130(2):570–581
Islam W, Adnan M, Shabbir A, Naveed H, Abubakar YS, Qasim M et al (2021) Insect-fungal-interactions: A detailed review on entomopathogenic fungi pathogenicity to combat insect pests. Microb Pathog 159:105122
Jacobsen ID, Große K, Hube B (2012) Embryonated chicken eggs as alternative infection model for pathogenic fungi. Host-Fungus Interactions: Methods Protoc:487–496
Jeger MJ (1999) Improved understanding of dispersal in crop pest and disease management: current status and future directions. Agric For Meteorol 97(4):331–349
Joop G, Vilcinskas A (2016) Coevolution of parasitic fungi and insect hosts. Zoology 119(4):350–358
Kang BR, Han JH, Kim JJ, Kim YC (2018) Dual biocontrol potential of the entomopathogenic fungus, Isaria javanica, for both aphids and plant fungal pathogens. Mycobiology 46(4):440–447
Kepler RM, Luangsa-Ard JJ, Hywel-Jones NL, Quandt CA, Sung GH, Rehner SA et al (2017) A phylogenetically-based nomenclature for Cordycipitaceae (Hypocreales). IMA Fungus 8:335–353
Khanday NY, Sofi SA (2021) Taxonomy, state-of-the-art, challenges and applications of visual understanding: a review. Comput Sci Rev 40:100374
Khandel P, Shahi SK (2018) Mycogenic nanoparticles and their bio-prospective applications: current status and future challenges. J Nanostructure Chem 8:369–391
Knickel K, Maréchal A (2018) Stimulating the social and environmental benefits of agriculture and forestry: an EU-based comparative analysis. Land Use Policy 73:320–330
Kurniawan K, Gustiano R, Kusmini II, Prakoso VA (2021) Genetic resources preservation and utilization of Indonesian native freshwater fish consumption. Ecol Environ Conserv 27(1):227–233
Landa Z, Osborne L, Lopez F, Eyal J (1994) A bioassay for determining pathogenicity of entomogenous fungi on whiteflies1. Biol Control 4(4):341–350
Leger RJS (2024) The evolution of complex Metarhizium-insect-plant interactions. Fungal Biol
Leger RJS, Joshi L, Roberts DW (1997) Adaptation of proteases and carbohydrases of saprophytic, phytopathogenic and entomopathogenic fungi to the requirements of their ecological niches. Microbiology 143(6):1983–1992
Li J, Xia Y (2022) Host–pathogen interactions between Metarhizium spp. and locusts. J Fungi 8(6):602
Li J, Cornelissen B, Rep M (2020) Host-specificity factors in plant pathogenic fungi. Fungal Genet Biol 144:103447
Lovett B, St Leger RJ (2017) The insect pathogens. Microbiol Spectrum 5(2):10–1128

Lundberg J, Moberg F (2003) Mobile link organisms and ecosystem functioning: implications for ecosystem resilience and management. Ecosystems 6:0087–0098

Madariaga-Mazón A, Hernández-Alvarado RB, Noriega-Colima KO, Osnaya-Hernández A, Martinez-Mayorga K (2019) Toxicity of secondary metabolites. Phys Sci Rev 4(12):20180116

Malavasi V, Soru S, Cao G (2020) Extremophile microalgae: the potential for biotechnological application. J Phycol 56(3):559–573

Mantzoukas S, Eliopoulos PA (2020) Endophytic entomopathogenic fungi: a valuable biological control tool against plant pests. Appl Sci 10(1):360

Martino E, Morin E, Grelet GA, Kuo A, Kohler A, Daghino S et al (2018) Comparative genomics and transcriptomics depict ericoid mycorrhizal fungi as versatile saprotrophs and plant mutualists. New Phytol 217(3):1213–1229

Matsuura HN, Fett-Neto AG (2015) Plant alkaloids: main features, toxicity, and mechanisms of action. Plant Toxins 2(7):1–15

Matthysen E, Clobert J (2012) Multicausality of dispersal: a review. Dispersal Ecol Evol 27:3–18

May RM, Gupta S, McLean AR (2001) Infectious disease dynamics: what characterizes a successful invader? Philos Trans Royal Soc Lond Ser B: Biol Sci 356(1410):901–910

McCarl BA (2010) Analysis of climate change implications for agriculture and forestry: an interdisciplinary effort. Clim Chang 100(1):119

McGarvey DJ, Croteau R (1995) Terpenoid metabolism. Plant Cell 7(7):1015

Meyer SL, Huettel RN, Sayre RM (1990) Isolation of fungi from Heterodera glycines and in vitro bioassays for their antagonism to eggs. J Nematol 22(4):532

Milutinović B, Schmitt T (2022) Chemical cues in disease recognition and their immunomodulatory role in insects. Curr Opin Insect Sci 50:100884

Minter DW, Brady BL (1980) Mononematous species of Hirsutella. Trans Br Mycol Soc 74(2):271–282

Mishra M (2023) The potential application of entomopathogenic fungi (EF) in insect pest management. In: Microbial biocontrol: molecular perspective in plant disease management. Singapore, Springer Nature Singapore, pp 323–347

Mohan M, Richardson G, Gopan G, Aghai MM, Bajaj S, Galgamuwa GP et al (2021) UAV-supported forest regeneration: current trends, challenges and implications. Remote Sens 13(13):2596

Moonjely S, Barelli L, Bidochka MJ (2016) Insect pathogenic fungi as endophytes. Adv Genet 94:107–135

Mora MAE, Castilho AMC, Fraga ME (2018) Classification and infection mechanism of entomopathogenic fungi. Arquivos Inst Biol 84:e0552015

Murphy BR, Batke SP, Doohan FM, Hodkinson TR (2015) Media manipulations and the culture of beneficial fungal root endophytes. Int J Biol 7(3):94–102

Naylor JR, McShane PE (1997) Predation by polychaete worms on larval and post-settlement abalone Haliotis iris (Mollusca: Gastropoda). J Exp Mar Biol Ecol 214(1–2):283–290

Ng TB, Wang HX (2005) Pharmacological actions of Cordyceps, a prized folk medicine. J Pharm Pharmacol 57(12):1509–1519

Nicoletti R, Becchimanzi A (2022) Ecological and molecular interactions between insects and fungi. Microorganisms 10(1):96

Orapiriyakul W, Young PS, Damiati L, Tsimbouri PM (2018) Antibacterial surface modification of titanium implants in orthopaedics. J Tissue Eng 9:2041731418789838

Ortiz-Urquiza A, Keyhani NO (2013) Action on the surface: entomopathogenic fungi versus the insect cuticle. Insects 4(3):357–374

Palma MS (2011) Peptides as toxins/defensins. Amino Acids 40(1):1–4

Parveen SS, Rashtrapal PS (2024) Integrated pest management strategies using endophytic entomopathogenic fungi. Plant Sci Today 11(1):568–574

Paschapur A, Subbanna ARNS, Singh AK, Jeevan B, Stanley J, Rajashekhar H, Mishra KK (2021) Unraveling the importance of metabolites from entomopathogenic fungi in insect pest management. Microbes Sustain Insect Pest Manage: Hydrol Enzyme Second Metab 2:89–120

Pathak A, Kothari R, Vinoba M, Habibi N, Tyagi VV (2021) Fungal bioleaching of metals from refinery spent catalysts: a critical review of current research, challenges, and future directions. J Environ Manag 280:111789

Pathan EK, Ghormade V, Tupe SG, Deshpande MV (2021) Insect pathogenic fungi and their applications: an Indian perspective. In: Satyanarayana T, Deshmukh SK, Deshpande MV (eds) Progress in mycology: an Indian perspective. Springer, Singapore, pp 311–327

Pautasso M, Dehnen-Schmutz K, Holdenrieder O, Pietravalle S, Salama N, Jeger MJ et al (2010) Plant health and global change–some implications for landscape management. Biol Rev 85(4):729–755

Perotto S, Angelini P, Bianciotto V, Bonfante P, Girlanda M, Kull T et al (2013) Interactions of fungi with other organisms. Plant Biosyst: Int J Deal All Aspects Plant Biol 147(1):208–218

Perumal V, Kannan S, Pittarate S, Krutmuang P (2024) A review of entomopathogenic fungi as a potential tool for mosquito vector control: a cost-effective and environmentally friendly approach. Entomol Res 54(3):e12717

Purty RS, Chatterjee S (2016) DNA barcoding: an effective technique in molecular taxonomy. Austin J Biotechnol Bioeng 3(1):1059

Quandt CA, Kepler RM, Gams W, Araújo JP, Ban S, Evans HC et al (2014) Phylogenetic-based nomenclatural proposals for Ophiocordycipitaceae (Hypocreales) with new combinations in Tolypocladium. IMA Fungus 5:121–134

Reddy N, Mahesh G, Priya M, Singh RS, Manjunatha L (2020) Hirsutella. In: Beneficial microbes in agro-ecology. Academic Press, pp 817–831

Rehner SA (2009) Molecular systematics of entomopathogenic fungi. In: Insect pathogens: molecular approaches and techniques. CAB International, Cambridge, MA, pp 145–165

Roberts DW, St Leger RJ (2004) Metarhizium spp., cosmopolitan insect-pathogenic fungi: mycological aspects. Adv Appl Microbiol 54(1):1–70

Rotchés-Ribalta R, Ruas S, Ahmed KD, Gormally M, Moran J, Stout J et al (2021) Assessment of semi-natural habitats and landscape features on Irish farmland: new insights to inform EU Common Agricultural Policy implementation. Ambio 50:346–359

Rouhan G, Gaudeul M (2014) Plant taxonomy: a historical perspective, current challenges, and perspectives. Mol Plant Taxon: Methods Protoc:1–37

Russell NJ (1998) Molecular adaptations in psychrophilic bacteria: potential for biotechnological applications. Adv Biochem Eng Biotechnol 61:1–21

Sahab A, Sabbour MM (2011) Virulence of four entomopathogenic fungi on some cotton pests with special reference to impact of some pesticides, nutritional and environmental factors on fungal growth. Egypt J Biol Pest Control 21(1)

Samal I, Bhoi TK, Majhi PK, Murmu S, Pradhan AK, Kumar D et al (2023) Combatting insects mediated biotic stress through plant associated endophytic entomopathogenic fungi in horticultural crops. Front Plant Sci 13:1098673

Sasan RK, Bidochka MJ (2012) The insect-pathogenic fungus Metarhizium robertsii (Clavicipitaceae) is also an endophyte that stimulates plant root development. Am J Bot 99(1):101–107

Sasmal SK, Bhowmick AR, Al-Khaled K, Bhattacharya S, Chattopadhyay J (2016) Interplay of functional responses and weak Allee effect on pest control via viral infection or natural predator: an eco-epidemiological study. Different Equat Dynam Syst 24:21–50

Sayed S, El-Shehawi A, Al-Otaibi S, El-Shazly S, Al-Otaibi S, Ibrahim R et al (2020) Isolation and efficacy of the endophytic fungus, Beauveria bassiana (Bals.) Vuillemin on grapevine aphid, Aphis illinoisensis Shimer (Hemiptera: Aphididae) under laboratory conditions. Egypt J Biol Pest Control 30(1):1–7

Schmid-Hempel P (2005) Evolutionary ecology of insect immune defenses. Annu Rev Entomol 50:529–551

Schowalter TD (1981) Insect herbivore relationship to the state of the host plant: biotic regulation of ecosystem nutrient cycling through ecological succession. Oikos:126–130

Segoli M, Kishinevsky M, Rozenberg T, Hoffmann I (2020) Parasitoid abundance and community composition in desert vineyards and their adjacent natural habitats. Insects 11(9):580

Seifert KA, Boulay H (2004) Hirsutella uncinata, a new hyphomycete from Australia. Mycologia 96(4):929–934

Shang Y, Feng P, Wang C (2015) Fungi that infect insects: altering host behavior and beyond. PLoS Pathog 11(8):e1005037

Sharma K, Parveen S (2011) Ecological study of fungi isolated from the surface water of Dudhawa Dam Dhamtari, Chhattisgarh, India. J Phytol 3(4)

Sharma A, Srivastava A, Shukla AK, Srivastava K, Srivastava AK, Saxena AK (2020) Entomopathogenic fungi: a potential source for biological control of insect pests. In: Phytobiomes: current insights and future vistas, pp 225–250

Singh AK (2012) Molecular taxonomy: use of modern methods in the identification of a species. Indian J Life Sci 2(1):143–147

Sleeman WC IV, Kapoor R, Ghosh P (2022) Multimodal classification: current landscape, taxonomy and future directions. ACM Comput Surv 55(7):1–31

Srivastava DS (2006) Habitat structure, trophic structure and ecosystem function: interactive effects in a bromeliad–insect community. Oecologia 149(3):493–504

Stamps WT, Linit MJ (1997) Plant diversity and arthropod communities: implications for temperate agroforestry. Agrofor Syst 39:73–89

Strauss SY (1987) Direct and indirect effects of host-plant fertilization on an insect community. Ecology 68(6):1670–1678

Stromberg BE (1997) Environmental factors influencing transmission. Vet Parasitol 72(3–4):247–264

Swift RJ, Rodewald AD, Senner NR (2017) Breeding habitat of a declining shorebird in a changing environment. Polar Biol 40:1777–1786

Toner P, Nelson D, Rao JR, Ennis M, Moore JE, Schock B (2021) Antimicrobial properties of phytohormone (gibberellins) against phytopathogens and clinical pathogens. Access Microbiol 3(10):000278

Torres MS, White JF (2009) Clavicipitaceae: free-living and saprotrophs to plant endophytes. Encycl Microbiol 1:422–430

Tóth BB, Barta Z (2010) Ecological studies of ectomycorrhizal fungi: an analysis of survey methods. Fungal Divers 45:3–19

Trizelia T, Rahma H, Syahrawati M (2023) Diversity of endophytic fungi of rice plants in Padang City, Indonesia, entomopathogenic to brown planthopper (Nilaparvata lugens). Biodivers J Biol Divers 24(4)

Tscharntke T, Rand TA, Bianchi FJ (2005) The landscape context of trophic interactions: insect spillover across the crop—noncrop interface. In: Annales Zoologici Fennici. Finnish Zoological and Botanical Publishing Board, pp 421–432

Turner AK, Begon M, Jackson JA, Bradley JE, Paterson S (2011) Genetic diversity in cytokines associated with immune variation and resistance to multiple pathogens in a natural rodent population. PLoS Genet 7(10):e1002343

Valero-Jiménez CA, Wiegers H, Zwaan BJ, Koenraadt CJ, van Kan JA (2016) Genes involved in virulence of the entomopathogenic fungus Beauveria bassiana. J Invertebr Pathol 133:41–49

van der Does HC, Rep M (2007) Virulence genes and the evolution of host specificity in plant-pathogenic fungi. Mol Plant-Microbe Interact 20(10):1175–1182

Van Klinken RD (1999) Host specificity testing: why do we do it and how we can do it better. In: Bozeman MT, Van Driesche RG, Heard T, McClay A, Reardon R (eds) Host specificity testing of exotic arthropod biological control agents: the biological basis for improvement in safety. USDA Forest Service, pp 54–68

van Veen FF, van Holland PD, Godfray HCJ (2005) Stable coexistence in insect communities due to density-and trait-mediated indirect effects. Ecology 86(12):3182–3189

Veliz, E. A., Martínez-Hidalgo, P., & Hirsch, A. M. (2017). Chitinase-producing bacteria and their role in biocontrol. AIMS microbiology, 3(3), 689.

Verma M, Mishra J, Arora NK (2019) Plant growth-promoting rhizobacteria: diversity and applications. Environ Biotechnol: For Sustain Future:129–173

Vilcinskas A, Jegorov A, Landa Z, Götz P, Matha V (1999) Effects of beauverolide L and cyclosporin A on humoral and cellular immune response of the greater wax moth, Galleria mellonella. Comp Biochem Physiol C: Pharmacol Toxicol Endocrinol 122(1):83–92
Wang YB, Wang Y, Fan Q, Duan DE, Zhang GD, Dai RQ et al (2020) Multigene phylogeny of the family Cordycipitaceae (Hypocreales): new taxa and the new systematic position of the Chinese cordycipitoid fungus Paecilomyces hepiali. Fungal Divers 103:1–46
Warcup JH (1951) The ecology of soil fungi. Trans Br Mycol Soc 34(3):376–399
Weiland-Bräuer N (2021) Friends or foes—microbial interactions in nature. Biology 10(6):496
Wilby KJ (2022) A scoping review of the clinical pharmacokinetics of bedaquiline. Clin Pharmacokinet 61(4):481–488
Wink M, Schimmer O (2010) Molecular modes of action of defensive secondary metabolites. In: Annual plant reviews volume 39: functions and biotechnology of plant secondary metabolites, 2nd edn, pp 21–161. https://doi.org/10.1002/9781444318876.ch2
Xiao YP, Wang YB, Hyde KD, Eleni G, Sun JZ, Yang Y et al (2023) Polycephalomycetaceae, a new family of clavicipitoid fungi segregates from Ophiocordycipitaceae. Fungal Divers 120(1):1–76
Yang LH, Postema EG, Hayes TE, Lippey MK, MacArthur-Waltz DJ (2021) The complexity of global change and its effects on insects. Curr Opin Insect Sci 47:90–102
Yerukala S, Butler DM, Bernard EC, Gwinn KD, Grewal PS, Ownley BH (2022) Colonization efficacy of the endophytic insect-pathogenic fungus, Beauveria bassiana, across the plant kingdom: a meta-analysis. Crit Rev Plant Sci 41(4):241–270
Zhang Y, Ma R, Wang Y, Sun W, Yang Z, Han M et al (2021) Viruses run: the evasion mechanisms of the antiviral innate immunity by hantavirus. Front Microbiol 12:759198
Zhao Q, Ye L, Wang Z, Li Y, Zhang Y, Keyhani NO, Huang Z (2021) Sustainable control of the rice pest, Nilaparvata lugens, using the entomopathogenic fungus Isaria javanica. Pest Manag Sci 77(3):1452–1464
Zhou D, Hyde KD (2001) Host-specificity, host-exclusivity, and host-recurrence in saprobic fungi. Mycol Res 105(12):1449–1457
Zimmermann G (2007) Review on safety of the entomopathogenic fungus Metarhizium anisopliae. Biocontrol Sci Tech 17(9):879–920

Endophytic Fungal Alkaloids: Production and Applications

12

Nihal Gören Sağlam, Fazilet Özlem Albayrak, and İbrahim Ortaş

Abstract

Endophytes, including bacteria and fungi, can colonize the inner tissues of all plants. Since some endophytes do not remain as endophytes throughout their life cycle, they are not fully understood. In recent years, endophyte fungi have been considered because of their potential reliable compounds and economic properties. Alkaloids are naturally occurring chemicals with useful physiological and pharmacodynamic effects. Alkaloid drugs exhibit bioactive properties such as fungicidal, antibacterial, and anticancer actions. The most important fungal alkaloids are indole, ergot, pyrrolizidine, and imidazole. Endophytic fungal alkaloids have several potential applications. Considering current understanding, this chapter provides detailed information regarding endophytic fungal alkaloids.

Keywords

Alkaloids · Fungal endophytes · Mycorrhiza · Secondary metabolites

N. Gören Sağlam (✉)
Biology Department, Science Faculty, Istanbul, Turkey
e-mail: gorenn@istanbul.edu.tr

F. Ö. Albayrak
Biology Department, Science Faculty, Mersin University, Mersin, Turkey

İ. Ortaş
Department of Soil Science and Plant Nutrition, Faculty of Agriculture, University of Cukurova, Adana, Turkey

K. A. Abd-Elsalam, S. F. AbuQamar (eds.), *Fungal Endophytes Volume I*,
https://doi.org/10.1007/978-981-97-7312-1_12

12.1 Introduction

Alkaloids are substances with more or less basic properties derived from amino acids, usually containing one or more nitrogen atoms in the ring. They are obtained from natural sources and can have strong physiological and pharmacodynamic effects at very low doses. These substances, generally found in salt form in nature, are called true alkaloids, while those without amino acid derivatives are called pseudoalkaloids, and those whose nitrogen atom is not part of the heterocyclic ring are called protoalkaloids. Alkaloids are mainly found in Angiosperms but rarely in bacteria, fungi, and Pteridophyta (Dewick 2002; Heinrich et al. 2012).

Endophytes are microorganisms, typically bacteria or fungi that live in the interior (including leaves, stems, roots, and seeds) tissues of plants and colonize them without causing any significant symptoms (Alam et al. 2021). They can be found in all parts of the plant, including the xylem and phloem (Tejesvi et al. 2007). Alam et al. (2021) defined endophytic fungi (EF) as an intriguing type of host-related fungal communities that colonize the intracellular or intercellular regions of the host tissues, benefiting both their hosts and themselves. In addition, some endophytes not only remain as endophytes throughout their life cycle but also show similar functions with some important microbial groups such as pathogens, epiphytes, and saprotrophic that colonize plant tissues.

As mentioned in Fig. 12.1, endophytic fungi could interact with their host plant via mutualistic, commensalistic, and pathogenic interactions (Jha et al. 2023). Endophytes are present in almost all plants worldwide because of their secondary metabolites, which have various potential uses in critical areas such as industry, medicine and agriculture. The fact that they can be used effectively in methods such as biological control increases the importance given to these microorganisms.

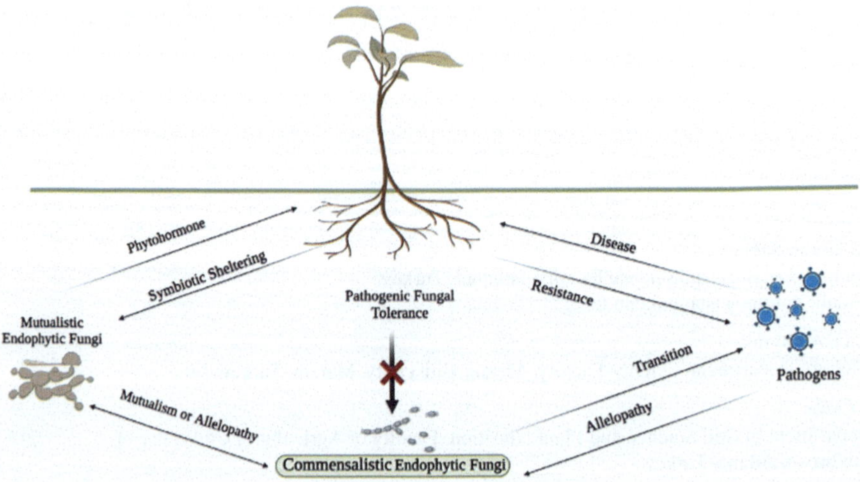

Fig. 12.1 Introduction of plant–fungi interactions (Jha et al. 2023)

Intricate biochemical interactions between endophytes and host plants are mediated by secondary metabolites, which direct complex interactions within the EF-host plant-pathogen community within a plant host. EFs are intensively related to host plants because they exist in a quiescent state as they perceive chemical changes in the host plants, which could be suffering injuries or various environmental stresses.

In addition to these definitions, endophytes are still not fully explained. The reason for this confusion is that some endophytes do not remain as endophytes throughout their life cycle. This means that in addition to mutualistic and commensalist life, it may include latent pathogens, latent saprotrophs, mycorrhizal fungi, and early stages of colonization by rhizobia bacteria. Therefore, endophytes may have one or more functions in a part of their life cycle (Porras-Alfaro and Bayman 2011). This chapter presents information about endophytic fungal alkaloids in detail, considering current knowledge.

12.2 Endophytic Fungal Alkaloids from Medicinal Plants

The following are some of the main traits and functions of endophytes:

Endophytes, which generally include bacteria and fungi, can enter plant tissue from roots, stems, flowers, leaves, or cotyledons and can spread to other plant tissues (Kobayashi and Palumbo 2000; Raghukumar 2008; Alam et al. 2021). European researchers first discovered endophyte fungi in the late nineteenth century. Following these discoveries, Bacon et al. (1977) reported that endophytes in the *Clavicipitaceae* family could cause toxic substance formation in cattle. Subsequent studies have shown that endophyte-related toxicity syndromes may become widespread and cause significant economic damage in the livestock sector (Hoveland 1993). Endophytes belonging to this family are one of the most interesting and economically significant examples of plant–fungi relationships and have inspired research on many endophytes in terms of the damage they caused. Members of classes such as *Dothideomycetes, Sordariomycetes, Leotiomycetes, Eurotiomycetes, and Pezizomycetes* form fungal endophyte colonies that are dominant in most plants (Jumpponen and Jones 2009). In addition to these classes, *Zygomycota* and *Basidiomycota* members and members of the order Agaricales, which are common in grasses, are also found as endophytes (Herrera et al. 2010; Khidir et al. 2010).

Endophytes constitute a symbiotic association with their host plants. Throughout plants' evolutionary history, they have established strategic and fine interactions with microorganisms in their environment. EF lives in the plant's tissues without causing any disease and often contributes to the overall health and well-being of the plant. The host plants' defenses can be boosted by rhizobia and mycorrhizal fungi, increasing their resistance to a variety of intruders. Most plant species are symbiotic with mycorrhizae, which are the largest symbiotic fungi. Approximately 90% of plants have mycorrhizal associations (Brundrett 2009). Plants that depend on mycorrhizae benefit greatly from mycorrhiza in protecting them from biotic and abiotic stressors.

AMF is one of the top interactions. AMF is the most common symbiotic relationship with plants in nature. AMF colonization of host plants improves root growth, length, root-shoot ratio, average diameter, surface area, lateral roots, and hydraulic properties. Mycorrhizal fungi bind plant cells together, increasing plant resistance to stressors (Della Mónica et al. 2023). There is considerable evidence that AMF and/or other fungal endophytes promote healthy plant growth by ensuring balanced nutrient intake and plant vitality. In addition, phosphate-dissolving fungi and other soil organisms work together with AMF infection, with notable effects on the physiology of the host.

Arbuscular and ectomycorrhizal fungi increase plant nutrient uptake in exchange for carbohydrate compounds in the rhizosphere, and endophytic fungi in roots, stems, and/or leaves provide resistance to biotic and abiotic stress conditions by using the same or some different mechanisms (Courty et al. 2010; Bastas 2021). Most endophytes maintain the potential for the synthesis of various bioactive metabolites used directly or indirectly against diseases, abiotic stress, and plant growth stimulation (Kusari and Spiteller 2012; Khare et al. 2018; Strobel 2018). The presence of more serpentine and ajmalicine in *C. roseus* roots after mycorrhizal inoculation suggests that mycorrhization had a higher impact on the root's alkaloids than on the shoots (Andrade et al. 2013). The effects of alkaloids in the interaction between *Epichloë*-infection and AMF can be summarized in Fig. 12.2, as mentioned by Zhong et al. (2022).

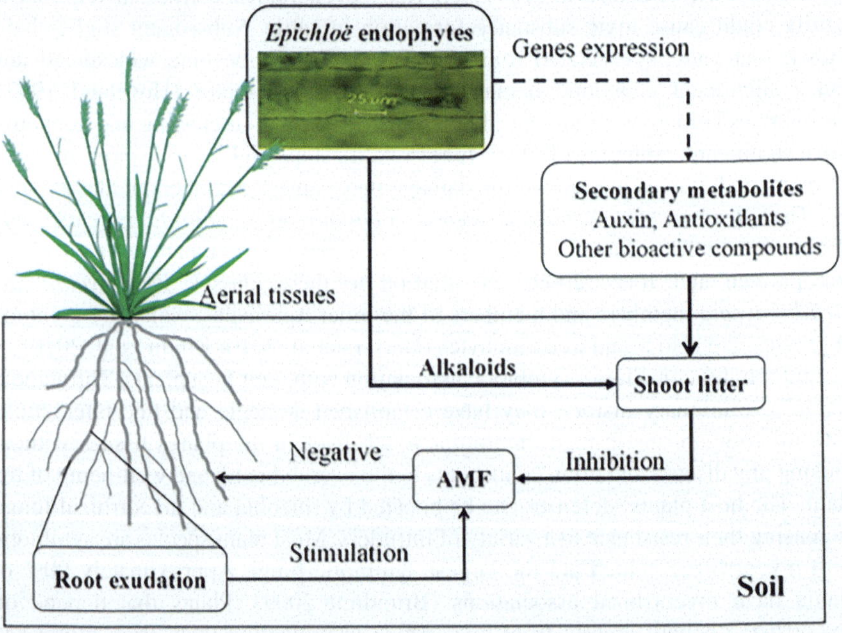

Fig. 12.2 Effects of alkaloids in the interaction between *Epichloë*-infection and AMF (Zhong et al. 2022)

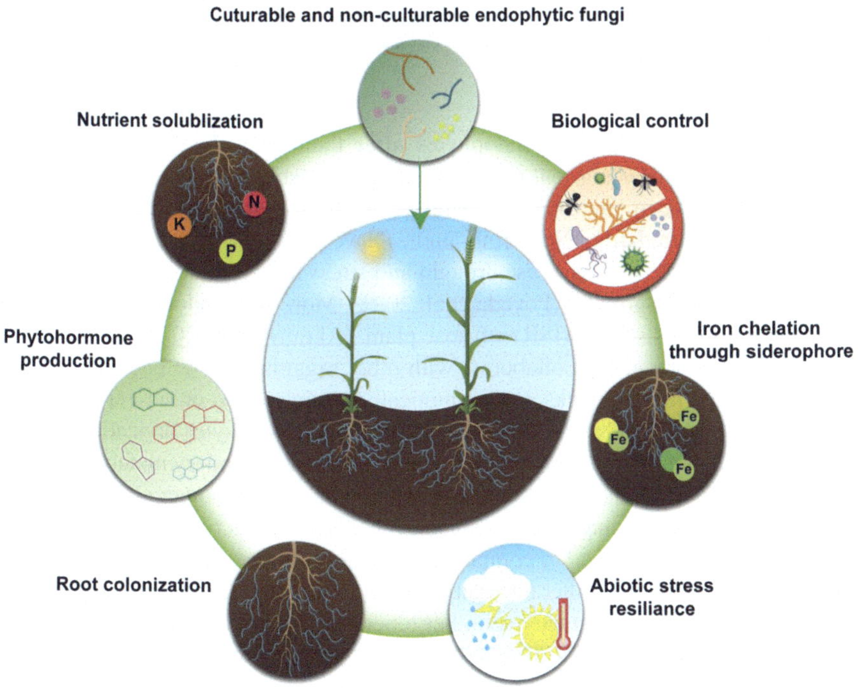

Fig. 12.3 Potential roles of endophytic fungi in sustainable agriculture (Akram et al. 2023)

Endophytes have formed an important basis for the discovery of drugs (Staniek et al. 2008). The bioactive metabolites of endophytes could be classified according to the alkaloids, cytochalazines, terpenoids, polyketides, steroids, and flavonoids they contain (Guo et al. 2008; Digra and Nonzom 2023). It has cellular targets such as cell division, glucose transfer, HIV-1 protease, and the cytoskeleton (Gimenez et al. 2007). This group of microorganisms, which offers an abundant and reliable source of new compounds, has wide use potential in important fields such as medicine, agriculture, and industry. Endophytic fungi could have a critical role in sustainable agriculture by enhancing plant growth and health through various mechanisms. The potential roles of endophytic fungi are listed in Fig. 12.3 (Akram et al. 2023).

Diverse Microorganisms A great variety of microorganisms, including fungi, bacteria, and occasionally other microorganisms. Fungal endophytes can be found in various plant species. Perennials are particularly colonized by various endophytic microorganisms (Demain 2014). It has been estimated by Strobel (2018) that only 1–2% of plant species have been studied, and little is known from the hydro-ecosystems, meaning that the bulk of EF interactions are still mostly unknown. Vandenkoornhuyse et al. (2002) postulated that little is known about biodiversity formation mechanisms, fungal strain-mediated processes, and EFs' crucial role in their habitats.

Beneficial Effects Numerous endophytes offer their host plants several advantages, including greater nutrient uptake, enhanced resistance to diseases and pests, increased tolerance to environmental stresses, including drought and salinity, and general growth stimulation. Numerous studies have found that EFs modify the host plant's metabolic process, increase drought and stress tolerance, accelerate nutrient uptake, and enhance growth. (Poveda et al. 2021). Moreover, endophytic fungi can induce systemic resistance (ISR) of host plant (Akram et al. 2023). In general, plants and microorganisms collaborate with other organisms to defend themselves against pest or pathogen attacks, to communicate warnings inter- or intra-specifically during development, and to create defiance exudates. Endophytes depend on many defensive mechanisms and regular life cycles to produce secondary metabolites and other bioactive chemicals. Rhizosphere creatures compete with other organisms and stress factors, similar to any other organism. When several microorganisms share an environment, they are forced to compete for resources to meet their needs for food, shelter, reproduction, and other necessities throughout their life cycles. Endophytes possess an additional advantage over other biocontrol agents because of their capability to colonize plant tissues. The multiple biological activities of AMF in ecosystems, their connections with native plants, and how these plants are better able to survive drought stress to complex physiological changes were evaluated and summarized by Madouh and Quoreshi (2023) after reviewing several studies. It is vital to remember that endophytes may have harmful impacts on the host plant depending on the environmental conditions. This is a typical biological mechanism for the operation of an ecosystem.

Secondary Metabolites The plant's defiance against stresses such as diseases and herbivores may depend heavily on these compounds, even though they may not directly contribute to the plant's primary metabolic activity. Endophytes create secondary metabolites that are beneficial to the host plant's health, such as anticancer agents, medicines, tastes, perfumes, and insecticides.

12.3 Endophytic Fungal Alkaloids from Medicinal Plants

In addition to the growing use of agricultural inputs, the environment and the safety of our food supply are negatively impacted by antibiotics and other chemicals used for animal health. Because the respiratory and nutritional systems of plants and animals both grow via synchronized cycles in nature, plant health directly improves animal health, directly improving human health. Animals, plants, and microorganisms all greatly depend on a strong immune system and adequate nutrition for their survival. As a result, the secondary metabolites they produce have grown in importance. The bodies of plants and animals produce secondary metabolites that are anticancer, anti-disease, and anti-pesticide drugs.

Endophytes can produce several secondary metabolites, such as alkaloids, terpenoids, diterpenoids, steroids, polyketides, and anthraquinones (Zheng et al. 2021). Fungal alkaloids refer to a broad class of compounds found among fungi metabolites. Fungi are an important resource for chemists and biologists, and studies on the biosynthesis and biological effects of such compounds target potential applications in drug development and agriculture. Alkaloids have significant roles not only for their chemical reasons but also for their various biological activities, such as anticancer, antiviral, and antifungal properties (Wang et al. 2011). The chemical structure of fungal alkaloids is diverse and consists of heterocyclic rings. The most important fungal alkaloids are ergot, indole, pyrrolizidine, and imidazole alkaloids. The biosynthesis of fungal alkaloids is quite complex and usually occurs through reactions catalyzed by more than one enzyme (Zhang et al. 2012). According to their chemical structures, alkaloids are grouped as follows (Ortega et al. 2021):

- Cytochalasans
- Indole alkaloids
- Isoindole alkaloids
- Pyrrolidone derivatives
- Pyridone and pyridinyl derivatives
- Diketopiperazine derivatives
- Quinoline and isoquinoline
- Quinazolines
- Amides and amines

Some important alkaloids produced by fungal endophytes included in these groups are listed below (Daley and Cordell 2021).

- Camptothecin
- Paclitaxel (taxol)
- Catharanthus alkaloids
- Cinchona alkaloids
- Huperzine A
- Swainsonine
- Piperine
- Berberine and related protoberberine alkaloids
- Opioid alkaloids
- Colchicine
- Tropane alkaloids
- Quinolizidine alkaloids

Camptothecine, which is a quinoline alkaloid, is produced by *Fusarium solani*, *Alternaria alternate*, *Phomposis* sp., and *Formitopsis* sp. It has a significant activity against some ovarian cancers. It could also have an active role in the treatment of leukemia (Archana and Nagadesi 2022; Cai et al. 2022). *Chaetomium* sp. has the ability to produce alkaloids such as chaetoglobosin, chaetominine, swainsonine, chaetofusin, and chaetoseminudin, which have antimicrobial and anticancer activities (An et al. 2020; Tian and Li 2022).

Penicillium sp. produces alkaloids such as paspaline, shearinines, meleagrine, paspalitrem, chrysogenamide, penicidones, glandicoline, trichistanin, spirotryprostatin, cerevisterol, chrysogine, peninsulfuranols, and penioxamide. These alkaloids have antimicrobial and anticancer activities (Archana and Nagadesi 2022; Nicoletti et al. 2023). Vinblastine and vincristine, which have anticancer activity, were produced by *Fusarium oxysporum* (Farzaei and Sayyari 2023). Moreover, *Aspergillus fumigatus* sp. could produce alkaloids such as pyripyropene, chaetominine, pseurotin A, tryptoquivaline, and asperfumoid, which have antifungal and anti-inflammatory activities (El-Hawary et al. 2020; Zou et al. 2022). Some important fungal alkaloids and their classes are listed in Table 12.1 (Archana and Nagadesi 2022).

Fungal alkaloids are of great importance in medicine, agriculture and ecology due to their biological activities. In this section, the potential pharmacological effects of different types of fungal alkaloids such as antifungal, antimicrobial, antitumor, and neurotoxicity can be examined (Uzma et al. 2018). Additionally, fungal alkaloids have important potential applications such as developing natural pesticides against harmful fungi in the agricultural sector, discovering new drugs in the medical field, and biotechnological uses in the industrial processes.

12.4 Endophytic Fungal Alkaloids with Unique Structures

Fungal alkaloids have unique bioactivities such as anti-neurodegenerative, anti-inflammatory, cytotoxic, and anti-depressive properties (Silva et al. 2023). As fungi can provide various sources of alkaloids (Archana and Nagadesi 2022), endophytic fungi are suggested as important resources because of their bioactive products (Zhang et al. 2022). There are several studies related to new endophytic fungal alkaloids findings and their potential (Li et al. 2023; Zhang et al. 2022). Moreover, endophytes are reported as excellent sources of bioactive and novel alkaloid products (Zhang et al. 2012). Endophytes have gained great attention because of their potential usage in forestry, agriculture, and pharmaceuticals. In the future, EF present promising opportunities for creating eco-friendly ecosystem management strategies and finding new bioactive substances.

Organic amines, ergot alkaloids, loline and indole diterpene derivatives are produced by plants with the effect of endophytic fungi. There are several compounds produced by various plants. For example, in gramineous plants endophytic fungi cause the production of pyrrolizidines, peramine, lolitrem B, and ergovaline (Han et al. 2022). These compounds can be varied according to the plant species. Some of the recently isolated endophytic fungal alkaloids are listed in Fig. 12.4 (Daley and Cordell 2021).

Table 12.1 The important fungal alkaloid names and classes

Class of alkaloids	Name of the alkaloid	Reference
Amines and amides	Phomoenamide	Zhang et al. (2012)
Amines and amides	p-aminoacetophenonic	Zhang et al. (2012)
Amines and amides	7-amino-4-methylcoumarin	Zhang et al. (2012)
Amines and amides	Ergot	Zhang et al. (2012)
Azaphilone alkaloids	Chaetofusins A and B	Kuklev and Dembitsky (2016)
Benzophenanthridine alkaloid	Sanguinarine	Wang et al. (2013)
Chromone alkaloid	Rohitukine, flavopiridol	Kumara et al. (2014)
Cinchona alkaloids	Quinine, quinidine, cinchonidine, cinchonine	Maehara et al. (2013)
Dibenzo–α–pyrone alkaloid	Rhizovagine A	Wang et al. (2020)
Diketopiperazine alkaloid	5S-hydroxynorvaline-S-Ile	Jiang et al. (2018)
Diketopiperazine alkaloid	Tryhistatin	Shan et al. (2010)
Diketopiperazine alkaloids	Piperazine-2,5-dione-1 and 2	Metwaly et al. (2015)
Dimeric pyrrolidine alkaloid	Collacyclumines A-D	Lin et al. (2020)
Dioxopiperazine alkaloids	Dehydrovariecolorin L and dehydroechinulin	Li et al. (2008)
Epipolythiodioxopiperazine (ETP)	Peninsulfuranols A-F	Zhu et al. (2016)
Ergot alkaloids	Fumigaclavine C and Pseurotin A	Pinheiro et al. (2013)
Flavipin–derived alkaloids	Azacoccones F-H	Harwoko et al. (2020)
Indole–diterpene alkaloids	Lolitrem B	Philippe (2016)
Indole alkaloid	Glandicoline B	Koolen et al. (2012)
Indole alkaloid	Piperine	Verma et al. (2011)
Indole alkaloids	Vincamine	Na et al. (2016)
Indole alkaloids	Vincristine and vinblastine	Kumar et al. (2013)
Indole alkaloids	Fumitremorgin B and C	Feng and Ma (2010)
Indole alkaloids	Cristatumins A-D	Du et al. (2012)
Indole alkaloids	Chetoseminudin F and G	Peng et al. (2019)
Indole derivatives	Chaetoglobosin B-D, F-G	Zhang et al. (2012)
Indole derivatives	Paspalitrem A, shearinines A, D–K, and paspaline	Zhang et al. (2012)
Indole diketopiperazine alkaloids	Isovariecolorin 1	Du et al. (2017)
Indole-diketopiperazine	Spirotryprostatin F	Gao et al. (2017)
Indolizidine alkaloids	Swainsonine	Grum et al. (2013)

(continued)

Table 12.1 (continued)

Class of alkaloids	Name of the alkaloid	Reference
Indolizidine alkaloids	Swainsonine	Song et al. (2019); Cook et al. (2014)
Indolosesquiterpenoid alkaloids	Mycoleptodiscins A and B	Ortega et al. (2013)
Isoquinolone alkaloid	3-O-methylviridicatin and viridicatol, 5-hydroxy-8-methoxy-4--phenylisoquinolin–1 (2H)-one	Ma et al. (2017)
Loline alkaloids	Amino pyrrolizidines	Blankenship et al. (2001)
Macfortine group of alkaloids	Chrysogenamide A	Lin et al. (2008)
Meleagrine alkaloid	Meleagrine and chrysogine	Yunianto et al. (2014)
Mycoleptodiscin alkaloids	Mycoleptodiscin B	Dissanayake et al. (2016)
Oxepine containing diketopiperazine type alkaloids	Varioloids A and B	Zhang et al. (2015a)
Polyketide isoquinoline alkaloid	Fusarimine	Yang et al. (2012)
Prenylated indole alkaloid	Penioxamide A and 18-hydroxydecaturin B	Zhang et al. (2015b)
Prenylated indole alkaloids	Amoenamide C and Sclerotiamide B	Zhang et al. (2019)
Protoberberine alkaloid	Palmatine 7–N–oxide	Agusta et al. (2014)
Pyridine alkaloid	Dideacetyl-pyripyropene A, pyripyropene A and E, 1,11-fumitremorgin C, tryptoquivaline J, chaetominine, 1-acetyl-carboline and nicotinic acid	Zou et al. (2022)
Pyridines	Penicidones A-C	Zhang et al. (2012)
Pyridone alkaloid	Campyridones A-D	Zhu et al. (2016)
Pyridone alkaloids	Fusapyridons A and B	Tsuchinari et al. (2007)
Pyridone alkaloids	Tolypyridone A	Li et al. (2015)
Pyrrole alkaloid	N-[4-(2-formyl-5-hydroxymethyl-pyrrol-1-yl)-butyl]-acetamide	Li et al. (2008)
Pyrrolidones alkaloids	Phomapyrrolidones A-C	Wijeratne et al. (2013)
Quinazoline alkaloid	(1R,4R)-1,4-(2,3)-indolmethane-1-methyl-2,4-dihydro-1H-pyrazino-[2,1-b]-quinazoline-3,6-dione	Zheng et al. (2012)
Quinazolines	Chaetominine	Zhang et al. (2012)
Quinazolinone alkaloids	Aniquinazolines A-D	An et al. (2013)
Quinoline	Camptothecine (CPT)	Zhang et al. (2012), Pund and Joshi (2017), Shweta et al. (2013)

(continued)

Table 12.1 (continued)

Class of alkaloids	Name of the alkaloid	Reference
Quinoline	Aspernigerin	Zhang et al. (2012)
Quinoline	Penicinoline	Zhang et al. (2012)
Quinoline and isoquinoline	Asperfumoid	Zhang et al. (2012)
Sesquiterpene alkaloids	Huperzine A	Su and Yang (2015)
Spiro–heterocyclic alkaloid	3'S-pestaloamide A	Daley and Cordell (2021)
Spiro-heterocyclic γ-lactam alkaloid	Pseurotin A	El-Hawary et al. (2020)
Spiroquinazoline alkaloids	Alanditrypinone, alantryphenone, alantrypinene, and alantryleunone	Barros and Rodrigues-Filho (2005)
Steroidal alkaloid	Peimisine and imperialine-3-D-glucoside	Pan et al. (2015)
Steroidal alkaloids	Peimisine and peiminine alkaloids	Pan et al. (2014)
Terpenoid indole alkaloids	Vincristine and vinblastine	Palem et al. (2015)

Additionally, several alkaloids have been reported from endophytic fungi, including quinoline, isoquinoline, alkaloids, quinazolines, miscellaneous, pyridine, amine, amide, and indole derivative alkaloids. Previous studies have shown that various alkaloids from EF have critical roles such as anticancer, cytotoxic, antimicrobial, and insecticidal activities (Zhang et al. 2012). Bastias et al. (2017) reported that certain fungi endophytes affect the host plant's defense signaling system, making it more resistant to intruders. Due to their effects, bioactive alkaloids can be summarized in four classes– peramine, ergot alkaloids, indole-diterpenes and lolines, which can be found in various combinations in endophyte taxa (Panaccione et al. 2014). Several chemicals are taken into consideration such as cyclobutane-containing alkaloids of fungal endophytes, which could have important functions in pharmacology and medicine. Fermentation of these alkaloids should also be taken into consideration (Dembitsky 2014). However, low yields of alkaloid compounds could restrict the clinical use of endophytic fungi. Moreover, the interactions between the host and the endophytic fungi among alkaloids should be evaluated (Zhang et al. 2012).

12.5 Endophytic Fungal Alkaloids with Novel Biological Activities

Endophytic fungi can produce complex secondary metabolites, which have significant properties in agricultural and pharmaceutical industries (Tousif et al. 2023). They have the effect of organizing several bioactive metabolites that may have an impact on the secondary metabolites pathways in plants. Endophytic fungi can also have an effect on several gene expressions related to the production of key enzymes in these pathways. They indicate the production of several phenolic compounds in

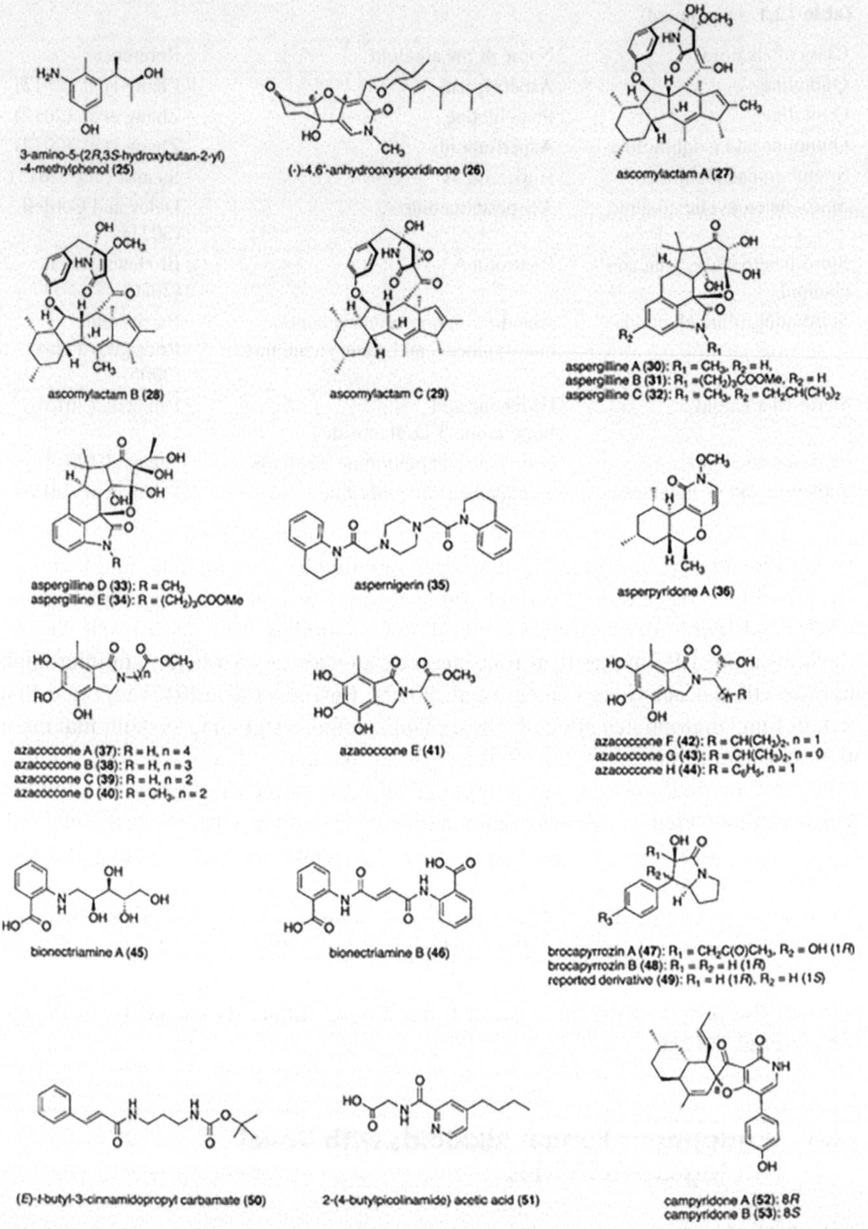

Fig. 12.4 Some of the recently isolated endophytic fungal alkaloids (Daley and Cordell 2021)

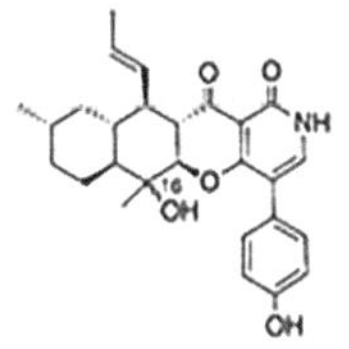

campyridone C (54): 16*R*
campyridone D (55): 16*S*

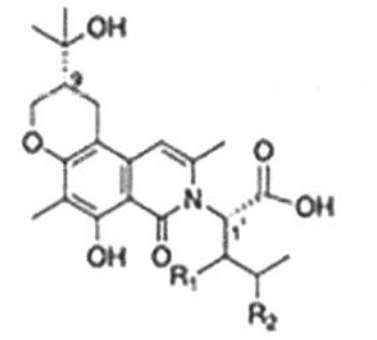

(9*R*, 1′*S*)-chaetoline A (56): R_1 = H, R_2 = CH_3
(9*R*, 1′*S*)-chaetoline B (57): R_1 = CH_3, R_2 = H

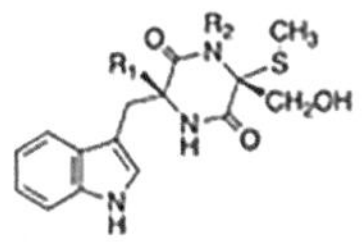

chetoseminudin E (58): R_1 = OH, R_2 = CH_3
chetoseminudin F (59): R_1 = OCH_3, R_2 = CH_3
chetoseminudin G (60): R_1 = OCH_3, R_2 = H

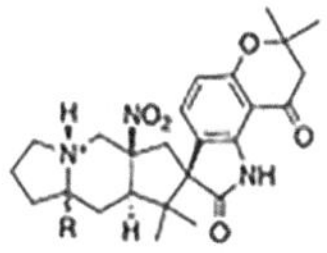

citranilin A (61): R = α-H
citranilin B (62): R = β-H

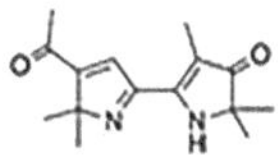

collacyclumine A (63)

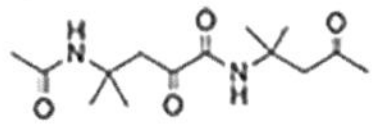

collacyclumine B (64)

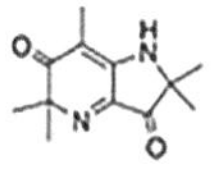

collacyclumine C (65)

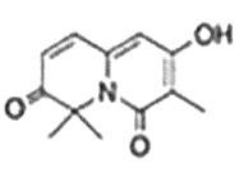

collacyclumine D (66)

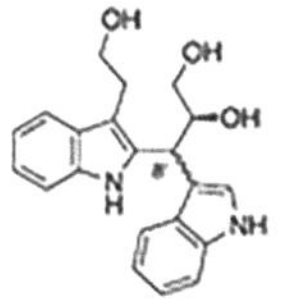

colletotryptin A (67): 8′*R*
colletotryptin B (68): 8′*S*

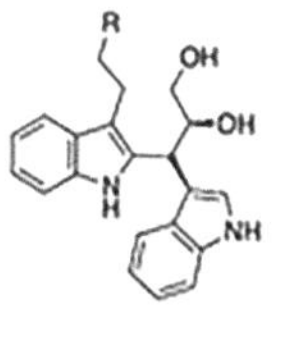

colletotryptin C (69): R = $OCOCH_3$
colletotryptin D (70): R = $NHCOCH_3$

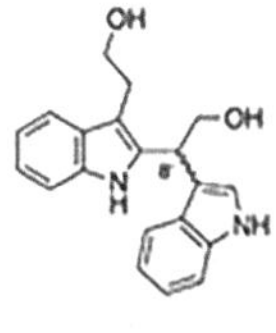

(-)-colletotryptin E (71): 8′*R*
(+)-colletotryptin F (72): 8′*S*

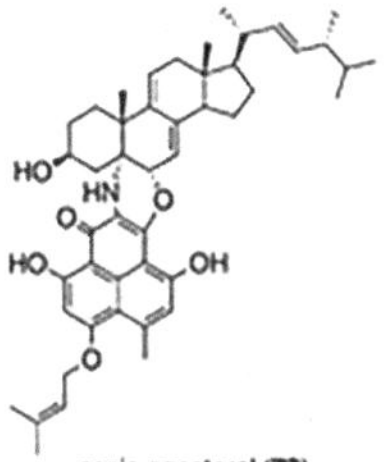

conio-azasterol (73)

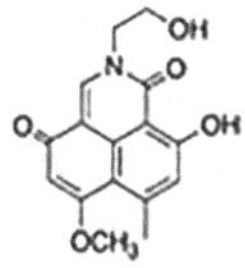

cylindrocarpyridone A (74)

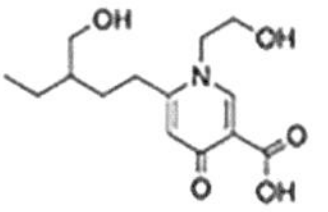

cylindrocarpyridone B (75)

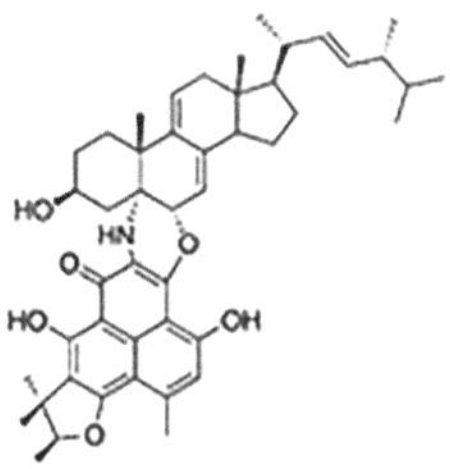

S-dehydroazasirosterol (76)

Fig. 12.4 (continued)

plants. Several industrially important metabolites can be produced due to the plant–endophyte interactions; therefore, this relationship has gained attention (Toppo et al. 2023).

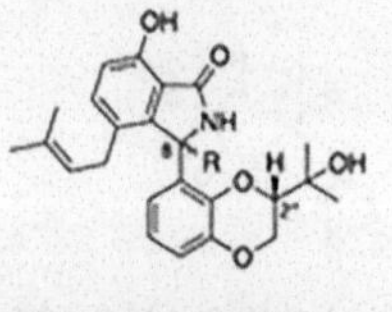

diaporisoindole A (77): R = H, 8S
diaporisoindole B (78): R = H, 8R
diaporisoindole D (80): R = OCH_3, 8S
diaporisoindole E (81): R = OCH_3, 8R

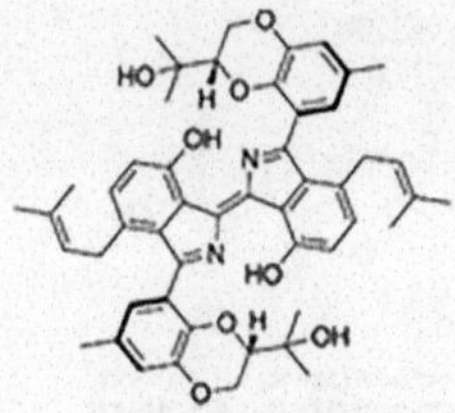

diaporisindole C (79)

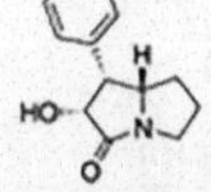

1,2-dihydrophenopyrrozin (82)

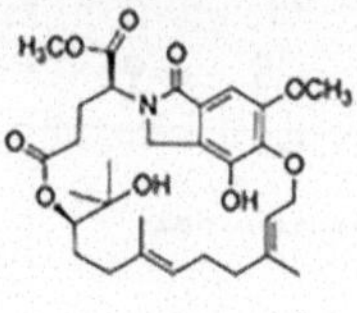

emericellolide A (83)

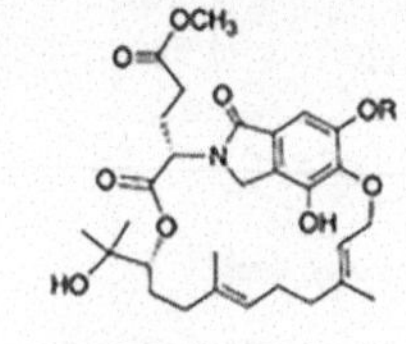

emericellolide B (84): R = CH_3
emericellolide C (85): R = H

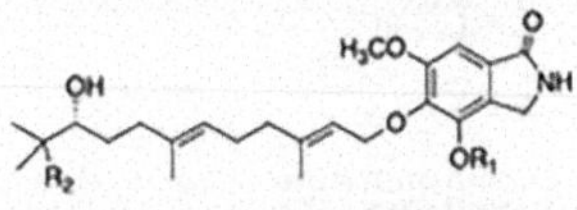

emeriphenolicin A (86): R_1 = CH_3, R_2 = Cl
emeriphenolicin B (87): R_1 = H, R_2 = Cl
emeriphenolicin C (88): R_1 = CH_3, R_2 = OH

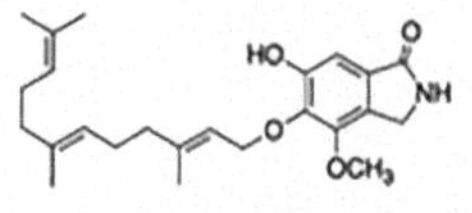

emeriphenolicin D (89)

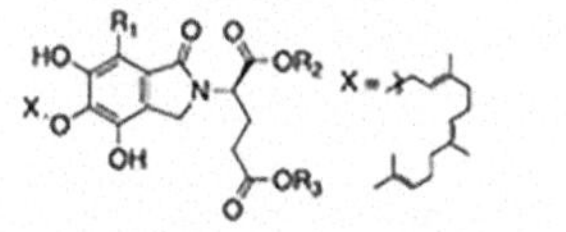

emeriphenolicin E (90): R_1 = H, R_2 = X, R_3 = CH_3
emeriphenolicin F (91): R_1 = H, R_2 = CH_3, R_3 = X
emeriphenolicin G (92): R_1 = X, R_2 = CH_3, R_3 = CH_3

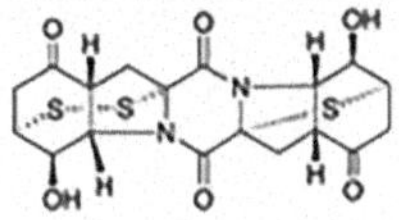

epicoccin A (93)

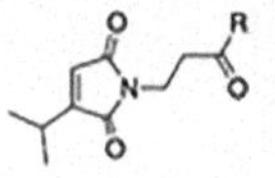

farinomalein A (94): R = OH
farinomalein C (95): R = $O(CH_2)_3CH_3$
farinomalein D (96): R = $OCH_2CHOHCH_2OH$
farinomalein E (97): R = OCH_2CCH_3COOH

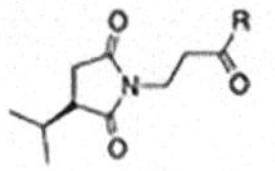

farinomalein F (98): R = OH
farinomalein G (99): R = OCH_3

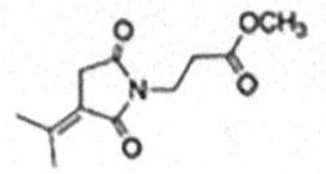

farinomalein H (100)

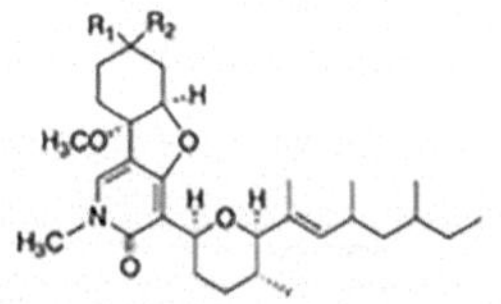

fusapyridon A (101): R_1 = R_2 = O
fusapyridon B (102): R_1 = R_2= OCH_3

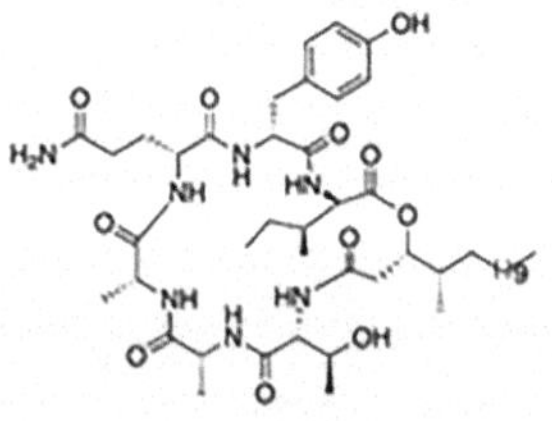

fusaripeptide A (103)

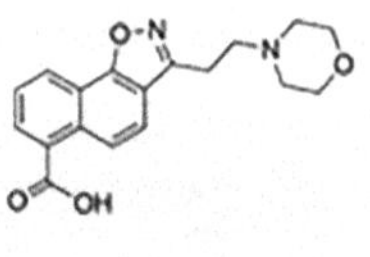

F. avenaceum alkaloid 1 (105)

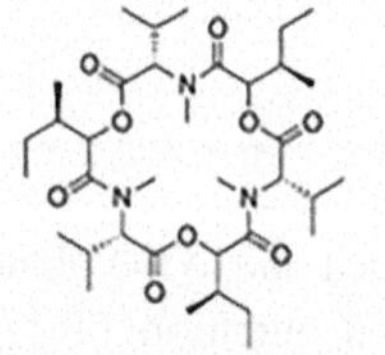

F. avenaceum alkaloid 2 (106)

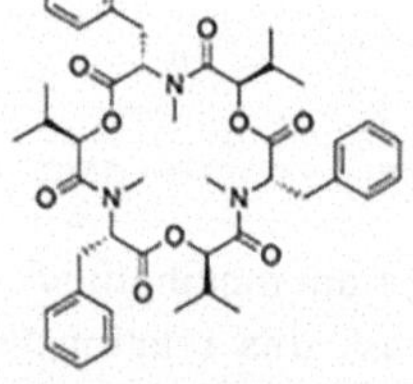

F. avenaceum alkaloid 3 (107)

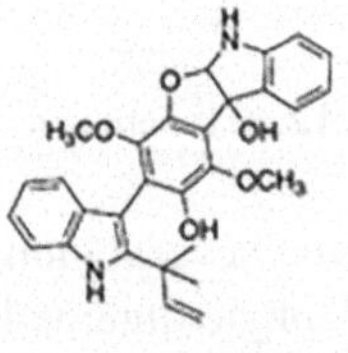

giluterrin (108)

Fig. 12.4 (continued)

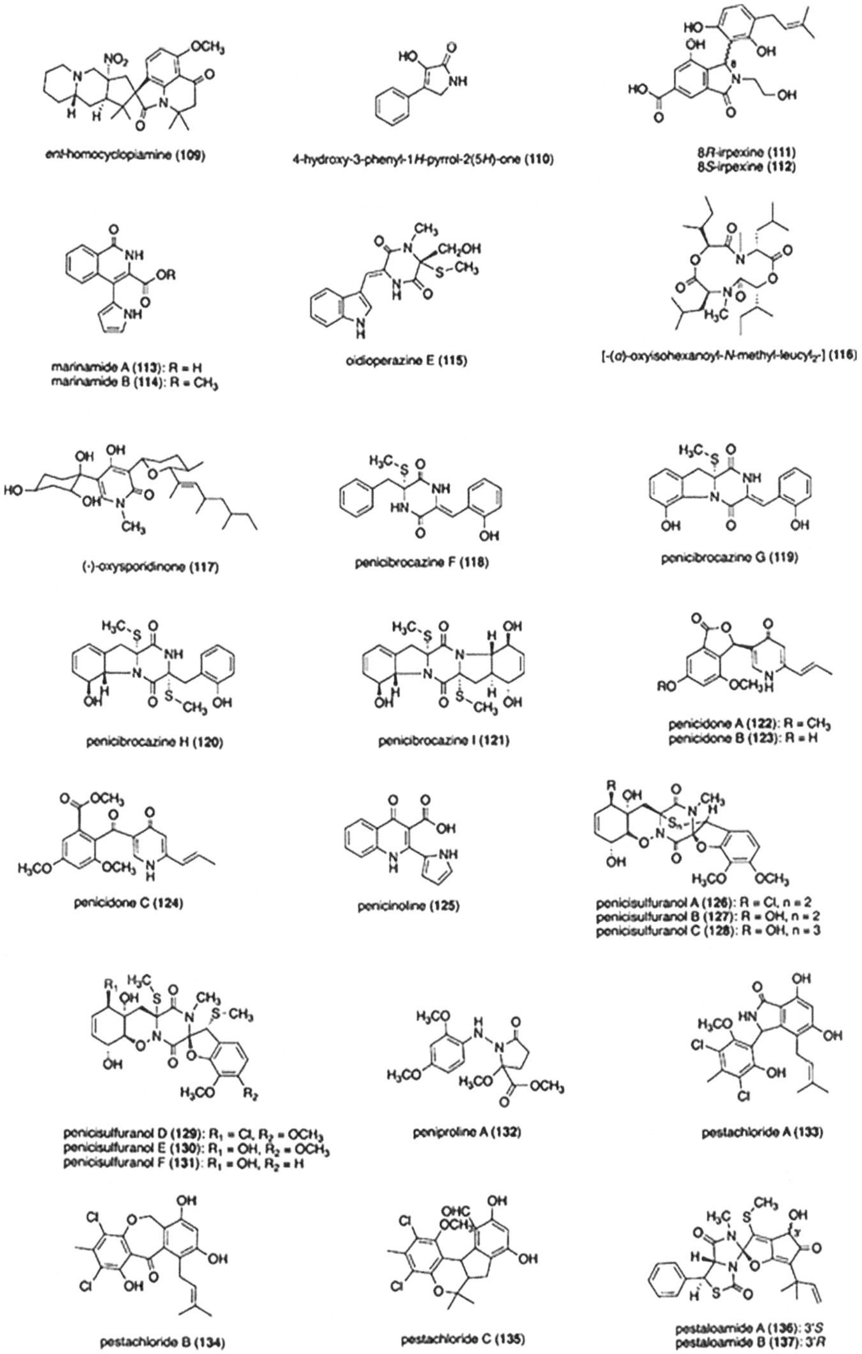

Fig. 12.4 (continued)

Endophytic fungi can grow in plant epidermis and aerial parts. They produce important secondary metabolites, including alkaloids, terpenes, monoterpenoids, and peptides (Hashem et al. 2023). Instead of chemicals, bioactive compounds isolated from endophytic fungi can be considered safe, cheap, easy, and more effective (Hashem et al. 2023). Several new bioactive compounds have been discovered via the relationship between endophytic fungi and plants (Khazaal et al. 2023). Fungal endophytes could increase plant resistance against insects by affecting the plant's immune system. In this response, it could be achieved by affecting salicylic acid and jasmonic acid pathways (Bastias et al. 2017). They can maintain a high level of resistance to biotic stress by increasing the amounts of salicylic acid, polyphenol contents and antioxidant levels (Farhat et al. 2023). EFs could supply a good antibacterial potential to the host plant in improving plant diseases (Song et al. 2023). Jha et al. (2023) summed up the interaction between the plant immune system and endophytic fungi in Fig. 12.5. Plants could recognize microbe-associated molecular patterns (MAMPs) inhibitors, encoded by endophytic fungi, which could enhance the plant's immune response (Jha et al. 2023).

It has been demonstrated that EFs produce bioactive secondary metabolites, and these compounds may contribute to the stimulation of plant development (García-Latorre et al. 2023). Moreover, the mutual relationship between endophytic fungi and plants could have positive effects on surviving in harsh environments (Farhat et al. 2023). Moreover, exploration of endophytic fungi could help to resource the ecological balance destruction by utilization of large numbers of secondary metabolites extractions from natural plants (Wang et al. 2023).

The bioactive substances that are obtained from endophytic fungi are categorized into eight groups: terpenoids, phenols, isocoumarins, xanthones, tetralones, enniatines, and benzopyranones. The most important difference between them is that each EF has a distinct secondary metabolite, which may make it possible to isolate and

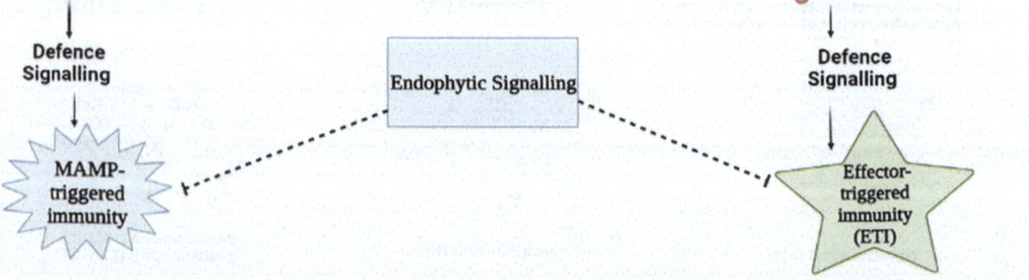

Fig. 12.5 The interaction between the plant immune system and endophytic fungi (Jha et al. 2023)

use different metabolites for medical purposes (Hashem et al. 2023). *Alternaria alternate*, an endophytic fungus obtained from Colocasia esculanta plant leaves, caused various metabolites that could have antiproliferative and antimicrobial activities (Khazaal et al. 2023). New alkaloids from endophytic fungus Penicillium citrinum isolated from *Ocimum tenuiflorum* were reported. In addition, these alkaloids could have effects on cancer cells (Lai et al. 2013).

It has been reported that endophytic fungi can be capable of producing a potential therapeutic drug for diabetes mellitus by the inhibitory potential properties of phenolic compounds towards α-glucosidase (Tousif et al. 2023). Silva et al. (2024) mentioned that secondary metabolites via the association of endophytic fungi and *Annona muricata* have biotechnological potential against different microorganisms and various cancer cells. According to the mentioned activities of endophytic fungi, we can summarize their potential activities as antioxidant, bacteriostatic, insecticidal, plant growth regulation, anticancer, and antitumor. Moreover, some problems related to endophytic fungi should be studied. First, the chemical structures of their secondary metabolites and the biological activities of endophytic fungi should be intensively analyzed, and their industrial production should be increased. The symbiosis mechanism and the molecular mechanism of the similar secondary metabolites between endophytic fungi and their host plants should be cleared (Wang et al. 2023).

12.6 Future Perspectives

Due to their innovative prospective applications in the pharmacological, medical, and agricultural fields, endophytic fungi should be taken into consideration/investigated for a sustainable, safe future. Alkaloids are among their naturally occurring substances and have several beneficial physiological and pharmacological qualities. Many ingredients could be found in the new era. This review has increased our understanding of how endophytes, especially mycorrhizal fungi and other microbial symbionts, affect the growth of the host plant. AMF symbiosis may be an essential organism for plants to maintain long-term health. Further research should be done about the significance of endophytic fungal bioactive compounds.

12.7 Conclusion

Fungal alkaloids are various natural metabolites produced by fungi and are of great importance because of their biosynthesis, pharmacological effects, and potential applications. Alkaloid drugs are the subject of current research because they exhibit bioactive properties like fungicidal, antibacterial, and anticancer action. Alkaloids are the chemicals that have received the most recent research. Alkaloids include derivatives of the indole molecule as well as pyrimidine, pyrrolizidine, quinoline, and isoquinoline, as well as amines and amides. Bioactive compounds make up alkaloids, which have complex chemical properties. Endophytes are used as effective new control agents against insect and herbivorous attacks that target their hosts because of the toxicity they produce when they are separated from various plant tissues. Focusing on the function of endophytes and their impact on sustainability and health is crucial.

References

Agusta A, Wulansari D, Nurkanto A, Fathoni A (2014) Biotransformation of protoberberine alkaloids by the endophytic fungus Coelomycetes AFKR-3 isolated from yellow moonsheed plant (*Archangelisia flava* (L.) Merr.). Procedia Chem 13:38–43

Akram S, Ahmed A, He P, Liu Y, Wu Y, Munir S, He Y (2023) Uniting the role of endophytic fungi against plant pathogens and their interaction. J Fungi 9(1):72

Alam B, Li JW, Ge Q, Khan MA, Gong JW, Mehmood S, Yuan YL, Gong WK (2021) Endophytic fungi: from symbiosis to secondary metabolite communications or vice versa? Front Plant Sci 12:791033

An CY, Li XM, Li CS, Wang MH, Xu GM, Wang BG (2013) Aniquinazolines A–D, four new quinazolinone alkaloids from marine-derived endophytic fungus *Aspergillus nidulans*. Mar Drugs 11(7):2682–2694

An L, Li CP, Zhang H, Wang ML, Kong LY, Yang MH (2020) Four cytochalasin alkaloids produced by *Chaetomium globosum*. Tetrahedron Lett 61(19):151838

Andrade SAL, Malik S, Sawaya A, Bottcher A, Mazzafera P (2013) Association with arbuscular mycorrhizal fungi influences alkaloid synthesis and accumulation in *Catharanthus roseus* and *Nicotiana tabacum* plants. Acta Physiol Plant 35:867–880

Archana O, Nagadesi PK (2022) Endophytic, non-endophytic fungal alkaloids and its applications. Saudi J Pathol Microbiol 7:4–19

Bacon CW, Porter JK, Robbins JD, Luttrell ES (1977) Epichloe typhina from toxic tall fescue grasses. Appl Environ Microbiol 34:76–81

Barros FAP, Rodrigues-Filho E (2005) Four spiroquinazoline alkaloids from *Eupenicillium* sp. isolated as an endophytic fungus from leaves of *Murraya paniculata* (Rutaceae). Biochem Syst Ecol 33(3):257–268

Bastas KK (2021) Impacts of climate changes on plant-beneficial microorganism interactions. Turk J Agric-Food Sci Technol 9:2594–2603

Bastias DA, Martínez-Ghersa MA, Ballaré CL, Gundel PE (2017) Epichloë fungal endophytes and plant defenses: not just alkaloids. Trends Plant Sci 22(11):939–948. https://doi.org/10.1016/j.tplants.2017.08.005

Blankenship JD, Spiering MJ, Wilkinson HH, Fannin FF, Bush LP, Schardl CL (2001) Production of loline alkaloids by the grass endophyte, *Neotyphodium uncinatum*, in defined media. Phytochemistry 58(3):395–401

Brundrett MC (2009) Mycorrhizal associations and other means of nutrition of vascular plants: understanding the global diversity of host plants by resolving conflicting information and developing reliable means of diagnosis. Plant Soil 320:37–77

Cai L, Wang J, Li Y, Qin M, Yin X, He Z, Kang J (2022) Global regulatory factor VeA upregulates the production of antitumor substances in endophytic *Fusarium solani*. Antonie Van Leeuwenhoek 115(8):1085–1100

Cook D, Gardner DR, Pfister JA (2014) Swainsonine-containing plants and their relationship to endophytic fungi. J Agric Food Chem 62(30):7326–7334

Courty PE, Buee M, Diedhiou AG, Frey-Klett P, Le Tacon F, Rineau F, Turpault MP, Uroz S, Garbaye J (2010) The role of ectomycorrhizal communities in forest ecosystem processes: new perspectives and emerging concepts. Soil Biol Biochem 42:679–698

Daley SK, Cordell GA (2021) Biologically significant and recently isolated alkaloids from endophytic fungi. J Nat Prod 84(3):871–897

Della Mónica IF, Vignale MV, Scervino JM, Iannone LJ, Novas MV (2023) Role of fungal endophytes on mycorrhizal-plant association and its impact on plant fitness. In: Microbial endophytes and plant growth. Academic Press, Cambridge, MA, pp 117–136

Demain AL (2014) Importance of microbial natural products and the need to revitalize their discovery. J Ind Microbiol Biotechnol 41:185–201

Dembitsky VM (2014) Naturally occurring bioactive Cyclobutane-containing (CBC) alkaloids in fungi, fungal endophytes, and plants. Phytomedicine 21(12):1559–1581. https://doi.org/10.1016/j.phymed.2014.07.005

Dewick PM (2002) Medicinal natural products: a biosynthetic approach. Wiley, Hoboken

Digra S, Nonzom S (2023) An insight into endophytic antimicrobial compounds: an updated analysis. In: Plant biotechnology reports. Springer, Cham, pp 1–31

Dissanayake RK, Ratnaweera PB, Williams DE, Wijayarathne CD, Wijesundera RLC, Andersen RJ, deSilva ED (2016) Antimicrobial activities of mycoleptodiscin B isolated from endophytic fungus *Mycoleptodiscus* sp. of *Calamus thwaitesii* Becc. J Appl Pharm Sci 6(1):001–006

Du FY, Li XM, Li CS, Shang Z, Wang BG (2012) Cristatumins A–D, new indole alkaloids from the marine-derived endophytic fungus *Eurotium cristatum* EN-220. Bioorg Med Chem Lett 22(14):4650–4653

Du FY, Li X, Li XM, Zhu LW, Wang BG (2017) Indolediketopiperazine alkaloids from Eurotium cristatum EN-220, an endophytic fungus isolated from the marine alga *Sargassum thunbergii*. Mar Drugs 15(2):24

El-Hawary SS, Moawad AS, Bahr HS, Abdelmohsen UR, Mohammed R (2020) Natural product diversity from the endophytic fungi of the genus *Aspergillus*. RSC Adv 10(37):22058–22079

Farhat H, Urooj F, Sohail N, Ullah S, Ali MS (2023) Plant disease management with amelioration of systemic resistance in soybean by endophytic fungi associated with GC-MS metabolic profiling of *Chaetomium* sp. Physiol Mol Plant Pathol:126. https://doi.org/10.1016/j.pmpp.2023.102049

Farzaei L, Sayyari M (2023) Trichoderma and sodium nitroprusside elicitation improves vinblastine and vincristine yield by increasing TIA pathway genes expression in cell suspension of *Catharanthus roseus*. Res Sq. https://doi.org/10.21203/rs.3.rs-2884766/v1

Feng C, Ma Y (2010) Isolation and anti-phytopathogenic activity of secondary metabolites from *Alternaria* sp. FL25, an endophytic fungus in *Ficus carica*. Chin J Appl Environ Biol 16:76–78

Gao N, Shang ZC, Yu P, Luo J, Jani KL, Kong LY, Yang MH (2017) Alkaloids from the endophytic fungus *Penicillium brefeldianum* and their cytotoxic activities. Chin Chem Lett 28(6):1194–1199

García-Latorre C, Rodrigo S, Marin-Felix Y, Stadler M, Santamaria O (2023) Plant-growth promoting activity of three fungal endophytes isolated from plants living in dehesas and their effect on *Lolium multiflorum*. Sci Rep 13(1):7354. https://doi.org/10.1038/s41598-023-34036-8

Gimenez C, Cabrera R, Reina M, Gonzalez-Coloma A (2007) Fungal endophytes and their role in plant protection. Curr Org Chem 11:707–720

Grum DS, Cook D, Baucom D, Mott IW, Gardner DR, Creamer R, Allen JG (2013) Production of the alkaloid swainsonine by a fungal endophyte in the host *Swainsona canescens*. J Nat Prod 76(10):1984–1988

Guo B, Wang Y, Sun X, Tang K (2008) Bioactive natural products from endophytes. Appl Biochem Microbiol 44:136–142

Han D, Wang K, Long F, Zhang W, Yao X, Chen S (2022) Effects of endophytic fungi on the secondary metabolites of *Hordeum bogdanii* under alkaline stress. AMB Express 12:73. https://doi.org/10.1186/s13568-022-01414-w

Harwoko H, Lee J, Hartmann R, Kurtán T, Müller WEG, Feldbrügge M, Kalscheuer R, Ancheeva E, Daletos G, Frank M, Liu Z, Proksch P, Azacoccones FH (2020) Azacoccones FH, new flavipin-derived alkaloids from an endop8hytic fungus *Epicoccum nigrum* MK214079. Fitoterapia 146:104698

Hashem AH, Attia MS, Kandil EK, Fawzi MM, Abdelrahman AS, Khader MS, Khodaira MA, Emam AE, Goma MA, Abdelaziz AM (2023) Bioactive compounds and biomedical applications of endophytic fungi: a recent review. Microb Cell Factories 22:107

Heinrich M, Barnes J, Gibbons S, Williamson EM (2012) Fundamentals of pharmacognosy and phytotherapy, 2nd edn. Churchhill Livingstone Elsevier, London

Herrera J, Khidir HH, Eudy DM, Porras-Alfaro A, Natvig DO, Sinsabaugh RL (2010) Variation in root-associated fungal endophytes: some taxonomic consistency at a transcontinental scale. Mycologia 102:1012–1026

Hoveland CS (1993) Importance and economic significance of the Acremonium endophytes to performance of animals and grass plant. Agric Ecosyst Environ 44:3–12

Jha P, Kaur T, Chhabra I, Panja A, Paul S, Kumar V, Malik T (2023) Endophytic fungi: hidden treasure chest of antimicrobial metabolites interrelationship of endophytes and metabolites. Front Microbiol 14:1227830

Jiang CS, Zhen-Fang Z, Xiao-Hong Y, Lan LF, Gu YC, Ye BP, Guo YW (2018) Antibacterial sorbicillin and diketopiperazines from the endogenous fungus Penicillium sp. GD6 associated Chinese mangrove *Bruguiera gymnorrhiza*. Chin J Nat Med 16(5):358–365

Jumpponen A, Jones KL (2009) Massively parallel 454 sequencing indicates hyperdiverse fungal communitiesin temperate *Quercus macrocarpa phyllosphere*. New Phytol 184:438–448

Khare E, Mishra J, Arora NV (2018) Multifaceted interactions between endophytes and plant: developments and prospects. Front Microbiol 9:2732

Khazaal HT, Khazaal MT, Abdel-Razek AS, Hamed AA, Ebrahim HY, Ibrahim RR, Bishr M, Mansour YE, el Dib RA, Soliman HSM (2023) Antimicrobial, antiproliferative activities and molecular docking of metabolites from *Alternaria alternata*. AMB Exp 13(1):68. https://doi.org/10.1186/s13568-023-01568-1

Khidir HK, Eudy DM, Porras-Alfaro A, Herrera J, Natvig DO, Sinsabaugh RL (2010) A general suite of fungal endophytes dominate the roots of two dominant grasses in a semiarid grassland. J Arid Environ 74:35–42

Kobayashi DY, Palumbo JD (2000) Bacterial endophytes and their effects on plants and uses in agriculture. In: Bacon CW, White JF (eds) Microbial endophytes. Marcel Dekker, New York, pp 199–236

Koolen HHF, Soares ER, da Silva FMA, de Almeida RA, de Souza ADL, de Medeiros LS, Filho ER, de Souza AQL (2012) An antimicrobial alkaloid and other metabolites produced by *Penicillium* sp. An endophytic fungus isolated from *Mauritia flexuosa* L. f. Quím Nova 35:771–774

Kuklev DV, Dembitsky VM (2016) Chemistry, origin, antitumor and other activities of fungal homo-dimeric alkaloids. Mathews J Pharm Sci 1(1):004

Kumar A, Patil D, Rajamohanan PR, Ahmad A (2013) Isolation, purification and characterization of vinblastine and vincristine from endophytic fungus *Fusarium oxysporum* isolated from *Catharanthus roseus*. PLoS One 8(9):e71805

Kumara PM, Soujanya KN, Ravikanth G, Vasudeva R, Ganeshaiah KN, Shaanker RU (2014) Rohitukine, a chromone alkaloid and a precursor of flavopiridol, is produced by endophytic fungi isolated from Dysoxylum binectariferum Hook. f and Amoora rohituka (Roxb). Wight & Arn. Phytomedicine 21(4):541–546

Kusari S, Spiteller M (2012) Metabolomics of endophytic fungi producing associated plant secondary metabolites: progress, challenges and opportunities. In: Roessner U (ed) Metabolomics. InTech, Rijeka, pp 241–266

Lai D, Brötz-Oesterhelt H, Müller WEG, Wray V, Proksch P (2013) Bioactive polyketides and alkaloids from *Penicillium citrinum*, a fungal endophyte isolated from *Ocimum tenuiflorum*. Fitoterapia 91:100–106. https://doi.org/10.1016/j.fitote.2013.08.017

Li DL, Li XM, Li TG, Dang HY, Wang BG (2008) Dioxopiperazine alkaloids produced by the marine mangrove derived endophytic fungus *Eurotium rubrum*. Helv Chim Acta 91(10):1888–1893

Li XB, Li L, Zhu RX et al (2015) Tetramic acids and pyridone alkaloids from the endolichenic fungus *Tolypocladium cylindrosporum*. J Nat Prod 78(9):2155–2160

Li Z, Zhang BW, Jiang L, Wang H, Ma QY, Wang HF, Zhang J, Chen FL, Zhao YX, Luo DQ (2023) Two new alkaloids from the endophytic fungus *Schizophyllum* sp. HM230 isolated from *Vincetoxicum mongolicum* Maxim. Nat Prod Res. https://doi.org/10.1080/14786419.2023.2176493

Lin Z, Wen J, Zhu T et al (2008) Chrysogenamide A from an endophytic fungus associated with *Cistanche deserticola* and its neuroprotective effect on SH-SY5Y cells. J Antibiot 61(2):81–85

Lin X, Ai W, Li M et al (2020) Collacyclumines A–D from the endophytic fungus *Colletotrichum salsolae* SCSIO 41021 isolated from the mangrove *Kandelia candel*. Phytochemistry 171:112237

Ma YM, Qiao K, Kong Y, Li MY et al (2017) A new isoquinolone alkaloid from an endophytic fungus R22 of *Nerium indicum*. Nat Prod Res 31(8):951–958

Madouh TA, Quoreshi AM (2023) The function of arbuscular mycorrhizal fungi associated with drought stress resistance in native plants of arid desert ecosystems: a review. Diversity-Basel 15(3):391

Maehara S, Simanjuntak P, Maetani Y et al (2013) Ability of endophytic filamentous fungi associated with *Cinchona ledgeriana* to produce Cinchona alkaloids. J Nat Med 67:421–423

Metwaly AM, Ghoneim MM, Musa A (2015) Two new antileishmanial diketopiperazine alkaloids from the endophytic fungus *Trichosporum* sp. Der Pharma Chem 7(11):322–327

Na R, Jiajia L, Dongliang Y, Yingzi P, Juan H et al (2016) Indentification of vincamine indole alkaloids producing endophytic fungi isolated from *Nerium indicum*, Apocynaceae. Microbiol Res 192:114–121

Nicoletti R, Andolfi A, Becchimanzi A, Salvatore MM (2023) Anti-insect properties of Penicillium secondary metabolites. Microorganisms 11(5):1302

Ortega HE, Graupner PR, Asai Y, Tendyke K, Qiu D, Shen YY, Rios N, Arnold AE, Coley PD, Kursar TA, Gerwick WH, Cubilla-Rios L (2013) Mycoleptodiscins A and B, cytotoxic alkaloids from the endophytic fungus *Mycoleptodiscus* sp. F0194. J Nat Prod 76(4):741–744

Ortega HE, Torres-Mendoza D, Caballero EZ, Cubilla-Rios L (2021) Structurally uncommon secondary metabolites derived from endophytic fungi. J Fungi 7(7):570

Palem PP, Kuriakose GC, Jayabaskaran C (2015) An endophytic fungus, Talaromyces radicus, isolated from *Catharanthus roseus*, produces vincristine and vinblastine, which induce apoptotic cell death. PLoS One 10(12):e0144476

Pan F, Hou K, Gao F, Hu B, Chen Q, Wu W (2014) Peimisine and peiminine production by endophytic fungus *Fusarium* sp. isolated from *Fritillaria unibracteata* var. wabensis. Phytomedicine 21(8–9):1104–1109

Pan F, Su X, Hu B, Yang N, Chen Q, Wu W (2015) *Fusarium redolens* 6WBY3, an endophytic fungus isolated from *Fritillaria unibracteata* var. wabuensis, produces peimisine and imperialine-3β-d-glucoside. Fitoterapia 103:213–221

Panaccione DG, Beaulieu WT, Cook D (2014) Bioactive alkaloids in vertically transmitted fungal endophytes. Funct Ecol 28:299–314. https://doi.org/10.1111/1365-2435.12076

Peng F, Hou SY, Zhang TY et al (2019) Cytotoxic and antimicrobial indole alkaloids from an endophytic fungus *Chaetomium* sp. SYP-F7950 of *Panax notoginseng*. RSC Adv 9(49):28754–28763

Philippe G (2016) Lolitrem B and indole diterpene alkaloids produced by endophytic fungi of the genus Epichloë and their toxic effects in livestock. Toxins 8(2):47

Pinheiro EAA, Carvalho JM, dos Santos DCP et al (2013) Antibacterial activity of alkaloids produced by endophytic fungus *Aspergillus* sp. EJC08 isolated from medicinal plant *Bauhinia guianensis*. Nat Prod Res 27(18):1633–1638

Porras-Alfaro A, Bayman P (2011) Hidden fungi, emergent properties: endophytes and microbiomes. Phytopathology 49(1):291

Poveda J, Eugui D, Abril-Urías P, Velasco P (2021) Endophytic fungi as direct plant growth promoters for sustainable agricultural production. Symbiosis 85:1–19

Pund S, Joshi A (2017) Nano-architectures for neglected tropical protozoal diseases: challenges and state of the art. Nano Microscale Drug Deliv Syst 1:439–480

Raghukumar C (2008) Marine fungal biotechnology: an ecological perspective. Fungal Divers 31:5–19

Shan W-G, Ying Y-M, Yu H-N, Yu H-N, Liu W-H, Zhan Z-J (2010) Diketopiperazine alkaloids from *Penicillium* spp. HS-3, an endophytic fungus in *Huperzia serrata*. Helv Chim Acta 93(4):772–776

Shweta S, Gurumurthy BR, Ravikanth G, Ramanan US, Shivanna MB (2013) Endophytic fungi from *Miquelia dentata* Bedd., produce the anticancer alkaloid, camptothecine. Phytomedicine 20(3–4):337–342

Silva J, Garcia J, Guimarães R et al (2023) Alkaloids from fungi. In: Carocho M, Heleno SA, Barros L (eds) Natural secondary metabolites. Springer, Cham. https://doi.org/10.1007/978-3-031-18587-8_16

Silva IMM, Silva RM, Paula VB, Estevinho LM (2024) Biological activities of endophytic fungi isolated from *Annona muricata* Linnaeus: a systematic review. Braz J Biol:84. https://doi.org/10.1590/1519-6984.259525

Song R, Wang J, Sun L, Zhang Y, Zhang Y, Ren Z, Zhao B, Lu H (2019) The study of metabolites from fermentation culture of *Alternaria oxytropis*. BMC Microbiol 19:1–8

Song L-S, Huo J, Wan L, Pan L, Jiang N, Fu J, Wei S, He L (2023) Differences and biocontrol potential of haustorial endophytic fungi from *Taxillus chinensis* on different host plants. BMC Microbiol 23(1). https://doi.org/10.1186/s12866-023-02878-x

Staniek A, Woerdenbag HJ, Kayser O (2008) Endophytes: exploiting biodiversity for the improvement of natural product–based drug discovery. J Plant Interact 3:75–93

Strobel G (2018) The emergence of endophytic microbes and their biological promise. J Fungi 4:57

Su J, Yang M (2015) Huperzine A production by *Paecilomyces tenuis* YS-13, an endophytic fungus isolated from Huperzia serrata. Nat Prod Res 29(11):1035–1041

Tejesvi MV, Kini KR, Prakash HS, Subbiah V, Shetty HS (2007) Genetic diversity and antifungal activity of species of *Pestalotiopsis* isolated as endophytes from medicinal plants. Fungal Divers 24:37–54

Tian Y, Li Y (2022) A review on bioactive compounds from marine-derived *Chaetomium* species. J Microbiol Biotechnol 32(5):541

Toppo P, Kagatay LL, Gurung A, Singla P, Chakraborty R, Roy S, Mathur P (2023) Endophytic fungi mediates production of bioactive secondary metabolites via modulation of genes involved in key metabolic pathways and their contribution in different biotechnological sector. 3 Biotech 1:191. https://doi.org/10.1007/s13205-023-03605-z

Tousif MI, Tauseef S, Nabeelah S, Sharmeen J, Zengin G, Legoabe L, Imran M, Mahomoodally MF (2023) Phenolics from endophytic fungi as natural α-glucosidase inhibitors: a comprehensive review. J Mol Struct 1291:135852. https://doi.org/10.1016/j.molstruc.2023.135852

Tsuchinari M, Shimanuki K, Hiramatsu F, Murayama F, Koseki T, Yoshihito S (2007) Fusapyridons A and B, novel pyridone alkaloids from an endophytic fungus, Fusarium sp. YG-45. Z Naturforsch B 62(9):1203–1207

Uzma F, Mohan CD, Hashem A, Konappa NM, Rangappa S, Kamath PV, Singh BP, Mudili V, Gupta VK, Siddaiah CN (2018) Endophytic fungi-alternative sources of cytotoxic compounds: a review. Front Pharmacol 9:309

Vandenkoornhuyse P, Baldauf SL, Leyval C, Straczek J, Young, & J.P.W. (2002) Extensive fungal diversity in plant roots. Science 295:2051–2051

Verma VC, Lobkovsky E, Gange AC (2011) Piperine production by endophytic fungus *Periconia* sp. isolated from *Piper longum* L. J Antibiot 64(6):427–431

Wang LW, Zhang YL, Lin FC, Hu YZ, Zhang CL (2011) Natural products with antitumor activity from endophytic fungi. Mini-Rev Med Chem 11:1056–1074

Wang H, Yin G, Yu CH, Wang Y, Sun ZL (2013) Inhibitory effect of sanguinarine on PKC-CPI-17 pathway mediating by muscarinic receptors in dispersed intestinal smooth muscle cells. Res Vet Sci 95(3):1125–1133

Wang A, Zhao S, Gu G, Xu D, Zhang X, Lai D, Zhou L (2020) Rhizovagine A, an unusual dibenzo-α-pyrone alkaloid from the endophytic fungus *Rhizopycnis vagum* Nitaf22. RSC Adv 10(47):27894–27898

Wang Z, Wang L, Pan Y, Zheng X, Liang X, Sheng L, Zhang D, Sun Q, Wang Q (2023) Research advances on endophytic fungi and their bioactive metabolites. Bioprocess Biosyst Eng 46:165–170

Wijeratne EK, He H, Franzblau SG, Hoffman AM, Gunatilaka AA (2013) Phomapyrrolidones A–C, antitubercular alkaloids from the endophytic fungus Phoma sp. NRRL 46751. J Nat Prod 76(10):1860–1865

Yang SX, Xiao J, Laatsch H, Holstein JJ, Dittrich B, Zhang Q, Gao J-M (2012) Fusarimine, a novel polyketide isoquinoline alkaloid, from the endophytic fungus *Fusarium* sp. LN12, isolated from *Melia azedarach*. Tetrahedron Lett 53(47):6372–6375

Yunianto P, Rusman Y, Saepudin E, Suwarso WP, Sumaryono W (2014) Alkaloid (Meleagrine and Chrysogine) from endophytic fungi (Penicillium sp.) of *Annona squamosa* L. Pak J Biol Sci 17(5):667–674

Zhang Y, Han T, Ming Q, Wu L, Rahman K, Qin L (2012) Alkaloids produced by endophytic fungi: a review. Nat Prod Commun 7(7):963–968

Zhang P, Li XM, Wang JN, Wang BG (2015a) Oxepine-containing diketopiperazine alkaloids from the algal-derived endophytic fungus *Paecilomyces variotii* EN-291. Helv Chim Acta 98(6):800–804

Zhang P, Li XM, Liu H, Li X, Wang BG (2015b) Two new alkaloids from *Penicillium oxalicum* EN-201, an endophytic fungus derived from the marine mangrove plant *Rhizophora stylosa*. Phytochem Lett 13:160–164

Zhang P, Yuan XL, Du YM, Zhang HB, Shen GM, Zhang ZF, Liang YJ, Zhao DL, Xu K (2019) Angularly prenylated indole alkaloids with antimicrobial and insecticidal activities from an endophytic fungus *Fusarium sambucinum* TE-6L. J Agric Food Chem 67(43):11994–12001

Zhang XL, Lu BG, Bai J, Li N (2022) Alkaloids isolated from the endophytic fungus *Fusarium concentricum*. Nat Prod Res. https://doi.org/10.1080/14786419.2022.2154346

Zheng CJ, Li L, Zou JP, Zou JP, Han T, Qin LP (2012) Identification of a quinazoline alkaloid produced by *Penicillium vinaceum*, an endophytic fungus from *Crocus sativus*. Pharm Biol 50(2):129–133

Zheng R, Li S, Zhang X, Zhao C (2021) Biological activities of some new secondary metabolites isolated from endophytic fungi: a review study. Int J Mol Sci 22(2):959

Zhong R, Zhang L, Zhang X (2022) Allelopathic Effects of foliar epichloë endophytes on belowground arbuscular mycorrhizal fungi: a meta-analysis. Agriculture 12:1768. https://doi.org/10.3390/agriculture12111768

Zhu M, Zhang X, Feng H, Dai J, Li J, Che Q, Gu Q, Zhu T, Li D (2016) Penicisulfuranols A–F, alkaloids from the mangrove endophytic fungus Penicillium janthinellum HDN13-309. J Nat Prod 80(1):71–75

Zou R, Wei C, Zhang X, Zhou D, Xu J (2022) Alkaloids from endophytic fungus Aspergillus fumigatus HQD24 isolated from the Chinese mangrove plant Rhizophora mucronata. Nat Prod Res 36(19):5069–5073

Endophytic Fungi as a Promising Source for Sulfur-Containing Compounds

13

Amer M. Abdelaziz, Mohamed S. Attia, Ahmed S. Doghish, and Amr H. Hashem

Abstract

Fungal endophytes are a diverse group of fungi that live inside plant tissues without causing apparent harm to the host plant. This symbiotic relationship can be beneficial for both the fungus and the plant. The fungi receive nutrients and a protected environment, while the plants may benefit from improved resistance to pathogens, enhanced growth, and increased tolerance to environmental stress. Fungal endophytes are known to produce a variety of bioactive compounds, which are molecules that can have a significant impact on living organisms, including potential medicinal and agricultural applications. Here are some examples of bioactive compounds produced by fungal endophytes. Research in this field is ongoing, and scientists are continually discovering new bioactive compounds from fungal endophytes. Sulfur-containing compounds are a diverse group of molecules that contain sulfur atoms in their chemical structure. Sulfur is an essential element for life, and it plays crucial roles in various biological processes. Thus, this chapter discusses the production of sulfur-containing compounds from endophytic fungi. Also, it shows different groups of sulfur-containing compounds that have different biological applications. Finally, displays different biological applications, e.g., antimicrobial as well as anticancer activities.

A. M. Abdelaziz (✉) · M. S. Attia · A. H. Hashem
Botany and Microbiology Department, Faculty of Science, Al-Azhar University, Cairo, Egypt
e-mail: amermorsy@azhar.edu.eg

A. S. Doghish
Department of Biochemistry, Faculty of Pharmacy, Badr University in Cairo (BUC), Badr, Cairo, Egypt

Biochemistry and Molecular Biology Department, Faculty of Pharmacy (Boys), Al-Azhar University, Nasr City, Cairo, Egypt

K. A. Abd-Elsalam, S. F. AbuQamar (eds.), *Fungal Endophytes Volume I*,
https://doi.org/10.1007/978-981-97-7312-1_13

Keywords

Fungal endophytes · Bioactive compounds · Sulfur-containing compounds · Antimicrobial activity · Anticancer activity

13.1 Introduction

Sulfur, a vital constituent in the realm of life, stands as a paramount element and ranks seventh in terms of abundance among the minerals found within the human body. Sulfur, an element classified as a chalcogen and nonmetal, is situated within the group of elements on the periodic table that display a wide range of characteristics in terms of redox states. The diverse array of chemical structures exhibited by sulfur-containing compounds, influenced by the element's multiple oxidation states, leads to the emergence of unique mechanisms of action for sulfur-containing natural products produced as secondary metabolites. Natural sulfur compounds are essential for the sustenance of all living organisms, as they are ubiquitously present across diverse taxonomic domains. Fungi represent a substantial and economically advantageous reservoir of these chemical compounds (Francioso et al. 2020).

Endophytes are defined as organisms that live in tissues beneath the epidermal cell layers and have no obvious negative effects on the host (Khalil et al. 2021). The most common microorganisms found as endophytes are fungi and bacteria. Fungi are the most commonly isolated endophytes. Fungal endophytes form a mutualistic or commensal relationship with their host plants. They colonize the intercellular spaces or live inside the plant cells, often within the roots, stems, leaves, or seeds (Alam et al. 2021). The presence of endophytes can confer various benefits to the host plant, such as enhanced nutrient uptake, stress tolerance, and protection against pathogens and herbivores (Chaudhary et al. 2022). Fungal endophytes belong to various fungal groups, including *Ascomycota* and *Basidiomycota*. Each plant species can harbor multiple endophyte species, and the composition of endophyte communities can vary depending on the plant species, tissue type, and environmental conditions (Zuo et al. 2021). Fungal endophytes play important ecological roles, which help plants adapt to environmental stresses, such as drought, high temperatures, or nutrient deficiencies. Endophytes can also contribute to plant defense against pathogens and herbivores by producing bioactive compounds that deter or inhibit their growth (Verma et al. 2022). Furthermore, fungal endophytes can influence plant community dynamics and nutrient cycling in ecosystems. Fungal endophytes are known to produce a wide array of secondary metabolites, including alkaloids, terpenoids, phenolics, and peptides (Rai et al. 2021). These bioactive compounds can have various biological activities, such as antimicrobial, antifungal, insecticidal, and anticancer properties. Some of these metabolites have been investigated for their potential pharmaceutical and agricultural applications (Hashem et al. 2023). This chapter explores the potential of endophytic fungi as a source for sulfur-containing compounds, discussing their importance in industries, diversity, bioactivity, isolation methods, and sustainability as an environmentally friendly alternative.

13.2 Biodiversity of Endophytic Fungi Producing Sulfur-Containing Compounds

Fungal endophytes can live inside terrestrial plants across global ecosystems. Endophytic microorganisms have been identified in various anatomical parts of plants, including the roots, stems, leaves, fruits, flowers, bark, and scales (Pirttilä et al. 2008). Fungal endophytes are found in both natural and stress habitats (Hashem et al. 2023; Selim et al. 2017; Yadav and Meena 2021). The distribution of plant microorganisms is influenced by a variety of factors, including types of agriculture, plant parts, climate, ecology, and geographical location (Lori et al. 2017; Russo et al. 2016). The presence and abundance of endophytic fungi can be influenced by various environmental factors, including atmospheric humidity and rainfall. It has been observed that the number of fungal isolates tends to decrease during the winter season in comparison to other seasons throughout the year (Zheng et al. 2016). Endophytic fungi have been documented in a wide range of plant species, exhibiting a significant diversity (Gouda et al. 2016; Sun and Guo 2012). The identification of endophytic fungi has revealed that the majority of these organisms belong to the phylum *Ascomycota* followed by *Basidiomycota* according to their prevalence. There are many genera of endophytic fungi that have the ability to produce sulfur-containing compounds, such as *Aspergillus, Penicillium, Chaetomium, Trichoderma, Epicoccum, Cladosporium, Fusarium, Alternaria, Phoma, Lasiodiplodia, Leptosphaeria,* and others.

13.2.1 *Aspergillus*

Aspergillus is considered the most isolated group among other fungal endophytes according to recent literature (Hashem et al. 2023). Also, the following genera are common, *Penicillium, Alternaria, Colletotrichum, Fusarium, Gibberella, Glomerella, Guignardia, Leptosphaerulina, Nigrospora, Phoma, Phomopsis, and Xylaria* (Zheng et al. 2016). Wu et al. (2019) isolated endophytic *Aspergillus micronesiensis,* which produced two cysteine residues containing merocytochalasans (cytochalasins A and B), which exhibited antibacterial activity toward Gram-positive bacteria. Ma et al. (2016) used the *Ficus carica* plant for the isolation of endophytic *A. tamari,* where results confirmed that a dipeptide compound (malformin E) was produced from this fungus. Moreover, *A. versicolor* 0312 and *A. versicolor* F210 were isolated from the stems of *Paris polyphylla var. yunnanensis* and bulbs of *Lycoris radiate,* respectively, which produced sulfur-containing compounds bisdethiobis(methylsulfanyl)apoaranotin and methylsulfonyl unit, 21-epi-taichunamide D (He et al. 2020; Li et al. 2020).

13.2.2 *Penicillium*

Penicillium is another group of endophytic ascomycetes, which are produced from different bioactive compounds. *Penicillium brocae* MA-231 was isolated from *Avicennia marina* and used for the production of many compounds penicibrocazines F–I, spirobrocazines A–B, penicibrocazines A–E, and Brocazines A-G in different studies (Meng et al. 2017, 2014, 2016, 2015). Guimarães et al. (2010) isolated *Penicillium crustosum* from *Viguiera robusta* which produced a compound containing sulfur named fusaperazine E. Likewise, *Penicillium crustosum* was isolated from the root of *Sonneratia caseolaris* and produced a compound named penixylarins D (Yu et al. 2019a). Also, the root of *Sonneratia caseolaris* was used for the isolation of endophytic *Penicillium janthinellum* HDN13-309, which produced many sulfur-containing compounds such as penispirozines A–D and penicisulfuranols A–D (Zhu et al. 2020; Zhu et al. 2017). Moreover, Kajula et al. (2016) isolated endophytic *Penicillium raciborskii* (TRT59) from *Rhododendron tomentosum* plant, where this fungus can produce bioactive compounds outovirins A–C.

13.2.3 *Trichoderma*

Endophytic *Trichoderma* has the ability to produce promising bioactive compounds. Endophytic *Trichoderma harzianum* was isolated from *Zingiber officinale* and exhibited a bioactive compound named pretrichodermamide G (Harwoko et al. 2021). Furthermore, Zhao et al. (2020) isolated *Trichoderma harzianum* D13 from the root of mangrove plant *Excoecaria agallocha Linn*, where this fungus had the ability to produce sulfur-containing compound called trichodermamide G. Moreover, bamboo leaf was used for isolation *Trichoderma* sp. BCC 5926, which produced pretrichodermamide A compound (Seephonkai et al. 2006).

13.2.4 *Chaetomium*

Chaetomium is one of the known endophytic fungi for the production of bioactive compounds, which are used in medical and pharmaceutical applications. *Chaetomium* sp. 88194 was isolated from *Cymbidium goeringii* plant, where compounds chaetocochin G, oidioperazine E, and chetoseminudin E were produced (Wang et al. 2015). Furthermore, Yu et al. (2018) used *Huperzia serrata Trev* for the isolation of endophytic *Chaetomium* sp. M336 and produced a new compound called 6-formamide-chetomin. Moreover, Peng et al. (2019) isolated *Chaetomium* sp. SYP-F7950 from *Panax notoginseng* plant, where results illustrated the production of two new compounds chetoseminudin F and G.

13.2.5 *Alternaria* and *Fusarium*

An endophytic strain of *Alternaria* sp. was successfully isolated from *Polygonum senegalense Meisn.*, a member of the Polygonaceae family. This particular strain exhibited the ability to produce two intriguing compounds, namely, alternariol 5-O-sulfate and alternariol 5-O-methyl ether-40-O-sulfate (Aly et al. 2008). Furthermore, Ibrahim et al. (2016) successfully isolated *Fusarium chlamydosporium* from the foliage of *Anvillea garcinii* (Burm.f.) DC. Subsequently, they observed that this particular fungal species exhibits the remarkable capability to synthesize a novel compound known as fusarithioamide B. Moreover, the *Davidia involucrata* plant was used for the isolation of endophytic *Fusarium* sp. (CTGU-ZL-34); a new compound called fusaodavinvin was produced from this promising strain.

13.2.6 Other Genera

Cladosporium cladosporioides MA-299 was isolated from *Bruguiera gymnorrhiza and produced new compounds named* thiocladospolides A–D (Zhang et al. 2019). Also, the root of *Avicennia marina (Forssk.) Vierh* was used for isolation endophytic *Cladosporium oxysporum, where it has the ability to produce* hiocladospolides F–J compounds (Wang et al. 2020). Furthermore, *Emericella* sp. AST0036 was isolated from *Astragalus lentiginosus* and produced the Secoemestrin D compound (Xu et al. 2013). Also, endophytic *Epicoccum nigrum was isolated from* the leaves of *Lysidice rhodostegia,* where it produced epicoccin I, ent-epicoccin G, and epicoccins J–T compounds (Wang et al. 2010b). Moreover, *Phoma* fungi have the ability to produce novel sirodesmin H (Pedras et al. 1990) and phomazines A–C (Kong et al. 2014). Furthermore, endophytic *Bipolaris sorokiniana* A606, *Botryosphaeria mamani*, which were isolated from *Huperzia serrate, Pogostemon cablin, and Bixa orellana* L., respectively, have the ability to produce different sulfur-containing compounds bionectin, E isocochlioquinones D, and botryosulfuranols A–C (Barakat et al. 2019; Wang et al. 2016). Guimarães et al. (2010) isolated endophytic *Colletotrichum gloeosporioides, which produced* colletopiperazine sulfur compound. Also, endophytic *Lasiodiplodia pseudotheobromae and Leptosphaeria* sp. XL026 were isolated from *Illigera rhodantha* and *Panax notoginseng,* respectively (Chen et al. 2019; Wei et al. 2014). Other endophytic *Neosartorya udagawae* HDN13-313 (Yu et al. 2019b), *Nigrospora sphaerica* (Centko et al. 2017), *Paraphaeosphaeria neglecta* FT462 (Li et al. 2017), *Pestalotiopsis* sp. HS30 (Wu et al. 2019), and *Setosphaeria rostrata* (Centko et al. 2017) were reported for the production of many sulfur-containing compounds (Table 13.1).

Table 13.1 Some examples for *Endophytic fungi* as a promising source for sulfur-containing compounds

Endophytic fungi	Plant	Compound	Ref.
Alternaria sp.	*Polygonum senegalense Meisn.* (Polygonaceae)	Alternariol 5-O-sulfate and alternariol 5-O-methyl ether-40 - O-sulfate	Aly et al. (2008)
Ascochyta sp. AJ 117309	*Taxus cuspidata var. nana Rehd*	FE399	Verekar et al. (2014)
Aspergillus micronesiensis	*Phyllanthus glaucus*	Cyschalasins A and cyschalasins B	Wu et al. (2019)
Aspergillus sydowii	*Scapania ciliata S. Lac*	Sydoxanthone A and sydoxanthone B	Song et al. (2013)
Aspergillus tamarii	*Ficus carica*	Malformin E	Ma et al. (2016)
Aspergillus terreus BCC 4651	Tree hole	Bisdethiobis(methylsulfanyl) apoaranotin	Haritakun et al. (2012)
Aspergillus versicolor 0312	*Paris polyphylla var. yunnanensis*	pentacyclic diketopiperazine with a 4,5-dihydrooxepine moiety versicolor A	He et al. (2020)
Aspergillus versicolor F210	*Lycoris radiate*	methylsulfonyl unit, 21-epi-taichunamide D	Li et al. (2020)
Bionectria sp. Y1085	*Huperzia serrata*	Bionectin E	Yang et al. (2019)
Bipolaris sorokiniana A606	*Pogostemon cablin*	isocochlioquinones D	Wang et al. (2016)
Botryosphaeria mamani	*Bixa orellana* L.	Botryosulfuranols A–C	Barakat et al. (2019)
Chaetomium sp. 88,194	*Cymbidium goeringii*	Chaetocochin G, oidioperazine E, and chetoseminudin E	Wang et al. (2015)
Chaetomium sp. M336	*Huperzia serrata Trev*	6-Formamide-chetomin	Yu et al. (2018)
Chaetomium sp. SYP-F7950	*Panax notoginseng*	Chetoseminudin F and G	Peng et al. (2019)
Cladosporium cladosporioides MA-299	*Bruguiera gymnorrhiza*	Thiocladospolides A–D	Zhang et al. (2019)
Cladosporium oxysporum	*Avicennia marina*	Thiocladospolides F–J	Wang et al. (2020)
Colletotrichum gloeosporioides,	*Viguiera robusta*	Colletopiperazine	Guimarães et al. (2010)
Emericella sp. AST0036	*Astragalus lentiginosus*	Secoemestrin D	Xu et al. (2013)
Epicoccum nigrum	*Lysidice rhodostegia*	epicoccin I, ent-epicoccin G, and epicoccins J–T	Wang et al. (2010b)
Exserohilum holmii	*Dactyloctenium aegyptium*	Exserohilone and 9,10-dihydroexserohilon	Sugawara et al. (1985)
Fusarium chlamydosporium	*Anvillea garcinii* (Burm.f.) DC.	Fusarithioamide B	Ibrahim et al. (2016)

(continued)

Table 13.1 (continued)

Endophytic fungi	Plant	Compound	Ref.
Fusarium sp. (CTGU-ZL-34).	*Davidia involucrata*	fusaodavinvin	Zhang et al. (2013)
Guignardia sp. IFB-E028	*Hopea hainanensis*	Guignasulfide	Wang et al. (2010a)
Lasiodiplodia pseudotheobromae	*Illigera rhodantha* (Hernandiaceae)	Lasiodiplines A–D	Wei et al. (2014)
Leptosphaeria sp. XL026	*Panax notoginseng*	Leptosphin A–C	Chen et al. (2019)
Neosartorya udagawae HDN13–313	*Avicennia marina*	Neosartoryonc A	Yu et al. (2019b)
Nigrospora sphaerica	Germinating fescue seed	Epoxyexserohilone	Centko et al. (2017)
Paraphaeosphaeria neglecta FT462	*Lycopodiella cernua*	Paraphaeosphaerides E–H	Li et al. (2017)
Penicillium brocae MA-231	*Avicennia marina*	Penicibrocazines F–I, Spirobrocazines A–B, Penicibrocazines A–E, and Brocazines A-G	Meng et al. (2017, 2014, 2016, 2015)
Penicillium crustosum	*Viguiera robusta*	Fusaperazine E	Guimarães et al. (2010)
Penicillium crustosum PRB-2	*Sonneratia caseolaris*	Penixylarins D	Yu et al. (2019a)
Penicillium janthinellum HDN13-309	*Sonneratia caseolaris*	Penispirozines A–D and penicisulfuranols A–D	Zhu et al. (2020); Zhu et al. (2017)
Penicillium raciborskii (TRT59)	*Rhododendron tomentosum*	Outovirin A–C	Kajula et al. (2016)
Pestalotiopsis sp. HS30	*Isodon xerophilus*	Pestaloamides A and B	Wu et al. (2019)
Phoma lingam isolate Leroy	Rapeseed	Sirodesmin H	Pedras et al. (1990)
Phoma sp. OUCMDZ-1847	*Kandelia candel*	phomazines A–C	Kong et al. (2014)
Phomopsis glabrae	*Pongamia pinnata*	PM181110	Su et al. (2020)
Setosphaeria rostrata	*Costus speciosus*	Rostratazines A–C	Centko et al. (2017)
Tolypocladium sp.	*Quercus virginiana Miller*	Scheme 54794 and Scheme 54796	Chu et al. (1993)
Trichoderma harzianum	*Zingiber officinale*	Pretrichodermamide G	Harwoko et al. (2021)
Trichoderma harzianum D13	*Excoecaria agallocha Linn*	Trichodermamide G	Zhao et al. (2020)
Trichoderma sp. BCC 5926	Bamboo leaf	Pretrichodermamide A	Seephonkai et al. (2006)

(continued)

Table 13.1 (continued)

Endophytic fungi	Plant	Compound	Ref.
Xylaria sp. YM 311647	*Azadirachta indica*	16-*O*-sulfo-18-norisopimar-7-en-4α,16-diol and 9-deoxy-hymatoxin A	Wu et al. (2014)

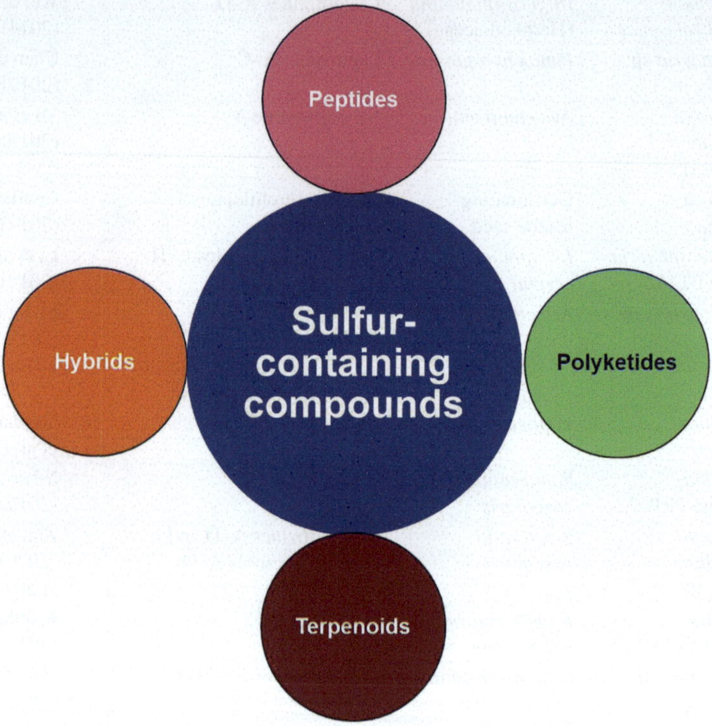

Fig. 13.1 Major groups of sulfur-containing compounds from endophytic fungi

13.3 Sulfur-Containing Compounds from Endophytic Fungi

There are many sulfur-containing compounds produced from endophytic fungi, and these compounds include peptides, polyketides, terpenoids, and others (Fig. 13.1).

13.3.1 Peptides

Diketopiperazine bionectin D was extracted from endophytic *Bionectria* sp. Y1085, which was isolated from Huperzia serrate plant (Yang et al. 2019). Furthermore, Lasiodiplines A–D are novel sulfur-containing diketopiperazines synthesized by *Lasiodiplodia pseudotheobromae* F2, which was extracted from *Illigera rhodantha*. Lasiodiplines E exhibited antibacterial activity against

Veillonella parvula, Actinomyces israelii, Streptococcus sp., *Bacteroides vulgates*, and *Peptostreptococcus* sp. The minimum inhibitory concentration (MIC) values for Lasiodiplines E against these bacteria were 0.25, 32.0, 0.12, 0.12, and 0.12 μg/mL, respectively (Wei et al. 2014). Furthermore, new spirocyclic thiodiketopiperazines botryosulfuranols A–C were extracted and identified from endophytic *Botryosphaeria mamani* that was isolated from *Bixa orellana* L. plant (Barakat et al. 2019). Furthermore, outovirin A was derived from endophytic Penicillium raciborskii (TRT59) extracted from the Rhododendron tomentosum plant. This compound exhibits a nitrogen-oxygen bond within the oxazinane ring, connecting the rings in a manner reminiscent of diketopiperazine and conduritol. In addition, it displays a unique sulfide bridge that connects the α- and β-carbons, setting it apart from the usual $\alpha - \alpha$ bridging (Kajula et al. 2016). Two recently discovered natural compounds, outovirins B and C, which bear a resemblance to the antifungal compound gliovirin, have been found in an extract of *Penicillium raciborskii* (Kajula et al. 2016). A variety of sulfur-containing compounds were generated by the endophytic fungus *Epicoccum nigrum*, derived from the leaves of *Lysidice rhodostegia*. The compounds were given the names epicoccin I, ent-epicoccin G, and epicoccins J-T (Wang et al. 2010b). Entepicoccin G and epicoccins M exhibited strong in vitro efficacy against the release of β-glucuronidase in rat polymorphonuclear leukocytes caused by the platelet-activating factor, with IC50 values of 3.07 and 4.16 μM, respectively. In addition, bisdethiobis(methylsulfanyl)apoaranotin was synthesized by the endophytic fungus *Aspergillus terreus* BCC 4651 (Haritakun et al. 2012). Furthermore, endophytic *Chaetomium* sp. 88194 was used for the production of new compounds chaetocochin G, oidioperazine E, and chetoseminudin E (Wang et al. 2015). Moreover, Peng et al. (2019) isolated *Chaetomium* sp. SYP-F7950 from *Panax notoginseng* plant, where the results illustrated the production of two new compounds chetoseminudin F and G. Many compounds penicibrocazines F–I, spirobrocazines A–B, penicibrocazines A–E, and Brocazines A-G were extracted from endophytic *Penicillium brocae* MA-231 (Meng et al. 2017, 2014, 2016, 2015). Also, two new compounds Scheme 54794 and Scheme 54796 were separated from endophytic *Tolypocladium* sp., which was isolated from *Quercus virginiana Miller* plant (Chu et al. 1993). Also, endophytic *Penicillium janthinellum* HDN13–309 was used for the production of many sulfur-containing compounds such as penispirozines A–D and penicisulfuranols A–D (Zhu et al. 2020; Zhu et al. 2017). Two new compounds exserohilone and 9,10-dihydroexserohilon were isolated from endophytic *Exserohilum holmii,* which was isolated from *Dactyloctenium aegyptium* (Sugawara et al. 1985). Pretrichodermamide G was produced from endophytic *Trichoderma harzianum,* which was isolated from *Zingiber officinale* (Harwoko et al. 2021). A novel cyclic pentapeptide, disulfide cyclo-(Leu-Val-Ile-Cys-Cys), known as malformin, was discovered from *Aspergillus tamarii* FR02, which was obtained from the root of *Ficus carica* (Ma et al. 2016). Five novel epipolysulfanyldioxopiperazines, namely, gliocladines A–E, were extracted from the endophytic fungus *Gliocladium roseum* 1A (Dong et al. 2005).

13.3.2 Polyketides

Polyketides are a class of organic compounds produced by the polymerization of acetyl and propionyl coenzyme A molecules. Wang et al. (2016) reported that endophytic *Bipolaris sorokiniana* A606 has the ability to produce new cytotoxic compound called isocochlioquinones D, where this compound exhibited antiproliferative activity. Li et al. (2017) isolated endophytic *Paraphaeosphaeria neglecta* FT462, which has ability to produced new compounds paraphaeosphaerides E–H. Furthermore, the endophytic fungus *Cladosporium cladosporioides* MA-299, obtained from *Bruguiera gymnorrhiza,* has the ability to synthesize four 12-membered macrolides, known as thiocladospolides A-D (Zhang et al. 2019). As well, five 12-membered macrolides, thiocladospolides F–J, were produced from endophytic *Cladosporium oxysporum,* which was isolated from *Avicennia marina* plant (Wang et al. 2020). Wu et al. (2019) isolated endophytic *Aspergillus micronesiensis,* which produced cyschalasins A and B, which exhibited antibacterial as well as anticancer activity. Trichodermamide G was produced from endophytic *Trichoderma harzianum* D13, which was isolated from the root of *Excoecaria agallocha* (Zhao et al. 2020). Moreover, Li et al. (2017) reported that endophytic *Paraphaeosphaeria neglecta* FT462 has the ability to produce many compounds containing sulfur called paraphaeosphaerides E–H. Also, endophytic *Alternaria* sp. was used for the production of two new compounds named alternariol 5-O-sulfate and alternariol 5-O-methyl ether-40 -O-sulfate (Aly et al. 2008). Zhang et al. (2013) produced a new compound called fusaodavinvin, which was produced from endophytic *Fusarium* sp. (CTGU-ZL-34).

13.3.3 Terpenoids

Terpenes are a broad and heterogeneous group of organic compounds that are synthesized by a diverse range of plant species and endophytic fungi. According to Masyita et al. (2022), the chemical structures and biological activity of fungal terpenes are extremely varied. Terpenes are classified into sesquiterpenes, diterpenes, and meroterpenes. Leptosphin A, a recently discovered compound containing a benzo[b]thiophene group, was obtained from the endophytic fungus *Leptosphaeria* sp. XL026, which was isolated from the leaves of *Panax notoginseng* (Chen et al. 2019). The bark of *Cephalotaxus hainanensis*, often known as Hainan plum-yew, yielded an endophytic fungus named S49. S49 yielded two novel sesquiterpenoids, namely, 1,10,11,12-guaianetetrol and 1,10,11,12-guaianetetrol (Mei et al. 2010). Two new compounds 16-O-sulfo-18-norisopimar-7-en-4α,16-diol and 9-deoxyhymatoxin A were produced from *Xylaria* sp. YM 311647, which was isolated from *Azadirachta indica* (Wu et al. 2014).

13.3.4 Hybrid Compounds

Two novel alkaloids Pestaloamides A and B were separated from endophytic *Pestalotiopsis* sp. HS30, which was isolated from *Isodon xerophilus* (Wu et al. 2019). PM181110 (133) is a novel depsipeptide derived from *Phomopsis glabrae*, a fungus found in the leaves of *Pongamia pinnata* (Fabaceae) (Su et al. 2020). Furthermore, a novel molecule, FE399, was isolated from the endophytic fungus *Ascochyta* sp. AJ 117309. This compound has shown significant potential as an anticancer agent (Verekar et al. 2014). Fusarithioamide B, derived from endophytic *Fusarium chlamydosporium*, exhibited potent antifungal properties against *C. albicans*, with a minimum inhibitory concentration (MIC) of 1.9 μg/mL. The study also has shown significant antibacterial efficacy against *E. coli, S. aureus, and B. cereus*, with minimum inhibitory concentration (MIC) values of 3.4, 2.9, and 3.9 μg/mL, respectively (Ibrahim et al. 2016).

13.4 Different Applications of Sulfur-Containing Compounds from Endophytic Fungi

13.4.1 Antimicrobial Activity

The issue of pathogenic bacteria and fungi developing resistance to commercially available drugs is a significant concern encountered by healthcare providers. This phenomenon occurs as these microorganisms acquire novel mechanisms to resist the effects of antimicrobial agents (Costelloe et al. 2010; Elbasuney et al. 2021). Hence, it is necessary to discover effective antimicrobial agents. Fungal endophytes are widely used to serve as important repositories of bioactive compounds, exhibiting a wide range of biological properties such as antimicrobial, antioxidant, anticancer, and antiviral activities, as well as antidiabetic activities (Table 13.2). Sulfur-containing compounds have recently garnered significant attention among these bioactive chemicals. The compound Diketopiperazine bionectin D, derived from the endophytic *Bionectria* sp. Y1085, exhibited antibacterial properties against *E. coli, S. aureus, and Salmonella typhimurium*, with a consistent minimum inhibitory concentration (MIC) value of 12.5 μg/mL (Yang et al. 2019). Lasiodiplines E is a recently discovered sulfurous diketopiperazine compound, which exhibited antibacterial activity, and this compound was purified from *Lasiodiplodia pseudotheobromae* extract (Wei et al. 2014). In addition, two novel compounds, Cyschalasins A and Cyschalasins B, which were isolated from *A. micronesiensis*, exhibited antibacterial properties against *S. aureus, MRSA, and Candida albicans* (Wu et al. 2019). A. tamari was utilized to produce the compound Malformin E, which exhibits antimicrobial and antifungal properties against *B. subtilis, S. aureus, P. aeruginosa, Escherichia coli, P. chrysogenum, C. albicans, and F. solani*. The minimum inhibitory concentration (MIC) values for malformin E range from 0.45 to 7.24 μM (Ma et al. 2016). Furthermore, bisdethiobis(methylsulfanyl)apoaranotin compound, which was produced from endophytic *Aspergillus terreus* BCC 4651exhibited weak

Table 13.2 Some antimicrobial sulfur-containing compounds from endophytic fungi

Endophytic fungi	Compound	Ref.
Aspergillus micronesiensis	Cyschalasins A and Cyschalasins B	Wu et al. (2019)
Aspergillus tamari	Malformin E	Ma et al. (2016)
Aspergillus terreus BCC 4651	Bisdethiobis(methylsulfanyl)apoaranotin	Haritakun et al. (2012)
Bionectria sp. Y1085	Bionectin E	Yang et al. (2019)
Chaetomium sp. M336	6-Formamide-chetomin	Yu et al. (2018)
Cladosporium cladosporioides MA-299	Thiocladospolides A–D	Zhang et al. (2019)
Cladosporium oxysporum	Thiocladospolides F–J	Wang et al. (2020)
Fusarium chlamydosporium	Fusarithioamide A–B	Ibrahim et al. (2016)
Guignardia sp. IFB-E028	Guignasulfide	Wang et al. (2010a)
Lasiodiplodia pseudotheobromae	Lasiodiplines E	Wei et al. (2014)
Leptosphaeria sp. XL026	Leptosphin A–C	Chen et al. (2019)
Paraphaeosphaeria neglecta FT462	paraphaeosphaerides E–H	Li et al. (2017)
Penicillium brocae MA-231	Penicibrocazines H–I, Spirobrocazines A	Meng et al. (2017, 2014, 2016, 2015)
Penicillium crustosum PRB-2	Penixylarins D	Yu et al. (2019a)
Trichoderma sp. BCC 5926	Pretrichodermamide A	Seephonkai et al. (2006)
Xylaria sp. YM 311647	16-*O*-sulfo-18-norisopimar-7-en-4α,16-diol and 9-deoxy-hymatoxin A	Wu et al. (2014)

antimycobacterial activity (Haritakun et al. 2012). In a study conducted by Yu et al. (2018), they utilized the endophytic fungus *Chaetomium* sp. M336 to generate 6-Formamide-chetomin. This compound showed potential as an antibacterial agent, exhibiting activity against *E. coli, S. aureus, S. typhimurium ATCC 6539, and E. faecalis*, all with a minimum inhibitory concentration (MIC) of 0.78 μg/mL. Moreover, Zhang et al. (2019) reported that thiocladospolide A displayed antimicrobial activity toward *Edwardsiella tarda* and Colletotrichum gloeosporioides with MIC values of 1, 8, and 2 μg/mL, respectively. Fusarithioamide A and fusarithioamide B were produced from endophytic *Fusarium chlamydosporium*, where both compounds showed antimicrobial activity toward *S. aureus, B. cereus, E. coli, P. aeruginosa, and C. albicans* (Ibrahim et al. 2016). Guignasulfide, which was produced from Guignardia sp. IFB-E028, exhibited antimicrobial activity against *Helicobacter pylori* with an MIC of 42.9 μM (Wang et al. 2010a). Chen et al. (2019) confirmed that Leptosphin A has promising antibacterial and antifungal activity with low MICs. Furthermore, endophytic Paraphaeosphaeria neglecta FT462 was used for the production of paraphaeosphaerides E-H compounds, where paraphaeosphaeride E displayed antibaterial activity against *E. coli* JW2496 at 20 μg/mL (Li et al. 2017). The endophytic *Penicillium brocae* MA-231 generated penicibrocazines H, which exhibited antibacterial activity against *V. harveyi*, *E. coli*, *A. hydrophilia*, and V. *parahaemolyticus*, with minimum inhibitory concentrations (MICs) of 16.0, 16.0, 32.0, and 16.0 μg/mL, respectively. Also, penicibrocazines I showed efficacy against *V. harveyi*, showing a minimum inhibitory concentration of 32.0 μg/mL. Additionally, with minimum inhibitory concentrations (MIC) of 32.0, 16.0, and 64.0 μg/mL for *E. coli, S. aureus, and Vibrio harveyi*, respectively, *Spirobrocazine A* demonstrated moderate antibacterial activity (Meng et al. 2017, 2014, 2016, 2015). Furtheromore, *Penicillium crustosum* PRB-2 produced penixylarins D, which has weak antibacterial activity *Bacillus subtilis* with an MIC50 of 100 μM (Yu et al. 2019a). Also, pretrichodermamide A, which was isolated from *Trichoderma sp.* BCC 5926, displayed antibacterial activity against *Mycobacterium tuberculosis* H37Ra with an MIC of 12.5 μg/mL (Seephonkai et al. 2006).

13.4.2 Anticancer Activity

Cancer involves a range of pathological conditions characterized by the unregulated and aberrant proliferation of various cell types, leading to the formation of an abnormal cellular mass (Khan et al. 2021). Cancer arises from a combination of external factors, such as tobacco use, alcohol consumption, smoking, an unhealthy diet, certain lifestyle choices, and exposure to external conditions like ultraviolet or ionizing and nonionizing radiation. Additionally, intrinsic factors including aging, DNA mutation, hormonal imbalances, and a weakened immune system contribute to the development of cancer. These factors influence the activation or inactivation of specific genes, leading to the abnormal proliferation of cells (Pandi et al. 2013). Cancer is a significant contributor to global mortality, with a rising incidence observed on an annual basis. The field of cancer treatment has witnessed notable progress in recent times, primarily attributed to the identification and enhancement of novel chemotherapeutic agents derived from natural origins (Uzma et al. 2018). Recent studies confirmed fungal endophytes as anticancer agents are alternative to traditional anticancer drugs (Table 13.3). Aly et al. (2008) reported that alternariol 5-O-sulfate and alternariol 5-O-methyl ether-40 -O-sulfate compound, which was separated from *Alternaria* sp., has anticancer activity against L5178Y with an EC50 of 4.5 μg/mL. Also, FE399 compound, which was produced from *Ascochyta* sp. AJ 117309, displayed promising anticancer activity against different cancerous cell lines (Verekar et al. 2014). Furthermore, two compounds Cyschalasins A and Cyschalasins B, which are isolated from *A. micronesiensis*, exhibited cytotoxic activity toward HL60, A549, Hep3B, MCF-7, and SW480 with IC50 values in the range of 3.0–19.9 μM (Wu et al. 2019). Moreover, malformin E compound showed exhibited anticancer activity toward MCF-7, A549, and HepG2 with IC50 values of 0.65, 2.42, and 36.02 μM, respectively (Ma et al. 2016). He et al. (2020) elucidated the presence of a sulfur-containing compound known as pentacyclic diketopiperazine, which exhibits a distinctive 4,5-dihydrooxepine moiety. This compound, aptly named versicolor, showcases intriguing properties worthy of further investigation. The compound demonstrated cytotoxic effects on the proliferation of the MOLT-4

Table 13.3 Some anticancer sulfur-containing compounds from endophytic fungi

Endophytic fungi	Compound	Ref.
Alternaria sp.	Alternariol 5-O-sulfate and alternariol 5-O-methyl ether-40-O-sulfate	Aly et al. (2008)
Ascochyta sp. AJ 117309	FE399	Verekar et al. (2014)
Aspergillus micronesiensis	Cyschalasins A and Cyschalasins B	Wu et al. (2019)
Aspergillus tamarii	Malformin E	Ma et al. (2016)
Aspergillus versicolor 0312	Pentacyclic diketopiperazine with a 4,5-dihydrooxepine moiety versicolor A	He et al. (2020)
Aspergillus versicolor F210	methylsulfonyl unit, 21-Epi-taichunamide D	Li et al. (2020)
Bipolaris sorokiniana A606	Isocochlioquinones D	Wang et al. (2016)
Botryosphaeria mamani	Botryosulfuranols A–B	Barakat et al. (2019)
Chaetomium sp. 88,194	Chaetocochin G	Wang et al. (2015)
Chaetomium sp. M336	6-Formamide-chetomin	Yu et al. (2018)
Fusarium chlamydosporium	Fusarithioamide B	Ibrahim et al. (2016)
Fusarium sp. (CTGU-ZL-34).	Fusaodavinvin	Zhang et al. (2013)
Guignardia sp. IFB-E028	Guignasulfide	Wang et al. (2010a)
Pestalotiopsis sp. HS30	Pestaloamides A and B	Wu et al. (2019)
Phoma sp. OUCMDZ-1847	Phomazines B	Kong et al. (2014)
Phomopsis glabrae	PM181110	Su et al. (2020)

cell line, exhibiting an IC50 value of 29.6 μM. In a separate investigation, the methylsulfonyl unit, specifically the 21-Epi-taichunamide D compound, which was isolated from the endophytic microorganism *Aspergillus* versicolor F210, exhibited notable anticancer properties against HL-60 and A549 cells. The half-maximal inhibitory concentration (IC50) values for HL-60 and A549 cells were determined to be 26.8 and 32.5 μM, respectively (Li et al. 2020). Furthermore, the antiproliferative activity of Isocochlioquinones D, derived from the endophytic fungus *Bipolaris sorokiniana* A606, was observed against SF-268, MCF-7, NCI-H460, and HepG-2 cell lines. The IC50 values for SF-268, MCF-7, NCI-H460, and HepG-2 were determined to be 32.8 μM, 28.3 μM, 42.6 μM, and 38.6 μM, respectively (Wang et al. 2016). Botryosulfuranols A exhibited activity against four distinct cancer cell lines, namely, HT-29, HepG2, Caco-2, and HeLa. The inhibitory concentration (IC50) values for these cell lines were measured to be 8.0, 11.4, 18.2, 23.5, and 9.3 μM, respectively. Botryosulfuranols B exhibited notable anticancer properties against three distinct cancer cell lines, namely, HT-29, HepG2, and HeLa. The inhibitory concentration (IC50) values for these cell lines were determined to be 63.2, 56.1, 61.2, 49.9, and 64.7 μM, respectively (Barakat et al. 2019). Chaetocochin G which was isolated from *Chaetomium* sp. 88194 showed anticancer activity toward MCF7 cancerous cell line (Wang et al. 2015). Yu et al. (2018) reported that 6-formamide-chetomin, which was separated from *Chaetomium* sp. M336, displayed cytotoxic activity toward HeLa, SGC-7901, and A549 cells with IC50 values of 21.6–27.1 μM. In a separate investigation, the compound known as fusarithioamide B was isolated from an extract derived from the microorganism *Fusarium chlamydosporium*. Notably, this compound exhibited cytotoxic effects against various cell lines, including SK-MEL, KB, BT-549, and SKOV-3 cells, as evidenced by their respective IC50 values of 9.3, 7.7, 0.4, and 0.8 μM, respectively (Ibrahim et al. 2016). Furthermore, the utilization of *Fusarium* sp. (CTGU-ZL-34) was employed in the synthesis of the fusaodavinvin compound, which exhibits noteworthy anticancer properties against various cell lines, including A549, HepG2, Caski, and MCF-7. The inhibitory concentration (IC50) values for these cell lines were determined to be 11.5, 15.3, 15.2, and 60.5 μg/mL, respectively (Zhang et al. 2013). Another, sulfur-containing compound, Guignasulfide, which separated from endophytic *Guignardia* sp. IFB-E028, exhibited cytotoxicity against HepG2 with an IC50 of 5.27 μM (Wang et al. 2010a). Also, Pestaloamides A and B from *Pestalotiopsis* sp. HS30 showed anticancer activity toward cancerous HCT116 cells at 40 μM (Wu et al. 2019). Also, phomazine B which was isolated from *Phoma* sp. OUCMDZ-1847 displayed anticancer activity against MGC-803 cells with an IC50 of 8.5 μM. Similarly, a novel depsipeptide PM181110 compound, derived from the endophytic *Phomopsis glabrae*, exhibited notable cytotoxicity against 40 human cancer cell lines at the nanomolar scale (mean IC50 = 89 nM). Additionally, it displayed significant inhibitory effects on 24 human tumor xenografts, with a mean IC50 value of 245 nM (Su et al. 2020).

13.5 Conclusion

Fungal endophytes play a crucial role in the production of bioactive compounds with various applications. Bioactive compounds include alkaloids, terpenoids, flavonoids, polyketides, peptides, and phenols, which are used in different biological applications. Sulfur-containing compounds from endophytic fungi have been of interest due to their diverse biological activities. There are many genera of fungi that have the ability to produce sulfur-containing compounds; these genera are as follows: *Aspergillus, Penicillium, Trichoderma, Chaetomium, Alternaria, Fusarium, Cladosporium, Epicoccum, Bipolaris, Colletotrichum, Lasiodiplodia,* and *Neosartorya*. Sulfur-containing compounds, which are produced from endophytic fungi, include terepenoids, polyketides, peptides, and hybrids. These compounds have antibacterial activity toward Gram-negative and Gram-positive bacteria and also have antifungal activity against unicellular and multicellular fungi. Moreover, these compounds have anticancer activity against different cancerous cell lines, which can be used later as anticancer agent for human after extensive studies in vivo.

References

Alam B et al (2021) Endophytic fungi: from symbiosis to secondary metabolite communications or vice versa? Front Plant Sci 12. https://doi.org/10.3389/fpls.2021.791033

Aly AH et al (2008) Cytotoxic metabolites from the fungal endophyte *Alternaria* sp. and their subsequent detection in its host plant *Polygonum senegalense*. J Nat Prod 71:972–980. https://doi.org/10.1021/np070447m

Barakat F et al (2019) Thiodiketopiperazines with two spirocyclic centers extracted from *Botryosphaeria mamane*, an endophytic fungus isolated from *Bixa orellana* L. Phytochemistry 158:142–148. https://doi.org/10.1016/j.phytochem.2018.11.007

Centko RM, Ratnaweera PB, Tysoe C, Withers SG, de Silva ED, Andersen RJ (2017) Alpha-glucosidase and alpha-amylase inhibiting thiodiketopiperazines from the endophytic fungus *Setosphaeria rostrata* isolated from the medicinal plant *Costus speciosus* in Sri Lanka. Phytochem Lett 22:76–80. https://doi.org/10.1016/j.phytol.2017.09.004

Chaudhary P, Agri U, Chaudhary A, Kumar A, Kumar G (2022) Endophytes and their potential in biotic stress management and crop production. Front Microbiol 13:933017. https://doi.org/10.3389/fmicb.2022.933017

Chen H-Y, Liu T-K, Shi Q, Yang X-L (2019) Sesquiterpenoids and diterpenes with antimicrobial activity from *Leptosphaeria sp.* XL026, an endophytic fungus in *Panax notoginseng*. Fitoterapia 137:104243. https://doi.org/10.1016/j.fitote.2019.104243

Chu M et al (1993) Two novel diketopiperazines isolated from the fungus *Tolypocladium sp.* Tetrahedron Lett 34:7537–7540. https://doi.org/10.1016/S0040-4039(00)60393-3

Costelloe C, Metcalfe C, Lovering A, Mant D, Hay AD (2010) Effect of antibiotic prescribing in primary care on antimicrobial resistance in individual patients: systematic review and meta-analysis. BMJ. 340:c2096

Dong J-Y, He H-P, Shen Y-M, Zhang K-Q (2005) Nematicidal epipolysulfanyldioxopiperazines from *Gliocladium roseum*. J Nat Prod 68:1510–1513. https://doi.org/10.1021/np0502241

Elbasuney S, El-Sayyad GS, Tantawy H, Hashem AH (2021) Promising antimicrobial and antibiofilm activities of reduced graphene oxide-metal oxide (RGO-NiO, RGO-AgO, and RGO-ZnO) nanocomposites. RSC Adv 11:25961–25975. https://doi.org/10.1039/D1RA04542C

Francioso A, Conrado AB, Mosca L, Fontana M (2020) Chemistry and biochemistry of sulfur natural compounds: key intermediates of metabolism and redox biology. Oxid Med Cell Longev 2020:8294158

Gouda S, Das G, Sen SK, Shin H-S, Patra JK (2016) Endophytes: a treasure house of bioactive compounds of medicinal importance. Front Microbiol 7:1538

Guimarães DO et al (2010) Diketopiperazines produced by endophytic fungi found in association with two Asteraceae species. Phytochemistry 71:1423–1429. https://doi.org/10.1016/j.phytochem.2010.05.012

Haritakun R, Rachtawee P, Komwijit S, Nithithanasilp S, Isaka M (2012) Highly conjugated ergostane-type steroids and aranotin-type diketopiperazines from the fungus *Aspergillus terreus* BCC 4651. Helv Chim Acta 95:308–313. https://doi.org/10.1002/hlca.201100335

Harwoko H et al (2021) Dithiodiketopiperazine derivatives from endophytic fungi *Trichoderma harzianum* and *Epicoccum nigrum*. Nat Prod Res 35:257–265. https://doi.org/10.1080/14786419.2019.1627348

Hashem AH et al (2023) Bioactive compounds and biomedical applications of endophytic fungi: a recent review. Microb Cell Fact 22:107. https://doi.org/10.1186/s12934-023-02118-x

He T, Wang Y, Du L, Li F, Hu Q, Cheng G, Wang W (2020) Overexpression of global regulator LaeA induced secondary metabolite production in Aspergillus versicolor 0312. Rec Nat Prod 14:387–394

Ibrahim SRM, Elkhayat ES, Mohamed GAA, Fat'hi SM, Ross SA (2016) Fusarithioamide A, a new antimicrobial and cytotoxic benzamide derivative from the endophytic fungus *Fusarium chlamydosporium*. Biochem Biophys Res Commun 479:211–216. https://doi.org/10.1016/j.bbrc.2016.09.041

Kajula M et al (2016) Bridged Epipolythiodiketopiperazines from Penicillium raciborskii, an Endophytic Fungus of *Rhododendron tomentosum* Harmaja. J Nat Prod 79:685–690. https://doi.org/10.1021/np500822k

Khalill AMA, Abdelaziz AM, Khaleil MM, Hashem AH (2021) Fungal endophytes from leaves of Avicennia marina growing in semi-arid environment as a promising source for bioactive compounds. Lett Appl Microbiol 72(3):263–274. https://doi.org/10.1111/lam.13414

Khan AU, Dagur HS, Khan M, Malik N, Alam M, Mushtaque M (2021) Therapeutic role of flavonoids and flavones in cancer prevention: current trends and future perspectives. Eur J Med Chem Rep 3:100010. https://doi.org/10.1016/j.ejmcr.2021.100010

Kong F, Wang Y, Liu P, Dong T, Zhu W (2014) Thiodiketopiperazines from the Marine-Derived Fungus *Phoma sp.* OUCMDZ-1847. J Nat Prod 77:132–137. https://doi.org/10.1021/np400802d

Li C-S et al (2017) NF-κB inhibitors, unique γ-pyranol-γ-lactams with sulfide and sulfoxide moieties from Hawaiian plant *Lycopodiella cernua* derived fungus *Paraphaeosphaeria neglecta* FT462. Sci Rep 7:10424. https://doi.org/10.1038/s41598-017-10537-1

Li H et al (2020) 21-Epi-taichunamide D and (±)-versicaline A, three unusual alkaloids from the endophytic *Aspergillus versicolor* F210. Tetrahedron Lett 61:152219. https://doi.org/10.1016/j.tetlet.2020.152219

Lori M, Symnaczik S, Mäder P, De Deyn G, Gattinger A (2017) Organic farming enhances soil microbial abundance and activity—a meta-analysis and meta-regression. PLoS One 12:e0180442

Ma Y-M, Liang X-A, Zhang H-C, Liu R (2016) Cytotoxic and antibiotic cyclic pentapeptide from an endophytic *Aspergillus tamarii* of Ficus carica. J Agric Food Chem 64:3789–3793. https://doi.org/10.1021/acs.jafc.6b01051

Masyita A et al (2022) Terpenes and terpenoids as main bioactive compounds of essential oils, their roles in human health and potential application as natural food preservatives. Food Chem X 13:100217. https://doi.org/10.1016/j.fochx.2022.100217

Mei W-L, Chen P, Wang H, Huang J-L, Dai H-F (2010) Two new sesquiterpenes from endophytic fungus S49 of *Cephalotaxus hainanensis*. J Asian Nat Prod Res 12:582–585

Meng L-H, Li X-M, Lv C-T, Huang C-G, Wang B-G (2014) Brocazines A–F, cytotoxic bisthiodiketopiperazine derivatives from *Penicillium brocae* MA-231, an endophytic fungus derived from the marine mangrove plant *Avicennia marina*. J Nat Prod 77:1921–1927

Meng L-H, Zhang P, Li X-M, Wang B-G (2015) Penicibrocazines A–E, five new sulfide diketopiperazines from the marine-derived endophytic fungus *Penicillium brocae*. Mar Drugs 13:276–287

Meng L-H et al (2016) Three diketopiperazine alkaloids with spirocyclic skeletons and one bisthiodiketopiperazine derivative from the mangrove-derived endophytic fungus *Penicillium brocae* MA-231. Org Lett 18:5304–5307. https://doi.org/10.1021/acs.orglett.6b02620

Meng L-H, Li X-M, Liu Y, Xu G-M, Wang B-G (2017) Antimicrobial alkaloids produced by the mangrove endophyte *Penicillium brocae* MA-231 using the OSMAC approach. RSC Adv 7:55026–55033. https://doi.org/10.1039/C7RA12081H

Pandi M, SenthilKumaran R, Rajapriya P, Yogeswari S, Muthumary J (2013) Taxol, a potential drug for the treatment of cancer. Biores Bull 2:1–9

Pedras MSC, Séguin-Swartz G, Abrams SR (1990) Minor phytotoxins from the blackleg fungus *Phoma lingam*. Phytochemistry 29:777–782. https://doi.org/10.1016/0031-9422(90)80017-B

Peng F et al (2019) Cytotoxic and antimicrobial indole alkaloids from an endophytic fungus *Chaetomium sp.* SYP-F7950 of Panax notoginseng. RSC Adv 9:28754–28763. https://doi.org/10.1039/C9RA04747F

Pirttilä AM, Podolich O, Koskimäki JJ, Hohtola E, Hohtola A (2008) Role of origin and endophyte infection in browning of bud-derived tissue cultures of Scots pine (Pinus sylvestris L.). Plant Cell Tissue Org Cult 95:47–55

Rai N et al (2021) Plant associated fungal endophytes as a source of natural bioactive compounds. Mycology 12:139–159. https://doi.org/10.1080/21501203.2020.1870579

Russo ML, Pelizza SA, Cabello MN, Stenglein SA, Vianna MF, Scorsetti AC (2016) Endophytic fungi from selected varieties of soybean (*Glycine max* L. Merr.) and corn (*Zea mays* L.) grown in an agricultural area of Argentina. Rev Argentina Microbiol 48:154–160

Seephonkai P, Kongsaeree P, Prabpai S, Isaka M, Thebtaranonth Y (2006) Transformation of an irregularly bridged epidithiodiketopiperazine to Trichodermamide A. Org Lett 8:3073–3075. https://doi.org/10.1021/ol0610461

Selim KA, Nagia M, Ghwas DEE (2017) Endophytic fungi are multifunctional biosynthesizers: ecological role and chemical diversity. In: Endophytic fungi: diversity, characterization and biocontrol. NOVA Science, New York, p 39

Song X-Q et al (2013) Xanthone derivatives from *Aspergillus sydowii*, an endophytic fungus from the liverwort *Scapania ciliata* S. Lac and their immunosuppressive activities. Phytochem Lett 6:318–321. https://doi.org/10.1016/j.phytol.2013.03.012

Su X-Z et al (2020) Pestaloamides A and B, two spiro-heterocyclic alkaloid epimers from the plant endophytic fungus *Pestalotiopsis sp.* HS30. Sci China Chem. 63:1208–1213. https://doi.org/10.1007/s11426-020-9762-0

Sugawara K, Sugawara F, Strobel GA, Fu Y, He CH, Clardy J (1985) Exserohilone: a novel phytotoxin produced by *Exserohilum holmii*. J Org Chem 50:5631–5633. https://doi.org/10.1021/jo00350a040

Sun X, Guo L-D (2012) Endophytic fungal diversity: review of traditional and molecular techniques. Mycology 3:65–76

Uzma F et al (2018) Endophytic fungi—alternative sources of cytotoxic compounds: a review. Front Pharmacol 9:309

Verekar SA, Mishra PD, Sreekumar ES, Deshmukh SK, Fiebig H-H, Kelter G, Maier A (2014) Anticancer activity of new depsipeptide compound isolated from an endophytic fungus. J Antibiot 67:697–701

Verma A, Shameem N, Jatav HS, Sathyanarayana E, Parray JA, Poczai P, Sayyed RZ (2022) Fungal endophytes to combat biotic and abiotic stresses for climate-smart and sustainable. Agric Front Plant Sci 13. https://doi.org/10.3389/fpls.2022.953836

Wang FW, Ye YH, Ding H, Chen YX, Tan RX, Song YC (2010a) Benzophenones from *Guignardia sp.* IFB-E028, an Endophyte on Hopea hainanensis. Chem Biodivers 7:216–220

Wang JM et al (2010b) Thiodiketopiperazines produced by the endophytic fungus *Epicoccum nigrum*. J Nat Prod 73:1240–1249. https://doi.org/10.1021/np1000895

Wang FQ et al (2015) Indole diketopiperazines from endophytic *Chaetomium sp* 88194 induce breast cancer cell apoptotic death. Sci Rep 5:9294. https://doi.org/10.1038/srep09294

Wang M et al (2016) Cytotoxic cochlioquinone derivatives from the endophytic fungus *Bipolaris sorokiniana* derived from Pogostemon cablin. Fitoterapia 110:77–82. https://doi.org/10.1016/j.fitote.2016.02.005

Wang W, Feng H, Sun C, Che Q, Zhang G, Zhu T, Li D (2020) Thiocladospolides F-J, antibacterial sulfur containing 12-membered macrolides from the mangrove endophytic fungus *Cladosporium oxysporum* HDN13-314. Phytochemistry 178:112462. https://doi.org/10.1016/j.phytochem.2020.112462

Wei W et al (2014) An antibacterial metabolite from Lasiodiplodia pseudotheobromae F2. Phytochemistry 100:103–109. https://doi.org/10.1016/j.phytochem.2014.01.003

Wu S-H, He J, Li X-N, Huang R, Song F, Chen Y-W, Miao C-P (2014) Guaiane sesquiterpenes and isopimarane diterpenes from an endophytic fungus *Xylaria sp.* Phytochemistry 105:197–204. https://doi.org/10.1016/j.phytochem.2014.04.016

Wu Z et al (2019) Cysteine residue containing merocytochalasans and 17,18-seco-Aspochalasins from *Aspergillus micronesiensis*. J Nat Prod 82:2653–2658. https://doi.org/10.1021/acs.jnatprod.9b00016

Xu Y-m, Espinosa-Artiles P, Liu MX, Arnold AE, Gunatilaka AAL (2013) Secoemestrin D, a cytotoxic epitetrathiodioxopiperizine, and emericellenes A–E, five sesterterpenoids from *Emericella sp* AST0036, a fungal endophyte of *Astragalus lentiginosus*1. J Nat Prod 76:2330–2336. https://doi.org/10.1021/np400762k

Yadav G, Meena M (2021) Bioprospecting of endophytes in medicinal plants of Thar Desert: an attractive resource for biopharmaceuticals. Biotechnol Rep 30:e00629. https://doi.org/10.1016/j.btre.2021.e00629

Yang Y-H, Yang D-S, Li G-H, Pu X-J, Mo M-H, Zhao P-J (2019) Antibacterial diketopiperazines from an endophytic fungus *Bionectria sp*. Y1085. J Antibiot 72:752–758

Yu F-X, Chen Y, Yang Y-H, Li G-H, Zhao P-J (2018) A new epipolythiodioxopiperazine with antibacterial and cytotoxic activities from the endophytic fungus *Chaetomium sp*. M336. Nat Prod Res 32:689–694. https://doi.org/10.1080/14786419.2017.1338285

Yu G et al (2019a) Secondary metabolites produced by combined culture of *Penicillium crustosum* and a *Xylaria sp*. J Nat Prod 82:2013–2017. https://doi.org/10.1021/acs.jnatprod.9b00345

Yu G et al (2019b) Methylsulfonylated polyketides produced by *Neosartorya udagawae* HDN13-313 via exogenous addition of small molecules. J Nat Prod 82:998–1001. https://doi.org/10.1021/acs.jnatprod.9b00035

Zhang L, Liu Y, Deng Z, Guo Z, Chen J, Tu X, Zou K (2013) Fusaodavinvin, a novel metabolite containing sulfur from the endophytic fungus *Fusarium sp*. (CTGU-ZL-34). Nat Prod Commun. 8:1934578X1300800120

Zhang F-Z, Li X-M, Yang S-Q, Meng L-H, Wang B-G (2019) Thiocladospolides A–D, 12-membered macrolides from the mangrove-derived endophytic fungus *Cladosporium cladosporioides* MA-299 and structure revision of pandangolide 3. J Nat Prod 82:1535–1541. https://doi.org/10.1021/acs.jnatprod.8b01091

Zhao D-L et al (2020) Antifungal nafuredin and epithiodiketopiperazine derivatives from the mangrove-derived fungus *Trichoderma harzianum* D13. Front Microbiol 11:1495

Zheng Y-K, Qiao X-G, Miao C-P, Liu K, Chen Y-W, Xu L-H, Zhao L-X (2016) Diversity, distribution and biotechnological potential of endophytic fungi. Ann Microbiol 66:529–542. https://doi.org/10.1007/s13213-015-1153-7

Zhu M et al (2017) Penicisulfuranols A–F, alkaloids from the Mangrove endophytic fungus *Penicillium janthinellum* HDN13-309. J Nat Prod 80:71–75. https://doi.org/10.1021/acs.jnatprod.6b00483

Zhu M et al (2020) Penispirozines A–H, three classes of dioxopiperazine alkaloids with spirocyclic skeletons isolated from the mangrove-derived *Penicillium janthinellum*. J Nat Prod 83:2647–2654. https://doi.org/10.1021/acs.jnatprod.0c00451

Zuo Y, Li X, Yang J, Liu J, Zhao L, He X (2021) Fungal endophytic community and diversity associated with desert shrubs driven by plant identity and organ differentiation in extremely arid desert ecosystem. J Fungi (Basel) 7:578. https://doi.org/10.3390/jof7070578

Phytohormone Production by Endophytic Fungi

14

Hajira Younas and Fatima Younas

Abstract

Due to the adverse impacts of climate change, diminishing agricultural land, and limited water resources, there is continuous pressure on researchers to improve agricultural productivity for the increasing population. Endophytic fungi (EF), a diverse group of microbes that colonize plant tissues without causing damage, have gained considerable attention due to their positive role in agriculture. This chapter explores the important phytohormones produced by EF, for instance, gibberellic acid (GA), indole-3-acetic acid (IAA), and abscisic acid (ABA), and their key roles in agronomy. IAA and GA production by EF is associated with enhanced plant growth, water uptake, and nutrient availability. Additionally, ABA production alleviates stress conditions such as nutrient deficiency, allowing plants to survive severe drought and salinity for short periods while also influencing secondary metabolites production such as flavonoids, phenols, terpenoids, and steroids. Despite these positive outcomes, the practical application of EF in agriculture is hindered by various factors such as large-scale inoculum production, regulatory standards, and field validation. This chapter highlights the diversity of EF, investigating their abundance and the rationale behind host plant selection. A major focus is to elaborate on the biosynthetic pathways for phytohormone production by EF, with insight into IAA and its multiple biosynthetic pathways. The pathway for GA production is also discussed. This chapter debates the role of essential factors in agriculture and highlights their potential applications in improving plant growth and stress tolerance.

H. Younas (✉)
School of Integrative Plant Sciences, Cornell University, Ithaca, NY, USA

F. Younas
School College of Pharmacy, University of the Punjab, Lahore, Pakistan

K. A. Abd-Elsalam, S. F. AbuQamar (eds.), *Fungal Endophytes Volume I*,
https://doi.org/10.1007/978-981-97-7312-1_14

Keywords

Endophytic fungi · Plant hormones · Abscisic acid · Gibberellic acid · Indole acetic acid · Sustainable agriculture

14.1 Introduction

Plants serve as natural habitats for a wide array of endophytes, mainly fungi and bacteria, forming mutualistic relationships (Chhipa and Deshmukh 2019; Baron and Rigobelo 2022). In recent years, the importance of endophytes has gained attention due to their potential use in industrial and commercial sectors. Endophytes are widespread in ecosystems, and it is extremely rare to find plant species without these associations (Lata et al. 2018; Martos et al. 2023). Endophytes have been discovered in various habitats, including fossilized leaves, stems, roots, seeds, and fruit tissues, providing evidence of endophyte-plant interactions dating back 300–400 million years (Li et al. 2023).

As habitats transitioned from aquatic to terrestrial, plants faced various challenging atmospheric conditions, such as nutrient-poor soils, temperature fluctuations, water availability, and elevated carbon dioxide levels. Endophytic microbes, comprising fungi, bacteria, and actinobacteria, are ubiquitous in nature due to their associations with diversified plant species (Khare et al. 2018). Endophytic bacteria and fungi residing within plant tissues play a vital role in enhancing the host plant's physiology by offering protective benefits. For example, they confer resistance to biotic stresses by producing compounds such as hormones, alkaloids, and mycotoxins. Moreover, they can induce the production of immunological or antimicrobial compounds in plants or compete with pathogenic invaders for space. The advantageous symbiotic relationship between endophytic fungi and their associated host plant is illustrated in Fig. 14.1.

Research suggests that over 1 million fungal endophytes are distributed worldwide in various environments, including lush tropical forests, scorching deserts, and temperate regions. These fungi have the capacity to colonize plant tissues (Lugtenberg et al. 2016). Fungal endophytes exhibit a versatile ability to inhabit various plant parts, i.e., roots, leaves, stems, twigs, bark, branches, seeds, and fruits (Derviş and Özer 2023; Priyashantha et al. 2023). Furthermore, they have been identified within resin canals, scales, meristems, petioles, and flowers (Kumar et al. 2019; Tripathy et al. 2023). Certain fungal species exhibit specificity to plant families and are found only in certain ones (Ali et al. 2018).

Endophytic fungi (EF) are valuable sources of various bioactive substances and are used in medicine, food, pharmaceutical industry, biotechnology, and agriculture. Many EF can produce phytohormones in higher quantities than their host plants (Baron and Rigobelo 2022). The literature reports that EF are effective in increasing the agricultural productivity of various plants for instance *Oryza sativa* (Seema et al. 2023), *Lolium multiflorum* (García-Latorre et al. 2023), *Cucumis sativus* (Tian et al. 2022), and *Brassica* spp. (Poveda et al. 2022). Like plants, EF can produce

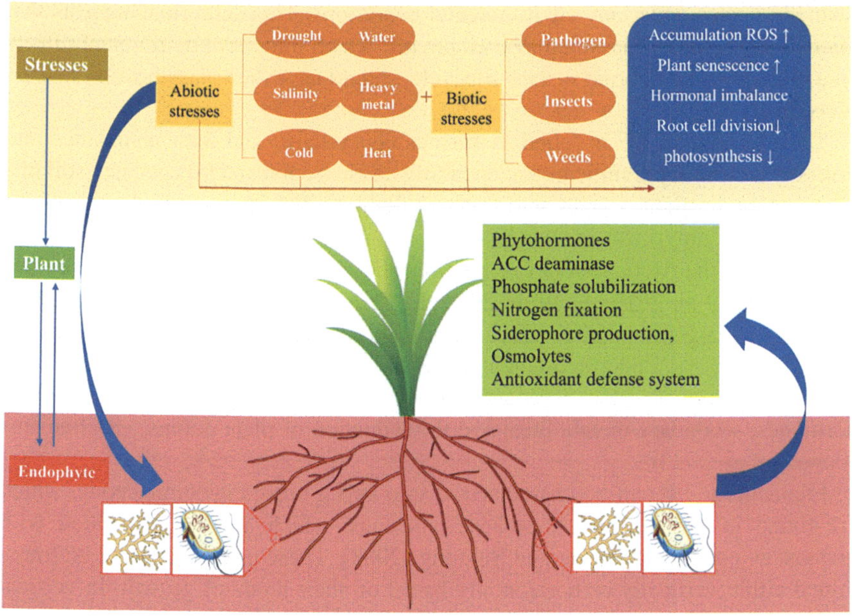

Fig. 14.1 The fungal endophyte and host plants interaction between abiotic and biotic stress environment (Lu et al. 2021)

significant phytohormones such as abscisic acid (Xu et al. 2018a, b), gibberellins (Wang et al. 2022), and indoleacetic acid (Ikram et al. 2023a, b; Mathew et al. 2023) in higher concentrations. However, the impact of these plant hormones on the metabolism and growth of EF, the host plant, and their roles in microbe-plant interactions are not well-documented. This chapter highlights the diversity of EF and their associated host plant ranges and the secretion of different types of phytohormones production by EF. Moreover, role of EF in agricultural sector will also be discussed in this chapter.

14.2 Diversity of EF

The term "endophyte," originally used by De Bary in 1886, refers to an organism residing within a plant (from "endo," meaning within, and "phyton," meaning plant). Currently, there is significant scientific debate regarding the use of the term "endophyte," primarily centered on whether it should exclusively designate habitat. One school of thought proposes that this term should be applied solely to habitat (Hardoim et al. 2015), while another suggests restricting its use and adopting a new one, "mutualistic endophytes," to describe microbes that confer assistances to their host plants (Card et al. 2016). A contemporary meaning for endophytes was put forth by Le Cocq et al. (2017), which describes "endophytes" as microbes that reside within plant tissues without causing harm to the host plant for at least part of their life cycles under any environmental conditions. This definition suggests that microbes previously classified as endophytes, which may have detrimental effects on a plant host at any point in their lifecycle, should have their classification revised accordingly.

The diversity of EF is found in various ecosystems, and their abundance and diversity could vary significantly, depending on environmental aspects, i.e., soil pH, temperature, nutrient content, and water availability. EF, particularly those belonging to the Ascomycetes, are widespread and prevalent across diverse ecosystems, ranging from the Arctic tundra to croplands and from tropical to temperate regions (Lugtenberg et al. 2016; Ali et al. 2018; El Mansy et al. 2020). Tropical environments host a larger number of EF, predominantly falling within a limited number of taxonomic classes. Endophytic filamentous fungi play a key part in improving plant growth by increasing plant resistance to biotic and abiotic stresses through phytohormones, secondary metabolites, and the induction of plant defense mechanisms (Poveda et al. 2022).

Filamentous EF can be characterized into two groups: clavicipitaceous, which are found in grasses, and non-clavicipitaceous, which are not associated with grasses, as outlined by Hyde and Soytong (2008). These endophytes can be transmitted either vertically or horizontally based on their mode of growth. In vertical transmission, the fungal endophytes enter developing seeds and colonize them during germination, conferring resistance to various abiotic stresses for instance high temperatures, heavy metals, salinity, and drought on the host plants. In horizontal transmission, EF generates vegetative spores or propagules that are subsequently disseminated to other plants within the population, mostly via air, animals, or insects (Lugtenberg et al. 2016; Samreen et al. 2021). The diverse EF belonging to the Ascomycota division is illustrated in Table 14.1.

Rodriguez et al. (2009) proposed a further classification of endophytes into four distinct classes. (1) Class 1 comprises non-clavicipitaceous endophytes capable of colonizing the roots and aerial parts of their host plants, with both vertical and horizontal transmission. (2) Class 2 consists of clavicipitaceous fungi that exclusively colonize grasses, inhabiting both belowground and aboveground parts of their host plants, and exhibit both vertical and horizontal transmission mechanisms. (3) Class 3 comprises endophytes primarily related with the leaves of tropical tree species, representing a diversified group characterized by horizontal transmission only. (4) Class 4 consists of the so-called septate EF, which are characterized by melanin in the septa and inhabit the root development of host plants, also exhibiting horizontal transmission.

In addition to ascomycetes, some basidiomycetes and yeasts also reside in plant tissues through various means without causing harm. Certain yeast species produce enzymes that enable them to enter or reside in plant tissues and reproduce within both intercellular and intracellular spaces, enhancing their survival and nutrient extraction from host plants (Kowalska et al. 2022). The comprehensive analysis of the EF distribution and diversity across extensive topographical regions is still in its nascent stages, with only a limited number of primary observations made (Chadha

Table 14.1 Fungal endophytes belonging to *Ascomycota* and *Basidiomycota* division along with their associate host plant

Endophytic fungi	Division	Plant	Family	Reference
Alternata sp.	Ascomycota	*Sophora alopecuroides*	Fabaceae	Ju et al. (2024)
Alternaria alternata	Ascomycota	*Solanum nigrum*	Solanaceae	Khan et al. (2015a, b)
Aspergillus sp.	Ascomycota	*Sophora alopecuroides*	Fabaceae	Ju et al. (2024)
Aspergillus awamori	Ascomycota	*Withania somnifera*	Solanaceae	Mehmood et al. (2019)
Aspergillus fumigatus	Ascomycota	*Oxalis corniculata*	Oxalidaceae	Bilal et al. (2018)
Aspergillus nidulans	Ascomycota	*Gibberella fujikuroi*	Nectriaceae	Xu et al. (2018a, b)
Aspergillus terreus	Ascomycota	*Sophora flavescens*	Fabaceae	Turbat et al. (2020)
Aspergillus welwitschiae	Ascomycota	*Glycine max*	Fabaceae	Husna et al. (2021)
Aspergillus welwitschiae	Ascomycota	*Fridericia chica*	Bignoniaceae	de Melo Pereira et al. (2024)
Aspergillus ustus	Ascomycota	*Oryza sativa*	Poaceae	Potshangbam et al. (2017)
Bipolaris sp.	Ascomycota	*Glycine max*	Fabaceae	Lubna et al. (2022)
Cladosporium angulosum	Ascomycota	*Sesuvium portulacastrum*	Aizoaceae	Yang et al. (2023)
Chaetomium globosum	Ascomycota	*Chromolaena odorata*	Asteraceae	Nongthombam et al. (2024)
Colletotrichum alienum	Ascomycota	*Anisomeles indica*	Lamiaceae	Toppo et al. (2024)
Colletotrichum cobbittiense	Ascomycota	*Anisomeles indica*	Lamiaceae	Toppo et al. (2024)
Colletotrichum cobbittiense	Ascomycota	*Anisomeles indica*	Lamiaceae	Toppo et al. (2024)
Colletotrichum yulongense	Ascomycota	*Anisomeles indica*	Lamiaceae	Toppo et al. (2024)
Curvularia sp.	Ascomycota	*Fridericia chica*	Bignoniaceae	de Melo Pereira et al. (2024)
Curvularia geniculata	Ascomycota	*Parthenium hysterophorus*	Asteraceae	Priyadharsini and Muthukumar (2017)
Curvularia lunata	Ascomycota	*Melia azedarach*	Meliaceae	Saad and Badry (2020)

(continued)

Table 14.1 (continued)

Endophytic fungi	Division	Plant	Family	Reference
Diaporthe cerradensis	Ascomycota	*Fridericia chica*	Bignoniaceae	de Melo Pereira et al. (2024)
Epulorhiza sp.	Ascomycota	*Neuwiedia singapureana*	Orchidaceae	Wang et al. (2024)
Fusarium avenaceum	Ascomycota	*Lolium multiflorum*	Poaceae	García-Latorre et al. (2023)
Fusarium equiseti	Ascomycota	*Anisomeles indica*	Lamiaceae	Toppo et al. (2024)
Fusarium oxysporum	Ascomycota	*Oryza sativa*	Poaceae	Potshangbam et al. (2017)
Fusarium proliferatum	Ascomycota	*Rhazya stricta*	Apocynaceae	Seema et al. (2023)
Fusarium proliferatum	Ascomycota	*Oxalis corniculata*	Oxalidaceae	Bilal et al. (2018)
Fusarium solani	Ascomycota	*Solanum lycopersicum*	Solanaceae	Pappas et al. (2018)
Fusarium tricinctum	Ascomycota	*Solanum nigrum*	Solanaceae	Khan et al. (2015a, b)
Galactomyces geotrichum	Ascomycota	*Oryza sativa*	Poaceae	Potshangbam et al. (2017)
Galactomyces geotrichum	Ascomycota	*Trapa japonica*	Lythraceae	Waqas et al. (2014)
Hypoxylon investiens	Ascomycota	*Fridericia chica*	Bignoniaceae	de Melo Pereira et al. (2024)
Neoscytalidium	Ascomycota	*Fridericia chica*	Bignoniaceae	de Melo Pereira et al. (2024)
Nigrospora sphaerica	Ascomycota	*Melia azedarach*	Meliaceae	Saad and Badry (2020)
Paecilomyces formosus	Ascomycota	*Glycine max*	Fabaceae	Bilal et al. (2018)
Paraconiothyrium archidendri	Ascomycota	*Betula utilis*	Betulaceae	Dasila et al. (2024)
Penicillium sp.	Ascomycota	*Cucumis sativus*	Cucurbitaceae	Waqas et al. (2012)
Penicillium sp.	Ascomycota	*Sophora flavescens*	Fabaceae	Turbat et al. (2020)
Penicillium funiculosum	Ascomycota	*Glycine max*	Fabaceae	Bilal et al. (2020)
Penicillium janthinellum	Ascomycota	*Solanum lycopersicum*	Solanaceae	Khan et al. (2015a, b)
Penicillium rubens	Ascomycota	*Fridericia chica*	Bignoniaceae	de Melo Pereira et al. (2024)

(continued)

Table 14.1 (continued)

Endophytic fungi	Division	Plant	Family	Reference
Penicillium simplicissimum	Ascomycota	*Oryza sativa*	Poaceae	Potshangbam et al. (2017)
Pezicula radicicola	Ascomycota	*Betula utilis*	Betulaceae	Dasila et al. (2024)
Phoma glomerata	Ascomycota	*Oryza sativa*	Poaceae	Potshangbam et al. (2017)
Phoma glomerata	Ascomycota	*Cucumis sativus*	Cucurbitaceae	Waqas et al. (2012)
Phoma glomerata	Ascomycota	*Citrullus colocynthis*	Cucurbitaceae	Ismaila et al. (2018)
Phomopsis liquidambaris	Ascomycota	*Arachis hypogaea*	Fabaceae	Jiang et al. (2022)
Sphingomonas sp.	Alphaproteobacteria	*Glycine max*	Fabaceae	Bilal et al. (2018)
Sarocladium terricola	Ascomycota	*Lolium multiflorum*	Poaceae	García-Latorre et al. (2023)
Talaromyces adpressus	Ascomycota	*Oryza coarctata*	Oryzinae	Airin et al. (2023)
Talaromyces argentinensis	Ascomycota	*Oryza coarctata*	Oryzinae	Airin et al. (2023)
Xylariaceae sp.	Ascomycota	*Lolium multiflorum*	Poaceae	García-Latorre et al. (2023)
Atractiella rhizophila	Basidiomycota	Quercus shumardii	Fagaceae	Jin et al. (2019)
Bjerkandera sp.	Basidiomycota	*Rubia* sp.	Rubiaceae	Feng et al. (2021)
Ceratobasidium sp.	Basidiomycota	*Rubia* sp.	Rubiaceae	Feng et al. (2021)
Funalia sp.	Basidiomycota	*Rubia* sp.	Rubiaceae	Feng et al. (2021)
Lentinus sp.	Basidiomycota	*Rubia* sp.	Rubiaceae	Feng et al. (2021)
Irpex sp.	Basidiomycota	*Rubia* sp.	Rubiaceae	Feng et al. (2021)
Piriformospora indica	Basidiomycota	*Glycine max*	Fabaceae	Bajaj et al. (2018)
Phlebia sp.	Basidiomycota	*Rubia* sp.	Rubiaceae	Feng et al. (2021)
Phlebiopsis sp.	Basidiomycota	*Rubia* sp.	Rubiaceae	Feng et al. (2021)
Porostereum spadiceum	Basidiomycota	*Glycine max*	Fabaceae	Hamayun et al. (2017)
Trametes sp.	Basidiomycota	*Rubia* sp.	Rubiaceae	Feng et al. (2021)
Tulasnella phuhinrongklaensis	Basidiomycota	*Phalaenopsis pulcherrima*	Orchidaceae	Srimuang and Lumyong (2018)

et al. 2015; Lugtenberg et al. 2016). Further research is required for exploring the EF diversity from different taxa and their associated host plants.

14.3 Rational for Plant Selection

A comprehensive understanding of the rationale and methods behind the choice of plants is crucial for optimizing the isolation of novel endophytes. Given the vast number and diversity of plants, strategic approaches are essential to narrow the search for EFs capable of producing phytohormones. The collection of endophytes for the discovery of phytohormones should target both fungal taxa and associated host plants (Torres et al. 2011). Several hypotheses, as outlined by various scholars, govern the plant selection strategies for endophyte discovery (Lugtenberg et al. 2016). The proposed hypotheses are as follows:

1. **Distinct Ecological Niche**: Plants with unique biological characteristics or those inhabiting unusual ecological niches are considered potential hosts for beneficial fungal endophytes.
2. **Ethnobotanical History:** Plants of ethnobotanical significance, especially those with a history of beneficial uses, are recognized as promising candidates for EF with phytohormone potential.
3. **Biodiversity-Rich Areas:** Plants that thrive in areas of high biodiversity are believed to harbor EF that reflect the richness in biodiversity.
4. **Endemism, Longevity, or Ancient Origins:** Plants that are endemic, possess exceptional longevity, or originate from ancient land masses are considered potential sources of endophytes.

The incorporation of these hypotheses into plant selection strategies aims to optimize the likelihood of discovering phytohormones produced by endophytic fungi. This approach recognizes the intricate interplay between endophytic fungi and plant taxa, offering a targeted methodology for endophyte isolation.

14.4 Phytohormones Production by EF

EF contribute to agriculture by producing plant hormones that aid in nutrient and water transport, activate biological defense mechanisms, and offer resistance against phytopathogens (Yan et al. 2019). The significant plant hormones produced by EF include indole acetic acid, abscisic acid, and gibberellic acid, which benefit plants to withstand biotic and abiotic stresses (Bilal et al. 2019). This section will discuss the phytohormones produced by EF and the metabolic pathways through which they are synthesized.

14.4.1 Indole Acetic Acid (IAA)

EF are well-known for their ability to synthesize a wide range of compounds, including volatile compounds, plant growth-regulating hormones, and iron-binding molecules, which can positively influence plant growth and development via both indirect and direct mechanisms. Among them, IAA, an important auxin, plays a pivotal part in improving various aspects of plant growth such as cell division, elongation, differentiation, root initiation, apical dominance, and resistance against pathogens (Zhang et al. 2022). IAA production is not limited to the Plantae kingdom but also occurs in plant-associated microbes, including endophytic, ectomycorrhizal, pathogenic, rhizospheric, and phyllosphere microbes (Khan et al. 2014; Kumla et al. 2014).

In the past few years, an increasing interest in the role of EF in producing IAA has been observed. IAA production is linked to root hair development, lateral root formation, improved nutrient absorption, and influence on fungal-plant interactions (Mehmood et al. 2019; Sarkar et al. 2021). IAA serves as a communication molecule between EF and their associated plants, aiding in growth and development. Furthermore, IAA can induce changes in the plant's innate defense mechanisms, providing resistance to pathogenic attacks (Fu et al. 2015). Research has highlighted the potential of IAA to serve as a signaling molecule within EF, thereby influencing gene expression in them (Jahn et al. 2021).

Several biosynthetic pathways for IAA production in microbes have been proposed. Specific fungi and yeasts associated with these pathways have been identified by researchers (Sun et al. 2014; Nutaratat et al. 2016). In plants, numerous pathways for IAA production exist, while in microbes, the known pathways include indole-3-pyruvate, indole-3-acetamide, tryptophan side-chain oxidase, indole-3-acetonitrile, and tryptamine. However, further research is needed to provide a detailed explanation of IAA production through all the proposed biosynthetic pathways.

IAA, a metabolic byproduct derived from L-tryptophan, is produced by a wide array of endophytes residing in plant roots. Several endophytic fungi (EF), including *Aureobasidium pullulans, C. acutatum, Cryptococcus flavus, Fusarium fujikuroi, F. oxysporum, F. proliferatum, Hannaella coprosmaensis, Muscodor cinnamomi, Pseudozyma aphidis, Piriformospora indica, Rhodosporidium paludigenum, Rhodotorula graminis, Sporisorium reilianum, Saccharomyces cerevisiae, Ustilago esculenta*, and *U. maydis*, are primarily associated with the IPyA and IAM biosynthetic pathways for IAA production (Sun et al. 2014; Nutaratat et al. 2016; Vrabka et al. 2019; Cosoveanu et al. 2021; Saber et al. 2023). Nonetheless, further detailed investigations into IAA production by EF, particularly those associated with plants, remain an essential area for future research. Table 14.2 shows the amounts of IAA produced by different endophytic fungal isolates.

Table 14.2 IAA production by endophytic fungi

Endophytic fungi	IAA (µg/mL)	Incubation time	Host plant	References
Alternaria alternata	319.24	7 days	*Elymus dahuricus* Turcz	Qiang et al. (2019a, b)
Arthrinium sp.	42	7 days	Chenopodium album	Ikram et al. (2023a, b)
Aspergillus awamori	24.2	7 days	*Withania somnifera*	Mehmood et al. (2019)
Aspergillus flavus	77.26	–	*Euphorbia indica*	Hamayun et al. (2019)
Aspergillus flavus	42.91	–	*Alpinia* sp.	Hartanto et al. (2019)
Aspergillus medius	58.02	72 h	*Triticum aestivum*	Prajapati et al. (2024)
Aspergillus niger	32.6	*3* days	*Axonopus purpusii*	Galeano et al. (2021)
Aspergillus ochraceus	176	14 days	*Avicennia marina*	Badawy et al. (2021)
Aspergillus pullulans	48.2	8 days	*Boswellia sacra*	Ali et al. (2019)
Aspergillus terreus	38	7 days	Chenopodium album	Ikram et al. (2023a, b)
Aspergillus terreus	40	–	*Chenopodium album*	Khan et al. (2022)
Cladosporium parahalotolerant	56.61	72 h	*Triticum aestivum*	Prajapati et al. (2024)
Colletotrichum alienum	122.5	6 days	*Anisomeles indica*	Toppo et al. (2024)
Colletotrichum cobbittiense (Ai3)	155.3	6 days	*Anisomeles indica*	Toppo et al. (2024)
Colletotrichum cobbittiense (Ai2)	129.2	6 days	*Anisomeles indica*	Toppo et al. (2024)
Colletotrichum gloeosporioides	160	7 days	*Jerusalem artichoke*	Khaekhum et al. (2021)
Colletotrichum. yulongense	179.2	6 days	*Anisomeles indica*	Toppo et al. (2024)
Comamonas antarctica (5ML1R)	38.24	4 days	*Indosasa shibataeoides*	Yue et al. (2024)
Comamonas Antarctica (D4N7)	33.3	4 days	*Indosasa shibataeoides*	Yue et al. (2024)
Curvularia lunata	45.17	8 days	*Elettaria* sp.	Munir et al. (2021)
Fusarium equiseti	170.3	6 days	*Anisomeles indica*	Toppo et al. (2024)
Hypocrea nigricans	40	7 days	*Chenopodium album*	Ikram et al. (2023a, b)

(continued)

Table 14.2 (continued)

Endophytic fungi	IAA (µg/mL)	Incubation time	Host plant	References
Microascus murinus	3.73	7 days	*Oryza sativa*	Pablo et al. (2020)
Nigrospora oryzae	351.01	–	*Oryza sativa*	Sodhi and Saxena (2023)
Penicillium citrinum	94.28	7 days	*Calotropis procera*	Jaiswal et al. (2024)
Penicillium chrysogenum	26.88	7 days	*Axonopus purpusii*	Galeano et al. (2023)
Penicillium roqueforti	36.9	7 days	*Solanum surattense*	Ikram et al. (2023a, b)
Penicillium roqueforti	51.3	–	*Solanum surattense*	Ikram et al. (2018)
Rhizopus oryzae	88.16	–	Adiantum capillus-veneris	Ismail et al. (2020)
Schizophyllum commune	11.7	8 days	*Elettaria* sp.	Munir et al. (2021)
Sordariomycetidae sp.	1638	–	*Arabidopsis thaliana*	Naureen et al. (2022)
Trichoderma atroviride	5.27	8 days	*Elettaria* sp.	Munir et al. (2021)
Talaromyces purpureogenus	405	8–10 days	*Triticum aestivum*	Kaur and Saxena (2023)
Trichoderma reesei	39.5	7 days	*Solanum surattense*	Ikram et al. (2023a, b)

14.4.2 IAA Production Pathway by EF

14.4.2.1 The Indole-3-Acetamide (IAM) Pathway

The IAA production through the IAM biosynthetic pathway has been thoroughly investigated. In this pathway, tryptophan is transformed into IAM, catalyzed via tryptophan monooxygenase. In the next step, IAM is hydrolyzed into ammonia and IAA, an active form of a phytohormone (Fig. 14.2). The enzymes involved in converting tryptophan to IAA vary among different microbes. Key genes such as IaaH and IaaM encode indole-3-acetamide hydrolase and tryptophan monooxygenase, respectively (Ahemad and Kibret 2014). This biosynthetic pathway has been reported for IAA biosynthesis as the predominant route in various fungal endophytes, including *Aspergillus awamori*, *Colletotrichum gloeosporioides*, *C. fructicola*, *C. acutatum*, and *F. oxysporum* (Tsavkelova et al. 2006; Numponsak et al. 2018; Mehmood et al. 2019). The transition of tryptophan to IAA is associated with microbe-plant interactions, where EF may affect the growth and development of the host plant via IAA production. Recent research has underscored the ubiquity of the IAM pathway across various EF, including cyanobacteria and actinomycetes (Tsavkelova et al. 2006). This pathway stands out as one of the most widely investigated tryptophan-dependent pathways.

14.4.2.2 The Indole-3-Pyruvic Acid (IPA) Pathway

The IPA pathway was initially revealed in plants and later found to be a prominent pathway for IAA production in bacteria. In this pathway, IPA is formed from tryptophan, which is catalyzed by aminotransferases, and then transferred into indole-3-acetaldehyde (IAAld) by indolepyruvate decarboxylase (IPDC). The final product, IAA, is obtained through the activity of aldehyde dehydrogenases (Sardar and Kempken 2018), as illustrated in Fig. 14.1. In microbes, key genes such as trpA, IPDC, and IAD are responsible for encoding aminotransferases, decarboxylases, and indolyl-3-acetaldoxime dehydratase, respectively.

In fungal endophytes, IPA is considered a major contributing pathway to IAA biosynthesis (Krause et al. 2015). Enzymes such as transaminase and

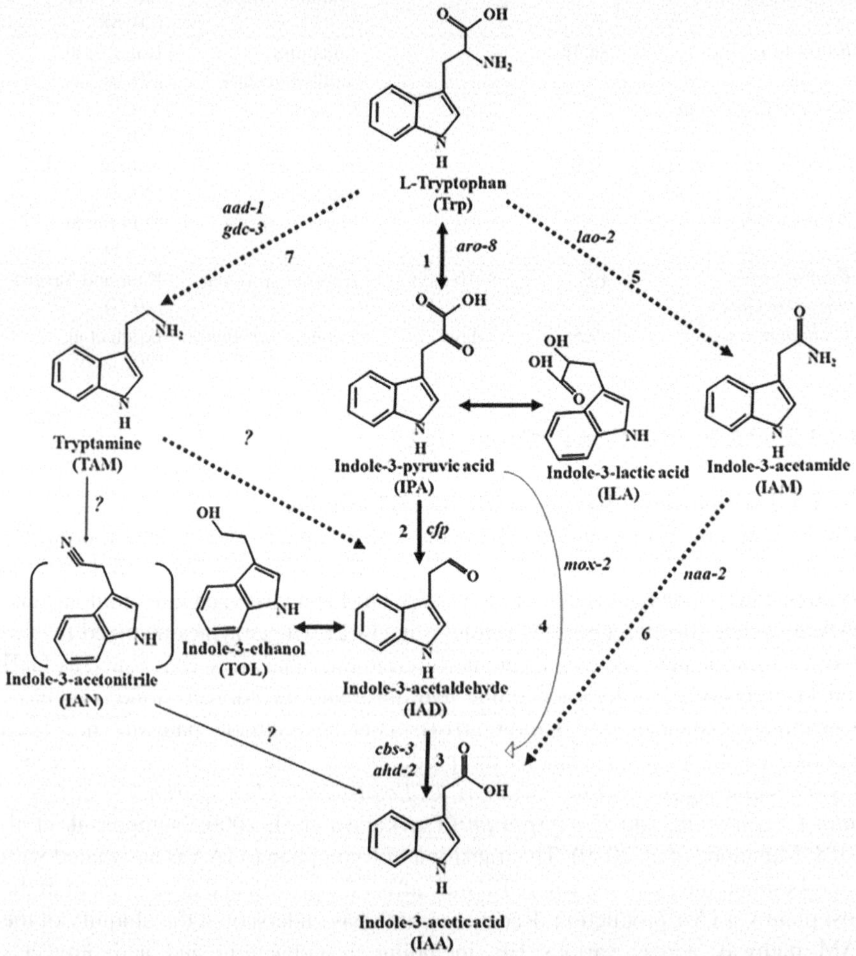

Fig. 14.2 Indole acetic acid production by EF involves the tryptophan precursor and utilizes TAM, IPA, LAM, and IAN biosynthetic pathways (Sardar and Kempken 2018)

indolyl-3-acetaldoxime dehydratase, used in this pathway, have been recognized in many endophytic fungi (EF) (Tivendale et al. 2014). These enzymes play an essential role in converting tryptophan to IPA and then to IAA (Cook et al. 2016). A genetic study of *Magnaporthe oryzae* revealed a comprehensive elucidation of the IPA biosynthetic pathway, including important enzymes like indole-3-pyruvic acid decarboxylase and tryptophan aminotransferase, as reported by Dong et al. (2022). The IPA pathway has emerged as a widespread and important route for IAA production in both fungi and bacteria; however, further knowledge is still required.

14.4.2.3 The Tryptamine (TAM) Pathway

The TAM pathway is another tryptophan-dependent pathway for IAA production, observed in a few fungi such as *Cyanodermella asteris* (Tang et al. 2023), *Colletotrichum* sp., and *Ustilago* sp. (Hilbert 2013). In this pathway, tryptophan is first transformed into tryptamine via tryptophan decarboxylase and then into IAAld with the help of amine oxidase (Tang et al. 2023). Lastly, IAAld is transformed into IAA through an important enzyme, i.e., aldehyde dehydrogenase (Fig. 14.1). Compared to the IPA and IAM pathways, the TAM biosynthetic pathway in *E. coli* is less explored, primarily due to limited data on the important amine oxidase involved in converting tryptamine to IAAld.

14.4.2.4 The Indole-3-Acetonitrile (IAN) Pathway

Research into the IAN biosynthetic pathway has primarily been conducted in plants. The mechanism behind IAA synthesis through this pathway includes the tryptophan transformation into indole-3-acetaldoxime (IAOx). This is followed by the subsequent transition of IAOx to indole-3-acetyl glucosinolate (IAG). The ending step includes the conversion of IAG to IAN and then to IAA via nitrilase (Zhao et al. 2001), as illustrated in Fig. 14.1.

Research on IAA production via the IAN in EF is not current enough to fully understand the mechanism. Researchers have not focused on IAA production through the IAN biosynthetic pathway. In this pathway, only a few enzyme-encoding genes have been identified, while the key enzyme responsible for converting tryptophan to IAOx remains unidentified. Investigations have revealed the presence of indole-3-acetaldehyde oxime dehydratase enzyme in *Bradyrhizobium* sp. and *Sclerotinia sclerotiorum*, consistently alongside genes encoding nitrile hydratases and nitrile hydrolases, confirming the presence of the IAN biosynthetic pathway. Rädisch et al. (2018) and Chen et al. (2021) indicate that a comprehensive understanding of IAA production via the IAN biosynthetic pathway in EF is necessary to elucidate the complete pathway.

14.5 Gibberellic Acid (GA)

Gibberellic acids are diterpenoid phytohormones first isolated in the 1920s from the phytopathogenic fungus *Gibberella fujikuroi* (Nguyen et al. 2023). GAs are isoprene derivatives with a four-ring structure. Among all isolated GAs, GA1, GA3,

GA4, and GA7 are well-understood and have been widely researched (Cui et al. 2020). GA plays a role in seed germination, stem elongation, leaf growth, floral initiation, and the formation of flowers and fruit, as well as senescence. Furthermore, it helps alleviate the negative effects of abiotic stresses on plant growth and development (Ali and Baloch 2020; Hossain et al. 2022). Research has reported that GA production by various endophytic fungal genera includes *Rhizoctonia* spp., *Fusarium* sp., *Rhizopus* sp., *Aspergillus* sp., *Absidia* spp., *Penicillium* spp., *Phoma* sp., *Monilia* sp., *Pythium* spp., *Verticillium* sp., *Trichoderma* spp., *Actinomucor* sp., and *Schizophyllum* spp. (Hamayun et al. 2010a, b; Waqas et al. 2012; Jaroszuk-Ściseł et al. 2014; Turaeva et al. 2020).

Meleigy and Khalaf (2009) reported GA production in 28 different strains of *Fusarium moniliforme*, with the highest amount estimated in only four strains. EF is considered significantly important for plant growth due to a wide variety of bioactive substances production, i.e., mycotoxins and phytohormones (Baron and Rigobelo 2022). The effect of various factors on GA production, including carbon source, temperature ranges (23, 25, 30, 37 °C), and pH of the medium (5.6–8), is still under investigation. Research reported the maximum production of GA, i.e., 105 mg/g, by *F. moniliforme* after 4 days of incubation at 30 °C in both liquid and solid Czapek's nutrient medium, which included *Jatropha curcas* plant residue as a carbon source (Ozimek et al. 2018).

14.5.1 Biosynthetic Pathway for GA Production

The biosynthetic pathway for GA production begins with the conversion of geranylgeranyl diphosphate (GGPP) into ent-copalyl diphosphate. Subsequent oxidation at position C-19 leads to the ent-kaurenoic acid formation, which then undergoes a series of reactions to produce ent-7α-hydroxykaurenoic acid, as illustrated in Fig. 14.3. A final oxidation at C-6β position results in GA_{12}-aldehyde formation. The key steps are similar among fungi, bacteria, and higher plants, with only minor variations occurring in the later steps.

In EF, GA_{12}-aldehyde is converted into GA_{14}-aldehyde, which is a distinctive feature where a single bifunctional terpene cyclase forms ent-kaurene from GGDP. In bacteria and higher plants, two enzymes, ent-kaurene synthase and ent-copalyl-diphosphate synthase, are involved in the later steps (Salazar-Cerezo et al. 2018). In bacteria, fungi, and higher plants, the initial stages share similarities, starting from GGDP and progressing to ent-7α-hydroxy-kaurenoic acid. In fungi, the biosynthesis of GA occurs via the mevalonic acid (MVA) pathway, which provides isopentenyl pyrophosphate (IPP) for terpenoid synthesis, including various forms of GAs (see Fig. 14.2).

In contrast to higher plants, IPP is synthesized via two pathways: the methyl erythritol phosphate pathway, carried out in plastids, and the MVA pathway, carried out in the cytoplasm (Hedden and Thomas 2012). These pathways are structured into three stages based on subcellular compartmentalization and enzymatic activities. The initial stage occurs in proplastids and involves soluble enzymes that lead

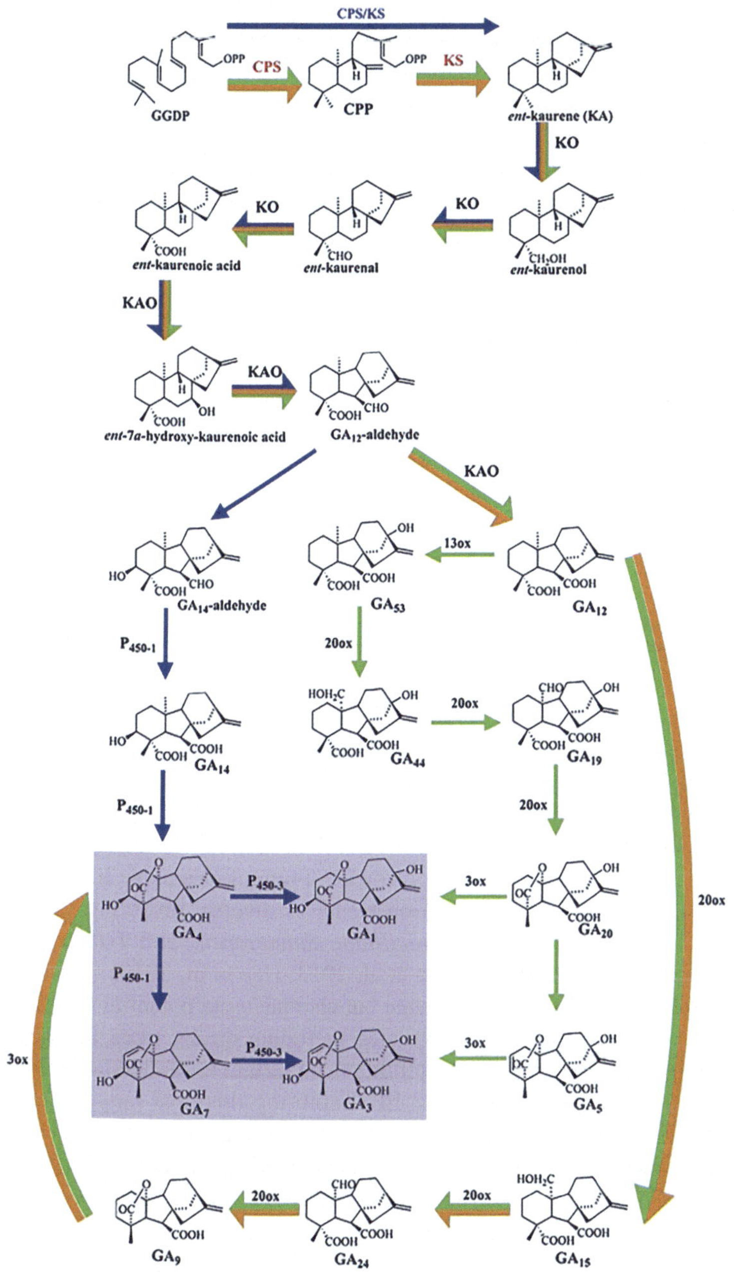

Fig. 14.3 GA production by EF via GGPP precursor (Salazar-Cerezo et al. 2018)

to ent-kaurene production. The next stage involves the ent-kaurene oxidation to GA_{12}-aldehyde. The final step takes place in the cytosol with the help of

2-oxoglutarate-dependent dioxygenases. In complex plants, GA_{12} serves as a branch point leading to the formation of GA_{12}-aldehyde, which is further changed to GA_{12} (a non-13-hydroxylated gibberellin). Subsequent 13-hydroxylation of GA_{12} results in GA_{53}. In the next step, GA_{12} and GA_{53} act as precursors for the 13-hydroxylation and non-13-hydroxylation pathways, respectively. With further oxidations at C-20 yields GA_9 and GA_{20}. The last step in GA formation involves the 3β-hydroxylation of two forms of gibberellins such as GA_9 and GA_{20} into GA_1 and GA_4, respectively (Bömke and Tudzynski 2009) (Table 14.3).

14.6 Abscisic Acid (ABA)

ABA, an important phytohormone, plays a vital part in various coordinating processes within plants. It is crucial in bud dormancy, seed germination, leaf senescence, and stress responses to salinity, heavy metals, and drought, highlighting its importance in medicine and agriculture (Rai et al. 2023; Hewage et al. 2020). Furthermore, increased levels of ABA stimulate stomatal closure, thereby minimizing transpiration (Lim et al. 2015). The literature has reported the negative effects of ABA on plant defense mechanisms against certain pathogen attacks. For instance, it hinders resistance in tobacco (*Nicotiana tabacum*) to *Ralstonia solanacearum* (Zhou et al. 2008), in barley (*Hordeum vulgare*) to *Magnaporthe oryzae* (Ulferts et al. 2015), and in tomato (*Solanum lycopersicum*) to *Botrytis cinerea* (Sivakumaran et al. 2016). Conversely, ABA increased resistance in *Arabidopsis* sp. against *Alternaria brassicicola* (Fan et al. 2009). The outcome of ABA can vary between species and associated host plants. In mutually beneficial plant-microbe interactions, ABA facilitates the establishment and colonization of compatible relationships. For instance, it enhances the interaction between tomato plants and the *Glomus intraradices*, an arbuscular mycorrhizal fungus (Xu et al. 2018a, b).

The host range for ABA production by EF is diverse, involving various plant genera for instance *Aspergillus, Penicillium, Sphaeropsis*, and *Trichoderma* (Xie et al. 2023; Xu et al. 2018a, b; Khan et al. 2013; Hao et al. 2010). These versatile microbes can be specific to a host range but also have the potential to form symbiotic associations with various plant species, including shrubs, trees, and herbaceous plants (Verma et al. 2022). This adaptability is important for understanding the complex interactions between plants and EF. Exploring this host range is crucial for elucidating the relationship between them and their potential use in agriculture. To the best of our knowledge, there is limited data on ABA production by EF and their associated biosynthetic pathways. Research primarily focuses on IAA and GA production by EF, because of their roles in enhancing plant development and growth.

In contrast, ABA, a stress hormone, has not received sufficient attention because its main functions are believed to be associated with stress regulation rather than growth promotion. Moreover, researchers encounter challenges in understanding the biosynthetic pathway of ABA production and its regulation in EF. As our understanding of the complex relationship between EF advances and host plants, elucidating the biosynthetic pathway for ABA production and its potential roles in

Table 14.3 GA production by endophytic fungi

Endophytic fungi	GA type	GA concentration	Host plant	References
Penicillium sp.	GA_3	2.8 ng/mL	*Monochoria vaginalis*	Ahmad et al. (2010)
	GA_4	2.6 ng/mL		
	GA_7	6.68 ng/mL		
	GA_9	1.61 ng/mL		
	GA_{24}	0.18 ng/mL		
Aspergillus sp.	GA_3	1.64 ng/mL	*Monochoria vaginalis*	Ahmad et al. (2010)
	GA_4	1.37 ng/mL		
	GA_7	6.29 ng/mL		
	GA_{12}	0.3 ng/mL		
	GA_{24}	0.59 ng/mL		
Aspergillus fumigatus	GA_1	0.091 ng/g	*Oxalis corniculata*	Bilal et al. (2018)
	GA_3	0.324 ng/g		
	GA_7	0.023 ng/g		
	GA_8	0.004 ng/g		
	GA_9	0.001 ng/g		
	GA_{12}	0.012 ng/g		
	GA_{24}	0.002 ng/g		
Fusarium proliferatum	GA_1	0.392 ng/g	*Oxalis corniculata*	Bilal et al. (2018)
	GA_3	0.089 ng/g		
	GA_7	0.492 ng/g		
	GA_8	0.003 ng/g		
	GA_9	0.195 ng/g		
	GA_{24}	0.018 ng/g		
Aspergillus fumigatus	GA4	24.8 ng/mL	*Glycine max*	Khan et al. (2011)
	GA9	1.2 ng/mL		
	GA12	9.8 ng/mL		
Phoma herbarum	GA_1	0.11 ng/mL	*Glycine max*	Hamayun et al. (2009)
	GA_3	2.91 ng/mL		
	GA_4	3.21 ng/mL		
	GA_7	1.4 ng/mL		
	GA_9	0.05 ng/mL		
	GA_{12}	0.23 ng/mL		
	GA_{15}	0.42 ng/mL		
	GA_{19}	0.53 ng/mL		
	GA_{20}	0.06 ng/mL		
Gliomastix murorum	GA1	0.32 ng/mL	*Elymus mollis*	Afzal khan et al. (2009)
	GA3	5.76 ng/mL		
	GA4	0.82 ng/mL		
	GA5	0.59 ng/mL		
	GA7	0.1 ng/mL		
	GA9	5.38 ng/mL		
	GA20	0.25 ng/mL		
	GA24	2.03 ng/mL		

(continued)

Table 14.3 (continued)

Endophytic fungi	GA type	GA concentration	Host plant	References
Penicillium trzebinski	GA1	1.2 ng/mL	*Hydrocharis dubia*	You et al. (2015)
	GA4	5 ng/mL		
Chaetomium globosum	GA1	0.67 ng/mL	*Capsicum annum*	Khan et al. (2012)
	GA4	21.8 ng/mL		
	GA9	0.51 ng/mL		
	GA12	12.4 ng/mL		
	GA20	1.11 ng/mL		
Cladosporium sp.	GA1	1.97 ng/mL	*Cucumis sativus*	Hamayun et al. (2010a, b)
	GA3	5.18 ng/mL		
	GA4	13.35 ng/mL		
	GA7	2.4 ng/mL		
	GA9	0.69 ng/mL		
	GA12	0.24 ng/mL		
	GA15	0.68 ng/mL		
	GA19	1.94 ng/mL		
	GA20	0.78 ng/mL		
Penicillium sp.	GA1	1.37 ng/mL	*Chrysanthemum coronarium*	Hamayun et al. (2010a, b)
	GA3	5.88 ng/mL		
	GA4	8.62 ng/mL		
	GA7	2.05 ng/mL		
	GA9	0.83 ng/mL		
	GA12	0.44 ng/mL		
	GA15	0.74 ng/mL		
	GA19	1.16 ng/mL		
	GA20	0.98 ng/mL		

microbe-plant interactions is urgent. Future studies are important to bridge this gap and provide a more in-depth understanding of the hormonal dynamics involved in endophytic symbiosis.

Xu et al. (2018a, b) reported the interaction between EF and host plants in terms of ABA. They suggested that EF might have receptors to detect ABA, which stimulates these receptors and triggers changes in the cellular machinery of EF. This affects gene functioning, protein synthesis, and degradation. Additionally, ABA enhances the citric acid cycle or Krebs cycle, an essential pathway for energy production in cells, potentially increasing the production of mycotoxins that provide protection against diseases and pest attacks. The interaction between the endophytic fungus *Aspergillus nidulans* and the mangrove plant via ABA is illustrated in Fig. 14.4.

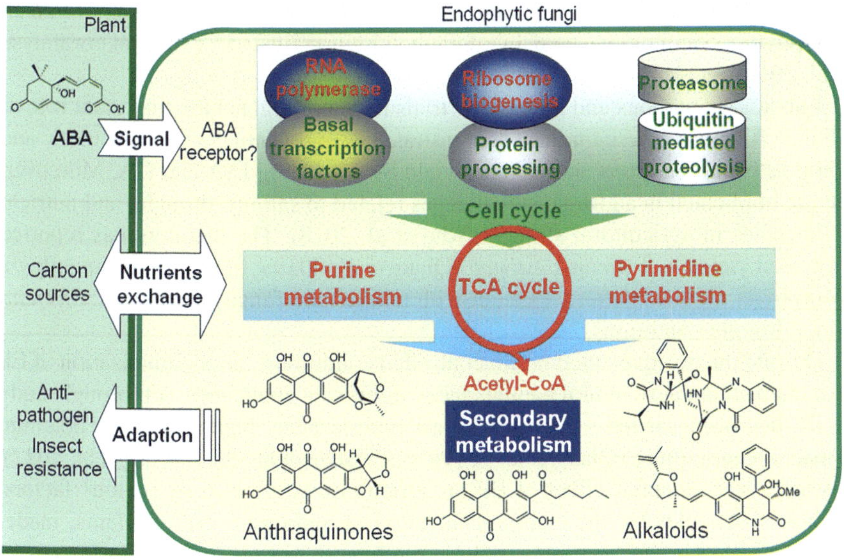

Fig. 14.4 ABA signaling between endophytic fungi and host plant improves nutrient exchange and provides protection against pests and insect attack (Xu et al. 2018a, b)

14.7 Role of EF in Agriculture

Recent studies have emphasized the significant contribution of EF to the development and growth of plants to environmental factors (Tsipinana et al. 2023; Tiwari and Bae 2022; Burragoni and Jeon 2021). These endophytes colonize different plant tissues, exhibiting distinct distribution patterns among various plant species (Aamir et al. 2020). In agricultural contexts, EF play advantageous and diverse roles as biofertilizers that can increase plant development by important phytohormones production, such as IAA, GA, and ABA. Phytohormones not only enhance plant growth but also protect against pests and pathogens and induce tolerance to abiotic stresses (Aamir et al. 2020). The significant effects of these phytohormones have been studied in various plant varieties, including tomato (Paradza et al. 2021), coffee (Lu et al. 2022), wheat (Asim et al. 2022), and barley (Murphy et al. 2018).

The presence of EF in plants is notable, but the mechanisms underlying the colonization of plants by EF need to be explored. For instance, the mechanisms behind the biosynthesis of IAA and GA are reported in the literature, but a detailed and well-explained mechanism for ABA production is still lacking. Some endophytic fungal genera, such as *Colletotrichum* spp., *Guignardia* sp., *Pestalotiopsis* sp., *Diaporthe* spp., and *Xylaria* spp., exhibit the ability to colonize a broad spectrum of plants as foliar endophytes without clear details of the inherent processes (Suryanarayanan et al. 2018). Understanding microbiome-plant interactions is crucial, considering that the plant and its associated microbes constitute a mini-ecosystem, the holobiont. The specific roles of keystone species, core (dominant ones), and satellite endophytic fungi (those with low tissue colonization) within an endophyte community remain largely uninvestigated (Vandenkoornhuyse et al. 2015).

Laboratory findings and glasshouse trials have highlighted the important role of EF in agriculture. These findings demonstrated EF's potential to alleviate biotic and abiotic stresses on crops and increase yield by producing IAA and GA. Moreover, EF are implicated in addressing challenges related to salinity, drought, and nutrient deficiencies in agricultural systems (Rho et al. 2018). The literature has reported increased yields in *Triticum aestivum* (Qiang et al. 2019a, b) and *Hordeum vulgare* (Murphy et al. 2018) after treatment with endophytic fungus *Alternaria alternata* under drought conditions.

Despite the positive outcomes in controlled conditions, the acclimatization of EF into mainstream agricultural habitats faces significant challenges. A thorough study of EF has been carried out over the past two decades, highlighting its potential impact on agriculture (Qiang et al. 2019a, b; Paradza et al. 2021; Asim et al. 2022). However, the potential effect of EF on agriculture is affected by various factors, such as strict rules for the commercialization of agronomic bioinoculants, inadequate validation of EF efficiency in the natural environment, and complexities associated with inoculum production on a large scale (Al-Ani 2019). There are research gaps between the reported benefits of EF and their application in agriculture under natural conditions.

A recent surge in patent registrations associated with EF suggests increasing interest across different domains, such as the environment, agriculture, industry, and medicine (Al-Ani 2019). More than 245 patents have been registered globally for utilizing EF for phytohormone production, secondary metabolites, and biotransforming agents between January 2001 and December 2019 (Torres-Mendoza et al. 2020). Additionally, over 185 patents were listed within the agriculture sector worldwide from January 1988 to December 2019, indicating the encouraging potential of EF in agriculture (Ortega et al. 2020).

One of the primary advantages conferred upon crops through the utilization of EF is the facilitation of plant advance, leading to increase agricultural yield. EF promotes plant performance via various mechanisms, such as enhanced synthesis of phytohormones, modulation of hormone levels, improved nutrient accessibility, and increased water uptake rates (Baron and Rigobelo 2022).

To study the growth-promoting effect of EF, it is important to determine the potential of the selected isolates to inhabit new plants upon reinoculation. A study by Paradza et al. (2021) studied the outcome of 16 isolated EF on pea (*Pisum sativum*) and tomato (*Solanum lycopersicum*) seedlings. Of these, seven isolates exhibited successful colonization of pea and nine colonized tomatoes. Furthermore, only four isolates, belonging to *Trichoderma* sp. and *Hypocrea* sp., could colonize both experimental plants. Out of four fungal endophytes, only *T. atroviride* showed a significant increase in tomato growth compared to pea. Research showed that the successful colonization of cotton (*Gossypium hirsutum*) and increased growth were achieved by inoculating with *T. lentiforme* and *T. asperelloides* under greenhouse conditions. Asim et al. (2022) demonstrated the improvement in seed germination, root and aerial biomass, and increased length of root and stem of *Triticum aestivum* upon bio-inoculation with *Fusarium oxysporum*. This beneficial effect is associated with increased levels of IAA and GA, along with a simultaneous decrease in jasmonic acid and ABA. In conclusion, research gaps remain in understanding the complex association between EF and host plant ranges and their potential role in agriculture. Further investigation is required to investigate the mechanisms governing these interactions and to optimize the cost-benefit ratio for plants hosting endophytes, facilitating their application in the agricultural sector.

14.8 Conclusion

The exploration of how EF produces phytohormones, such as IAA, GA, and ABA, opens various possibilities for improving agriculture. EF and host plant interaction has the potential to increase yield and alleviate stress conditions, making food production more sustainable.

Literature reports that EF, by producing two important phytohormones such as IAA and GA, can optimize nutrient availability and improve plant growth and water uptake. Additionally, ABA production helps host plants and EF respond better to stressful situations. This contributes to making crops more adaptable and resilient to various environmental concerns.

Despite their positive impacts in lab conditions, inoculating EF on a large scale in mainstream agriculture may face challenges such as regulatory constraints, logistical issues, and validation requirements. However, the rising number of patents in recent years indicates growing attention to capitalizing on the agricultural profits offered by EF.

Exploring the biosynthetic pathways for phytohormone production by EF is important for gaining insights. Further research is required to unveil the details of these pathways. Moreover, there is a necessity to explore other important phytohormones common in both plants and EF, demanding detailed studies in this domain. Moving ahead, further investigation is significant to understand the processes behind the interactions between EF and plants. This knowledge will be helpful in efficiently incorporating these fungi into agricultural practices, thus ensuring a sustainable future for agriculture.

References

Aamir M, Rai KK, Zehra A, Kumar S, Yadav M, Shukla V et al (2020) Fungal endophytes: classification, diversity, ecological role, and their relevance in sustainable agriculture. In: Kumar A, Singh VK (eds) Microbial endophytes. Woodhead Publishing, Sawston, pp 291–323

Afzal Khan S, Hamayun M, Kim HY, Yoon HJ, Lee IJ, Kim JG (2009) Gibberellin production and plant growth promotion by a newly isolated strain of *Gliomastix murorum*. World J Microbiol Biotechnol 25:829–833

Ahemad M, Kibret M (2014) Mechanisms and applications of plant growth promoting rhizobacteria: current perspective. J King Saud Univ Sci 26:1–20

Ahmad N, Hamayun M, Khan SA, Khan AL, Lee IJ, Shin DH (2010) Gibberellin-producing endophytic fungi isolated from *Monochoria vaginalis*. J Microbiol Biotechnol 20(12):1744–1749

Airin AA, Arafat MI, Begum RA, Islam MR, Seraj ZI (2023) Plant growth-promoting endophytic fungi of the wild halophytic rice *Oryza coarctata*. Ann Microbiol 73(1):36

Al-Ani LKT (2019) Recent patents on endophytic fungi and their international market. In: Singh HB, Keswani C, Singh SP (eds) Intellectual property issues in microbiology. Springer, Singapore, pp 271–284

Ali S, Baloch AM (2020) Overview of sustainable plant growth and differentiation and the role of hormones in controlling growth and development of plants under various stresses. Recent Pat Food Nutr Agric 11(2):105–114

Ali AH, Radwan U, El-Zayat S, El-Sayed MA (2018) Desert plant-fungal endophytic association: the beneficial aspects to their hosts. Biol Forum Int J 10(1):138–145

Ali A, Bilal S, Khan AL, Mabood F, Al-Harrasi A, Lee IJ (2019) Endophytic *Aureobasidium pullulans* BSS6 assisted developments in phytoremediation potentials of *Cucumis sativus* under Cd and Pb stress. J Plant Interact 14(1):303–313

Asim S, Hussain A, Murad W, Hamayun M, Iqbal A, Rehman H, Tawab A, Irshad M, Alataway A, Dewidar AZ, Elansary HO (2022) Endophytic *Fusarium oxysporum* GW controlling weed and an effective biostimulant for wheat growth. Front Plant Sci 13:922343

Badawy AA, Alotaibi MO, Abdelaziz AM, Osman MS, Khalil AM, Saleh AM, Mohammed AE, Hashem AH (2021) Enhancement of seawater stress tolerance in barley by the endophytic fungus *Aspergillus ochraceus*. Metabolites 11(7):428

Bajaj R, Huang Y, Gebrechristos S, Mikolajczyk B, Brown H, Prasad R, Varma A, Bushley KE (2018) Transcriptional responses of soybean roots to colonization with the root endophytic fungus *Piriformospora indica* reveals altered phenylpropanoid and secondary metabolism. Sci Rep 8(1):10227

Baron NC, Rigobelo EC (2022) Endophytic fungi: a tool for plant growth promotion and sustainable agriculture. Mycology 13(1):39–55

Bilal L, Asaf S, Hamayun M, Gul H, Iqbal A, Ullah I, Lee IJ, Hussain A (2018) Plant growth promoting endophytic fungi *Aspergillus fumigatus* TS1 and *Fusarium proliferatum* BRL1 produce gibberellins and regulate plant endogenous hormones. Symbiosis 76:117–127

Bilal S, Shahzad R, Khan AL, Al-Harrasi A, Kim CK, Lee IJ (2019) Phytohormones enabled endophytic *Penicillium funiculosum* LHL06 *protects Glycine max* L. from synergistic toxicity of heavy metals by hormonal and stress-responsive proteins modulation. J Hazard Mater 379:120824

Bilal S, Shahzad R, Imran M, Jan R, Kim KM, Lee IJ (2020) Synergistic association of endophytic fungi enhances *Glycine max* L. resilience to combined abiotic stresses: heavy metals, high temperature and drought stress. Ind Crops Prod 143:111931

Bömke C, Tudzynski B (2009) Diversity, regulation, and evolution of the gibberellin biosynthetic pathway in fungi compared to plants and bacteria. Phytochemistry 70(15–16):1876–1893

Burragoni SG, Jeon J (2021) Applications of endophytic microbes in agriculture, biotechnology, medicine, and beyond. Microbiol Res 245:126691

Card S, Johnson L, Teasdale S, Caradus J (2016) Deciphering endophyte behavior: the link between endophyte biology and efficacious biological control agents. FEMS Microbiol Ecol 92(8):fiw114

Chadha N, Mishra M, Rajpal K, Bajaj R, Choudhary DK, Varma A (2015) An ecological role of fungal endophytes to ameliorate plants under biotic stress. Arch Microbiol 197(7):869–881

Chen K, Wang Z, Ding K, Chen Y, Asano Y (2021) Recent progress on discovery and research of aldoxime dehydratases. Green Synth Catal 2(2):179–186

Chhipa H, Deshmukh SK (2019) Fungal endophytes: rising tools in sustainable agriculture production. In: Endophytes secondary metabolites. Springer, Cham, pp 1–24

Cook SD, Nichols DS, Smith J, Chourey PS, McAdam EL, Quittenden L, Ross JJ (2016) Auxin biosynthesis: are the indole-3-acetic acid and phenylacetic acid biosynthesis pathways mirror images?. Plant Physiol 171(2):1230-1241.

Cosoveanu A, Chowdhary K, Cabrera R, Sharma S (2021) Role of phytohormones-producing fungal endophytes in plant–microbial interactions under stress. In: Endophytes: potential source of compounds of commercial and therapeutic applications, pp 195–223

Cui W, Song Q, Zuo B, Han Q, Jia Z (2020) Effects of gibberellin (GA4+ 7) in grain filling, hormonal behavior, and antioxidants in high-density maize (*Zea mays* L.). Plants 9(8):978

Dasila K, Pandey A, Sharma A, Samant SS, Singh M (2024) Endophytic fungi from Himalayan silver birch as potential source of plant growth enhancement and secondary metabolite production. Braz J Microbiol 55(1):557-570.

de Melo Pereira DÍ, Gurgel RS, de Souza ATF, Matias RR, de Souza Falcão L, Chaves FCM, da Silva GF, Martínez JG, de Lima Procópio RE, Fantin C, Albuquerque PM (2024) Isolation and identification of pigment-producing endophytic fungi from the Amazonian species *Fridericia chica*. J Fungi 10(1):77

Derviş S, Özer G (2023) Plant-associated *Neoscytalidium dimidiatum*—taxonomy, host range, epidemiology, virulence, and management strategies: a comprehensive review. J Fungi 9(11):1048

Dong L, Ma Y, Chen CY, Shen L, Sun W, Cui G et al (2022) Identification and characterization of auxin/IAA biosynthesis pathway in the Rice blast fungus *Magnaporthe oryzae*. J Fungi 8(2):208

El Mansy SM, Nouh FAA, Mousa MK, Abdel-Azeem AM (2020) Endophytic fungi: diversity, abundance, and plant growth-promoting attributes. In: Agriculturally important fungi for sustainable agriculture: volume 1: perspective for diversity and crop productivity. Springer, Cham, pp 21–59

Fan J, Hill L, Crooks C, Doerner P, Lamb C (2009) Abscisic acid has a key role in modulating diverse plant-pathogen interactions. Plant Physiol 150(4):1750–1761

Feng L, Zhang AX, Li L, Zhang XJ, Wang Z, Tan NH (2021) Diversity of cultivable endophytic fungi in two Rubia plants and their potential for production of anti-tumour Rubiaceae-type cyclopeptides. Lett Appl Microbiol 73(6):759–769

Fu SF, Wei JY, Chen HW, Liu YY, Lu HY, Chou JY (2015) Indole-3-acetic acid: a widespread physiological code in interactions of fungi with other organisms. Plant Sig Behav 10(8):e1048052

Galeano RMS, Franco DG, Chaves PO, Giannesi GC, Masui DC, Ruller R, Corrêa BO, da Silva Brasil M, Zanoelo FF (2021) Plant growth promoting potential of endophytic Aspergillus niger 9-p isolated from native forage grass in Pantanal of Nhecolândia region, Brazil. Rhizosphere 18:100332

Galeano RMS, Silva SM, Yonekawa MKA, de Alencar Guimarães NC, Giannesi GC, Masui DC, Corrêa BO, da Silva Brasil M, Zanoelo FF (2023) *Penicillium chrysogenum* strain 34-P promotes plant growth and improves initial development of maize under saline conditions. Rhizosphere 26:100710

García-Latorre C, Rodrigo S, Marin-Felix Y, Stadler M, Santamaria O (2023) Plant-growth promoting activity of three fungal endophytes isolated from plants living in dehesas and their effect on *Lolium multiflorum*. Sci Rep 13(1):7354

Hamayun M, Khan SA, Khan AL, Rehman G, Sohn EY, Shah AA, Kim SK, Joo GJ, Lee IJ (2009) *Phoma herbarum* as a new gibberellin-producing and plant growth-promoting fungus. J Microbiol Biotechnol 19(10):1244–1249

Hamayun M, Khan SA, Iqbal I, Ahmad B, Lee IJ (2010a) Isolation of a gibberellin-producing fungus (*Penicillium* sp. MH7) and growth promotion of crown daisy (*Chrysanthemum coronarium*). J Microbiol Biotechnol 20(1):202–207

Hamayun M, Khan SA, Khan AL, Rehman G, Kim YH, Iqbal I, Hussain J, Sohn EY, Lee IJ (2010b) Gibberellin production and plant growth promotion from pure cultures of *Cladosporium* sp. MH-6 isolated from cucumber (*Cucumis sativus* L.). Mycologia 102(5):989–995

Hamayun M, Hussain A, Khan SA, Kim HY, Khan AL, Waqas M, Irshad M, Iqbal A, Rehman G, Jan S, Lee IJ (2017) Gibberellins producing endophytic fungus *Porostereum spadiceum* AGH786 rescues growth of salt affected soybean. Front Microbiol 8:686

Hamayun M, Hussain A, Afzal Khan S, Iqbal A, Lee IJ (2019) *Aspergillus flavus* promoted the growth of soybean and sunflower seedlings at elevated temperature. BioMed Res Int 2019:1295457

Hao G, Du X, Zhao F, Ji H (2010) Fungal endophytes-induced abscisic acid is required for flavonoid accumulation in suspension cells of *Ginkgo biloba*. Biotechnol Lett 32:305–314

Hardoim PR, Van Overbeek LS, Berg G, Pirttilä AM, Compant S, Campisano A, Döring M, Sessitsch A (2015) The hidden world within plants: ecological and evolutionary considerations for defining functioning of microbial endophytes. Microbiol Mol Biol Rev 79(3):293–320

Hartanto A, Lutfia A, Munir E (2019) Identification of a potential IAA-producing fungus isolated from *Alpinia* sp Rhizome in Hutan Sibayak, North Sumatera. J Phys Conf Ser 1351(1):012024. IOP Publishing

Hedden P, Thomas SG (2012) Gibberellin biosynthesis and its regulation. Biochem J 444(1):11–25

Hewage KAH, Yang JF, Wang D, Hao GF, Yang GF, Zhu JK (2020) Chemical manipulation of abscisic acid signaling: a new approach to abiotic and biotic stress management in agriculture. Adv Sci 7(18):2001265

Hilbert M (2013) Biochemical and molecular analyses of the biosynthesis pathway of the indole derivatives in Piriformospora indica

Hossain A, Pamanick B, Venugopalan VK, Ibrahimova U, Rahman MA, Siyal AL, Maitra S, Chatterjee S, Aftab T (2022) Emerging roles of plant growth regulators for plants adaptation to abiotic stress–induced oxidative stress. In: Emerging plant growth regulators in agriculture. Academic Press, Cambridge, MA, pp 1–72

Husna, Hussain A, Shah M, Hamayun M, Qadir M, Iqbal A (2021) Heavy metal tolerant endophytic fungi *Aspergillus welwitschiae* improves growth, ceasing metal uptake and strengthening antioxidant system in *Glycine max* L. Environ Sci Pollut Res:1–15. https://doi.org/10.1007/s11356-021-16640-1

Hyde KD, Soytong K (2008) The fungal endophyte dilemma. Fungal Divers 33(163):e173

Ikram M, Ali N, Jan G, Jan FG, Rahman IU, Iqbal A, Hamayun M (2018) IAA producing fungal endophyte *Penicillium roqueforti* Thom., enhances stress tolerance and nutrients uptake in wheat plants grown on heavy metal contaminated soils. PLoS One 13(11):e0208150

Ikram M, Ali N, Jan G, Jan FG, Pervez R, Romman M, Zainab R, Yasmin H, Khan N (2023a) Isolation of endophytic fungi from halophytic plants and their identification and screening for auxin production and other plant growth promoting traits. J Plant Growth Reg 42(8):4707–4723

Ikram M, Ali N, Jan G, Jan FG, Pervez R, Romman M, Zainab R, Yasmin H, Khan N (2023b) Isolation of endophytic fungi from halophytic plants and their identification and screening for auxin production and other plant growth promoting traits. J Plant Growth Regul 42(8):4707–4723

Ismail AH, Mehmood ASIF, Qadir MUHAMMAD, Husna AI, Hamayun MUHAMMAD, Khan NAEEM (2020) Thermal stress alleviating potential of endophytic fungus rhizopus oryzae inoculated to sunflower (Helianthus annuus L.) and soybean (*Glycine max* L.). Pak J Bot 52(5):1857–1865

Ismaila AH, Qadira M, Husnaa MI, Ahmadb A, Hamayuna M (2018) Endophytic fungi isolated from Citrullus colocynthis. Leaves and Their potential for secretion of indole acetic acid and gibberellin. J Appl Environ Biol Sci 8(6):80–84

Jahn L, Hofmann U, Ludwig-Müller J (2021) Indole-3-acetic acid is synthesized by the endophyte Cyanodermella asteris via a tryptophan-dependent and-independent way and mediates the interaction with a non-host plant. Int J Mol Sci 22(5):2651

Jaiswal S, Ojha A, Mishra SK (2024) Assessment of plant growth-promoting parameters of endophytes isolated from *Calotropis procera* and their performance under irrigated and non-irrigated conditions. Curr Microbiol 81(1):1–13

Jaroszuk-Ściseł J, Kurek E, Trytek M (2014) Efficiency of indoleacetic acid, gibberellic acid and ethylene synthesized in vitro by *Fusarium culmorum* strains with different effects on cereal growth. Biologia 69:281–292

Jiang HJ, Zhao YY, Pan YT, Sun K, Xie XG, Dai CC (2022) The endophytic fungus *Phomopsis liquidambaris* promotes phosphorus uptake by *Arachis hypogaea* L. by regulating host auxin, gibberellins, and cytokinins signaling pathways. J Soil Sci Plant Nutr 22(4):4913–4927

Jin W, Peng L, Zhang X, Sun H, Yuan Z (2019) Effects of endophytic and ectomycorrhizal basidiomycetes on *Quercus virginiana* seedling growth and nutrient absorption. J Sustain For 38(5):457–470

Ju M, Zhang Q, Wang R, Yan S, Zhang Q, Li P, Hao F, Gu P (2024) Community ecological succession of endophytic fungi associates with medicinal compound accumulation in Sophora alopecuroides. Microbiol Spectr 12:e0307623

Kaur R, Saxena S (2023) Evaluation of drought-tolerant endophytic fungus *Talaromyces purpureogenus* as a bioinoculant for wheat seedlings under normal and drought-stressed circumstances. Folia Microbiol:1–19. https://doi.org/10.1007/s12223-023-01051-1

Khaekhum S, Ekprasert J, Suebrasri T, Mongkolthanaruk W, Riddech N, Jogloy S, Boonlue S (2021) The first member of *Exserohilum rostratum* beneficial for promoting growth and yield of sunchoke (*Helianthus tuberosus* L.). Rhizosphere 19:100379

Khan AL, Hamayun M, Kim YH, Kang SM, Lee JH, Lee IJ (2011) Gibberellins producing endophytic *Aspergillus fumigatus* sp. LH02 influenced endogenous phytohormonal levels, isoflavonoids production and plant growth in salinity stress. Process Biochem 46(2):440–447

Khan AL, Shinwari ZK, Kim YH, Waqas M, Hamayun M, Kamran M, Lee IJ (2012) Role of endophyte *Chaetomium globosum* LK4 in growth of *Capsicum annuum* by production of gibberellins and indole acetic acid. Pak J Bot 44(5):1601–1607

Khan AL, Waqas M, Khan AR, Hussain J, Kang SM, Gilani SA, Hamayun M, Shin JH, Kamran M, Al-Harrasi A, Yun BW (2013) Fungal endophyte *Penicillium janthinellum* LK5 improves growth of ABA-deficient tomato under salinity. World J Microbiol Biotechnol 29:2133–2144

Khan AL, Waqas M, Hussain J, Al-Harrasi A, Al-Rawahi A, Al-Hosni K, Kim MJ, Adnan M, Lee IJ (2014) Endophytes *Aspergillus caespitosus* LK12 and *Phoma* sp. LK13 of *Moringa peregrina* produce gibberellins and improve rice plant growth. J Plant Interact 9(1):731–737

Khan AR, Ullah I, Waqas M, Shahzad R, Hong SJ, Park GS, Jung BK, Lee IJ, Shin JH (2015a) Plant growth-promoting potential of endophytic fungi isolated from Solanum nigrum leaves. World J of Microbiol Biotechnol 31:1461–1466

Khan AL, Waqas M, Hussain J, Al-Harrasi A, Hamayun M, Lee IJ (2015b) Phytohormones enabled endophytic fungal symbiosis improve aluminum phytoextraction in tolerant *Solanum lycopersicum*: an example of *Penicillium janthinellum* LK5 and comparison with exogenous GA3. J Hazard Mater 295:70–78

Khan MI, Ali N, Jan G, Hamayun M, Jan FG, Iqbal A, Hussain A, Lee IJ (2022) Salt stress alleviation in *Triticum aestivum* through primary and secondary metabolites modulation by *Aspergillus terreus* BTK-1. Front Plant Sci 13:779623

Kowalska J, Krzymińska J, Tyburski J (2022) Yeasts as a potential biological agent in plant disease protection and yield improvement—A short review. Agric 12(9):1404.

Khare E, Mishra J, Arora NK (2018) Multifaceted interactions between endophytes and plant: developments and prospects. Front Microbiol 9:2732

Krause K, Henke C, Asiimwe T, Ulbricht A, Klemmer S, Schachtschabel D, Boland W, Kothe E (2015) Biosynthesis and secretion of indole-3-acetic acid and its morphological effects on *Tricholoma vaccinum*-spruce ectomycorrhiza. Appl Environ Microbiol 81(20):7003–7011

Kumar V, Soni R, Jain L, Dash B, Goel R (2019) Endophytic fungi: recent advances in identification and exploration. In: Advances in endophytic fungal research: present status and future challenges. Springer, Cham, pp 267–281

Kumla J, Suwannarach N, Bussaban B, Matsui K, Lumyong S (2014) Indole-3-acetic acid production, solubilization of insoluble metal minerals and metal tolerance of some sclerodermatoid fungi collected from northern Thailand. Ann Microbiol 64:707–720

Lata R, Chowdhury S, Gond SK, White JF Jr (2018) Induction of abiotic stress tolerance in plants by endophytic microbes. Lett Appl Microbiol 66(4):268–276

Le Cocq K, Gurr SJ, Hirsch PR, Mauchline TH (2017) Exploitation of endophytes for sustainable agricultural intensification. Mol Plant Pathol 18(3):469–473

Li Z, Xiong K, Wen W, Li L, Xu D (2023) Functional endophytes regulating plant secondary metabolism: current status, prospects and applications. Int J Mol Sci 24(2):1153

Lim CW, Baek W, Jung J, Kim JH, Lee SC (2015) Function of ABA in stomatal defense against biotic and drought stresses. Int J Mol Sci 16(7):15251–15270

Lu H, Wei T, Lou H, Shu X, Chen Q (2021) A critical review on communication mechanism within plant-endophytic fungi interactions to cope with biotic and abiotic stresses. J Fungi 7(9):719

Lu L, Karunarathna SC, Hyde KD, Suwannarach N, Elgorban AM, Stephenson SL, Al-Rejaie S, Jayawardena RS, Tibpromma S (2022) Endophytic fungi associated with coffee leaves in China exhibited in vitro antagonism against fungal and bacterial pathogens. J Fungi 8(7):698

Lubna, Khan MA, Asaf S, Jan R, Waqas M, Kim KM, Lee IJ (2022) Endophytic fungus *Bipolaris* sp. CSL-1 induces salt tolerance in *Glycine max* L. via modulating its endogenous hormones, antioxidative system and gene expression. J Plant Int 17(1):319–332

Lugtenberg BJ, Caradus JR, Johnson LJ (2016) Fungal endophytes for sustainable crop production. FEMS Microbiol Ecol 92(12):fiw194

Martos F, Hobbhahn N, de Vega C, Johnson SD (2023) Phylogeography and host-shift speciation in the vampirecups, an enigmatic clade of endophytic holoparasitic plants. J Biogeogr 50(11):1852–1865

Mathew SA, Helander M, Saikkonen K, Vankova R, Dobrev PI, Dirihan S, Fuchs B (2023) Epichloë endophytes shape the foliar endophytic fungal microbiome and alter the auxin and salicylic acid phytohormone levels in two meadow fescue cultivars. J Fungi 9(1):90

Mehmood A, Hussain A, Irshad M, Hamayun M, Iqbal A, Khan N (2019) In vitro production of IAA by endophytic fungus *Aspergillus awamori* and its growth promoting activities in *Zea mays*. Symbiosis 77:225–235

Meleigy SA, Khalaf MA (2009) Biosynthesis of gibberellic acid from milk permeate in repeated batch operation by a mutant *Fusarium moniliforme* cells immobilized on loofa sponge. Bioresour Technol 100(1):374–379

Munir E, Yurnaliza Y, Lutfia A, Hartanto A (2021) Antifungal activity and IAA production by endophytic fungi isolated from *Elettaria* sp. IOP Conf Ser Earth Environ Sci 782(4):042037. IOP Publishing

Murphy BR, Doohan FM, Hodkinson TR (2018) From concept to commerce: developing a successful fungal endophyte inoculant for agricultural crops. J Fungi 4:24

Naureen A, Nasim FUH, Choudhary MS, Ashraf M, Grundler FM, Schleker ASS (2022) A new endophytic fungus CJAN1179 isolated from the Cholistan desert promotes lateral root growth in *Arabidopsis* and produces IAA through tryptophan-dependent pathway. Arch Microbiol 204(3):181

Nguyen TAN, Higa T, Shiina A, Utami YD, Hiruma K (2023) Exploring the roles of fungal-derived secondary metabolites in plant-fungal interactions. Physiol Mol Plant Pathol 125:102021

Nongthombam KS, Mutum SS, Pandey RR (2024) Distribution, diversity and biochemical analysis of endophytic fungi associated with chromolaena odorata. J Appl Life Sci Int 27(1):17–29

Numponsak T, Kumla J, Suwannarach N, Matsui K, Lumyong S (2018) Biosynthetic pathway and optimal conditions for the production of indole-3-acetic acid by an endophytic fungus, *Colletotrichum fructicola* CMU-A109. PLoS One 13(10):e0205070

Nutaratat P, Srisuk N, Arunrattiyakorn P, Limtong S (2016) Indole-3-acetic acid biosynthetic pathways in the basidiomycetous yeast *Rhodosporidium paludigenum*. Arch Microb 198:429–437

Ortega HE, Torres-Mendoza D, Cubilla-Rios L (2020) Patents on endophytic fungi for agriculture and bio-and phytoremediation applications. Microorganisms 8:1237

Ozimek E, Jaroszuk-Ściseł J, Bohacz J, Korniłłowicz-Kowalska T, Tyśkiewicz R, Słomka A, Nowak A, Hanaka A (2018) Synthesis of indoleacetic acid, gibberellic acid and ACC-deaminase by *Mortierella strains* promotes winter wheat seedlings growth under different conditions. Int J Mol Sci 19(10):3218

Pablo CHD, Pagaduan JRR, Langres HKC, Hipol RM (2020) Plant growth-promoting characteristics of root fungal endophytes isolated from a traditional Cordillera rice landrace. Stud Fungi 5(1):536–549

Pappas ML, Liapoura M, Papantoniou D, Avramidou M, Kavroulakis N, Weinhold A, Broufas GD, Papadopoulou KK (2018) The beneficial endophytic fungus Fusarium solani strain K alters tomato responses against spider mites to the benefit of the plant. Front Plant Sci 9:1603

Paradza VM, Khamis FM, Yusuf AA, Subramanian S, Ekesi S, Akutse KS (2021) Endophytic colonisation of *Solanum lycopersicum* and *Phaseolus vulgaris* by fungal endophytes promotes seedlings growth and hampers the reproductive traits, development, and survival of the greenhouse whitefly, *Trialeurodes vaporariorum*. Front Plant Sci 12:771534

Potshangbam M, Devi SI, Sahoo D, Strobel GA (2017) Functional characterization of endophytic fungal community associated with *Oryza sativa* L. and *Zea mays* L. Front Microbiol 8:325

Poveda J, Rodríguez VM, Díaz-Urbano M, Sklenář F, Saati-Santamaría Z, Menéndez E, Velasco P (2022) Endophytic fungi from kale (*Brassica oleracea* var. acephala) modify roots-glucosinolate profile and promote plant growth in cultivated *Brassica* species. First description of *Pyrenophora gallaeciana*. Front Microbiol 13:981507

Prajapati P, Yadav M, Nishad JH, Gautam VS, Kharwar RN (2024) Salt tolerant fungal endophytes alleviate the growth and yield of saline-affected wheat genotype PBW-343. Microbiol Res 278:127514

Priyadharsini P, Muthukumar T (2017) The root endophytic fungus *Curvularia geniculata* from *Parthenium hysterophorus* roots improves plant growth through phosphate solubilization and phytohormone production. Fungal Ecol 27:69–77

Priyashantha AKH, Karunarathna SC, Lu L, Tibpromma S (2023) Fungal endophytes: an alternative biocontrol agent against phytopathogenic fungi. Encyclopedia 3(2):759–780

Qiang X, Ding J, Lin W, Li Q, Xu C, Zheng Q, Li Y (2019a) Alleviation of the detrimental effect of water deficit on wheat (*Triticum aestivum* L.) growth by an indole acetic acid-producing endophytic fungus. Plant Soil 439:73–391

Qiang X, Ding J, Lin W, Li Q, Xu C, Zheng Q, Li Y (2019b) Alleviation of the detrimental effect of water deficit on wheat (*Triticum aestivum* L.) growth by an indole acetic acid-producing endophytic fungus. Plant Soil 439:373–391

Rädisch R, Chmátal M, Rucká L, Novotný P, Petrásková L, Halada P, Kotik M, Pátek M, Martínková L (2018) Overproduction and characterization of the first enzyme of a new aldoxime dehydratase family in Bradyrhizobium sp. Int J Biol Macromol 115:746–753

Rai GK, Mishra S, Chouhan R, Mushtaq M, Chowdhary AA, Rai PK, Kumar RR, Kumar P, Perez-Alfocea F, Colla G, Cardarelli M (2023) Plant salinity stress, sensing, and its mitigation through WRKY. Front Plant Sci 14:1238507

Rho H, Hsieh M, Kandel SL, Cantillo J, Doty SL, Kim SH (2018) Do endophytes promote growth of host plants under stress? A meta-analysis on plant stress mitigation by endophytes. Microb Ecol 75:407–418

Rodriguez RJ, White JF, Arnold AE, Redman RS (2009) Fungal endophytes: diversity and functional roles. New Phytol 182(2):314–330

Saad MMG, Badry HH (2020) Phytohormones producing fungal endophytes enhance nutritional status and suppress pathogenic fungal infection in tomato. J Agric Sci Technol 22(5):1383–1395

Saber WI, Al-Askar AA, Ghoneem KM (2023) Exclusive biosynthesis of pullulan using taguchi's approach and decision tree learning algorithm by a novel endophytic *Aureobasidium pullulans* strain. Polymers 15(6):1419

Salazar-Cerezo S, Martínez-Montiel N, García-Sánchez J, Pérez-y-Terrón R, Martínez-Contreras RD (2018) Gibberellin biosynthesis and metabolism: a convergent route for plants, fungi and bacteria. Microbiol Res 208:85–98

Samreen T, Naveed M, Nazir MZ, Asghar HN, Khan MI, Zahir ZA, Kanwal S, Jeevan B, Sharma D, Meena VS, Meena SK (2021) Seed associated bacterial and fungal endophytes: diversity, life cycle, transmission, and application potential. Appl Soil Ecol 168:104191

Sardar P, Kempken F (2018) Characterization of indole-3-pyruvic acid pathway-mediated biosynthesis of auxin in *Neurospora crassa*. PLoS One 13(2):e0192293

Sarkar S, Dey A, Kumar V, Batiha GES, El-Esawi MA, Tomczyk M, Ray P (2021) Fungal endophyte: an interactive endosymbiont with the capability of modulating host physiology in myriad ways. Front Plant Sci 12:701800

Seema N, Hamayun M, Hussain A, Shah M, Irshad M, Qadir M, Iqbal A, Alrefaei AF, Ali S (2023) Endophytic *Fusarium proliferatum* reprogrammed phytohormone production and antioxidant system of *Oryza sativa* under drought stress. Agronomy 13(3):873

Sivakumaran A, Akinyemi A, Mandon J, Cristescu SM, Hall MA, Harren FJ, Mur LA (2016) ABA suppresses Botrytis cinerea elicited NO production in tomato to influence H2O2 generation and increase host susceptibility. Front Plant Sci 7:709

Sodhi GK, Saxena S (2023) Plant growth-promoting endophyte Nigrospora oryzae mitigates abiotic stress in rice (*Oryza sativa* L.). FEMS Microbiol Ecol 99(9):fiad094

Srimuang EHM, Lumyong S (2018) A new endophytic fungus, *Tulasnella phuhinrongklaensis* (Cantharellales, Basidiomycota) isolated from roots of the terrestrial orchid, *Phalaenopsis pulcherrima*. Phytotaxa 374(2):099–109

Sun PF, Fang WT, Shin LY, Wei JY, Fu SF, Chou JY (2014) Indole-3-acetic acid-producing yeasts in the phyllosphere of the carnivorous plant *Drosera indica* L. PLoS One 9(12):e114196

Suryanarayanan TS, Devarajan PT, Girivasan KP, Govindarajulu MB, Kumaresan V, Murali TS, Rajamani T, Thirunavukkarasn V, Venkatesan G (2018) The host range of multi-host endophytic fungi. Curr Sci 115:1963–1969

Tang J, Li Y, Zhang L, Mu J, Jiang Y, Fu H, Zhang Y, Cui H, Yu X, Ye Z (2023) Biosynthetic pathways and functions of Indole-3-Acetic acid in microorganisms. Microorganisms 11(8):2077

Tian Y, Fu X, Zhang G, Zhang R, Kang Z, Gao K, Mendgen K (2022) Mechanisms in growth-promoting of cucumber by the endophytic fungus *Chaetomium globosum* strain ND35. J Fungi 8(2):180

Tivendale ND, Ross JJ, Cohen JD (2014) The shifting paradigms of auxin biosynthesis. Trends Plant Sci 19(1):44–51

Tiwari P, Bae H (2022) Endophytic fungi: key insights, emerging prospects, and challenges in natural product drug discovery. Microorganisms 10(2):360

Toppo P, Jangir P, Mehra N, Kapoor R, Mathur P (2024) Bioprospecting of endophytic fungi from medicinal plant *Anisomeles indica* L. for their diverse role in agricultural and industrial sectors. Sci Rep 14(1):588

Torres M, Tadych M, White JF, Bills GF, Pirttila A, Sorvari S (2011) Isolation and identification of fungal endophytes. In: Prospects and applications for plant-associated microbes: a laboratory manual Part B: fungi. BBi, Paimio, pp 153–164

Torres-Mendoza D, Ortega HE, Cubilla-Rios L (2020) Patents on endophytic fungi related to secondary metabolites and biotransformation applications. J Fungi 6:58

Tripathy P, Sethi S, Panchal D, Prakash O, Sharma A, Mondal RB, Pal S (2023) Biogenic synthesis of nanoparticles by amalgamating microbial endophytes: potential environmental applications and future perspectives. In: Microbial endophytes and plant growth. Academic Press, Cambridge, MA, pp 215–231

Tsavkelova EA, Klimova SY, Cherdyntseva TA, Netrusov AI (2006) Microbial producers of plant growth stimulators and their practical use: a review. Appl Biochem Microbiol 42:117–126

Tsipinana S, Husseiny S, Alayande KA, Raslan M, Amoo S, Adeleke R (2023) Contribution of endophytes towards improving plant bioactive metabolites: a rescue option against red-taping of medicinal plants. Front Plant Sci 14:1248319

Turaeva B, Soliev A, Eshboev F, Kamolov L, Azimova N, Karimov H, Zukhritdinova N, Khamidova K (2020) The use of three fungal strains in producing of indole-3-acetic acid and gibberellic acid. Plant Cell Biotechol Mol Biol 21(35–36):32–43

Turbat A, Rakk D, Vigneshwari A, Kocsubé S, Thu H, Szepesi Á, Bakacsy LD, Škrbić B, Jigjiddorj EA, Vágvölgyi C, Szekeres A (2020) Characterization of the plant growth-promoting activities of endophytic fungi isolated from *Sophora flavescens*. Microorganisms 8(5):683

Ulferts S, Delventhal R, Splivallo R, Karlovsky P, Schaffrath U (2015) Abscisic acid negatively interferes with basal defence of barley against *Magnaporthe oryzae*. BMC Plant Biol 15:1–13

Vandenkoornhuyse P, Quaiser A, Duhamel M, Le Van A, Dufresne A (2015) The importance of the microbiome of the plant holobiont. New Phytol 206:1196–1206

Verma A, Shameem N, Jatav HS, Sathyanarayana E, Parray JA, Poczai P, Sayyed RZ (2022) Fungal endophytes combat biotic and abiotic stresses for climate-smart and sustainable agriculture. Front Plant Sci 13:953836

Vrabka J, Niehaus EM, Münsterkötter M, Proctor RH, Brown DW, Novák O, Pěnčik A, Tarkowská D, Hromadová K, Hradilová M, Oklešt'ková, J. (2019) Production and role of hormones during interaction of *Fusarium* species with maize (*Zea mays* L.) seedlings. Front Plant Sci 9:1936

Wang K, Peng XP, Feng D, Tian C, Wan WL, Lou HX, Li G (2022) Metabolic profiling for the discovery of structurally diverse gibberellins and their precursors from the endophytic fungus Fusarium sp. NJ-F5. ACS Omega 7(49):45624–45635

Wang T, Chi M, Chen J, Liang L, Wang Y, Chen Y (2024) The diversity and growth-promoting potential of the endophytic fungi of *Neuwiedia singapureana* (Orchidaceae) in China. Diversity 16(1):34

Waqas M, Khan AL, Kamran M, Hamayun M, Kang SM, Kim YH, Lee IJ (2012) Endophytic fungi produce gibberellins and indoleacetic acid and promote host-plant growth during stress. Molecules 17(9):10754–10773

Waqas M, Khan AL, Kang SM, Kim YH, Lee IJ (2014) Phytohormone-producing fungal endophytes and hardwood-derived biochar interact to ameliorate heavy metal stress in soybeans. Biol Fertil Soils 50:1155–1167

Xie XG, Sheng KX, Lu WL, Wang HR, Rahman K, Feng KM, Han T (2023) Endophytic fungus *Trichoderma atroviride* D16 increases tanshinone accumulation in *Salvia miltiorrhiza* plantlets by enhancing abscisic acid and salicylic acid signaling. Symbiosis 90(2):171–181

Xu L, Li T, Wu Z, Feng H, Yu M, Zhang X, Chen B (2018a) Arbuscular mycorrhiza enhances drought tolerance of tomato plants by regulating the 14-3-3 genes in the ABA signaling pathway. Appl Soil Ecol 125:213–221

Xu G, Yang S, Meng L, Wang BG (2018b) The plant hormone abscisic acid regulates the growth and metabolism of endophytic fungus *Aspergillus nidulans*. Sci Rep 8(1):6504

Yan L, Zhu J, Zhao X, Shi J, Jiang C, Shao D (2019) Beneficial effects of endophytic fungi colonization on plants. Appl Microbiol Biotechnol 103:3327–3340

Yang N, Zhang W, Wang D, Cao D, Cao Y, He W, Lin Z, Chen X, Ye G, Chen Z, Wei X (2023) A novel endophytic fungus strain of Cladosporium: its identification, genomic analysis, and effects on plant growth. Front Microbiol 14:1287582

You YH, Kang SM, Choi YM, Lee MC, Kim JG (2015) Gibberellins production and identification of endophytic fungi isolated from aquatic plants in fresh water. Korean J Mycol 43(1):71–76

Yue J, Yang F, Xiao Y, Lin S, He Z, Wang S, Zhao J, Yuan J, Li L, Liu L (2024) *Comamonas endophytica* sp. nov., a novel indole acetic acid producing endophyte isolated from bamboo in China. Int J Syst Evol Microbiol 74(1):006217

Zhang M, Gao C, Xu L, Niu H, Liu Q, Huang Y, Lv G, Yang H, Li M (2022) Melatonin and indole-3-acetic acid synergistically regulate plant growth and stress resistance. Cell 11(20):3250

Zhao Y, Christensen SK, Fankhauser C, Cashman JR, Cohen JD, Weigel D, Chory J (2001) A role for flavin monooxygenase-like enzymes in auxin biosynthesis. Science 291(5502):306–309

Zhou J, Zhang H, Yang Y, Zhang Z, Zhang H, Hu X, Chen J, Wang XC, Huang R (2008) Abscisic acid regulates TSRF1-mediated resistance to *Ralstonia solanacearum* by modifying the expression of GCC box-containing genes in tobacco. J Exp Bot 59(3):645–652

Recent Trends in Phytotoxic Compounds from Endophytic Fungi

15

Faijuddin Ahammad, Saidul Islam, Arghya Banerjee, Sourav Mondal, Bholanath Mondal, Sk Naim Aktar, Maksud Hasan Shah, and Mahatab Ali

Abstract

Plant endophytes are thought to represent new sources of naturally occurring substances that have a range of biological effects, such as antibacterial, anti-inflammatory, anticancer, herbicidal, antileishmanial, and antioxidant properties.

F. Ahammad
Department of Plant Pathology, Sam Higgingbottom University of Agriculture, Technology and Science, Prayagraj, Uttar Pradesh, India

S. Islam (✉)
Nadia Krishi Vigyan Kendra, Bidhan Chandra Krishi Viswavidyalaya (BCKV), Gayeshpur, Nadia, West Bengal, India

A. Banerjee
Department of Plant Pathology, School of Agriculture and Allied Science, The Neotia, University, Jhinga, South 24 Parganas, West Bengal, India

S. Mondal
Rathindra Krishi Vigyan Kendra, Palli Siksha Bhavana, Visva Bharati University, Sriniketan, Bolpur, West Bengal, India

B. Mondal
Department of Plant Pathology, Palli Siksha Bhavana, Visva Bharati University, Sriniketan, Bolpur, West Bengal, India

S. N. Aktar
Department of Agronomy, Palli Siksha Bhavana, Visva Bharati University, Sriniketan, Bolpur, West Bengal, India

M. H. Shah
Department of Agronomy, Bidhan Chandra Krishi Viswavidyalaya (BCKV), Mohanpur, Nadia, West Bengal, India

M. Ali
Department of Genetics and Plant Breeding, Bidhan Chandra Krishi Viswavidyalaya (BCKV), Mohanpur, Nadia, West Bengal, India

K. A. Abd-Elsalam, S. F. AbuQamar (eds.), *Fungal Endophytes Volume I*,
https://doi.org/10.1007/978-981-97-7312-1_15

Numerous strains have been reported to produce a wide range of specialized products, including terpenoids, alkaloids, polyketides, phenolic compounds, coumarins, and quinone derivatives. Various plant-microbe relationships have an impact on the environment and agriculture. The ability to create a wide range of proteins and metabolites has been developed by both plants and microorganisms to facilitate or regulate their interactions with other organisms and alter their environmental conditions. A number of these high-value metabolites that endophytes produce have found extensive usage in medicine and other socioeconomic fields. Quick progress in the field of genomics has made it possible to unravel the biology and ecology of endophytes, including their potential for metabolism, by analyzing their genetic blueprints and examining the ways in which they carry out particular tasks in diverse ecological and environmental settings. In this article, we examine current developments in our knowledge of the biology and ecology of endophytes and how these discoveries have been applied to promote environmentally friendly farming practices, create novel medicinal metabolites, and enhance environmental health. Advances in gene modification technologies and a deeper understanding of endophyte biology have led to the promise that endophytes can be genetically engineered to produce "value-added" compounds and be used to address environmental and agricultural issues. To reach their full potential, endophyte biology still has a few knowledge gaps that need to be filled.

Keywords

Endophyte fungi · Biomedical applications · Transmission modes · Potential applications in agriculture

15.1 Introduction

In 1866, Bary first used the term "endophyte" to describe any creature that lives inside the tissues of plants. Since then, endophytes have been classified as asymptomatic bacteria living inside plants (Carroll 1986) and microorganisms that spend a portion of their life cycle inside plant tissues and organs without obviously harming the host plant (Petrini 1991). Endophytic fungi have been found in a wide variety of plant species, according to many reports (Rajamanikyam et al. 2017; Souza and Santos 2017; Kouipou Toghueo and Boyom 2019). Over time, the idea of endophytes has changed (Hallmann et al. 1997; Hardoim et al. 2015). These microbes can be taken out of the inside of plants or separated from plant tissues that have been damaged on the surface (Hallmann et al. 1997).

Endophytic fungi are an abundant source of bioactive substances that are essential for the survival and upkeep of endophytes in plants as well as for improving host plant health and resilience to stressful conditions. These compounds include hormones (such as auxin, gibberellins), cellulases, proteases, and chitinases and antimicrobial agents (Eid et al. 2019). Numerous biotechnological applications could benefit greatly from these metabolites (Rana et al. 2019; Rustamova et al. 2020). The advantages of endophytic fungi for their host plants, their capacity to generate natural products with a wide range of biological activities, and the role of omics tools in expanding our knowledge of symbiotic relationships to promote plant health are all covered in this review.

The first endophytic strain was identified from ryegrass seeds (*Lolium temulentum* L.). Mycologists and pharmaceutical chemists have become interested in this field after discovering the paclitaxel-producing endophytic fungus (called *Taxomyces* and renamed) from the Pacific yew and other discoveries from other plant species. This finding sparked interest in endophytic fungus as a viable source of plant metabolites and sparked study into them as a novel source of bioactive compounds. A fascinating class of microorganisms essential to the symbiotic connections between plants and fungi is endophytic fungi. Endophytes live inside plant tissues and don't seem to harm them, in contrast to pathogens that cause illnesses in plants. Understanding the variety, abundance, and roles of endophytic fungi, therefore, is a dynamic area of multidisciplinary study that crosses the fields of microbiology, ecology, plant biology, and natural product chemistry. Endophytic fungi play a crucial role in determining the sustainability and well-being of plant ecosystems, as scientists continue to elucidate the complex mechanisms that control their interactions with their host plants.

To address the shortcomings of current fungal bio-prospecting techniques, new screening approaches that are capable of detecting fungi in different ecological settings must be used. One such approach is to focus on fungi that form mutualistic relationships with plants and live inside their tissues without causing any visible symptoms; these fungi are referred to as endophytes, and they have great potential as a source of bioactive compounds (Digra and Nonzom 2023; Ul-Hassan et al. 2012; Chowdhary et al. 2012; Meshram et al. 2022). Endophytic fungi secrete a variety of bioactive compounds that have a variety of purposes, including promoting plant growth, triggering defense mechanisms against pathogens, and acting as agents to mitigate the effects of salt and drought (Digra and Nonzom 2023; Khare et al. 2018; Strobel 2018; Meshram and Gupta 2019; Meshram et al. 2023). Bioactive chemicals are produced as a result of the coevolution of endophytic fungi and their host plants. These compounds can boost the host plant's fitness and play a range of roles in plant-microbe interactions (Fig. 15.1) (Aly et al. 2013; Saxena et al. 2015; Venugopalan and Srivastava 2015; Kusari and Spiteller 2016). Through metabolic interactions, endophytic fungi create communication with their host plants (Krohn et al. 2007; Kusari et al. 2012). The xenohormesis theory (Howitz and Sinclair 2008) states that under selective evolutionary pressure, heterotrophic organisms like fungi learn to detect chemical cues from host plants that are caused by stress and begin to produce similar compounds themselves. They essentially reproduce the host plant's biological characteristics (Kusari et al. 2014; Venugopalan and Srivastava 2015). Endophytic fungi not only synthesize chemicals that are similar to those of their host plants, but they also display a wide range of secondary metabolites that have interesting biological and/or medicinal qualities. Over the past three decades, a diverse array of bioactive substances with potential use in medicine and healthcare have been identified in endophytic fungus. Antimicrobial (Digra and Nonzom 2023; Caruso et al. 1990; Deshmukh et al. 2018), anticancer (Bedi et al.

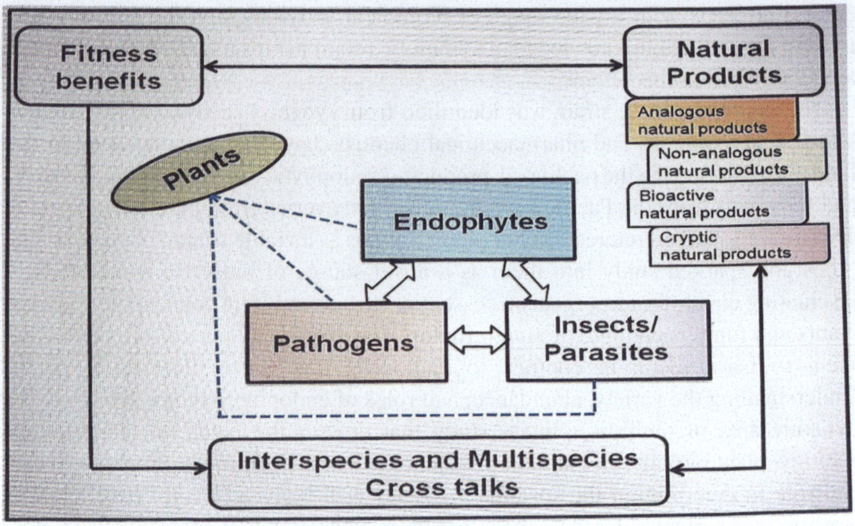

Fig. 15.1 Diagram shows the cost-benefit of interactions between plants and fungi, including those involving the host plant, pathogens, endophytic fungi, and insects that cause fungal endophytes to create a variety of metabolites with significant bioactivities (Gupta et al. 2023)

2018; Deshmukh et al. 2019), antioxidant (Li et al. 2016; Kaur et al. 2020a, b; Manganyi and Ateba 2020; Nischitha and Shivanna 2022), anti-inflammatory (Pal et al. 2020; Mazumder et al. 2021), antidiabetic (Artanti et al. 2012; Indrianingsih and Tachibana 2017; Agrawal et al. 2021), and immunosuppressive actions (Song et al. 2013; Zhang et al. 2021; Deng et al. 2021; Wei et al. 2022) are only a few of the qualities these compounds display. Such physiologically active metabolites from endophytic fungi are abundant, which emphasizes their significance as a valuable source of possible medicinal compounds (Digra and Nonzom 2023; Deshmukh et al. 2015; Adeleke and Babalola 2021; Strobel 2018; Deshmukh et al. 2018). The goal of this chapter is to outline the main biomedical uses of endophytic fungi for plant protection in agriculture.

15.2 History

The history of endophytic fungi is closely intertwined with the broader exploration of plant-microbe interactions and the study of symbiosis. While interest in endophytic fungi dates back to the nineteenth century, their systematic investigation and recognition of significance have evolved gradually over time. Here's an overview of the historical progression:

Early Observations (Nineteenth Century): The earliest references to endophytic fungi can be traced back to the nineteenth century. Scientists observed fungal structures within plant tissues, although their significance remained unclear. These observations often centered on fungal growth within plant roots and stems.

Discovery and Initial Studies (Twentieth Century): The twentieth century marked the beginning of more focused investigations into endophytic fungi. In the mid-twentieth century, researchers started recognizing that certain fungi could inhabit plant tissues without causing visible harm. However, the full extent of their diversity and ecological roles remained largely unexplored. 1980s–1990s:

Recognition of Mutualistic Relationships: During the 1980s and 1990s, researchers began acknowledging that many endophytic fungi established mutualistic relationships with their host plants. These fungi were found to confer various benefits to plants, including enhanced growth, stress tolerance, and protection against pathogens and herbivores. This realization shifted the perception of endophytes from being regarded as mere saprophytic colonizers to being valued as essential partners in maintaining plant health.

Advancements in Molecular Techniques (Late Twentieth Century–Present): The late twentieth century witnessed a significant leap forward in the study of endophytic fungi with the advent of molecular techniques, particularly DNA sequencing. These powerful tools revolutionized our understanding of endophytic fungal communities by enabling precise identification and classification of species within plant tissues. As a result, researchers uncovered a much greater diversity of endophytic fungi than previously known. Moreover, molecular techniques facilitated investigations into the genetic and functional aspects of the symbiotic interactions between endophytes and their host plants.

Twenty-First Century: Ecological and Biotechnological Insights: The twenty-first century has witnessed a surge in research on endophytic fungi, focusing on their ecological roles, functional diversity, and potential biotechnological applications. Scientists have delved into understanding how endophytic fungi contribute to plant growth promotion, stress tolerance enhancement, and nutrient cycling in ecosystems. Furthermore, the discovery of bioactive compounds produced by endophytes has spurred interest in their applications across agriculture, medicine, and industry.

Current Trends: Symbiosis and Sustainability: Recent studies underscore the critical role of endophytic fungi in promoting sustainable agriculture and supporting ecosystem stability. Researchers are increasingly exploring the intricate signaling and metabolic interactions between endophytes and host plants, unveiling the mechanisms underpinning these beneficial relationships. Emphasizing symbiosis and sustainability, current trends in endophytic fungal research highlight the importance of these hidden partners in enhancing plant health and resilience. Throughout its history, the study of endophytic fungi has evolved from initial observations to a sophisticated discipline with profound implications for plant biology, ecology, and biotechnology. As our understanding deepens, the vital contributions of these symbiotic organisms to the health and vigor of plants become increasingly apparent.

15.3 Biomedical Applications of Fungal Endophytes

15.3.1 Antimicrobial Activity

Due to their constant development of new defense mechanisms against antimicrobial medicines, pathogenic bacteria and fungi that are resistant to commercial treatments pose a serious threat to healthcare services (Elbasuney et al. 2021; Costelloe et al. 2010). Therefore, the search for potent antibacterial medicines is urgent. Fungal endophytes are invisible to their hosts; they live inside plant tissues and are important sources of bioactive substances with a range of biological activities, such as antioxidant, antiviral, antibacterial, and anticancer effects (Sharaf et al. 2022). Numerous active secondary metabolites, including steroids, flavonoids, terpenoids, peptides, quinones, lignans, alkaloids, phenylpropanoids, phenolics, and isocoumarins, are produced by these endophytic fungi and have been shown to have strong antimicrobial effects on a range of pathogenic microorganisms (Elbasuney et al. 2021). In order to combat antimicrobial resistance, it is crucial to investigate novel antimicrobials derived from recently identified fungal endophytes (Costelloe et al. 2010). *Aspergillus* species is a common fungal endophyte; leaf tissues of medicinal plants such as *Callistemon subulatus* have yielded several isolates of this species. As per Kamel et al. (2020), isoshamixanthone, which was extracted from *Aspergillus* species, had antibacterial properties against microorganisms that cause disease. Additionally, research conducted by Sharaf et al. (2022) revealed the isolation of *A. favus* from *Ocimum basilicum*, which at 1000 μg/mL demonstrated antibacterial and antifungal activities against a variety of pathogens, including *Salmonella typhimurium, Bacillus cereus, B. subtilis, Escherichia coli, Pseudomonas aeruginosa, Klebsiella pneumonia,* and *Candida albicans*. Similar to this, *A. fumigatus* showed encouraging antibacterial and antifungal properties against common drug-resistant microorganisms when it was isolated from the leaves of *Albizia lucidior* and *Ocimum basilicum* (Hussein et al. 2022; Sharaf et al. 2022). These results demonstrate the potential of *Aspergillus* species in particular as fungal endophytes as sources of new antimicrobial compounds to counteract antibiotic resistance.

The potential antibacterial characteristics of *Penicillium*, a group of endophytic ascomycetes often found within plants, have drawn interest, particularly in relation to its ability to combat resistant microbes. The antifungal and antibacterial properties of *Penicillium* species against a range of diseases have been emphasized by recent research. In 2020, Wu and colleagues identified an endophytic *Penicillium* species from *Stephania dielsiana*, the host plant, and found that it exhibited promising antibacterial efficacy against seven distinct pathogenic bacteria. Similarly, *Penicillium citrinum* isolated from *Azadirachta indica* has been shown to have promising antibacterial action against human pathogenic bacteria and fungi by Kumari et al. (2021). Furthermore, with minimum inhibitory concentrations (MIC) ranging from 12.5 to 100 μg/ml, Zhao et al. (2019) discovered three novel α-pyrone derivatives from *Penicillium ochrochloron* isolated from Taxus medium that showed antimicrobial efficacy against specific pathogenic bacterial and fungal strains. Furthermore, *Penicillium pinophilum* from *Alloteropsis cimicina* and *Penicillium funiculosum* from Ficus elastic were isolated in investigations conducted by Ding et al. (2019) and Nischitha and Shivanna (2021), respectively. Potential antibacterial activity was shown by extracts from these *Penicillium* species against strains of clinical bacteria. These results highlight *Penicillium* endophytes' potential as important suppliers of antimicrobial compounds to combat resistant microorganisms.

15.3.2 Antiviral Activity

Viruses pose significant threats worldwide, causing outbreaks with severe symptoms, high mortality rates, and substantial economic burdens on society. With the continual emergence of new viruses, the limited availability of vaccines and antiviral drugs underscores the urgent need for novel antiviral compounds to address drug resistance issues. The escalating resistance of human diseases to conventional antibiotics further emphasizes the necessity of discovering alternative agents for managing these ailments. Bioactive compounds derived from natural sources offer a vast reservoir of chemical diversity, surpassing even the largest combinatorial databases. Recent studies have highlighted the potential of endophytic fungi as sources of novel antiviral agents. For instance, Han et al. (2019) reported the isolation of an endophytic strain of *Trichoderma harzianum* from Kadsura angustifolia, yielding nigranoic acid, which exhibited potent antiviral activity by inhibiting HIV-1 reverse transcriptase. Similarly, Khiralla et al. (2020) isolated *Curvularia papendorfii* from *Vernonia amygdalina*, demonstrating that the crude extract of *C. papendorfii* displayed antiviral effects against coronaviruses, reducing virus-induced cytopathogenic effects by 40% at lower multiplicity of infection. Furthermore, Liu et al. (2019) isolated *Phoma* sp. from *Aconitum vilmorinianum*, leading to the discovery of a rare 14-nordrimane sesquiterpenoid called phomanolide. These findings underscore the potential of endophytic fungi and their bioactive compounds as promising sources of novel antiviral agents for combating viral diseases and addressing drug resistance challenges.

15.3.3 Antioxidant Activity

Antioxidant-active compounds are essential for protecting cells from the oxidative damage caused by free radicals that are produced from oxygen and reactive oxygen species (ROS). These dangerous substances are linked to the development and course of numerous diseases, including cancer, cardiovascular conditions, Alzheimer's, diabetes mellitus, hypertension, and aging. They also cause DNA damage, carcinogenesis, and cellular degeneration. Antioxidants are recognized to be present in natural sources such fruits, vegetables, and medicinal plants. Additionally, metabolites produced by fungal endophytes have emerged as a promising reservoir of novel antioxidants. Among fungal endophytes, *Aspergilli* have been particularly noted for their antioxidant production. For instance, Brissow et al. (2017) demonstrated the antioxidant potential of an extract from endophytic

Aspergillus nidulans isolated from *Passiflora incarnata*. Similarly, *Penicillium* spp. have been recognized for their antioxidant activity. Ikram et al. (2019) identified ferulic acid, cinnamic acid, quercetin, and rutin in the extract of *Penicillium roqueforti*, all of which exhibit promising antioxidant properties.

Moreover, research by Nischitha and Shivanna (2021) highlighted endophytic *Penicillium pinophilum* as a natural source of novel bioactive compounds with both antimicrobial and antioxidant properties. *Chaetomium*, another fungal endophyte, has also been acknowledged for its antioxidant potential. Zhao et al. (2021) isolated flavonoids from *Chaetomium cruentum*, extracted from *Conyza blinii* H. Lev, demonstrating significant antioxidant activity. Additionally, endophytic *Diaporthe* species have been reported for their antioxidant activity, as evidenced by research conducted by Tan et al. (2022). Furthermore, genera such as *Epicoccum* and *Nigrospora* have shown promise in terms of antioxidant activity. Examples include *Epicoccum nigrum* (Yan et al. 2020), *Nigrospora sphaerica* (Gautam et al. 2022), and *Nigrospora oryzae* (Vig et al. 2022). These findings underscore the diverse and valuable contributions of fungal endophytes to the pool of natural antioxidants.

15.3.4 Anticancer Activity

Cancer, characterized by uncontrolled and abnormal cell proliferation, poses a significant global health challenge and remains a leading cause of mortality. Its multifactorial etiology involves both extrinsic factors, such as tobacco use, alcohol consumption, unhealthy diet, and environmental exposures, and intrinsic factors, like aging, DNA mutations, hormonal imbalances, and compromised immune function. These factors contribute to the dysregulation of specific genes, culminating in aberrant cell growth and tumor formation. In recent years, there has been a growing interest in the exploration of natural bioactive compounds derived from endophytic fungi as potential sources of novel anticancer agents. Among fungal endophytes, *Aspergilli* have garnered particular attention due to their abundance and promising anticancer activity. For instance, *Pulchranin*, an anticancer compound, was isolated from the endophytic *Aspergillus* strain TRL1 obtained from *Tabebuia rosea*, exhibiting effective inhibition against human liver (Hep-G2) and breast (MCF-7) cancer cell lines.

Similarly, *Pyrano xanthones* with anticancer properties were identified from *Aspergillus* ASCLA, demonstrating activity against human cervix carcinoma. Additionally, gartryprostatins A, B, and C, isolated from endophytic *Aspergillus* sp. GZWMJZ-258, exhibited anticancer activity against human FLT3-ITD mutant AML.

Another endophytic fungus, *A. candidus* LDJ-5 isolated from *Rhizophora apiculata* Blume, yielded prenylterphenyllins as potential anticancer compounds. *Alternaria* sp. sb23 produced anticancer compounds alterchothecenes and trichothecenes, which showed activity against colon and breast cancer. Similarly, compounds including dibenzoα-pyrone derivatives and benzo-γ-pyrone derivatives, isolated from endophytic *A. alternata*, demonstrated anticancer effects against various human cancer cell lines. Taxol, a well-known anticancer compound, is produced by fungal endophytes such as *Taxus brevifolia*, commonly used in cancer treatment. Overall, the discovery of novel anticancer compounds from endophytic fungi holds promise for the development of new therapeutic agents against cancer, offering potential alternatives to conventional chemotherapy.

15.3.5 Criteria for Plant Selection

Understanding the methodologies and rationales behind the isolation of novel endophytic microorganisms, especially those producing unique bioactive compounds, is crucial. With the vast number of plant species globally, innovative strategies are essential to efficiently identify endophytes with desired bioactivities. Specific criteria guide the selection of plants for endophyte isolation and natural product discovery. Plants thriving in distinctive environmental niches, particularly those with uncommon biological features and innovative survival mechanisms, are prioritized for study. Additionally, plants with a rich ethnobotanical history, often utilized by indigenous communities for specific purposes, are chosen based on direct interactions with locals or through relevant literature. Interestingly, the healing properties attributed to certain plants may actually stem from bioactive compounds produced by their endophytes rather than the plants themselves.

Endemic plants, those with exceptional longevity, or species associated with ancient landmasses are also targeted. Such plants may host endophytic communities that have coevolved unique traits over time. Furthermore, regions with high biodiversity are explored, as they offer a greater likelihood of encountering diverse endophyte populations. Upon selection, each plant is meticulously identified, and its precise location is recorded using GPS technology. Small portions of stems and leaves, known as explants, are carefully collected, ensuring the removal of excess moisture, and stored in sealed plastic bags. These systematic collection methods facilitate the isolation of novel endophytic microorganisms and enhance the discovery of bioactive compounds with potential applications in various fields.

Efforts are made to preserve collected plant materials at 4 °C until isolation procedures commence. Upon arrival in the laboratory, the plant samples undergo thorough surface treatment using 70% ethanol, occasionally followed by flaming, and then air drying under a laminar-flow hood. These steps aim to eradicate surface contaminants effectively. The interior tissues are then painstakingly dissected and put onto water agar plates after the outer tissues have been carefully removed from the samples using a sterile knife blade. The hyphal ends of the fungi are removed after many days of incubation and placed onto potato dextrose agar. Here, it is common for the fungal endophytes to sporulate on particular plant components. Finally, common morphological and molecular biological methods and procedures are employed to identify these endophytic fungi. A distinct class of microorganisms known as endophytic fungus lives in plant tissues without evidently harming their host plants. Mutual benefits characterize this symbiotic relationship: the plant provides the fungus with nourishment and shelter, and the fungus may also have growth-promoting or protective effects on the plant. These endophytic fungi live in the leaves, stems, roots, and seeds of plants, among other plant components.

15.3.6 Relationship Between Endophytic Fungi and Plant

Throughout time, the interaction between an endophyte and its host plant has changed, marked by particular adaptations on the parts of the fungal and the plant (Verma et al. 2012). This complex and multifaceted relationship depends on the particular traits of the plant, the fungus, or both. The external and endogenous hypotheses are the two primary theories put forth to explain the precise origin of endophytes; however, this remains unclear. The endogenous theory states that endophytes have a similar genetic background to their hosts and are derived from plant chloroplasts and mitochondria. On the other hand, the exogenous hypothesis postulates that endophytes penetrate their host plants using a variety of techniques, including root wounds, induced channels, and surface contact (Hondelmann et al. 2020). Endophytic fungus and their host plants have evolved a variety of relationships over time, including mutualism, antagonism, and neutralism. When fungal endophytes exhibit neutralism, they usually stay dormant in their host tissues for the remainder of the host's life. Conversely, certain endophytes might stay inactive until their environment is right for them to wake up, which would suggest mutualistic or antagonistic relationships (Verma et al. 2016).

15.4 Host Plant and Endophytic Fungi Transmission Modes

Endophytic fungi exhibit diverse modes of reproduction, encompassing three primary pathways:

1. Via seeds or hyphae, diseased plants can transmit their infection to their progeny vertically, a process also referred to as intercellular transmission. One example of this approach is *Neotyphodium* sp.
2. Horizontal transmission, also known as intracellular transmission, happens when sexual spores from infected plants spread. It is known that *Epichloe* spp. use this technique.
3. Mixed transmission involves a combination of vertical and horizontal transmission routes (Gebretsadik et al. 2022).

Vertical transmission represents an asexual reproductive mechanism wherein intercellular hyphae are transferred from an infected plant to initiate infection in non-plants or to germinate infected seeds (Daley et al. 2017).

Horizontal transmission, on the other hand, involves the production of infectious sexual spores as part of the fungus's sexual life cycle. These spores are typically found in the stroma around the leaf sheath of grass flags, as seen in certain *Epichloe* spp. Upon emergence, fungal propagules develop and infect leaves by penetrating the leaf cuticle through stomata, eventually expanding intercellularly within healthy tissues (Verma et al. 2016). However, many *Epichloe* spp. utilize a combination of vertical and horizontal transmission, where both systemic and nonsystemic endophytes are transmitted intracellularly.

15.4.1 Plant's Biotic Stress Management Through Endophytic Fungi

Endophytes have become increasingly appealing to researchers in both industrial and agricultural fields due to their capacity to enhance a plant's resistance to a wide array of stresses encountered in its natural environment. These stressors encompass biotic challenges like insect pests, herbivores, nematodes, and other pathogens, as well as abiotic stressors such as drought and chemical exposure (Seleiman et al. 2021).

15.4.2 Production of Biocontrol Agents by Endophytic Fungi

Fungal endophytes play a crucial role in producing biocontrol agents that bestow antimicrobial and pesticide properties upon their host plants (Gong et al. 2020). Among these protective compounds are antixenosis and antibiosis agents, which deter insects from feeding or reduce their feeding efficiency (Hondelmann et al. 2020). Endophytic fungi are responsible for synthesizing and releasing specific bioactive molecules within the plant tissue, which contribute to its defense mechanisms without harming the host (Liarzi et al. 2016). These compounds, including various plant phytochemicals, belong to different functional classes and collectively serve to protect the plant against various threats (Verma et al. 2016).

15.4.3 The Nematicidal Ability of Endophytic Fungi for Host Plant

Fungal endophytes exhibit nematicidal properties due to their containment of toxic compounds targeting nematodes (Hondelmann et al. 2020). However, Diedhiou et al. (2003) observed reduced nematicidal activity of the endophytic fungus *Fusarium oxysporum* against *Meloidogyne incognita*, a plant-parasitic nematode affecting tomato plants. In contrast, Zum et al. (2009) discovered that by reducing the population of *Radopholus similis* (Cobb), a prominent parasitic nematode on bananas, the combination of endophytes *Trichoderma atroviride* and *F. oxysporum* increased banana output.

15.4.4 Phytohormones Production Ability of Endophytic Fungi for Host Plant

By aiding in the solubilization of phosphate and encouraging the absorption of vital nutrients, plant hormones, and metabolites, endophytic fungi are necessary in augmenting plant development (Firáková et al. 2007). Among these, endophytes that produce metabolites and enzymes that promote plant growth, especially in harsh environmental conditions—like *Colletotrichum, Piriformospora* indica, *Sebacina vermifera*, and *Penicillium*—have attracted a lot of interest (Redman et al. 2011). For example, it has been observed that the endophyte *Cladosporium sphaerospermum,* which is present in Glycine max, releases gibberellic acid, which stimulates the growth of soybean and rice plants (Hamayun et al. 2010). Similar to this, the microspore of *Pestalotiopsis* creates a metabolite called a pestalotin analogue, which has gibberellin-like properties and increases germination rates (Li et al. 2018). According to Khan et al. (2014), endophytes that create derivatives of indole acetic acid, such as *Fusarium tricinctum* and *Alternaria alternata*, aid in the development of plants. Moreover, Johnson et al. (2014) demonstrated that endophytes associated with *Piriformospora indica* roots regulate phytohormones crucial for host plant growth, thereby enhancing nutrient absorption and translocation, particularly phosphorus and nitrogen, from the soil.

15.4.5 Enzyme Production of Endophytic Fungi

The breakdown of soil biomass by endophytic fungi plays a crucial role in nutrient recycling, contributing to soil fertility and quality. Endophytic fungi are known to produce and secrete various extracellular hydrolases, including pectinase, cellulase, xylanase, and proteinase, which serve multiple functions such as defense against pathogens and nutrient acquisition from the host (Leo et al. 2016). These enzymes facilitate symbiotic interactions by breaking down complex macromolecules like polysaccharides, proteins, and organic phosphates into smaller molecules that can be transported across cells for metabolic processes (Strong et al. 2015).

More than 50 endophytic fungal strains, including hydrolases and proteases, were discovered by Sunitha et al. (2013). Similar to this, Chathurdevi and Gowrie (2016) found that endophytic fungi isolated from the medicinal plant *Cardiospermum halicacabum* promote plant growth by generating a variety of extracellular enzymes that allow the plant to flourish in unfavorable environments. Notably, fungal enzymes like chitinases play a crucial role in breaking down carbonaceous and nitrogenous compounds derived from chitin, thereby contributing to nutrient cycling and soil health.

15.5 Management of Abiotic Stress by Plants Through Endophytic Fungi

Global agricultural production is seriously threatened by abiotic stresses, such as drought, salinity, and temperature swings, which frequently result in yield losses greater than 50% (Wang et al. 2010). Plants have evolved mechanisms to cope with abiotic stress, including (i) promptly initiating response mechanisms upon encountering stress and (ii) harnessing endophytic bioactive compounds synthesized by endophytes to act as stress-alleviating agents (Gong et al. 2020).

15.5.1 Plant's Drought Stress Management by Endophytic Fungi

Drought poses a significant threat to global crop productivity, exerting adverse effects on plant life such as impaired germination, membrane damage, reduced photosynthesis, and altered production of reactive oxygen species (Seleiman et al. 2021). However, research indicates that plants harboring fungal endophytes exhibit greater resilience to drought compared to noninfected plants. These endophyte-infected plants demonstrate enhanced drought tolerance by mechanisms such as increased tissue solute accumulation, reduced leaf water conduction, slowed transpiration, or the development of a thicker cuticle. For instance, studies have shown that endophytic *Neotyphodium* spp. enhance drought tolerance in grass plants by regulating stomatal function and osmoregulation (Chhipa and Deshmukh 2019). This was supported by the findings of Redman et al. (2011), who observed increased biomass in rice plants inoculated with *Fusarium culmorum* and *Curvularia protuberata* under drought conditions compared to non-inoculated plants.

15.5.2 Plant's Salinity Stress Management Through Endophytic Fungi

Extreme weather patterns and intensive farming methods have raised soil salinity significantly in recent decades, endangering agricultural productivity worldwide (Wicke et al. 2011). Soil salinity, which is defined by the build-up of soluble salts, modifies the physical and chemical characteristics of soil, making it less suitable for agricultural use (Hu et al. 2011). In response to salt stress, plants produce reactive oxygen species (ROS) that can harm cells, including superoxide dismutase (SOD). Nonetheless, by releasing antioxidant metabolites into the plant tissues, fungus endophytes such as *Piriformospora indica* improve the tolerance of plants to salt stress (Baltruschat et al. 2008). These metabolites play a crucial role in scavenging ROS from the cells, primarily through the regeneration of antioxidants like ascorbate and glutathione. For instance, Rodriguez et al. (2008) found that nonsymbiotic plants such as dune grass (*Leymus mollis*) exhibited severe wilting and desiccation within 7 days of exposure to a salt solution of about 500 mmol/L, often resulting in plant death after 14 days. In contrast, plants in symbiosis with *F. culmorum* demonstrated significantly higher salt tolerance, enduring the same salt concentration for well over 14 days before exhibiting signs of wilting.

15.5.3 Plant's Temperature Stress Management by Endophytic Fungi

High temperatures can hinder crop development by denaturing proteins and inducing protein aggregation (Hussain et al. 2018). Conversely, low temperatures can disrupt metabolic processes by inhibiting enzyme reactions and altering intermolecular interactions, thereby affecting protein structure and cell membrane integrity (Andreas et al. 2016). Certain plants, such as tomatoes, watermelon, and wheat, have been found to benefit from the presence of endophytic *Curvularia* spp. within their tissues. In *Dichanthelium lanuginosum*, a grass species, the presence of *Curvularia* sp. endophytes has been shown to protect the plant from temperature-related stress, particularly in areas where soil temperatures exceed 57 °C. *D. lanuginosum* exhibits remarkable temperature tolerance, ranging from 38 to 65 °C, attributed to its endophytic association with the fungus *C. protuberata* and a mycovirus known as *Curvularia thermal tolerance virus* (CThTV). *C. protuberata* is known to thrive in geothermal soils and has been observed to confer cold stress tolerance in germinated rice seeds under laboratory conditions (Redman et al. 2011).

15.5.4 Management of Heavy Metal Stress by the Plant Through Endophytic Fungi

Increased industrialization has led to a growing concern over heavy metal pollution, as these pollutants are nondegradable and pose significant risks to the environment. Heavy metal toxicity can severely impact crop productivity, with potential losses ranging from 25% to 80% across various cultivated crops due to inhibited root growth caused by heavy metal exposure (Singh et al. 2011). Fungal endophytes have been identified as having mechanisms for chelating or accumulating heavy metals, thereby enhancing plant resistance to heavy metal stress. These mechanisms bolster the plant's antioxidative defenses, facilitating detoxification and enabling plant growth in polluted soil (Wang et al. 2017). Siderophores, which are metal-chelating compounds produced by plant roots in the rhizosphere, play a crucial role in preventing heavy metal absorption by plants. By forming complexes with heavy metals that are less readily absorbed by plant roots, siderophores effectively reduce heavy metal uptake by plants. *P. fortinii* and *Rhizodermea veluwensis* are known to synthesize siderophores, which contribute to heavy metal removal from the rhizosphere (Yamaji et al. 2016).

15.5.5 Characteristics of Endophytic Fungi

1. Endophytic fungi typically maintain a stealthy presence within their host plants, often without causing any visible signs of disease. Unlike pathogens that instigate harmful infections, these fungi establish a harmonious relationship with the plant, maintaining a delicate balance without inducing adverse effects.
2. Endophytic fungi exhibit remarkable adaptability in their colonization strategies, occupying intercellular spaces and even residing within plant cells themselves. Their distribution spans across various plant tissues, with variations influenced by the specific fungal species and the host plant's characteristics.
3. The biodiversity of endophytic fungi is extensive, encompassing a plethora of species spanning diverse taxonomic groups. This rich diversity is shaped by factors such as the host plant's species, geographical location, and prevailing environmental conditions.
4. A significant attribute of endophytic fungi is their capacity to produce bioactive compounds, including a myriad of secondary metabolites with diverse biological activities. These compounds hold promise for applications in agriculture, medicine, and various industrial sectors.
5. Endophytic fungi engage in multifaceted interactions with their host plants, exhibiting both beneficial and neutral effects. They contribute to plant health by promoting growth, enhancing nutrient uptake, bolstering tolerance to environmental stresses, and providing defense mechanisms against herbivores and pathogens.
6. In ecological terms, endophytic fungi wield considerable influence on ecosystem dynamics by impacting plant vitality, nutrient recycling processes, and overall plant community compositions. They also mediate interactions between plants and other organisms within their habitats.
7. The bioprospecting potential of endophytic fungi is substantial, driven by their ability to produce bioactive compounds of pharmaceutical, agricultural, and biotechnological significance. Exploration of these fungal sources has unveiled compounds with promising antimicrobial, anticancer, and pesticidal properties, offering avenues for drug discovery and sustainable agricultural practices.

15.6 Potential Applications in Agriculture and Plant Protection

It is well-known that endophytic fungi can survive inside plant tissues without endangering their host plants. Numerous bioactive substances, including phytotoxic substances, that may find use in plant protection and agriculture are produced by these fungi. A well-researched case study of endophyte application emphasizes how endophyte-mediated phytoremediation is facilitated by the breakdown of environmental contaminants, protection of plants from stress, stimulation of plant growth, and direct and indirect support of plants to withstand high concentrations of pollutants (He et al. 2020; Tiwari et al. 2022). In soil phytoremediation, the breakdown of

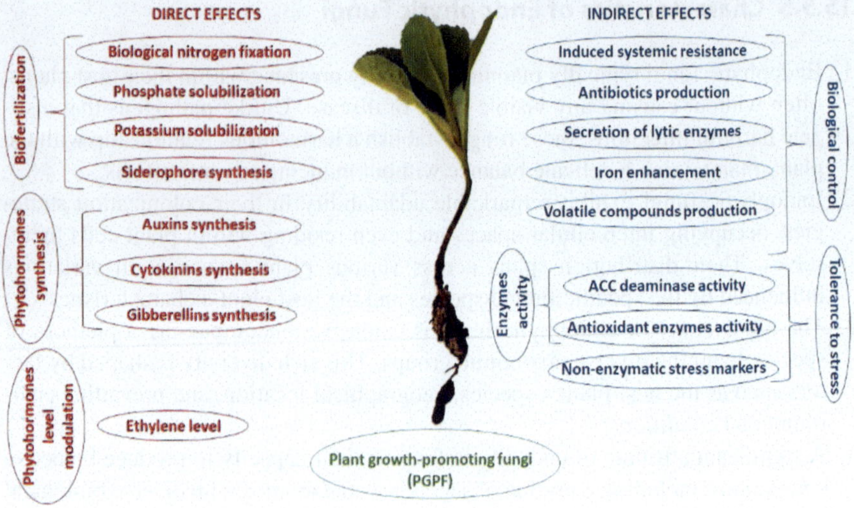

Fig. 15.2 Impacts of fungi that promote plant development directly and indirectly (Ozimek and Hanaka 2021)

contaminants can be a useful use of endophytes as bio-resources to replenish the soil's lost fertility. Currently, efforts to find new metabolites with pharmacological and/or biotechnological utility are driving endophyte research projects (Fig. 15.2). The diverse functions of endophytes in promoting plant growth and promoting sustainable agriculture. Endophytes play important ecological responsibilities, but they are also becoming more significant as biological resources for the synthesis of valuable compounds with medicinal significance. According to Hettiarachchige et al. (2019), endophytes exhibit host mimicking in their autonomous synthesis of metabolites that are anticipated to be excellent candidates for genetic engineering. Bioprospection of endophytes for natural product-based drug discovery has produced high-value metabolites, including anticancer drugs like taxol and camptothecin, anti-HIV drugs like lariatins A and B, HIV-1 protease inhibitors like hinnuliquinone, and cholinesterase inhibitors like huperzine A (Tiwari and Bae 2020, 2022). The discovery and successful commercialization of taxol, a billion-dollar anticancer medicine, show that the rich yet little-studied metabolic diversity of endophytes has significant potential to impact the pharmaceutical industry. It is possible to extract and analyze high-value metabolites from a variety of endophytes, which opens up new possibilities for drug discovery based on natural products. The comprehensive study of endophytes makes it necessary to comprehend the dynamics between plants and endophytes, as well as the mechanisms underlying metabolite induction, in order to comprehend the molecular underpinnings of this interaction and the wide range of applications it has from a socioeconomic standpoint. Some recent trends in this field include:

1. Biopesticide Development: Researchers have been investigating the potential of phytotoxic compounds produced by endophytic fungi as biopesticides. These compounds can be used to control pests and diseases in crops without causing harm to humans, animals, or the environment. Various studies have focused on identifying endophytic fungi that produce phytotoxic metabolites with pesticidal properties.
2. Stress Tolerance and Plant Growth Promotion: It has been shown that certain phytotoxic substances made by endophytic fungi enhance plant development and stress tolerance. These substances have the ability to improve nutrient uptake, promote plant growth, and protect plants from environmental stresses such disease invasions, salt, and drought.
3. Identification of Novel Compounds: Advances in analytical techniques and genomics have facilitated the identification of novel phytotoxic compounds from endophytic fungi. Researchers are using tools like mass spectrometry, nuclear magnetic resonance (NMR), and genome mining to discover previously unknown bioactive compounds with potential agricultural applications.
4. Mode of Action Studies: Understanding the mechanisms of action of phytotoxic compounds is crucial for their practical application. Recent trends involve elucidating how these compounds affect plant physiology, cellular processes, and the interactions between plants and other organisms in the ecosystem.
5. Bioprospecting in Biodiversity Hotspots: Many endophytic fungi are found in biodiversity-rich areas such as rainforests. Researchers are exploring these regions to discover new fungal strains and their associated phytotoxic compounds. This bioprospecting approach aims to identify unique and potent bioactive molecules for various applications.
6. Fungal-Plant Interactions: The study of the interactions between endophytic fungi and their host plants is an evolving field. Researchers are investigating how plants perceive and respond to phytotoxic compounds produced by endophytic fungi, which can provide insights into the potential benefits of these compounds for plant health.
7. Biotechnological Applications: Advances in biotechnology have enabled the production of phytotoxic compounds through fermentation and genetic engineering. This could potentially lead to the scalable production of these compounds for commercial purposes.

15.6.1 Phytotoxic Compounds

Phytotoxic compounds are substances that have a toxic or harmful effect on plants. These compounds can interfere with various physiological and biochemical processes in plants, leading to reduced growth, development, and overall health. Phytotoxic compounds can be produced by a variety of sources, including certain plants themselves, microorganisms like fungi and bacteria, and even synthetic chemicals. Here are a few examples:

1. Allelopathic Compounds: Some plants release allelopathic compounds into their environment, which inhibit the growth of nearby plants. These compounds can be found in various plant parts such as leaves, stems, roots, and even in the soil around the plant. They play a role in competition between plant species for resources.
2. Microbial Metabolites: Microorganisms, including endophytic fungi and bacteria, can produce phytotoxic compounds as part of their metabolic processes. These compounds might help the microorganisms establish dominance or provide a competitive advantage over other plant species.
3. Herbicides: Synthetic herbicides are chemicals designed to control or kill unwanted plants, often referred to as weeds. These chemicals are purposely used to be phytotoxic and are applied in agricultural and landscaping contexts to manage vegetation.
4. Plant Disease Toxins: Some pathogens, including fungi and bacteria, produce toxins that harm plants and contribute to disease development. These toxins can interfere with cellular processes, leading to cell death and tissue damage.
5. Natural Products: Various natural products, such as secondary metabolites produced by plants, can have phytotoxic effects on other plants. These compounds might be involved in plant defense against herbivores or pathogens, deterring their consumption.
6. Environmental Stress Compounds: Environmental stressors like heavy metals, high salinity, or pollutants can produce phytotoxic effects on plants. These stressors can disrupt cellular processes and lead to plant damage or death.
7. Overproduction of Ethylene: Ethylene is a natural plant hormone that can have phytotoxic effects when produced in excess. High levels of ethylene can lead to leaf wilting, premature senescence (aging), and other detrimental effects on plants.

15.6.2 Effects of Phytotoxic Compounds

It's important to note that the effects of phytotoxic compounds can vary based on factors such as concentration, exposure duration, plant species, and environmental conditions. Some phytotoxic compounds have been explored for their potential applications in agriculture, as they could be used as natural herbicides or plant growth regulators. However, the use of such compounds requires careful consideration of their specific effects and potential impacts on both target and nontarget plants.

Phytotoxic compounds can have a range of effects on plants, depending on factors such as the type of compound, concentration, exposure duration, and the specific plant species involved. These effects can vary from subtle changes in growth and development to severe damage or even death of the plant. Here are some common effects of phytotoxic compounds on plants:

1. Inhibition of Germination: Phytotoxic compounds can hinder seed germination by interfering with the enzymatic processes and cellular activities required for the seed to sprout and develop into a seedling.
2. Stunted Growth: Many phytotoxic compounds can slow down or inhibit the growth of plants. This can result in reduced stem and root growth, leading to stunted overall development.
3. Leaf Chlorosis: Chlorosis refers to the yellowing of leaves due to reduced chlorophyll production. Phytotoxic compounds can disrupt the photosynthetic process, causing leaves to lose their green color and reducing the plant's ability to produce energy.
4. Necrosis: Necrosis is the death of plant cells or tissues. Phytotoxic compounds can trigger necrotic lesions on leaves, stems, or other plant parts, leading to tissue death and potential infection sites for pathogens.
5. Abnormal Leaf Morphology: Exposure to phytotoxic compounds can result in the development of abnormal or distorted leaf shapes, sizes, and structures.
6. Reduced Reproductive Success: Phytotoxic compounds can negatively impact the reproductive processes of plants, including the formation of flowers, pollen, and seeds. This can lead to decreased fruit production and reproductive success.
7. Altered Hormone Regulation: Some phytotoxic compounds interfere with the regulation of plant hormones, disrupting the balance of growth and development processes.
8. Water Stress and Wilting: Phytotoxic compounds can disrupt the water balance within plant cells, leading to water stress, wilting, and loss of turgor pressure.
9. Root Damage: Phytotoxic compounds can affect root growth and development, impacting nutrient and water uptake. This can further stress the plant and impair overall health.
10. Allelopathy: Certain phytotoxic compounds released by plants or microorganisms can have allelopathic effects, inhibiting the growth and development of neighboring plants, thereby reducing competition for resources.
11. Changes in Secondary Metabolite Production: Exposure to phytotoxic compounds can influence the production of secondary metabolites in plants, which can affect their defense mechanisms against herbivores or pathogens.

15.6.3 Omics Approaches to Explore Endophytic Fungi

According to Hardoim et al. (2015), endophytic fungi and host plants have complex interactions that are impacted by biotic, abiotic, and genetic variables. In order to fully utilize the potential of these microorganisms in agriculture, that is, to improve plant growth, increase crop yields, control pests, reduce pathogen virulence, and strengthen plant resistance to environmental stresses like extreme temperatures, pH fluctuations, drought, heavy metal contamination, and nutrient deficiencies—it is imperative to comprehend this symbiotic relationship (Shankar Naik 2019; Lugtenberg et al. 2016).

Table 15.1 Benefits of endophytic fungi to host plants revealed by omics-based approaches

Host plant	Endophyte	Benefits	Omics approaches	References
Zea mays	*Exophiala pisciphila*	Heavy metal tolerance by the remodeled host cell walls	Transcriptomic	Shen et al. (2020)
Brassica napus	*Piriformospora indica*	Stress responses; energy production	Proteomic	Shrivastava et al. (2018)
Hordeum vulgare	*Piriformospora indica*	Salt stress tolerance	Metabolomic Transcriptomic Lonomic	Ghaffari et al. (2016)
Eucalyptus globulus	*Chaetomium cupreum*	Heavy metal tolerance	Transcriptomic	Ortiz et al. (2019)
Lolium arundinaceum	*Epichloe coenophiala*	Disease resistance; abiotic stress response	Transcriptomic	Dinkins et al. (2017)
Theobroma cacao	*Colletotrichum tropicale*	Changes in host physiology, metabolism, and anatomy; resistance to pathogens and herbivores	Transcriptomic	Mejía et al. (2014)
Zea mays	*Fusarium verticillioides*	Reduction of harmful effects of phytopathogen	Transcriptomic Metabolomic	Jonkers et al. (2012)
Hordeum vulgare	*Piriformospora indica*	Drought stress tolerance through photosynthesis stimulation, energy releasing, and enhanced antioxidative defense system	Proteomic	Ghabooli et al. (2013)
Lolium perenne L. cv Samson	*Epichloe festucae*	Changes in host development, particularly trichome formation and cell wall biogenesis; resistance to drought and infection by fungal pathogens	Transcriptomic Metabolomic	Dupont et al. (2015)

Understanding the critical role endophytic fungi play in supporting plant health and enabling environmentally friendly, sustainable agricultural methods, a great deal of research has been devoted to elucidating the numerous facets of this symbiotic relationship (Kaur 2020). The investigation of endophyte-plant interactions has been transformed by recent technical developments and the use of bioinformatic techniques. The genetic architecture, taxonomy, and phylogeny of endophytes can be better understood by genomic studies that make use of next-generation sequencing (NGS) technologies (Bosamia et al. 2020). Fungal endophytes' impacts on plant gene expression under various situations can be fully understood by the integration of omics data from genomes, transcriptomics, proteomics, and metabolomics (Table 15.1). While transcriptomics, proteomics, and metabolomics describe the whole range of transcribed RNAs, proteins, and metabolites, respectively, in response to certain environmental stimuli, genomics offers a comprehensive picture of an organism's genetic repertoire (Kaul et al. 2016; Bosamia et al. 2020). Plant defense is a complex system of defenses, and endophytic fungi are important because they affect plant metabolism, which increases tolerance to infections and induces systemic resistance (Gao et al. 2010). Ambrose and Belanger (2012) examined the differential expression of genes linked to the endophyte *Epichloe festucae* colonization of *Festuca rubra* using quantitative transcriptome analysis. More than 200 plant genes involved in various physiological processes were found to express differently in the colonized and non-colonized samples. The transcript abundance and properties of a secreted protein in endophyte-infected *F. rubra* were noteworthy because they indicated the protein's possible role in disease resistance. Compiling transcriptome data with genomic information was essential to comprehending how *E. festucae*'s distinct pattern of gene expression could provide resistance to the host.

Moreover, there is a wealth of evidence supporting the role of fungal endophytes in promoting plant growth (Bilal et al. 2018; Khalil et al. 2021). Yuan et al. (2019) used comparative transcriptomics and proteomics to show how the endophyte *Gilmaniella* sp. AL12 affects the control of the metabolic process in the medicinal herb *Atractylodes lancea*. According to their research, endophytes influenced the immunological response of plants, which may have a positive impact on interactions between plants and endophytes. Furthermore, higher biomass and sesquiterpenoid content were observed in the shoots of *A. lancea* due to the presence of *Gilmaniella* sp. AL12, which activated plant genes linked to the synthesis of proteins involved in carbon fixation, carbohydrate metabolism, and energy production.

Abiotic stressors that hinder plant growth and development, including saline stress, can have an effect on crop productivity (Kumar 2014). Alikhani et al. (2013) evaluated the impact of the endophyte *Piriformospora indica* on *Hordeum vulgare* L.'s ability to withstand salt by using a proteomic approach. According to their findings, endophytic fungi changed the plant host's physiological and proteomic reactions, causing a systemic response to salt stress and providing chances to improve plant resilience to environmental stressors. Knowledge metabolism, physiology, gene expression, and other aspects of endophytic-plant interactions has been greatly improved by omics-based technologies (Chetia et al. 2019). This has led to a better knowledge of the beneficial impacts of endophytic fungi on plant health.

15.7 Conclusions

In conclusion, the study of phytotoxic compounds from endophytic fungi has emerged as a dynamic and rapidly evolving field with promising implications for various sectors. Recent trends have shed light on the multifaceted roles these compounds play in plant-microbe interactions, agriculture, biotechnology, and ecosystem health. Through a combination of advanced techniques, innovative research approaches, and interdisciplinary collaboration, researchers have made significant strides in understanding and harnessing the potential of phytotoxic compounds from endophytic fungi. The biotechnological potential of phytotoxic compounds is becoming increasingly evident. These compounds, with their diverse biological activities, offer opportunities for drug discovery, as well as applications in the fields of medicine, pharmaceuticals, and industrial processes. Moreover, the intricate interactions between endophytic fungi and their host plants have far-reaching ecological implications. Recent research has highlighted their roles in plant community dynamics, nutrient cycling, and ecosystem resilience, underscoring their importance in maintaining healthy ecosystems.

In conclusion, recent trends in the study of phytotoxic compounds from endophytic fungi have illuminated a path toward sustainable solutions for agriculture, enhanced plant health, and a broader understanding of the intricate relationships that shape our ecosystems. As researchers, scientists, and stakeholders collectively contribute to this expanding field, the potential for transformative discoveries and applications remains boundless.

References

Adeleke BS, Babalola OO (2021) Pharmacological potential of fungal endophytes associated with medicinal plant: review. J Fungi 7:147

Agrawal S, Samanta S, Deshmukh SK (2021) The antidiabetic potential of endophytic fungi: future prospects as therapeutic agents. Biotechnol Appl Biochem 69:1159–1165

Alikhani M, Khatabi B, Sepehri M, Nekouei MK, Mardi M, Salekdeh GH (2013) A proteomics approach to study the molecular basis of enhanced salt tolerance in barley (*Hordeum vulgare* L.) conferred by the root mutualistic fungus *Piriformospora indica*. Mol BioSyst 9(6):1498–1510. https://doi.org/10.1039/c3mb70069k

Aly AH, Debbab A, Proksch P (2013) Fungal endophytes-Secret producers of bioactive plant metabolites. Int J Pharm Sci 68:499–505

Ambrose KV, Belanger FC (2012) SOLiD-SAGE of endophyte-infected red fescue reveals numerous effects on host transcriptome and an abundance of highly expressed fungal secreted proteins. PLoS One 7(12):e53214. https://doi.org/10.1371/journal.pone.0053214

Andreas LB, Jaudzems K, Stanek J et al (2016) Structure of fully protonated proteins by proton-detected magic-angle spinning NMR. Proc Natl Acad Sci U S A 113(33):9187–9192. https://doi.org/10.1073/pnas.1602248113

Artanti N, Tachibana S, Kardono LB, Sukiman H (2012) Isolation of alpha-glucosidase inhibitors produced by an endophytic fungus, *Colletotrichum* sp. TSC13 from Taxus sumatrana. Pak J Biol Sci 15:673–679

Baltruschat H, Fodor J, Harrach BD et al (2008) Salt tolerance of barley induced by the root endophyte *Piriformospora indica* is associated with a strong increase in antioxidants. New Phytol 180(2):501–510. https://doi.org/10.1111/j.1469-8137.2008.02583.x

Bedi A, Adholeya A, Deshmukh SK (2018) Novel anticancer compounds from endophytic fungi. Curr Biotechnol 7:168–184

Bilal L, Asaf S, Hamayun M et al (2018) Plant growth promoting endophytic fungi *Aspergillus fumigatus* TS1 and *Fusarium proliferatum* BRL1 produce gibberellins and regulates plant endogenous hormones. Symbiosis 76(2):117–127. https://doi.org/10.1007/s13199-018-0545-4

Bosamia TC, Barbadikar KM, Modi A (2020) Genomic insights of plant endophyte interaction: prospective and impact on plant fitness. In: Kumar A, Radhakrishnan EK (eds) Microbial endophytes: functional biology and applications. Elsevier, Amsterdam, pp 227–249. https://doi.org/10.1016/B978-0-12-819654-0.00009-0

Brissow ER, da Silva IP, de Siqueira KA et al (2017) 18-Deshydroxy cytochalasin: an antiparasitic compound of Diaporthe phaseolorum-92C, an endophytic fungus isolated from *Combretum lanceolatum* Pohl ex Eichler. Parasitol Res 116(7):1823–1830. https://doi.org/10.1007/s00436-017-5451-9

Carroll G (1986) The biology of endophytism in plants with particular reference to woody perennials. In: Fokkema NJ, Van Den Heuvek J (eds) Microbiology of the phyllosphere. Cambridge University Press, Cambridge, pp 205–222

Caruso DJ, Palombo EA, Moulton SE, Zaferanloo B (1990) Exploring the promise of endophytic fungi: a review of novel antimicrobial compounds. Microorganisms 2022:10

Chathurdevi G, Gowrie S (2016) Endophytic fungi isolated from medical plant-a promising source of potential bioactive metabolites. Int J Curr Pharm Res 8:50–56

Chetia H, Kabiraj D, Bharali B et al (2019) Exploring the benefits of endophytic fungi via Omics. In: Singh B (ed) Advances in endophytic fungal research: present status and future challenges. Springer, Cham, pp 51–81. https://doi.org/10.1007/978-3-030-03589-1_4

Chhipa H, Deshmukh SK (2019) Fungal endophytes: rising tools in sustainable agriculture production. In: Endophytes and secondary metabolites. Springer, Cham. https://doi.org/10.1007/978-3-319-76900-4_26-1

Chowdhary K, Kaushik N, Coloma AG, Raimundo CM (2012) Endophytic fungi and their metabolites isolated from Indian medicinal plant. Phytochem Rev 11:467–485

Costelloe C, Metcalfe C, Lovering A, Mant D, Hay AD (2010) Effect of antibiotic prescribing in primary care on antimicrobial resistance in individual patients: systematic review and meta-analysis. BMJ 340:c2096. https://doi.org/10.1136/bmj.c2096

Daley DK, Brown KJ, Badal S (2017) Fungal metabolites. In: Pharmacognosy. Elsevier, Amsterdam, pp 413–421

Deng M, Chen X, Shi Z, Xie S, Qiao Y, Chen G, Tan X, Lu Y, Qi C, Zhang Y (2021) New immunosuppressive secondary metabolites from the endophytic fungus *Aspergillus* sp. Fitoterapia 151:104882

Deshmukh SK, Verekar SA, Bhave SV (2015) Endophytic fungi: a reservoir of antibacterials. Front Microbiol 5:715

Deshmukh SK, Gupta MK, Prakash V, Saxena S (2018) Endophytic fungi: a source of potential antifungal compounds. J Fungi 4:77

Deshmukh SK, Gupta MK, Prakash V, Reddy MS (2019) Fungal endophytes: a novel source of cytotoxic compounds. In: Jha S (ed) Endophytes and secondary metabolites; reference series in phytochemistry. Springer, Cham, pp 365–426

Diedhiou PM, Hallmann J, Oerke EC, Dehne HW (2003) Effects of arbuscular mycorrhizal fungi and a non-pathogenic *Fusarium oxysporum* on *Meloidogyne incognita* infestation of tomato. Mycorrhiza 13(4):199–204. https://doi.org/10.1007/s00572-002-0215-4

Digra S, Nonzom S (2023) An insight into endophytic antimicrobial compounds: an updated analysis. Plant Biotechnol Rep:1–31. https://doi.org/10.1007/s11816-023-00824-x

Ding Z, Tao T, Wang L et al (2019) Bioprospecting of novel and bioactive metabolites from endophytic fungi isolated from rubber tree *Ficus elastica* leaves. J Microbiol Biotechnol 29(5):731–738. https://doi.org/10.4014/jmb.1901.01015

Dinkins RD, Nagabhyru P, Graham MA, Boykin D, Schardl CL (2017) Transcriptome response of *Lolium arundinaceum* to its fungal endophyte *Epichloe coenophiala*. New Phytol 213(1):324–337. https://doi.org/10.1111/nph.14103

Dupont PY, Eaton CJ, Wargent JJ et al (2015) Fungal endophyte infection of ryegrass reprograms host metabolism and alters development. New Phytol 208(4):1227–1240. https://doi.org/10.1111/nph.13614

Eid AM, Salim SS, Hassan SD, Ismail MA, Fouda A (2019) Role of endophytes in plant health and abiotic stress management. In: Kumar V, Prasad R, Kumar M, Choudhary DK (eds) Microbiome in plant health and disease: challenges and opportunities. Springer, Cham, pp 119–144. https://doi.org/10.1007/978-981-13-8495-0_6

Elbasuney S, El-Sayyad GS, Tantawy H, Hashem AH (2021) Promising antimicrobial and antibioflm activities of reduced graphene oxide-metal oxide (RGO-NiO, RGO-AgO, and RGO-ZnO) nano composites. RSC Adv 11(42):25961–25975. https://doi.org/10.1039/D1RA04542C

Firáková S, Šturdíková M, Múcková M (2007) Bioactive secondary metabolites produced by microorganisms associated with plants. Biologia 62(3):251–257. https://doi.org/10.2478/s11756-007-0044-1

Gao F-K, Dai C-C, Liu X-Z (2010) Mechanisms of fungal endophytes in plant protection against pathogens. Afr J Microbiol Res 4(13):1346–1351

Gautam VS, Singh A, Kumari P et al (2022) Phenolic and flavonoid contents and antioxidant activity of an endophytic fungus *Nigrospora sphaerica* (EHL2), inhabiting the medicinal plant Euphorbia hirta (dudhi) L. Arch Microbiol 204(2):140. https://doi.org/10.1007/s00203-021-02650-7

Gebretsadik KG, Zhang Y, Chen J (2022) Screening and evaluation for antixenosis resistance in wheat accessions and varieties to grain aphid, Sitobion miscanthi (Takahashi) (Hemiptera: Aphididae). Plants (Basel) 11(8):1094. https://doi.org/10.3390/plants11081094

Ghabooli M, Khatabi B, Ahmadi FS et al (2013) Proteomics study reveals the molecular mechanisms underlying water stress tolerance induced by *Piriformospora indica* in barley. J Proteome 94:289–301. https://doi.org/10.1016/j.jprot.2013.09.017

Ghaffari MR, Ghabooli M, Khatabi B, Hajirezaei MR, Schweizer P, Salekdeh GH (2016) Metabolic and transcriptional response of central metabolism affected by root endophytic fungus *Piriformospora indica* under salinity in barley. Plant Mol Biol 90(6):699–717. https://doi.org/10.1007/s11103-016-0461-z

Gong Z, Xiong L, Shi H et al (2020) Plant abiotic stress response and nutrient use efficiency. Sci China Life Sci 63(5):635–674. https://doi.org/10.1007/s11427-020-1683-x

Gupta A, Meshram V, Gupta M, Goyal S, Qureshi KA, Jaremko M, Shukla KK (2023) Fungal endophytes: microfactories of novel bioactive compounds with therapeutic interventions; a comprehensive review on the biotechnological developments in the field of fungal endophytic biology over the last decade. Biomolecules 13:1038. https://doi.org/10.3390/biom13071038

Hallmann J, Quadt-Hallmann A, Mahaffee WF, Kloepper JW (1997) Bacterial endophytes in agricultural crops. Can J Microbiol 43(10):895–914. https://doi.org/10.1139/m97-131

Hamayun M, Khan SA, Khan AL et al (2010) Gibberellin production and plant growth promotion from pure cultures of *Cladosporium* sp. MH-6 isolated from cucumber (*Cucumis sativus* L.). Mycologia 102(5):989–995. https://doi.org/10.3852/09-261

Han M, Qin D, Ye T et al (2019) An endophytic fungus from *Trichoderma harzianum* SWUKD3.1610 that produces nigranoic acid and its analogues. Nat Prod Res 33(14):2079–2087. https://doi.org/10.1080/14786419.2018.1486311

Hardoim PR, Van Overbeek LS, Berg G et al (2015) The hidden world within plants: ecological and evolutionary considerations for defining functioning of microbial endophytes. Microbiol Mol Biol Rev 79(3):293–320. https://doi.org/10.1128/MMBR.00050-14

He W, Megharaj M, Wu C-Y, Subashchandrabose SR, Dai C-C (2020) Endophyte- assisted phytoremediation: mechanisms and current application strategies for soil mixed pollutants. Crit Rev Biotechnol 40(1):31–45

Hettiarachchige IK, Elkins AC, Reddy P, Mann RC, Guthridge KM, Sawbridge TI, Forster JW, Spangenberg GC (2019) Genetic modification of asexual Epichloë endophytes with the perA gene for peramine biosynthesis. Mol Genet Genomics 294:315–328

Hondelmann P, Paul C, Schreiner M, Meyhöfer R (2020) Importance of antixenosis and antibiosis resistance to the cabbage whitefly (*Aleyrodes proletella*) in brussels sprout cultivars. Insects 11(1):56

Howitz KT, Sinclair DA (2008) Xenohormesis: sensing the chemical cues of other spices. Cell 133:387–391

Hu HC, Tian F, Hu H (2011) Soil particle size distribution and its relationship with soil water and salt under mulched drip irrigation in Xinjiang of China. Sci China Technol Sci 54(6):1568–1574. https://doi.org/10.1007/s11431-010-4276-x

Hussain HA, Hussain S, Khaliq A et al (2018) Chilling and drought stresses in crop plants: implications, cross talk, and potential management opportunities. Front Plant Sci 9:393. https://doi.org/10.3389/fpls.2018.00393

Hussein ME, Mohamed OG, El-Fishawy AM et al (2022) Identification of antibacterial metabolites from endophytic fungus *Aspergillus fumigatus*, isolated from *Albizia lucidior* Leaves (Fabaceae), utilizing metabolomic and molecular docking techniques. Molecules 27(3):1117. https://doi.org/10.3390/molecules27031117

Ikram M, Ali N, Jan G, Hamayun M, Jan FG, Iqbal A (2019) Novel antimicrobial and antioxidative activity by endophytic *Penicillium roqueforti* and *Trichoderma reesei* isolated from *Solanum surattense*. Acta Physiol Plant 41(9):164. https://doi.org/10.1007/s11738-019-2957-z

Indrianingsih AW, Tachibana S (2017) Glucosidase inhibitor produced by an endophytic fungus, *Xylariaceae* sp QGS 01 from *Quercus gilva* Blume. Food Sci Hum Wellness 6:88–95

Johnson JM, Alex T, Oelmüller R (2014) *Piriformospora indica*: the versatile and multifunctional root endophytic fungus for enhanced yield and tolerance to biotic and abiotic stress in crop plants. J Trop Agric 52:103–122

Jonkers W, Rodriguez Estrada AER, Lee K, Breakspear A, May G, Kistler HC (2012) Metabolome and transcriptome of the interaction between *Ustilago maydis* and *Fusarium verticillioides* in vitro. Appl Environ Microbiol 78(10):3656–3667. https://doi.org/10.1128/AEM.07841-11

Kamel RA, Abdel-Razek AS, Hamed A et al (2020) Isoshamixanthone: a new pyrano xanthone from endophytic *Aspergillus* sp. ASCLA and absolute configuration of epiisoshamixanthone. Nat Prod Res 34(8):1080–1090. https://doi.org/10.1080/14786419.2018.1548458

Kaul S, Sharma T, Dhar MK (2016) 'Omics' tools for better understanding the plant-endophyte interactions. Front Plant Sci 7:955. https://doi.org/10.3389/fpls.2016.00955

Kaur T (2020) Fungal endophyte-host plant interactions: role in sustainable agriculture. In: Hasanuzzaman M, Teixeira Filho MCM, Fujita M, Nogueira TAR (eds) Sustainable crop production. IntechOpen, London, pp 211–228. https://doi.org/10.5772/intechopen.92367

Kaur N, Arora DS, Kalia N, Kaur M (2020a) Bioactive potential of endophytic fungus *Chaetomium globosum* and GC-MS analysis of its responsible components. Sci Rep 10(1):18792. https://doi.org/10.1038/s41598-020-75722-1

Kaur N, Arora DS, Kalia N, Kaur M (2020b) Antibioflm, antiproliferative, antioxidant and antimutagenic activities of an endophytic fungus *Aspergillus fumigatus* from *Moringa oleifera*. Mol Biol Rep 47:2901–2911

Khalil AMA, Hassan SE-D, Alsharif SM et al (2021) Isolation and characterization of fungal endophytes isolated from medicinal plant *Ephedra pachyclada* as plant growth-promoting. Biomol Ther 11(2):140. https://doi.org/10.3390/biom11020140

Khan NA, Khan MIR, Asgher M, Fatma M, Masood A, Syeed S (2014) Salinity tolerance in plants. revisiting the role of sulfur metabolites. J Plant Biochem Physiol 2:120–131

Khare E, Mishra J, Arora NV (2018) Multifaceted interactions between endophytes and plant: developments and prospects. Front Microbiol 9:2732

Khiralla A, Spina R, Varbanov M et al (2020) Evaluation of antiviral, antibacterial and antiproliferative activities of the endophytic fungus *Curvularia papendorfii*, and isolation of a new polyhydroxy acid. Microorganisms 8(9):1353. https://doi.org/10.3390/microorganisms8091353

Kouipou Toghueo RMK, Boyom FF (2019) Endophytic fungi from *Terminalia* species: a comprehensive review. J Fungi (Basel) 5(2):43. https://doi.org/10.3390/jof5020043

Krohn K, Ullah Z, Hussain H, Florke U, Schulz B, Draeger S, Pescitelli G, Salvadori P, Antus S, Kurtan T (2007) Massari lactones e-g, new metabolites from the endophytic fungus *Coniothyrium* sp., associated with the plant *Artemisia maritima*. Chirality 19:464–470

Kumar M (2014) Crop plants and abiotic stresses. J Biomol Res Ther 03(1):1000e125. https://doi.org/10.4172/2167-7956.1000e125

Kumari P, Singh A, Singh DK et al (2021) Isolation and purification of bioactive metabolites from an endophytic fungus *Penicillium citrinum* of *Azadirachta indica*. S Afr J Bot 139:449–457. https://doi.org/10.1016/j.sajb.2021.02.020

Kusari S, Spiteller M (2016) The promise of endophytic fungi as sustainable resource of biologically relevant pro-drugs: a focus on Cameroon. In: Fungi applications and management strategies. CRC Press, Boca Raton, pp 1–13

Kusari S, Hertweck C, Spiteller M (2012) Chemical ecology of endophytic fungi: origins of secondary metabolites. Chem Biol 19:792–798

Kusari S, Singh S, Jayabaskaran C (2014) Biotechnological potential of plant-associated endophytic fungi: hope versus hype. Trends Biotechnol 32:297–303

Leo VV, Passari AK et al (2016) A novel triculture system (CC3) for simultaneous Enzyme Production and hydrolysis of common grasses through submerged fermentation. Front Microbiol 7:44

Li HL, Li XM, Li X, Wang CY, Liu H, Kassack MU, Meng LH, Wang BG (2016) Antioxidant hydroanthraquinones from the marine algal-derived endophytic fungus *Talaromyces islandicus* EN-501. J Nat Prod 80:162–168

Li S-J, Zhang X, Wang X-H, Zhao C-Q (2018) Novel natural compounds from endophytic fungi with anticancer activity. Eur J Med Chem 156:316–343. https://doi.org/10.1016/j.ejmech.2018.07.015

Liarzi O, Bucki P, Braun Miyara SB, Ezra D (2016) Bioactive volatiles from an endophytic *Daldinia* cf. *concentrica* isolate affect the viability of the plant parasitic nematode *meloidogyne javanica*. PLoS One 11(12):e0168437. https://doi.org/10.1371/journal.pone.0168437

Liu S-S, Jiang J-X, Huang R et al (2019) A new antiviral 14-nordrimane sesquiterpenoid from an endophytic fungus Phoma sp. Phytochem Lett 29:75–78. https://doi.org/10.1016/j.phytol.2018.11.005

Lugtenberg BJJ, Caradus JR, Johnson LJ (2016) Fungal endophytes for sustainable crop production. FEMS Microbiol Ecol 92(12):fiw194. https://doi.org/10.1093/femsec/fiw194

Manganyi MC, Ateba CN (2020) Untapped potentials of endophytic fungi: a review of novel bioactive compounds with biological applications. Microorganisms 8:1934

Mazumder K, Ruma YN, Akter R, Aktar A, Hossain MM, Shahina Z, Mazumdar S, Kerr PG (2021) Identification of bioactive metabolites and evaluation of in vitro anti-inflammatory and in vivo antinociceptive and antiarthritic activities of endophyte fungi isolated from *Elaeocarpus floribundus* blume. J Ethnopharmacol 273:113975

Mejía LC, Herre EA, Sparks JP et al (2014) Pervasive effects of a dominant foliar endophytic fungus on host genetic and phenotypic expression in a tropical tree. Front Microbiol 5:479. https://doi.org/10.3389/fmicb.2014.00479

Meshram V, Gupta M (2019) Endophytic fungi: a quintessential source of potential bioactive compounds. In: Hodkinson T, Doohan F, Saunders M, Murphy B (eds) Sustain: endophytes for a growing world, vol 13. Cambridge University Press, Cambridge, pp 277–309

Meshram V, Kapoor N, Dwibedi V, Srivastava A, Saxena S (2022) Extracellular resveratrol producing endophytic fungus, Quambalaria cyanescens. S Afr J Bot 146:409–416

Meshram V, Elazar M, Maymon M, Sharma G, Shawahna R, Charuvi D, Freeman S (2023) Endophytic *Fusarium clavum* confers growth and salt tolerance in *Cucumis melo*. Environ Exp Bot 206:105–153

Nischitha R, Shivanna MB (2021) Metabolite fngerprinting, in vitro antimicrobial and antioxidant activities and in-silico docking in *Alloteropsis cimicina* and its endophytic fungus *Penicillium pinophilum*. Mol Biol Rep 48(5):4021–4037. https://doi.org/10.1007/s11033-021-06410-0

Nischitha R, Shivanna MB (2022) Screening of secondary metabolites and antioxidant potential of endophytic fungus *Penicillium citrinum* and host *Digitaria bicornis* by spectrophotometric and electrochemical methods. Arch Biol 204:206

Ortiz J, Soto J, Fuentes A, Herrera H, Meneses C, Arriagada C (2019) The endophytic fungus *Chaetomium cupreum* regulates expression of genes involved in the tolerance to metals and plant growth promotion in *Eucalyptus globulus* roots. Microorganisms 7(11):490. https://doi.org/10.3390/microorganisms7110490

Ozimek E, Hanaka A (2021) *Mortierella* species as the plant growth-promoting fungi present in the agricultural soils. Agriculture (Switzerland) 11(1):1–18. https://doi.org/10.3390/agriculture11010007

Pal PP, Shaik AB, Begum AS (2020) Prospective leads from endophytic fungi from anti-inflammatory drug discovery. Planta Med 86:941–959

Petrini O (1991) Fungal endophytes of tree leaves. In: Andrews JH, Hirano SS (eds) Microbial ecology of leaves. Springer, Cham, pp 179–197. https://doi.org/10.1007/978-1-4612-3168-4_9

Rajamanikyam M, Vadlapudi V, Amanchy R, Upadhyayula SM (2017) Endophytic fungi as novel resources of natural therapeutics. Braz Arch Biol Technol 60(0):e17160542. https://doi.org/10.1590/1678-4324-2017160542

Rana KL, Kour KL, Sheikh I et al (2019) Endophytic fungi: biodiversity, ecology significance, and potential industrial applications. In: Yadav A, Mishra S, Singh S, Gupta A (eds) Recent advancement in white biotechnology through fungi, vol 1. Springer, Cham, pp 1–62. https://doi.org/10.1007/978-3-030-10480-1_1

Redman RS, Kim YO, Woodward DACJ et al (2011) Increased fitness of rice plants to abiotic stress via habitat adapted symbiosis: a strategy for mitigating impacts of climate change. PLoS One 6(7):e14823. https://doi.org/10.1371/journal.pone.0014823

Rodriguez RJ, Henson J, Van Volkenburgh E et al (2008) Stress tolerance in plants via habitat-adapted symbiosis. ISME J 2(4):404.e416.R

Rustamova N, Bozorov K, Efferth T, Yili A (2020) Novel secondary metabolites from endophytic fungi: synthesis and biological properties. Phytochem Rev 19(2):425–448. https://doi.org/10.1007/s11101-020-09672-x

Saxena S, Meshram V, Kapoor N (2015) *Muscodor tigerii* sp. non-volatile antibiotic producing endophytic fungus from the Northeastern Himalayas. Ann Microbiol 65:47–57

Seleiman MF, Al-Suhaibani N, Ali N et al (2021) Drought stress impacts on plants and different approaches to alleviate its adverse effects. Plants (Basel) 10(2). https://doi.org/10.3390/plants10020259

Shankar Naik BS (2019) Functional roles of fungal endophytes in host fitness during stress conditions. Symbiosis 79(2):99–115. https://doi.org/10.1007/s13199-019-00635-1

Sharaf MH, Abdelaziz AM, Kalaba MH, Radwan AA, Hashem AH (2022) Antimicrobial, antioxidant, cytotoxic activities and phytochemical analysis of fungal endophytes isolated from *Ocimum basilicum*. Appl Biochem Biotechnol 194(3):1271–1289. https://doi.org/10.1007/s12010-021-03702-w

Shen M, Schneider H, Xu R et al (2020) Dark septate endophyte enhances maize cadmium (Cd) tolerance by the remodeled host cell walls and the altered Cd subcellular distribution. Environ Exp Bot 172:104000. https://doi.org/10.1016/j.envexpbot.2020.104000

Shrivastava N, Jiang L, Li PL et al (2018) Proteomic approach to understand the molecular physiology of symbiotic interaction between *Piriformospora indica* and *Brassica napus*. Sci Rep 8(1):5773. https://doi.org/10.1038/s41598-018-23994-z

Singh HP, Kaur G, Batish DR, Kohli RK (2011) Lead (Pb)-inhibited radicle emergence in Brassica campestris involves alterations in starch-metabolizing enzymes. Biol Trace Elem Res 144(1–3):1295–1301. https://doi.org/10.1007/s12011-011-9129-3

Song XQ, Zhang X, Han QJ, Li XB, Li G, Li RJ, Jiao Y, Zhou JC, Lou HX (2013) *Xanthone derivatives* from *Aspergillus sydowii*, an endophytic fungus from the liverwort *Scapania ciliata S. Lac* and their immunosuppressive activities. Phytochem Lett 6:318–321

Souza BS, Santos TT (2017) Endophytic fungi in economically important plants: ecological aspects, diversity and potential biotechnological applications. J Bioenergy Food Sci 4(2):113–126. https://doi.org/10.18067/jbfs.v4i2.121

Strobel G (2018) The emergence of endophytic microbes and their biological promise. J Fungi 4:57

Strong PJ, Xie S, Clarke WP (2015) Methane as a resource: can the methanotrophs add value. Environ Sci Technol 49(7):4001–4018. https://doi.org/10.1021/es504242n

Sunitha VH, Devi DN, Srinivas C (2013) Extracellular enzymatic activity of endophytic Fungal Strains isolated from medicinal plants. World J Agric Sci 9(1):01–09

Tan W-N, Nagarajan K, Lim V et al (2022) Metabolomics analysis and antioxidant potential of endophytic *Diaporthe fraxini* ED2 grown in different culture media. J Fungi 8(5):519. https://doi.org/10.3390/jof8050519

Tiwari P, Bae H (2020) Horizontal gene transfer and endophytes: an implication for the acquisition of novel traits. Plants 9:305. https://doi.org/10.3390/plants9030305

Tiwari P, Bae H (2022) Endophytic fungi: key insights, emerging prospects, and challenges in natural product drug discovery. Microorganisms 10:360. https://doi.org/10.3390/microorganisms10020360

Tiwari P, Adil M, Basavegowda N, Chen J (2022) Plant-associated endophytes: molecular mechanisms and significance in promoting sustainable agriculture. In: Tiwari P (ed) Endophytes: types, potential uses and mechanisms of action. Nova Publishers, New York

Ul-Hassan UR, Strobel GA, Booth E, Knighton B, Floerchinger C, Sears J (2012) Modulation of volatile organic compound formation in the mycodiesel-producing endophyte *Hypoxylon* sp. CI-4. Microbiology 158:465–473

Venugopalan A, Srivastava S (2015) Endophytes as in vitro production platforms of high value plant secondary metabolites. Biotechnol Adv 33:873–887

Verma VC, Singh SK, Kharwar RN (2012) Histological investigation of fungal Endophytes in healthy tissues of *Azadirachta indica* A, Juss. Kasetsart J (Nat Sci) 46:229–237

Verma V, Ravindran P, Kumar PP (2016) Plant hormone-mediated regulation of stress responses. BMC Plant Biol 16:86. https://doi.org/10.1186/s12870-016-0771-y

Vig R, Bhadra F, Gupta SK, Sairam K, Vasundhara M (2022) Neuroprotective effects of quercetin produced by an endophytic fungus *Nigrospora oryzae* isolated from *Tinospora cordifolia*. J Appl Microbiol 132(1):365–380. https://doi.org/10.1111/jam.15174

Wang ZN, Xiao Y, Chen WS, Tang KX, Zhang L (2010) Increased vitamin C content accompanied by an enhanced recycling pathway confers oxidative stress tolerance in *Arabidopsis*. J Integr Plant Biol 52(4):400–409. https://doi.org/10.1111/j.1744-7909.2010.00921.x

Wang L-W, Wang J-L, Chen J et al (2017) A novel derivative of (-) mycousnine produced by the endophytic fungus *Mycosphaerella nawae*, exhibits high and selective immunosuppressive activity on T cells. Front Microbiol 8:1251. https://doi.org/10.3389/fmicb.2017.01251

Wei PP, Ji JC, Ma XJ, Li ZH, Ai HL, Lei XX, Liu JK (2022) Three new pyrrole alkaloids from the endophytic fungus *Albifimbria viridis*. Nat Prod Bioprospect 12:5

Wicke B, Smeets E, Dornburg V et al (2011) The global technical and economic potential of bioenergy from salt-affected soils. Energy Environ Sci 4(8):2669–2681. https://doi.org/10.1039/C1EE01029H

Yamaji K, Watanabe Y, Masuya H, Shigeto A, Yui H, Haruma T (2016) Root fungal endophytes enhance heavy-metal stress tolerance of *Clethra barbinervis* growing naturally at mining sites via growth enhancement, promotion of nutrient uptake and decrease of heavy-metal concentration. PLoS One 11(12):e0169089. https://doi.org/10.1371/journal.pone.0169089

Yan Z, Huang C, Guo H et al (2020) Isobenzofuranone monomer and dimer derivatives from the mangrove endophytic fungus *Epicoccum nigrum* SCNU-F0002 possess α-glucosidase inhibitory and antioxidant activity. Bioorg Chem 94:103407. https://doi.org/10.1016/j.bioorg.2019.103407

Yuan J, Zhang W, Sun K et al (2019) Comparative transcriptomic of *Atractylodes lancea* in response to endophytic fungus *Gilmaniella* sp. AL12 reveals regulation in plant metabolism. Front Microbiol 10:1208. https://doi.org/10.3389/fmicb.2019.01208

Zhang W, Li J, Wei C, Deng X, Xu J (2021) Chemical epigenetic modifiers enhance the production of immunosuppressants from the endophytic fungus *Aspergillus fumigatus* isolated from *Cynodon dactylon*. Nat Prod Res 36:4481–4485

Zhao T, Xu LL, Zhang Y et al (2019) Three new α-pyrone derivatives from the plant endophytic fungus *Penicillium ochrochloronthe* and their antibacterial, antifungal, and cytotoxic activities. J Asian Nat Prod Res 21(9):851–858. https://doi.org/10.1080/10286020.2018.1495197

Zhao S, Wu X, Duan X et al (2021) Optimal extraction, purification and antioxidant activity of total flavonoids from endophytic fungi of *Conyza blinii* H. Lév. PeerJ 9:e11223

Zum FA, Mendoza A, Cabrera JA et al (2009) The burrowing nematode of banana: strategies for controlling the uncontrollable. Acta Hortic 828:101–108. https://doi.org/10.17660/ActaHortic.2009.828.9

Endophytic Fungi: A Biofactories for the Synthesis of Nanomaterials

16

Rishabh Anand Omar, Divya Chauhan, Neetu Talreja, and Mohammad Ashfaq

Abstract

Researchers continue to focus on synthesizing novel nanomaterials (NMs) by using environmentally friendly, nontoxic, and facile processes, like biological systems, to enhance the utilization of NMs. Preferential adoption of biological systems for NMs synthesis surpasses traditional chemical methods, which can pose environmental risks. Endophytes are organisms in micron size that sustain inside plant tissues, displaying a non-harmful relationship with the plants. Endophytic microorganisms (EMs) encompass bacteria, archaea, yeast, fungi, and viruses, and they can produce a diverse array of secondary metabolites, rendering them well-suited agents for the production of NPs. Recently, a novel development in research at the intersection of biology and nanotechnology has emerged through an endophyte-mediated synthesis of NPs. The green or endophytic driven NPs synthesis is one of the most ecological and eco-friendly methods regarded as potential biofactories. These fungi help synthesize NPs and improve stability, controlled size, and biocompatibility, significantly affecting the NPs' efficacy at end applications.

R. A. Omar
Centre for Environmental Science and Engineering, Indian Institute of Technology Kanpur, Kanpur, India

D. Chauhan
Department of Drinking Water and Sanitation, Ministry of Jal Shakti, New Delhi, India

N. Talreja (✉)
Department of Science, Faculty of Science and Technology, Alliance University, Anekal, Bengaluru, Karnataka, India

M. Ashfaq (✉)
Department of Biotechnology, University Centre for Research & Development, Chandigarh University, Gharaun, Mohali, Punjab, India

K. A. Abd-Elsalam, S. F. AbuQamar (eds.), *Fungal Endophytes Volume I*,
https://doi.org/10.1007/978-981-97-7312-1_16

Keywords

Nanomaterials · Green chemistry · Endophytic fungi · Antibacterial · Anticancer

16.1 Introduction

The advent of nanomaterials (NMs) and their applications in biological systems revolutionized numerous applications. NMs applications are spread as biomarkers, diagnosing tools, nano drugs, bioimaging agents, antimicrobial agents, and drug delivery systems (Malik et al. 2023; Patra et al. 2018; Anjum et al. 2021; Singh and Nalwa 2011). Nanotechnology has recently been used in the food agricultural sector. It plays a vital role in fertilizing crops and promoting plant growth by providing micro/macro elements to the plants. It also protects plants against various fungal and bacterial diseases and preserves food quality (Ashfaq et al. 2017; Gahoi et al. 2021). Researchers continuously focused on synthesizing NMs with new perspectives by environmentally friendly processes.

Researchers are focusing on synthesizing NMs using environmentally friendly, nontoxic, and facile processes like biological systems, to enhance the utilization of NMs. Preferential acceptance of biological systems for NMs synthesis surpasses traditional chemical methods, which can pose environmental risks (Ahmed et al. 2022; Peralta-Videa et al. 2016; Fadiji et al. 2022). The biological systems are often termed "green chemistry," as they integrate microbial biotechnology and nanotechnology approaches to produce NMs. Embracing these eco-friendly alternatives for NMs production will lead to more sustainable agriculture and food production systems in the long run (Fadiji et al. 2022; Iavicoli et al. 2017). Utilizing biological processes provides several advantages compared to chemical and physical methods, including the absence of toxic waste and low consumption of energy, making them cost-effective, environmentally responsive, and straightforward (Shah et al. 2018). NMs have diverse applications in various industries and technologies. These NMs haves at least one dimension below 100 nm (Bogas et al. 2022; Baker et al. 2015; Azmath et al. 2016). EMs can synthesize NMs through intracellular or extracellular processes, leading to the classification of NMs as intra/extracellular NMs (Rana et al. 2020).

Numerous NMs, including Se, Ag, Te, Au, Si, Pt, Ti, Pd, QDs, Zr, and magnetite-NPs, have been successfully biosynthesized using EMs (Bogas et al. 2022; Gupta et al. 2022). The versatility of microorganisms is evident in their rapid multiplication and adaptability to diverse environments, making them ideal candidates for the biological production of NMs, particularly when compared to plants or other larger organisms (Azmath et al. 2016). Moreover, microorganisms offer a cost-effective and efficient approach by biosynthesizing NMs in solution, simplifying filtration and separation processes for the NMs (Hulkoti and Taranath 2014). In this aspect, endophytes can produce various kinds of NMs. Undoubtedly, we can say that endophytic fungi emerged as biofactories for synthesizing nanoparticles (NPs).

Endophytes are plant tissue residing in micron-sized organisms, displaying a non-harmful relationship with the plants. However, recent studies indicate that some can undergo changes in their life modes and become pathogenic (Fadiji and Babalola 2020; Brader et al. 2017). These EMs encompass bacteria, archaea, yeast, fungi, and viruses, and they can produce a diverse type of secondary metabolites, rendering them well-suited agents for the production of NPs (Baker and Satish 2015). Several plant endophytes, including bacteria like *Penicillium citrinum, Colletotrichum* sp., *Bacillus cereus, Pseudomonas veronii,* and fungi like *Saccharomonospora* sp. and *Aspergillus fumigatus,* were identified as having the capability to synthesize NPs (Sunkar and Nachiyar 2012; Syed et al. 2016; Ahmad et al. 2003; Verma et al. 2013). Recently, a novel development in research at the intersection of biology and nanotechnology has emerged through an endophyte-mediated synthesis of NPs. EMs establish symbiotic relationships with plants (Rahman et al. 2017) and are cultured to isolate novel chemical compounds of significance for medicinal and industrial purposes (Schulz et al. 2002). Various plant species (~300,000) on Earth harbor EMs. The fungal endophytes are capable of producing substantial amounts of biomolecules and proteins that are crucial in preventing agglomeration and stabilizing NPs during their synthesis (Netala et al. 2016b). Leveraging this capability, it becomes feasible to manufacture various NPs efficiently with the potential for multiple applications (Kathiresan et al. 2009). Figure 16.1 shows the graphical illustration of the synthesis of Ag-NPs using endophytic fungi (EFs) and their application. This chapter discusses endophytic microorganisms, especially EFs, the biosynthesis of EFs-driven NPs, their applications, and future prospects.

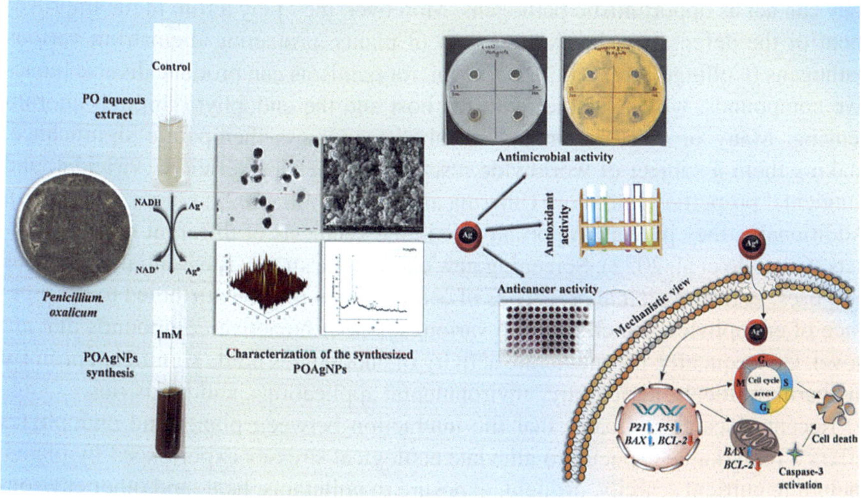

Fig. 16.1 Graphical illustration of the synthesis of Ag-NPs using EFs and their application. (The image was reproduced with permission (Gupta et al. 2022))

16.2 Endophytes

Over the past few decades, botanists have come to realize that plants harbor an incredibly vast array of organisms known as endophytes. Endophytic microbes, including bacteria and fungi, exist within plants either in intercellular or intracellular spaces, engaging in a symbiotic relationship between them. The term "endophytes" was coined by Bary in 1866, derived from the Greek words "endon," meaning "within," and "phyton," meaning "plants" (Collinge et al. 2022; Schulz 2006). Endophytes can colonize in various types of plants therefore they are widespread. Interactions of endophytes with the host plants involve both mutualistic and antagonistic relationships (Asaf et al. 2023; Huang et al. 2017; Bilański and Kowalski 2022). Notably, endophytes exhibit remarkable adaptability, aligning with the characteristics of their compounds that stimulate plant growth and aid in adapting to the plant's environment (Das and Varma 2009).

Endophytes have been extracted from various parts of plants, such as leaf, stem, root, petiole, bark, seeds, flowers, duds, resin ducts, scale primordia, meristem, and fruits (De Melo Pereira et al. 2012). Typically, a higher concentration of endophytes is found in the belowground sections of plants, particularly in the roots, compared to aboveground parts. It has been demonstrated that these microorganisms can translocate from the roots to the leaves through the xylem (Frank et al. 2017). The diversity of endophytes is vast, differing from species of one plant to species of another plant or even within the same species of plant having different environments (Chareprasert et al. 2006). Endophytes exhibit various relationships with their host plants; commonly, they engage in a mutualistic relationship, but in some instances, they can act as opportunistic pathogens. Moreover, they play a role in the improvement of the defense mechanisms system of plants, protecting them from various pathogens (Collinge et al. 2022). These microorganisms can produce diverse bioactive compounds, which interact with the host and the endophyte through quorum sensing. Many of these secondary metabolites possess therapeutic significance, making them a subject of worldwide research for their bactericidal, viricidal, and fungicidal properties, and more (Sharma and Mallubhotla 2022; Gupta et al. 2023). Additionally, they play an important role in the synthesis of different natural products (Conrado et al. 2022). Screening new endophytic strains is essential for various purposes, as the medicinal properties of certain plants may be attributed to the presence of endophytes, which produce various types of bioactive compounds that are novel. Consequently, the metabolic activity of endophytes holds significant promise in pharmaceuticals, agriculture, environmental applications, and industries.

Recent research indicates that the interaction between plants and endophytes offers a remarkable capacity to alleviate ecological stresses experienced by plants, including nutrient scarcity, drought, exposure to pollutants, heat, and other environmental pressures (Rho et al. 2018; Byregowda et al. 2022). Both plants and animals host diverse types of microbes that play various roles, such as stimulating the immune system and promoting development (Attia et al. 2022). Endophytes play an important role in plant growth promotion by producing plant-regulating hormones, modifying the production of endogenous hormones, and enhancing the availability of essential nutrients like nitrogen and phosphorus (Glick 2012; Ahemad 2015). Moreover, they can trigger a latent defense mechanism known as induced systemic resistance, significantly increasing protection against a wide range of pathogens (Sharma et al. 2023). Despite the potential benefits of endophytes, certain limitations exist in this field, such as challenges related to the reinoculation of endophytes into plants, utilizing endophytes for their nutritional benefits, determining whether endophytes are cultivar-specific, and the need for extensive research to implement practical applications (Moronta-Barrios et al. 2018; Afzal et al. 2019). One of the main issues lies in evaluating and classifying endophytes understanding their multiple interactions, nutrient exchange processes, and systemic effects (Bonfante 2018).

16.3 Endophytic Fungi

Many fungal species exist on Earth, estimated to be around 1 million (Blackwell 2011). In the last century, it was demonstrated that the association of fungi with various higher plants is approximately 0.1 million in count., which are associated either as saprophytes or parasites, found on decaying or deceased organic matter (Rochín-Hernández et al. 2022). Growing evidence highlights the significance of mycorrhizal fungi in promoting plant growth. EFs play a crucial role in biologically active compound production that could be another source for manufacturing plant-consequent biologically active substances. These compounds possess diverse properties, including anticancer, antioxidant, antimicrobial, and immunomodulating activities (Netala et al. 2016a). EFs have garnered significant attention in applied microbiology, prompting numerous experiments to explore their important role in enhancing the survival and development of plants (Hiruma et al. 2018; Toju et al. 2016). These experiments reveal that many root EFs promote plant growth by either suppressing pathogens or enhancing the nutritional conditions in the rhizosphere (Newsham 2011). The influence of various factors, such as host plants, fungal strains, soil type, and adeptness, plays an important role in these interactions (Svenningsen et al. 2018). Furthermore, EFs aid plants in tolerating both abiotic and biotic stresses while enhancing the nutrition of plants, particularly in nutrient-limiting conditions. On the basis, o of transmission the endophytes are further categorized into two categories, namely, horizontally transmitted endophytes or seed-borne endophytes (Perotto et al. 2018).

16.4 Nanomaterials (NMs)

The term "nano" is derived from the Greek word "nanos," which means dwarf or extremely small. In the context of nanotechnology, "nano" represents one billionth of a meter. Nanotechnology involves innovative approaches to synthesize, design, and manipulate elements on a small scale ranging from 1 to 100 nm (Khatami et al. 2018). NPs possess unique physicochemical properties due to their high surface-to-volume ratio and nano-size. NPs research has become integral to various applications, including chemical imaging photonics, electronics, catalysis, information storage, sensing, ecological remediation, nanomedicines, biological labeling, and drug delivery (Afreen et al. 2018). Multiple processes have been employed for the synthesis of NPs, which are classified into three processes: (1) physical, (2) chemical, and (3) biological (Singla et al. 2019; Omar et al. 2019). With the increment in environmental pollution in the past few years, biological approaches are considered safe since they do not involve harmful chemicals that can contaminate waste streams (Omar et al. 2022). Chemical methods utilize or produce highly toxic chemicals and hazardous materials, which limits their applicability in biomedicine (Ying et al. 2022). Physical methods require substantial energy, making them less environmentally friendly (Abid et al. 2022). Due to concerns about the environment and energy efficiency, biological methods have gained popularity. Additionally, physical and chemical means of NPs synthesis are costly and often have a low production rate. Large-scale production using these methods can lead to instability and polydispersity, especially during a reduction in liquid media. In contrast, the bio-based method offers numerous pros, such as being economically efficient, easily scalable for large-scale production, and not requiring high temperature, pressure, or toxic chemicals (Yang et al. 2022). Microbes play a crucial role in biological methods, as they can absorb and accumulate metals (Singh et al. 2017).

16.5 Classification of EFs Used for the Synthesis of Nanomaterials

Endophytes, often fungi and bacteria, reside within living plant cells. EFs, an endosymbiont, coexist with plants with a commensalism relationship. These relatively unexplored organisms produce valuable metabolic products beneficial for agricultural and drug industries (Petrini et al. 1992). EFs are widespread and have been identified in various plant types (Arnold et al. 2000). Even within the same host, different strains of these fungi can display varying capabilities in utilizing different substances. Consequently, it is possible to isolate endophytes from diverse plants belonging to different classes or families and cultivate them in various environments and habitats (Petrini et al. 1992). The genetic makeup of both organisms influences the type of relationship between an endophyte and a plant and is further influenced by the surrounding environment (Moricca and Ragazzi 2008).

Fungal endophytes are categorized into four classes based on their host range, tissue colonization, planta presence, transmission, diversity, and the significance they provide to their hosts (Rodriguez et al. 2009). (1) Class 1 endophytes, typically *Clavicipitaceous* endophytes, consist of a limited number of *Clavicipitaceous* species which is phylogenetically related. They are delicate and challenging to culture, mainly found in certain cool and warm season grasses (White et al. 2004). (2) Class II endophytes encompass diverse species, all belonging to the Dikarya group (*Ascomycota* or *Basidiomycota*). These endophytes can bestow specific stress tolerance adaptations to their host plants (Rodriguez et al. 2008). (3) Endophytes in

Class III are differentiated on the basis of their transmission patterns and occurrence. They are found in vascular and nonvascular plants, both woody and herbaceous angiosperms, particularly in Antarctic communities. These endophytes are notable for their diversity within individual host tissues, plants, and populations (Rodriguez et al. 2009; Harrison and Griffin 2020; Zhang and Yao 2015). (4) in the IVth class, the endophytes are limited to the roots of the plants and show dark-colored melanized septa. *Ascomycetous* fungi belonging to class IV can be sterile or conidial and inside inter- and intracellular forms hyphae melanized-like structures, as well as microsclerotia inside the roots. In this class, endophytes are found plants act as host, particularly nonmycorrhizal ones, in various ecosystems such as Antarctic, alpine, subalpine temperate zones, and tropical regions (Jumpponen 2001).

EFs have been considered promising sources of natural products with significant biotechnological importance, capable of producing substances that safeguard plants from insect attacks and diseases, and possess potential medicinal applications (Borges et al. 2009). Several species of EFs were determined as potential sources of several compounds used for the treatment of diabetes, cancer, and immunosuppressive conditions, and also as an insect (Strobel and Daisy 2003). Some fungi among them generate secondary metabolites exhibiting antimicrobial anticancer properties (Xie et al. 2018). For instance, *Fusarium oxysporum*, an endophyte isolated from medicinal plants, has demonstrated antimicrobial and anticancer activity (Xie et al. 2018; Strobel and Daisy 2003). Additionally, certain EFs produce metabolites similar to those found in the host plant, which possess therapeutic functions, such as steroids, alkaloids, flavonoids, terpenoids, quinones, phenolic acids, and peptides (Hashem et al. 2023; Kaur et al. 2020). Shankar et al. (Shankar et al. 2003) have described a procedure of gold nanomaterial synthesis using EFs, which is as follows: The first step involved disinfection of the plant leaf surface from which endophytes were isolated. For disinfection, leaves were cut and immersed in 0.01% sterilized mercuric chloride, and successive washes with sterilized DI water were made. Further isolation from the sterilized leaf was done on potato dextrose agar plates under sterilized conditions. The plates were incubated for 8 days at 25 ± 2 °C. After the fungal growth on potato dextrose agar plates, some of the fungal colonies were transferred to the liquid medium containing potato dextrose broth/malt, glucose, peptone, and yeast extract. The liquid medium was incubated for another 8 days under shaking conditions (200 rpm) at the same temperature used previously. After incubation, the mycelia of the grown EFs were separated by centrifuging the medium at 5000 rpm for 20 min at 20 °C temperature. The separated mycelia of fungi were washed three times with sterilized DI water. Further, the synthesis of nanoparticles from the isolated fungi was done. For the synthesis, approximately 10 g of harvested wet fungal biomass was taken. 1 mM of gold chloride solution was mixed with the wet biomass. The mixture was incubated for several days at 25–27 °C under stirring conditions at 200 rpm. Samples were collected at different time intervals from the solution. The solution confirmation was made by determining the physicochemical properties of the synthesized NPs (Raghupathi et al. 2011).

16.6 EFs Involve in NPs Synthesis

Using mycoendophytes as a biosynthetic technique represents a groundbreaking approach to developing efficient and cost-effective NPs production methods. EFs are involved extensively in NPs synthesis because of their wide range of enzyme production, significantly enhancing NPs productivity (Raghupathi et al. 2011). Recently, mycoendophytes have been harnessed to produce NPs from various parts of medicinal plants, such as stems, leaves, and roots (Ramalingmam et al. 2015). The most commonly utilized method for NPs production is the extracellular approach. Although intracellular synthesis allows for better control over size and shape, it is time-consuming and expensive to collect and recover the NPs. In the extracellular approach, the enzyme nitrate reductase, essential for creating Ag-NPs, is released into the medium, converting Ag^+ ions into Ag-NPs (Tran et al. 2013). Therefore, when it comes to producing Ag-NPs from fungi, the extracellular approach has proven to be the most successful method.

16.6.1 Zinc Oxide (ZnO)-NPs

The microorganism-derived synthesis of ZnO-NPs is relatively not explored till now. Only some studies have indicated that certain microorganisms, including yeast, bacterial, and fungal strains, can produce ZnO-NPs. Among fungi, only a limited number of genera have been found to have this NPs synthesis ability (Mohd Yusof et al. 2019). ZnO-NPs possess excellent antibacterial and antifungal properties because of their large surface-to-volume ratio, which enhances their interaction with pathogens (Raghunath and Perumal 2017). Fungal extracts with secondary metabolites act as a surrounding material, polymerizing to form ZnO-NPs (Ganesan et al. 2020). The use of fungi in NPs production offers the advantage of easy scalability, as fungi are efficient producers of extracellular enzymes (Ganesan et al. 2020). In specific instances, zinc sulfide quantum dots were synthesized by a fungus, namely, *A. flavus*. The average diameter of dots was determined ~18.5–58.7 nm (Uddandarao and Balakrishnan 2017). Another study by Uddandarao et al. (2019) showed biosynthesis of zinc sulfide nanoparticles (12–24 nm) and a composite of $ZnSo_4$ sulfide and gold semiconductor (10–18 nm) NPs using *A. flavus* as the endophytic fungus (Uddandarao et al. 2019). *Cochliobolus geniculatus*, an endophytic fungus tolerant to Zn, separated from *Nothapodytes foetida* which is a medicinal plant, and showed potential in generating proteins that aid in ZnO-NPs production (Kadam et al. 2019). Similarly, *Periconium* sp. extract also effectively mediated the sol-gel synthesis of ZnO-NPs, producing nanoparticles which having a quasi-spherical shape with an approximate size of 16–78 nm (Ganesan et al. 2020). Some hexagonal zinc oxide NPs (34–55 nm) were also synthesized using *Xylaria acuta* extract (Sumanth et al. 2020). Figure 16.2 Shows the TEM and SAED patterns of zinc oxide NPs. The TEM micrograph confirms that the synthesized ZnO-NPs ranged from 30 to 50 nm.

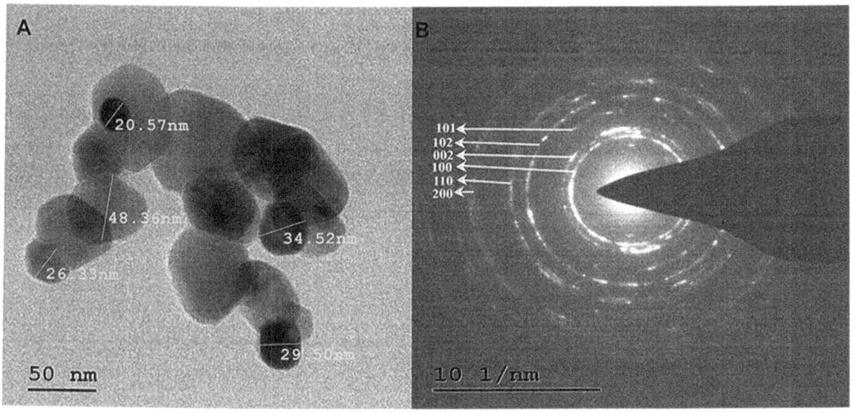

Fig. 16.2 (**a**) TEM and (**b**) SAED patterns of ZnO-NPs prepared by *Xylaria acuta* fungal extract. (The image was taken with the permission (Sumanth et al. 2020))

16.6.2 Gold (Au)-NPs

Gold NPs have gained popularity in medicine because of their nontoxic nature and exceptional absorption properties. They are widely recognized as valuable biosensors and potential agents for fighting cancer (Dykman and Khlebtsov 2011; Hammami et al. 2021; Lim et al. 2011). *Penicillium citrinum*, isolated from brown algae, produced Au-NPs, with most particles falling within the 60–80 nm size range. Cytosolic extract from the endophytic fungus was vital in reducing Au ions. Fourier-transform infrared spectra (FTIR) showed distinct peaks corresponding to various surface functional groups (Manjunath et al. 2017). Similarly, another endophytic fungus, *C. cladosporioides*, found in a plant, namely, *Solanum wightii*, used to produce Au-NPs biologically with an average size of 60 nm (Manjunath Hulikere et al. 2017). Employing a green technique of synthesis with *Fusarium solani* ATLOY-8 (the endophytic strain) residing in the *Chonemorpha fragrans* plant, Au-NPs were synthesized, displaying a pink (ruby red) color with a 510–560 nm range plasmon band between. The XRD pattern confirmed the crystalline nature of the NPs, and SEM analysis revealed an approximate size between 40 and 45 nm (Clarance et al. 2020). Moreover, Au-NPs were produced using *Cladosporium* sp. Extract in liquid form resides in *Commiphora wightii* plant. These NPs have approximate sizes of 5–10 nm formed in a spherical shape and peaking at 524 nm absorbance. XRD analysis confirmed their crystalline structure and EDX analysis indicated the Au and O atoms were present (Munawer et al. 2020).

16.6.3 Silver (Ag)-NPs

Like zinc oxide and gold-NPs, silver-NPs have been frequently used in numerous end applications and synthesized by using EFs (Guilger-Casagrande and de Lima

2019). For instance, Musarrat et al. reported a filamentous endophytic fungus, namely, *Amylomyces rouxii*. The fungus can produce Ag-NPs from a plant, namely, *Phoenix dactylifera*. Ag-NPs have been recognized for their microbicidal efficacy against both Gram-positive and negative bacteria and also plant fungal pathogens (Musarrat et al. 2010). Devi et al. reported that *Penicillium* sp., *Centella asiatica* plant's endophye, can synthesize Ag-NPs with a size of 100 nm, as observed through SEM. Two fungal strains, namely, *Aspergillus niger* PFR6 and *Aspergillus tamarii* PFL2 were used by Devi and Joshi isolated from *P. fulgens* L., to synthesize Ag-NPs, which showing peaks at 419 nm absorbance and 430 nm absorbance, respectively, and appeared as spherical particles. Singh et al. (Singh et al. 2013) used roots of Aloe vera and plants of *Curcuma longa* to prepare Ag-NPs from various species of *Penicillium*, with particle size 15–40 nm. Ag-NPs were generated from *Penicillium* species isolated from *Glycosmis mauritiana*, appearing as spherical NPs with various sizes determined at 50000 magnification at two places (Govindappa et al. 2016). *A. terreus*, an endophytic fungus isolated from *Calotropis procera*, synthesized oval to spherical Ag-NPs with an average size of 16.54 nm (Rani et al. 2017). *Phomopsis liquidambaris* (an endophytic fungus) was employed to produce Ag-NPs, with a sharp absorption spectrum at 430 nm, indicating the formation of spherical, polydispersed, and stable NPs with an approximate diameter of 18.7 nm (Seetharaman et al. 2018). *Cladosporium cladosporioides* biogenically produced Ag-NPs from *Sargassum wightii*, brown algae, with particle sizes below 100 nm (Manjunath Hulikere and Joshi 2019). Popli et al. prepared and characterized Ag-NPs from *Loranthus micranthus* residing endophyte's liquid extract, observing an absorbance of Ag-NPs at 438 nm (Popli et al. 2018). The endophytic fungus *Nemania* sp. was a good source of Ag-NPs, with a surface plasmon resonance band at 460 nm and a size range of 5–70 nm (Farsi and Farokhi 2018). Ag-NPs were produced in a one-step process using the endophytic fungus *Botryosphaeria rhodina*, resulting in well-distributed Ag-NPs with spherical shape (Akther et al. 2019b). *Talaromyces purpureogenus* mycelial extract (MEEF) was used in the green synthesis of Ag-NPs, showing an absorption peak from 390 to 420 nm and confirming the successful formation of Tp-Ag-NPs (Hu et al. 2019). Ag-NPs were synthesized from an endophyte, namely, *Cladosporium perangustum*, that appeared 30–40 nm spherical particles (Govindappa et al. 2020). *Curculigo orchioides* residing endophyte, namely, *Penicillium cinnamopurpureum*, produced spherical and polydispersed Ag-NPs with an average size of 30 nm (Dinesh et al. 2022). *Aspergillus austroafricanus* CGJ-B3 endophyte, synthesized s2–51 nm spherical Ag-NPs (Danagoudar et al. 2021). *Penicillium oxalicum*, another endophytic fungus, was shown to produce Ag-NPs with a strong peak 430 nm absorbance and a cubic crystalline structure (face-centered) (Seetharaman et al. 2021). Ag-NPs (66–88 nm) synthesized from the *Dendrophthoe falcata* plant residing endophytic fungus *Alternaria alternata* (Govindappa et al. 2022). It is essential to mention here that these fungi help synthesize NPs and improve stability, controlled size, and biocompatibility, which significantly affect the efficacy of the NPs in end applications. Figure 16.3 shows the capped and uncapped Ag-NPs.

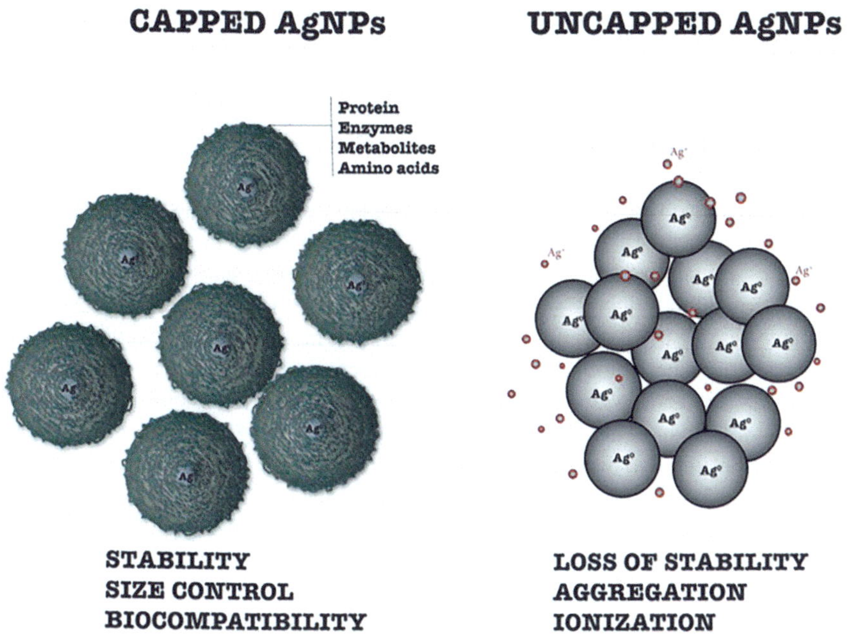

Fig. 16.3 Capped and uncapped Ag-NPs. (The image was taken with the permission (Guilger-Casagrande and de Lima 2019))

16.6.4 Iron (Fe)-NPs

To synthesize Fe-NPs, *Tecomella undulata* residing *P. oxalicum* (an endophytic fungus), was utilized. The prepared Fe-NPs underwent various characterization methods. A distinct absorption band at 300 nm, typical of Fe-NPs, was observed by the absorption spectra of the UV-vis spectrophotometer. The FTIR spectrum exhibited absorption peaks ranging from 3430.70 to 466.83/cm, indicating Fe-NPs. The SEM images revealed spherical particles with 140 nm of diameter. Additionally, AFM pictures indicated a particle diameter of 3.9 nm and a height of 100 nm. The mycosynthesized Fe-NPs were employed to decolorize methylene blue dye in H_2O_2. Within 6 h, approximately 99.17% dye decolorization was determined. This process, arbitrated by the mycosynthesized Fe-NPs, offers a promising solution to eliminate the hazardous effluents from textiles for a longer period (Mathur et al. 2021).

16.6.5 Cupric Oxide (CuO)-NPs

CuO-NPs were produced through the utilization *Aegle marmelos* plant (a medicinal tree) residing endophyte *A. terreus*. CuO-NPs indicated peaks within the range of 200–700 nm in UV-vis analysis. Spectral analysis revealed an absorption band of

CuO-NPs around 551 nm. The size of the CuO-NPs was below 100 nm (Mani et al. 2021).

16.7 Mechanism of Fungal-Driven NPs Synthesis

Despite extensive research on the microorganisms-derived green synthesis of NPs, the specific processes and biochemical pathways involved in metal nanoparticle production remain relatively unknown. Microbial enzymes and secondary metabolites are crucial in reducing metal ions into NPs, both the inner side of the cell and the outer side of the extracellular environment. When exposed to metal ion solutions, bacteria release biomolecules and enzymes that possess the potential to reduce metal salts, converting toxic metallic ions into less harmful metal NPs (Kazemi et al. 2023; Osman et al. 2024; Bahrulolum et al. 2021). Microorganisms can produce NPs through two distinct processes. Intracellular biosynthesis involves the use of proteins, enzymes, and cofactors within the cell. Also, extracellular biogenesis involves the secretion of biomolecules that interact with the cell wall with functional groups to facilitate NPs formation (Li et al. 2011; Lahiri et al. 2021). Both intracellular and extracellular pathways can be mediated by various organic compounds, including enzymes released inner or outer side of the cell, which enhances metal reduction. Although the exact mechanisms by which NPs produced from biological sources are not fully understood, it is believed that key routes include reduction, enzymatic oxidation, adsorption, and chelation with cell walls peptides or polysaccharides (Ramrakhiani and Ghosh 2018; Sharma et al. 2019; Slavin et al. 2017). The NPs biosynthesis can be considered a bottom-up strategy since NPs are produced through oxidation or reduction interactions among metal ions and cellular chemical derivatives. The process involves the metal ion uptake through biomass, followed by their conversion into elemental form, which is a dynamic process. The ions included into the cell, at the time of intracellular biosynthesis, are regulated by activities of the plasma membrane and cell wall, and facilitated by reductive enzymes, cofactors, and carrier proteins (Salunke et al. 2016; Ghosh et al. 2021; Zhang et al. 2011). Metabolic processes, including enzyme-mediated reductions, take place in the cytoplasm, resulting in the formation of metal NPs. Metal ions can undergo extracellular biosynthesis by interacting with the cell wall's functional chemical groups or through the release of metabolites. Intracellular NPs biosynthesis is usually dependent on the metabolism of the cell, requiring specific circumstances for bioactivity and growth, which means that microbes should be alive. Parameters, such as pressure, culture media, contamination, salinity, metal concentrations, and temperature, are crucial factors during biological synthesis. An advantage is that the approach is capable to produce metal sulfide and metal oxide NPs. Also, in extracellular biosynthesis, there is no requirement for microorganism to be alive during the process, making it a more feasible option for industrial metal reduction processes. The control of process variables is better and simpler with dead cells, as they are less harmful, and metal recovery and biomass renewal and recycling are easier (Koul et al. 2021; Atalah et al. 2022). The advantages of obtaining metal NPs from biological sources lie in the ease of producing dedicated products in cleaner and more cost-effective ways. Biological synthesis is simpler than the other processes (chemical or physical) due to operating parameters, such as the use of biomass (living or dead) and ambient or near ambient pressure and temperature for biological (Karunakaran et al. 2023; Osman et al. 2024; Mukherjee and Patra 2017). For intracellular biosynthesis, cell lysis is used to recover nanoparticles depending on the metabolism of the microorganisms. In contrast, when extracellular mechanisms are responsible for biosynthesis, the recovery of nanoparticles is simplified, involving solid-liquid separation methods like centrifugation and filtration (Gahlawat and Choudhury 2019; Salvadori et al. 2014). The versatility of the method is exhibited by the ability to synthesize various metal-NPs (metals, sulfides, and oxides of metal) using microbes and different metal salts. The stability of NPs is crucial for their application, as their minor size tends to cause them to clump together, affecting their behavior. Microorganisms or plant extract-derived NPs utilize covering and stabilizing compounds to inhibit agglomeration (Pandit et al. 2022; Singh et al. 2018).

16.8 Applications of EFs-Driven NPs

16.8.1 Antibacterial

Ag-NPs possess antimicrobial mechanisms, including the suppression of DNA and mRNA synthesis, rapture to the cell membrane resulting in cell constituent leakage, suppression of protein formation, cell synthesis disruption, dysfunction of mitochondria, and suppression of the ETS (electron transport system) chain (Bogas et al. 2022). The endophytic fungus *A. rouxii*, residing in *P. dactylifera* plant, produces Ag-NPs with wide bactericidal effect against both Gram-positive and negative bacteria, and also plant fungal pathogens (Musarrat et al. 2010). Similarly, *A. clavatus*, an endophytic fungus, generates Ag-NPs that inhibit *Pseudomonas fluorescens* and *Escherichia coli* growth (Verma et al. 2010). Ag-NPs produced by other EFs, such as *Fusarium* sp., *Cryptosporiopsis ericae* PS4 and *A. terreus*, show antibacterial efficacy against various bacterial strains (Devi and Joshi 2014; Rani et al. 2017). Furthermore, Au-NPs produced by *C. cladosporioides* demonstrate significant antibacterial activity against certain bacterial species (Manjunath Hulikere et al. 2017). In addition Ag-NPs, and CuO-NPs produced by *A. terreus* exhibit high bactericidal activity against specific bacterial strains (Mani et al. 2021). CuO-NPs exert their antibacterial effect through the generation of reactive species of oxygen species (ROS), causing cellular membrane disruption, oxidation of proteins, peroxidation of lipids, and DNA degradation in bacterial cells (Angelé-Martínez et al. 2017). Figure 16.4 shows the schematic representation of the microbicidal mechanism of metal-NPs synthesized by EFs. Overall, biogenic nanoparticles produced by EFs show promising antimicrobial properties against various bacterial species and have the potential to be used as alternatives to traditional antibiotics in certain applications.

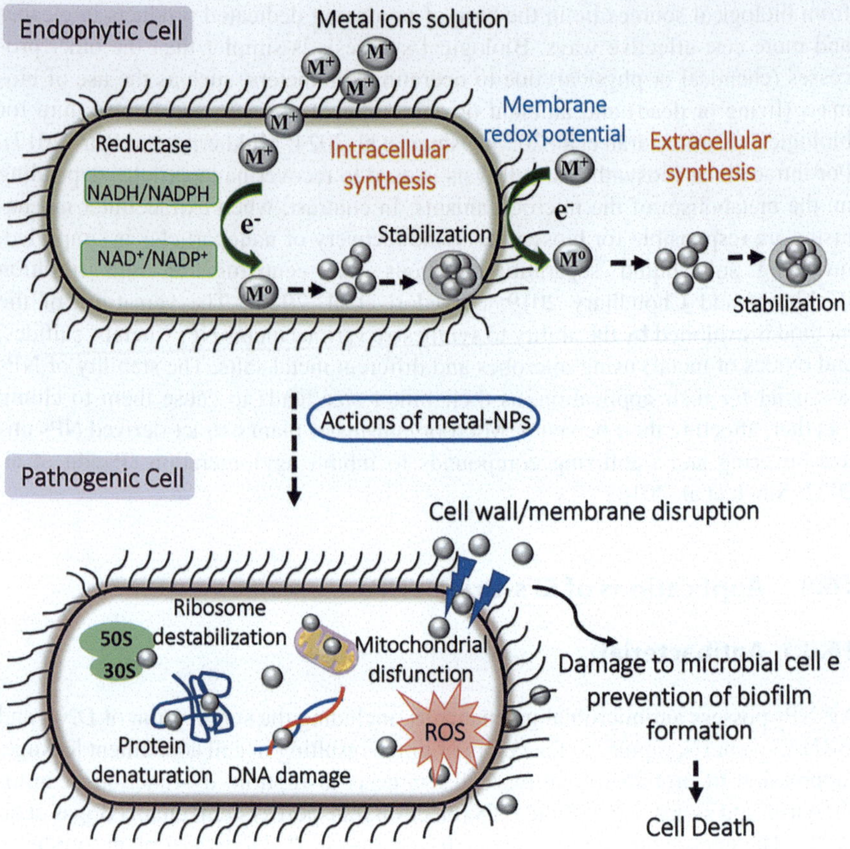

Fig. 16.4 A schematic representation of the microbicidal mechanism of metal-NPs synthesized by EFs. (The figure was reproduced with permission (Bogas et al. 2022))

16.8.2 Antifungal

NPs have emerged as a promising approach for managing phytopathogenic fungi in agriculture due to their excellent antifungal efficacy against various types of fungi (Arciniegas-Grijalba et al. 2017). The antifungal activity of NPs is affected by various factors, such as their dimension, shape, allocation, agglomeration, content, surface chemistry, and crystallinity (Koduru et al. 2018). Smaller NPs with a higher surface area-to-volume ratio tend to exhibit enhanced antifungal activity (Rai et al. 2018). Synthesis methods play a crucial role in controlling these parameters (Rafique et al. 2017), but the presence of metal precursors or surfactants in the final product can impact the surface chemistry of NPs and their antifungal activity (Alghuthaymi et al. 2015). The antifungal mechanisms of metal NPs on phytopathogenic fungi involve various processes. These include the release of ions from NPs that bind to specific proteins, affecting cell permeability and essential membrane proteins. NPs can hinder the germination and development of conidia of fungi intervening with ETS, oxidation of protein, and potential of the membrane. Additionally, they may affect mitochondrial membrane potential, induce ROS generation, and cause severe protein damage, disruption of membranes, and rapturing DNA, ultimately leading to cell death (Zhao et al. 2018; Mikhailova 2020). While the antifungal potential of NPs has been demonstrated, there is still limited research on their influence on fungal infections. For example, Ag-NPs have shown bactericidal effect *Candida albicans* and have potential for the development of antifungal medicines (Verma et al. 2010). Au-NPs derived from the endophytic fungus *C. cladosporioides* have displayed antifungal activity against *A. niger* (MTCC 281) (Manjunath Hulikere et al. 2017). ZnO-NPs synthesized from extract endophytic fungi have also showed significant antifungal efficacy against various fungal pathogen of plants (Sumanth et al. 2020). In conclusion, NPs offer a promising avenue for managing phytopathogenic fungi, and their antifungal mechanisms warrant further investigation to fully understand their potential applications in agriculture and medicine. CuO-NPs synthesized from the endophytic fungus *A. terreus* exhibit lower antifungal efficacy against *C. albicans* and *A. niger* contrasted with their bactericidal activity against *V. cholerae, E. coli.*, and *Proteus mirabilis*. This disparity can be attributed to the fungal cell wall's rigidity, primarily composed of N-acetylglucosamine, which creates a challenge for metal-NPs to penetrate. However, with repeated doses over time, the cell wall can be damaged, allowing the pathogenic fungus to be susceptible to the CuO-NPs. The study found that CuO-NPs effectively prevent or delay the bacterial growth of Gram-positive and negative bacteria due to their ability to modify the potential of cell membranes, interrupting transport and promoting cell death (Mani et al. 2021). In fungus, CuO-NPs absorb into the cells of fungi and release ions of copper, which in turn degrade the DNA structure, leading to cell death. The cell membrane's functional groups mainly amine and carboxylic groups can interact with Cu ions, providing a route for entry of Cu++ ions to the cell and exert their effect (Qamar et al. 2020). Also, ZnO-NPs synthesized using an endophytic fungal extract (*Periconium* sp.) demonstrate potent fungicidal activity against *C. albicans* and showed an approximately 24 mm inhibition zone (Ganesan et al. 2020).

16.8.3 Antiviral

The entry and attachment of a virus to host cells rely on interactions between viral surface components and ligands/proteins on the cell membrane. Developing antiviral medications that disrupt these connections, thereby reducing the attachment of the virus and its cellular penetration, is a promising strategy. Among the metal NPs studied for their action against microorganisms, Ag-NPs have emerged as a particularly promising antiviral agent. The complexity of virus structures has made it challenging to understand how nanoparticles interact with viruses. Ag-NPs have been shown to combat viruses through two main mechanisms, which are binding to the external coat proteins, preventing the virus from attaching to cell receptors, and binding to the nucleic acid of the virus, inhibiting replication or multiplication inside host cells (Salleh et al. 2020). Ag-NPs can modify the surface proteins' shape, decreasing their detectability and preventing them from adhering to host receptors. Despite significant efforts to develop vaccines and drugs against infections caused by viruses, complete prevention of viral diseases remains elusive. Therefore, there is a requirement of the synthesis of effective drug for abatement of a variety of viruses. Limited cases have shown the viricidal activity of NPs, synthesized biologically using bacteria and fungi. *Aspergillus* sp. derived Ag-NPs were showed bacteriophage virus abated, and *A. niger* derived Ag-NPs efficiently reduce the development of virus in the *E. coli* as a host bacterium. (Galdiero et al. 2011). Gaikwad et al. also reported that Ag-NPs can protect against human parainfluenza virus type 3 and herpes simplex virus. Thus, the interactions between biosynthesized NPs and viruses are gaining importance. Although reports of EFs producing antiviral biosynthesized NPs are scarce, future efforts can target EFs for NPs production and the development of antiviral medicines, presenting a unique approach for treating viral diseases (Gaikwad et al. 2013).

16.8.4 Antilarval

India, due to its favorable ecological conditions, is highly endemic to diseases caused by mosquitos. They are notorious for transmitting various diseases. To combat insectoids diseases, different pesticides and larvicides are used. However, the excessive use of synthetic insect and larva killing materials has led to increased resistance in them, posing health and environmental concerns. Consequently, alternative strategies, such as biological control, are being explored, which offer long-term effectiveness, cost-effectiveness, and safety for living organisms and ecosystems (Salunkhe et al. 2011; Najitha Banu et al. 2014; Dhanasekaran and Thangaraj 2013). Therefore, NPs synthesized by microorganisms are gaining attention. NPs derived from various sources like *Bacillus thuringiensis*, *Agaricus bisporus*, *Cochliobolus lunatus*, *A. niger*, *Pseudomonas mandelii*, *E. coli*, *Streptomyces* sp., *Penicillium* sp., and *Vibrio* sp. have demonstrated larvicidal activity (Najitha Banu et al. 2014). However, there is limited data on the larvicidal activity of endophytic microorganism-derived NPs. Notably, Ag-NPs was recently identified as effective insecticides for controlling mosquito vectors (Patil et al. 2012). For instance, Seetharaman et al. investigated the larvicidal activity of Ag-NPs synthesized by the endophytic fungus *P. liquidambaris*, finding that the synthesized NPs inhibited the growth of *Culex quinquefasciatus* and *Aedes aegypti* larvae in a dose-reliant manner (Seetharaman et al. 2018). Additionally, Danagoudar et al. (2020) discovered that Ag-NPs biologically synthesized from *P. citrinum* CGJ-C1 which resided in a *Tragia involucrata* L. (a medicinal plant) with potential larvicidal action, highlighting the possibility of using mycoAg-NPs to reduce both cancer and mosquito-caused diseases (Danagoudar et al. 2020). Moreover, Seetharaman et al. reported that *P. oxalicum* strain LA-1 liquid extract-derived Ag-NPs were farther effective in abating *C. quinquefasciatus* larvae (Seetharaman et al. 2021).

16.8.5 Anticancer

Cancer is an indicating cause of death worldwide, and some drugs have been associated with toxicity in normal human cells. Therefore, requirement of newer drugs molecules for breast and cervical cancers. However, there has been limited research on the cellular toxicology of NPs on cell lines of cervical cancer and human breast (Krishnaraj et al. 2014; Kaler et al. 2013; Alimardani et al. 2023; Ashfaq et al. 2022). Fungi have gained attention for their ability to produce NPs, and they are considered ideal for NPs generation due to various factors. Exploring EFs for metallic NPs production can be an innovative strategy since there are limited reports on their use for manufacturing anticancer medications. The selective pH-based cellular toxicity of ZnO-NPs against carcinoma cells involves the rapid solubility of these NPs into ions of Zn^{2+} within acidic intracellular compartments. This induces oxidative stress and cell injury in cancer cells due to ROS generation, leading to mitochondrial dysfunction and apoptosis (Sana et al. 2020). Fungal extracts from *Cryptendoxyla hypophloia* CGJ D2, *P. citrinum* CGJ C1, *Cladosporium* sp., and *P. citrinum* have shown antioxidant and cytotoxic activities, particularly *P. citrinum* CGJ C2 strain's ethyl acetate extract (Ananda et al. 2018). Research on Ag-NPs synthesized by the EFs *B. rhodina* demonstrated their potential for capturing free radicals and leading cell death in lung cancer cell lines (A549) (Akther et al. 2019b). Similarly, *P. citrinum* CGJ-C1-derived AgNPs had a toxic effect against multiple carcinoma cell lines but were harmless to noncancerous cells (Danagoudar et al. 2020). AgNPs from *P. oxalicum* demonstrated dose-based toxicity for cell lines of human breast cancer (MDA-MB-231) (Seetharaman et al. 2021). ZnONPs produced by *A. tenuissima* culture filtrate also showed cytotoxic effects against various cancer cells (Abdelhakim et al. 2020). Gold NPs synthesized with the help of an endophytic bacterium *F. solani* from *C. fragrans* exhibited cytotoxicity against cervical and breast cancer cells (Clarance et al. 2020). CuONPs produced from endophyte *A. terreus* showed a concentration-dependent inhibition of cell line proliferation, especially in HT-29 colon cancer cells (Mani et al. 2021).

16.8.6 Antioxidant

Botryosphaeria rhodian-derived Ag-NPs exhibited enhanced antioxidant activity, surpassing that of ascorbic acid (Bogas et al. 2022; Akther et al. 2019a). Similarly, Ag-NPs produced from the *Exserohilum rostrata*, residing in leaf of *Ocimum tenuiflorum*, showed antioxidant abilities comparable to regular antioxidants like quercetin and ascorbic acid (Bagur et al. 2020). *C. perangustum*-derived Ag-NPs displayed excellent antioxidant activity in a variety of assays, outperforming ascorbic acid and butylated hydroxytoluene (Govindappa et al. 2016). *A. austroafricanus* CGJ-B3-derived AgNPs shows antioxidant, cellular toxicity, and anticline esterase activities (Danagoudar et al. 2021). Additionally, *A. alternata*-derived Ag-NPs shows enhanced antioxidant activity than the standard in various assays (Govindappa et al. 2022). Au-NPs produced by the endophytic fungus *P. citrinum* separated from brown algae demonstrated antioxidant activity related to the quantity of Au-NPs in mixture used for reaction (Manjunath et al. 2017). Similarly, Au-NPs, synthesized by the endophytic fungus *C. cladosporioides*, showed antioxidant activity comparable to ascorbic acid and dose-dependent radical scavenging capacity (Manjunath Hulikere et al. 2017). Au-NPs (MycoAu-NPs) synthesized by using *Cladosporium* sp., which effectively shows the anticarcinoma effect against cancer cell line of human breast MCF-7, inducing apoptosis (IC50 38.23/g/L) (Munawer et al. 2020). ZnO-NPs produced by using *Periconium* sp. extract demonstrated strong antioxidant capabilities at 100 g/mL concentration (Ganesan et al. 2020). *A. tenuissima* liquid filtrate-derived ZnO-NPs also exhibited antioxidant activity at various dosages, comparable to ascorbic acid (Abdelhakim et al. 2020). Overall, the production and applications of metal-NPs derived from EFs have shown promising potential in agroecosystems and biomedical domains, which are also briefly described in Table 16.1.

16.8.7 Biosensor Development

Biosensors are widely used in various fields, including biomedical diagnostics, disease monitoring, environmental surveillance, discovery of drugs, and forensics. Different techniques could be employed to raise biosensors, utilizing biomolecules with superior affinity to detect various analytes with great sensitivity and selectivity. Algae-derived NPs have shown promise in biosensing applications. For instance,

Table 16.1 EFs-driven NPs and their various products

S. No.	EFs	NPs	Applications	Ref
1.	*Exserohilum rostrata*	Ag-NPs	Antibacterial and antimitotic activity	Bagur et al. (2020)
2.	*Alternaria tenuissima*	Zn-NPs	Antimicrobial and anticancer activity	Abdelhakim et al. (2020)
3.	*Saccharomyces boulardii*	Ag-NPs	Anticancer activity	Kaler et al. (2013)
4.	*Penicillium citrinum*	Ag-NPs	Anticancer (MGF-7) activity	Danagoudar et al. (2020)
5.	*Cochliobolus lunatus*	Ag-NPs	Larvicidal activity	Salunke et al. (2016)
6.	*F oxysporum,*	Ag-NPs	Antiviral against herpes simplex virus and herpes simplex virus	Gaikwad et al. (2013)
7.	*Penicillium* sp.	Ag-NPs	Antibacterial activity	Singh et al. (2014)
8.	*Aspergillus terreus*	Cu-NPs	Anticancer activity	Mani et al. (2021)
9.	*Penicillium oxalicum*	Fe-NPs	Methyl blue decolorization	Mathur et al. (2021)
10.	*Alternaria alternata*	Ag-NPs	Antioxidant activity	Govindappa et al. (2022)

Au-NPs have been effectively utilized in pregnancy tests to detect hormones like hCG (Kuppusamy et al. 2014). NPs of platinum serve as a novel biosensor with exceptional sensitivity for detecting adrenaline, beneficial for addressing allergies, asthma, cardiac operations, and heart attacks. The optical properties of fungal-derived Au-NPs are advantageous in identifying hormone types and quantities, aiding in cancer diagnosis (Chaudhary et al. 2020). Micro-AuAg alloys produced from algae have demonstrated considerable electrocatalytic potential for the detection of initial cancer cells and 2-butanone. Additionally, Ag-NPs produced by *Noctiluca scintillans* have been tested as a hydrogen peroxide (H_2O_2) sensor, with temperature, duration, and pH influencing the catalytic degradation of H_2O_2 and causing a color shift (Elgamouz et al. 2020; Michael et al. 2022). In another study, Au-NPs derived from *Hypnea valencia* were found able to detect levels of human chorionic gonadotropin hormone in pregnant women's urine samples (Kuppusamy et al. 2014).

16.9 Conclusion

Green synthesis, or endophytic-driven NPs synthesis, is one of the most ecological and eco-friendly methods observed as efficient biofactories. Filamentous fungal species are considered efficient in NPs production due to their higher production of enzymes and secondary metabolites. EFs, which reside in components of plants through a mutualistic relationship, have been identified as potential sources for NPs manufacturing because of their ability to intake, tolerate, and accumulate metals. The current chapter examines recent research on endophytes, nanomaterials, and EFs, and their role in producing functional-NMs for applications in agroecosystems and biomedical sectors. In various fields, such as biology and agriculture, NMs have captured interest because of their size-reliant properties. Ag-NPs have been known for their fungicidal and antibacterial effects, while nano-pesticides have shown promise as less harmful environmental treatments for various conditions. Endophytic fungal NPs have emerged as a fascinating aspect of nanotechnology, offering the potential for the synthesis of nano-weapons and nanomedicines against plant pathogens in the future. However, further research is required to understand the specific mechanisms of molecules underlying endophytic fungal NPs production to get better control over nanoparticle size and shape.

References

Abdelhakim HK, El-Sayed ER, Rashidi FB (2020) Biosynthesis of zinc oxide nanoparticles with antimicrobial, anticancer, antioxidant and photocatalytic activities by the endophytic Alternaria tenuissima. J Appl Microbiol 128:1634–1646

Abid N, Khan AM, Shujait S, Chaudhary K, Ikram M, Imran M, Haider J, Khan M, Khan Q, Maqbool M (2022) Synthesis of nanomaterials using various top-down and bottom-up approaches, influencing factors, advantages, and disadvantages: a review. Adv Colloid Interf Sci 300:102597

Afreen S, Omar RA, Talreja N, Chauhan D, Ashfaq M (2018) Carbon-based nanostructured materials for energy and environmental remediation applications. In: Prasad R, Aranda E (eds) Approaches in bioremediation: the new era of environmental microbiology and nanobiotechnology. Springer, Cham

Afzal I, Shinwari ZK, Sikandar S, Shahzad S (2019) Plant beneficial endophytic bacteria: mechanisms, diversity, host range and genetic determinants. Microbiol Res 221:36–49

Ahemad M (2015) Enhancing phytoremediation of chromium-stressed soils through plant-growth-promoting bacteria. J Genet Eng Biotechnol 13:51–58

Ahmad A, Senapati S, Khan MI, Kumar R, Ramani R, Srinivas V, Sastry M (2003) Intracellular synthesis of gold nanoparticles by a novel alkalotolerant actinomycete, Rhodococcus species. Nanotechnology 14:824

Ahmed SF, Mofijur M, Rafa N, Chowdhury AT, Chowdhury S, Nahrin M, Islam ABMS, Ong HC (2022) Green approaches in synthesising nanomaterials for environmental nanobioremediation: technological advancements, applications, benefits and challenges. Environ Res 204:111967

Akther T, Vabeiryureilai M, Nachimuthu Senthil K, Davoodbasha M, Srinivasan H (2019a) Fungal-mediated synthesis of pharmaceutically active silver nanoparticles and anticancer property against A549 cells through apoptosis. Environ Sci Pollut Res Int 26:13649–13657

Akther T, Vabeiryureilai M, Nachimuthu Senthil K, Davoodbasha M, Srinivasan H (2019b) Fungal-mediated synthesis of pharmaceutically active silver nanoparticles and anticancer property against A549 cells through apoptosis. Environ Sci Pollut Res 26:13649–13657

Alghuthaymi MA, Almoammar H, Rai M, Said-Galiev E, Abd-Elsalam KA (2015) Myconanoparticles: synthesis and their role in phytopathogens management. Biotechnol Biotechnol Equip 29:221–236

Alimardani V, Rahiminezhad Z, Dehghankhold M, Farahavar G, Jafari M, Abedi M, Moradi L, Niroumand U, Ashfaq M, Abolmaali SS, Yousefi G (2023) Nanotechnology-based cell-mediated delivery systems for cancer therapy and diagnosis. Drug Deliv Transl Res 13:189–221

Ananda D, Chandrashekhar GJ, Sunil Kumar R, Heggodu GRK, Balenahalli Narasingappa R (2018) Antioxidant and cytotoxic potential of endophytic fungi isolated from medicinal plant Tragia involucrata L. Pharmacogn Res 10. https://doi.org/10.4103/pr.pr_137_17

Angelé-Martínez C, Nguyen K v, Ameer FS, Anker JN, Brumaghim JL (2017) Reactive oxygen species generation by copper(II) oxide nanoparticles determined by DNA damage assays and EPR spectroscopy. Nanotoxicology 11:278–288

Anjum S, Ishaque S, Fatima H, Farooq W, Hano C, Abbasi BH, Anjum I (2021) Emerging applications of nanotechnology in healthcare systems: grand challenges and perspectives. Pharmaceuticals (Basel) 14:707

Arciniegas-Grijalba PA, Patiño-Portela MC, Mosquera-Sánchez LP, Guerrero-Vargas JA, Rodríguez-Páez JE (2017) ZnO nanoparticles (ZnO-NPs) and their antifungal activity against coffee fungus Erythricium salmonicolor. Appl Nanosci 7:225–241

Arnold AE, Maynard Z, Gilbert GS, Coley PD, Kursar TA (2000) Are tropical fungal endophytes hyperdiverse? Ecol Lett 3:267–274

Asaf S, Jan R, Khan MA, Lubna KA, Lubna, Asif S, Bilal S, Ahmad W, Waqas M, Kim KM, Al-Harrasi A, Lee IJ (2023) Unraveling the mutualistic interaction between endophytic Curvularia lunata CSL1 and tomato to mitigate cadmium (Cd) toxicity via transcriptomic insights. Sci Total Environ 861:160542

Ashfaq M, Verma N, Khan S (2017) Carbon nanofibers as a micronutrient carrier in plants: efficient translocation and controlled release of Cu nanoparticles. Environ Sci Nano 4:138–148

Ashfaq M, Talreja N, Chauhan D, Afreen S, Sultana A, Srituravanich W (2022) Two-dimensional (2D) hybrid nanomaterials for diagnosis and treatment of cancer. J Drug Deliv Sci Technol 70:103268

Atalah J, Espina G, Blamey L, Muñoz-Ibacache SA, Blamey JM (2022) Advantages of using extremophilic bacteria for the biosynthesis of metallic nanoparticles and its potential for rare earth element recovery. Front Microbiol 13:855077

Attia MS, Hashem AH, Badawy AA, Abdelaziz AM (2022) Biocontrol of early blight disease of eggplant using endophytic Aspergillus terreus: improving plant immunological, physiological and antifungal activities. Bot Stud 63:26

Azmath P, Baker S, Rakshith D, Satish S (2016) Mycosynthesis of silver nanoparticles bearing antibacterial activity. Saudi Pharm J 24:140–146

Bagur H, Poojari CC, Melappa G, Rangappa R, Chandrasekhar N, Somu P (2020) Biogenically synthesized silver nanoparticles using endophyte fungal extract of ocimum tenuiflorum and evaluation of biomedical properties. J Clust Sci 31:1241–1255

Bahrulolum H, Nooraei S, Javanshir N, Tarrahimofrad H, Mirbagheri, v. S., Easton, A. J. & Ahmadian, G. (2021) Green synthesis of metal nanoparticles using microorganisms and their application in the agrifood sector. J Nanobiotechnol 19:86

Baker S, Satish S (2015) Biosynthesis of gold nanoparticles by Pseudomonas veronii AS41G inhabiting Annona squamosa L. Spectrochim Acta A Mol Biomol Spectrosc 150:691–695

Baker S, Mohan Kumar K, Santosh P, Rakshith D, Satish S (2015) Extracellular synthesis of silver nanoparticles by novel Pseudomonas veronii AS41G inhabiting Annona squamosa L. and their bactericidal activity. Spectrochim Acta A Mol Biomol Spectrosc 136 Pt C:1434–1440

Bilański P, Kowalski T (2022) Fungal endophytes in Fraxinus excelsior petioles and their in vitro antagonistic potential against the ash dieback pathogen Hymenoscyphus fraxineus. Microbiol Res 257:126961

Blackwell M (2011) The fungi: 1, 2, 3 … 5.1 million species? Am J Bot 98:426–438

Bogas AC, Henrique Rodrigues S, Gonçalves MO, de Assis M, Longo E, Paiva de Sousa C (2022) Endophytic microorganisms from the tropics as biofactories for the synthesis of metal-based nanoparticles: healthcare applications. Front Nanotechnol:4. https://doi.org/10.3389/fnano.2022.823236

Bonfante P (2018) The future has roots in the past: the ideas and scientists that shaped mycorrhizal research. New Phytol 220:982–995

Borges WDS, Borges KB, Bonato PS, Said S, Pupo MT (2009) Endophytic fungi: natural products, enzymes and biotransformation reactions. Curr Org Chem 13:1137–1163

Brader G, Corretto E, Sessitsch A (2017) Metagenomics of plant microbiomes. In: Charles TC, Liles MR, Sessitsch A (eds) Functional metagenomics: tools and applications. Springer, Cham

Byregowda R, Prasad SR, Oelmüller R, Nataraja KN, Prasanna Kumar MK (2022) Is endophytic colonization of host plants a method of alleviating drought stress? Conceptualizing the hidden world of endophytes. Int J Mol Sci 23:9194

Chareprasert S, Piapukiew J, Thienhirun S, Whalley AJS, Sihanonth P (2006) Endophytic fungi of teak leaves Tectona grandis L. and rain tree leaves Samanea saman Merr. World J Microbiol Biotechnol 22:481–486

Chaudhary R, Nawaz K, Khan AK, Hano C, Abbasi BH, Anjum S (2020) An overview of the algae-mediated biosynthesis of nanoparticles and their biomedical applications. Biomolecules 10:1498

Clarance P, Luvankar B, Sales J, Khusro A, Agastian P, Tack JC, Al Khulaifi MM, Al-Shwaiman HA, Elgorban AM, Syed A, Kim HJ (2020) Green synthesis and characterization of gold nanoparticles using endophytic fungi Fusarium solani and its in-vitro anticancer and biomedical applications. Saudi J Biol Sci 27:706–712

Collinge DB, Jensen B, Jørgensen HJ (2022) Fungal endophytes in plants and their relationship to plant disease. Curr Opin Microbiol 69:102177

Conrado R, Gomes TC, Roque GSC, de Souza AO (2022) Overview of bioactive fungal secondary metabolites: cytotoxic and antimicrobial compounds. Antibiotics (Basel) 11:1604

Danagoudar A, Pratap GK, Shantaram M, Ghosh K, Kanade SR, Joshi CG (2020) Characterization, cytotoxic and antioxidant potential of silver nanoparticles biosynthesised using endophytic fungus (Penicillium citrinum CGJ-C1). Mater Today Commun 25:101385

Danagoudar A, Pratap GK, Shantaram M, Ghosh K, Kanade SR, Sannegowda LK, Joshi CG (2021) Antioxidant, cytotoxic and anti-choline esterase activity of green silver nanoparticles synthesized using Aspergillus austroafricanus CGJ-B3 (Endophytic Fungus). Anal Chem Lett 11:15–28

Das A, Varma A (2009) Symbiosis: the art of living. In: Varma A, Kharkwal AC (eds) Symbiotic gungi: principles and practice. Springer, Berlin
De Melo Pereira GV, Magalhães KT, Lorenzetii ER, Souza TP, Schwan RF (2012) A multiphasic approach for the identification of endophytic bacterial in strawberry fruit and their potential for plant growth promotion. Microb Ecol 63:405–417
Devi LS, Joshi SR (2014) Evaluation of the antimicrobial potency of silver nanoparticles biosynthesized by using an endophytic fungus, Cryptosporiopsis ericae PS4. J Microbiol 52:667–674
Dhanasekaran D, Thangaraj R (2013) Evaluation of larvicidal activity of biogenic nanoparticles against filariasis causing Culex mosquito vector. Asian Pac J Trop Dis 3:174–179
Dinesh B, Monisha N, Shalini HR, Prathap GK, Poyya J, Shantaram M, Hampapura JS, Karigar CS, Joshi CG (2022) Antibacterial activity of silver nanoparticles synthesized using endophytic fungus—Penicillium cinnamopurpureum. Spectrosc Lett 55:20–34
Dykman LA, Khlebtsov NG (2011) Gold nanoparticles in biology and medicine: recent advances and prospects. Acta Nat 3:34–55
Elgamouz A, Idriss H, Nassab C, Bihi A, Bajou K, Hasan K, Abu Haija M, Patole SP (2020) Green synthesis, characterization, antimicrobial, anti-cancer, and optimization of colorimetric sensing of hydrogen peroxide of algae extract capped silver nanoparticles. Nanomaterials 10:1861
Fadiji AE, Babalola OO (2020) Metagenomics methods for the study of plant-associated microbial communities: a review. J Microbiol Methods 170:105860
Fadiji AE, Mortimer PE, Xu J, Ebenso EE, Babalola OO (2022) Biosynthesis of nanoparticles using endophytes: a novel approach for enhancing plant growth and sustainable agriculture. Sustainability 14:10839
Farsi M, Farokhi S (2018) Biosynthesis of antibacterial silver nanoparticles by endophytic fungus nemania sp. Isolated from Taxus baccata L. (Iranian Yew). Zahedan J Res Med Sci. 20:e57916
Frank AC, Saldierna Guzmán JP, Shay JE (2017) Transmission of bacterial endophytes. Microorganisms 5:70
Gahlawat G, Choudhury AR (2019) A review on the biosynthesis of metal and metal salt nanoparticles by microbes. RSC Adv 9:12944–12967
Gahoi P, Omar RA, Verma N, Gupta GS (2021) Rhizobacteria and acylated homoserine lactone-based nanobiofertilizer to improve growth and pathogen defense in cicer arietinum and triticum aestivum plants. ACS Agric Sci Technol 1:240–252
Gaikwad S, Ingle A, Gade A, Rai M, Falanga A, Incoronato N, Russo L, Galdiero S, Galdiero M (2013) Antiviral activity of mycosynthesized silver nanoparticles against herpes simplex virus and human parainfluenza virus type 3. Int J Nanomedicine 8:4303–4314
Galdiero S, Falanga A, Vitiello M, Cantisani M, Marra, v. & Galdiero, M. (2011) Silver nanoparticles as potential antiviral agents. Molecules 16:8894–8918
Ganesan V, Hariram M, Vivekanandhan S, Muthuramkumar S (2020) Periconium sp. (endophytic fungi) extract mediated sol-gel synthesis of ZnO nanoparticles for antimicrobial and antioxidant applications. Mater Sci Semicond Process 105:104739
Ghosh S, Ahmad R, Zeyaullah M, Khare SK (2021) Microbial nano-factories: synthesis and biomedical applications. Front Chem 9:626834
Glick BR (2012) Plant growth-promoting bacteria: mechanisms and applications. Scientifica 2012:963401
Govindappa M, Farheen H, Chandrappa CP, Channabasava, Rai RV, Raghavendra VB (2016) Mycosynthesis of silver nanoparticles using extract of endophytic fungi, Penicillium species of Glycosmis mauritiana, and its antioxidant, antimicrobial, anti-inflammatory and tyrokinase inhibitory activity. Adv Nat Sci Nanosci Nanotechnol 7:035014
Govindappa M, Lavanya M, Aishwarya P, Pai K, Lunked P, Hemashekhar B, Arpitha BM, Ramachandra YL, Raghavendra, v. B. (2020) Synthesis and characterization of endophytic fungi, cladosporium perangustum mediated silver nanoparticles and their antioxidant, anticancer and nano-toxicological study. BioNanoScience 10:928–941
Govindappa M, Manasa DJ, Vinaykiya V, Bhoomika V, Dutta S, Pawar R, Raghavendra VB (2022) Screening of antibacterial and antioxidant activity of biogenically synthesized silver nanopar-

ticles from alternaria alternata, endophytic fungus of dendrophthoe falcata-a parasitic plant. BioNanoScience 12:128–141

Guilger-Casagrande M, de Lima R (2019) Synthesis of silver nanoparticles mediated by fungi: a review. Front Bioeng Biotechnol 7:287

Gupta P, Rai N, Verma A, Saikia D, Singh SP, Kumar R, Singh SK, Kumar D, Gautam, v. (2022) Green-based approach to synthesize silver nanoparticles using the fungal endophyte penicillium oxalicum and their antimicrobial, antioxidant, and in vitro anticancer potential. ACS Omega 7:46653–46673

Gupta A, Meshram V, Gupta M, Goyal S, Qureshi KA, Jaremko M, Shukla KK (2023) Fungal endophytes: microfactories of novel bioactive compounds with therapeutic interventions a comprehensive review on the biotechnological developments in the field of fungal endophytic biology over the last decade. Biomolecules 13:1038

Hammami I, Alabdallah NM, Jomaa AA, Kamoun M (2021) Gold nanoparticles: synthesis properties and applications. J King Saud Univ Sci 33:101560

Harrison JG, Griffin EA (2020) The diversity and distribution of endophytes across biomes, plant phylogeny and host tissues: how far have we come and where do we go from here? Environ Microbiol 22:2107–2123

Hashem AH, Attia MS, Kandil EK, Fawzi MM, Abdelrahman AS, Khader MS, Khodaira MA, Emam AE, Goma MA, Abdelaziz AM (2023) Bioactive compounds and biomedical applications of endophytic fungi: a recent review. Microb Cell Factories 22:107

Hiruma K, Kobae Y, Toju H (2018) Beneficial associations between Brassicaceae plants and fungal endophytes under nutrient-limiting conditions: evolutionary origins and host–symbiont molecular mechanisms. Curr Opin Plant Biol 44:145–154

Hu X, Saravanakumar K, Jin T, Wang MH (2019) Mycosynthesis, characterization, anticancer and antibacterial activity of silver nanoparticles from endophytic fungus Talaromyces purpureogenus. Int J Nanomedicine 14:3427–3438

Huang X, Ho S-H, Zhu S, Ma F, Wu J, Yang J, Wang L (2017) Adaptive response of arbuscular mycorrhizal symbiosis to accumulation of elements and translocation in Phragmites australis affected by cadmium stress. J Environ Manag 197:448–455

Hulkoti NI, Taranath TC (2014) Biosynthesis of nanoparticles using microbes—a review. Colloids Surf B: Biointerfaces 121:474–483

Iavicoli I, Leso V, Beezhold DH, Shvedova AA (2017) Nanotechnology in agriculture: opportunities, toxicological implications, and occupational risks. Toxicol Appl Pharmacol 329:96–111

Jumpponen A (2001) Dark septate endophytes—are they mycorrhizal? Mycorrhiza 11:207–211

Kadam VV, Ettiyappan JP, Mohan Balakrishnan R (2019) Mechanistic insight into the endophytic fungus mediated synthesis of protein capped ZnO nanoparticles. Mater Sci Eng B 243:214–221

Kaler A, Jain S, Banerjee UC (2013) Green and rapid synthesis of anticancerous silver nanoparticles by Saccharomyces boulardii and insight into mechanism of nanoparticle synthesis. Biomed Res Int 2013:872940

Karunakaran G, Sudha KG, Ali S, Cho E-B (2023) Biosynthesis of nanoparticles from various biological sources and its biomedical applications. Molecules 28:4527

Kathiresan K, Manivannan S, Nabeel MA, Dhivya B (2009) Studies on silver nanoparticles synthesized by a marine fungus, Penicillium fellutanum isolated from coastal mangrove sediment. Colloids Surf B Biointerfaces 71:133–137

Kaur N, Arora DS, Kalia N, Kaur M (2020) Bioactive potential of endophytic fungus Chaetomium globosum and GC–MS analysis of its responsible components. Sci Rep 10:18792

Kazemi S, Hosseingholian A, Gohari SD, Feirahi F, Moammeri F, Mesbahian G, Moghaddam ZS, Ren Q (2023) Recent advances in green synthesized nanoparticles: from production to application. Mater Today Sustain 24:100500

Khatami M, Alijani HQ, Sharifi I (2018) Biosynthesis of bimetallic and core-shell nanoparticles: their biomedical applications—a review. IET Nanobiotechnol 12:879–887

Koduru JR, Kailasa SK, Bhamore JR, Kim K-H, Dutta T, Vellingiri K (2018) Phytochemical-assisted synthetic approaches for silver nanoparticles antimicrobial applications: a review. Adv Colloid Interf Sci 256:326–339

Koul B, Poonia AK, Yadav D, Jin JO (2021) Microbe-mediated biosynthesis of nanoparticles: applications and future prospects. Biomolecules 11:886

Krishnaraj C, Muthukumaran P, Ramachandran R, Balakumaran MD, Kalaichelvan PT (2014) Acalypha indica Linn: biogenic synthesis of silver and gold nanoparticles and their cytotoxic effects against MDA-MB-231, human breast cancer cells. Biotechnol Rep 4:42–49

Kuppusamy P, Mashitah MY, Maniam GP, Govindan N (2014) Biosynthesized gold nanoparticle developed as a tool for detection of HCG hormone in pregnant women urine sample. Asian Pac J Trop Dis 4:237

Lahiri D, Nag M, Sheikh HI, Sarkar T, Edinur HA, Pati S, Ray RR (2021) Microbiologically-synthesized nanoparticles and their role in silencing the biofilm signaling cascade. Front Microbiol 12:636588

Li X, Xu H, Chen Z-S, Chen G (2011) Biosynthesis of nanoparticles by microorganisms and their applications. J Nanomater 2011:270974

Lim Z-ZJ, Li J-EJ, NG C-T, Yung L-YL, Bay B-H (2011) Gold nanoparticles in cancer therapy. Acta Pharmacol Sin 32:983–990

Malik S, Muhammad K, Waheed Y (2023) Emerging applications of nanotechnology in healthcare and medicine. Molecules 28:6624

Mani VM, Kalaivani S, Sabarathinam S, Vasuki M, Soundari AJPG, Ayyappa Das MP, Elfasakhany A, Pugazhendhi A (2021) Copper oxide nanoparticles synthesized from an endophytic fungus Aspergillus terreus: bioactivity and anti-cancer evaluations. Environ Res 201:111502

Manjunath Hulikere M, Joshi CG (2019) Characterization, antioxidant and antimicrobial activity of silver nanoparticles synthesized using marine endophytic fungus- Cladosporium cladosporioides. Process Biochem 82:199–204

Manjunath Hulikere M, Joshi CG, Danagoudar A, Poyya J, Kudva AK, Dhananjaya BL (2017) Biogenic synthesis of gold nanoparticles by marine endophytic fungus-Cladosporium cladosporioides isolated from seaweed and evaluation of their antioxidant and antimicrobial properties. Process Biochem 63:137–144

Manjunath HM, Joshi CG, Raju NG (2017) Biofabrication of gold nanoparticles using marine endophytic fungus—Penicillium citrinum. IET Nanobiotechnol 11:40–44

Mathur P, Saini S, Paul E, Sharma C, Mehtani P (2021) Endophytic fungi mediated synthesis of iron nanoparticles: characterization and application in methylene blue decolorization. Curr Res Green Sustain Chem 4:100053

Michael A, Singh A, Roy A, Islam MR (2022) Fungal- and algal-derived synthesis of various nanoparticles and their applications. Bioinorg Chem Appl 2022:3142674

Mikhailova EO (2020) Silver nanoparticles: mechanism of action and probable bio-application. J Funct Biomater 11:84

Mohd Yusof H, Mohamad R, Zaidan UH, Abdul Rahman NA (2019) Microbial synthesis of zinc oxide nanoparticles and their potential application as an antimicrobial agent and a feed supplement in animal industry: a review. J Anim Sci Biotechnol 10:57

Moricca S, Ragazzi A (2008) Fungal endophytes in Mediterranean oak forests: a lesson from Discula quercina. Phytopathology 98:380–386

Moronta-Barrios F, Gionechetti F, Pallavicini A, Marys E, Venturi V (2018) Bacterial microbiota of rice roots: 16s-based taxonomic profiling of endophytic and rhizospheric diversity, endophytes isolation and simplified endophytic community. Microorganisms 6:14

Mukherjee S, Patra CR (2017) Biologically synthesized metal nanoparticles: recent advancement and future perspectives in cancer theranostics. Future Sci OA 3:Fso203

Munawer U, Raghavendra VB, Ningaraju S, Krishna KL, Ghosh AR, Melappa G, Pugazhendhi A (2020) Biofabrication of gold nanoparticles mediated by the endophytic Cladosporium species: photodegradation, in vitro anticancer activity and in vivo antitumor studies. Int J Pharm 588:119729

Musarrat J, Dwivedi S, Singh BR, Al-Khedhairy AA, Azam A, Naqvi A (2010) Production of antimicrobial silver nanoparticles in water extracts of the fungus Amylomyces rouxii strain KSU-09. Bioresour Technol 101:8772–8776

Najitha Banu A, Balasubramanian C, Moorthi P v (2014) Biosynthesis of silver nanoparticles using Bacillus thuringiensis against dengue vector, Aedes aegypti (Diptera: Culicidae). Parasitol Res 113:311–316

Netala VR, Kotakadi VS, Bobbu P, Gaddam SA, Tartte V (2016a) Endophytic fungal isolate mediated biosynthesis of silver nanoparticles and their free radical scavenging activity and anti microbial studies. 3 Biotech 6:132

Netala VR, Bethu MS, Pushpalatha B, Baki VB, Aishwarya S, Rao JV, Tartte V (2016b) Biogenesis of silver nanoparticles using endophytic fungus Pestalotiopsis microspora and evaluation of their antioxidant and anticancer activities. Int J Nanomedicine 11:5683–5696

Newsham KK (2011) A meta-analysis of plant responses to dark septate root endophytes. New Phytol 190:783–793

Omar RA, Afreen S, Talreja N, Chauhan D, Ashfaq M (2019) Impact of nanomaterials in plant systems. In: PRASAD R (ed) Plant nanobionics: volume 1, advances in the understanding of nanomaterials research and applications. Springer International Publishing, Cham

Omar RA, Chauhan D, Talreja N, Mangalaraja R v, Ashfaq M (2022) Chapter 12—Vegetables waste for biosynthesis of various nanoparticles. In: Abd-Elsalam KA, Periakaruppan R, Rajeshkumar S (eds) Agri-waste and microbes for production of sustainable nanomaterials. Elsevier, Amsterdam

Osman AI, Zhang Y, Farghali M, Rashwan AK, Eltaweil AS, Abd El-Monaem EM, Mohamed IMA, Badr MM, Ihara I, Rooney DW, Yap P-S (2024) Synthesis of green nanoparticles for energy, biomedical, environmental, agricultural, and food applications: a review. Environ Chem Lett. https://doi.org/10.1007/s10311-023-01682-3

Pandit C, Roy A, Ghotekar S, Khusro A, Islam MN, Emran TB, Lam SE, Khandaker MU, Bradley DA (2022) Biological agents for synthesis of nanoparticles and their applications. J King Saud Univ Sci 34:101869

Patil CD, Borase HP, Patil S v, Salunkhe RB, Salunke BK (2012) Larvicidal activity of silver nanoparticles synthesized using Pergularia daemia plant latex against Aedes aegypti and Anopheles stephensi and nontarget fish Poecillia reticulata. Parasitol Res 111:555–562

Patra JK, Das G, Fraceto LF, Campos EVR, Rodriguez-Torres MDP, Acosta-Torres LS, Diaz-Torres LA, Grillo R, Swamy MK, Sharma S, Habtemariam S, Shin H-S (2018) Nano based drug delivery systems: recent developments and future prospects. J Nanobiotechnol 16:71

Peralta-Videa JR, Huang Y, Parsons JG, Zhao L, Lopez-Moreno L, Hernandez-Viezcas JA, Gardea-Torresdey JL (2016) Plant-based green synthesis of metallic nanoparticles: scientific curiosity or a realistic alternative to chemical synthesis? Nanotechnol Environ Eng 1:4

Perotto S, Daghino S, Martino E (2018) Ericoid mycorrhizal fungi and their genomes: another side to the mycorrhizal symbiosis? New Phytol 220:1141–1147

Petrini O, Sieber TN, Toti L, Viret O (1992) Ecology, metabolite production, and substrate utilization in endophytic fungi. Nat Toxins 1:185–196

Popli D, Anil V, Subramanyam AB, Namratha MN, Ranjitha VR, Rao SN, Rai RV, Govindappa M (2018) Endophyte fungi, Cladosporium species-mediated synthesis of silver nanoparticles possessing in vitro antioxidant, anti-diabetic and anti-Alzheimer activity. Artif Cells Nanomed Biotechnol 46:676–683

Qamar H, Rehman S, Chauhan DK, Tiwari AK, Upmanyu, v. (2020) Green synthesis, characterization and antimicrobial activity of copper oxide nanomaterial derived from momordica charantia. Int J Nanomedicine 15:2541–2553

Rafique M, Sadaf I, Rafique MS, Tahir MB (2017) A review on green synthesis of silver nanoparticles and their applications. Artif Cells Nanomed Biotechnol 45:1272–1291

Raghunath A, Perumal E (2017) Metal oxide nanoparticles as antimicrobial agents: a promise for the future. Int J Antimicrob Agents 49:137–152

Raghupathi KR, Koodali RT, Manna AC (2011) Size-dependent bacterial growth inhibition and mechanism of antibacterial activity of zinc oxide nanoparticles. Langmuir 27:4020–4028

Rahman L, Shinwari ZK, Iqrar I, Rahman L, Tanveer F (2017) An assessment on the role of endophytic microbes in the therapeutic potential of Fagonia indica. Ann Clin Microbiol Antimicrob 16:53

Rai M, Ingle AP, Paralikar P, Anasane N, Gade R, Ingle P (2018) Effective management of soft rot of ginger caused by Pythium spp. and Fusarium spp.: emerging role of nanotechnology. Appl Microbiol Biotechnol 102:6827–6839

Ramalingmam P, Muthukrishnan S, Thangaraj PR (2015) Biosynthesis of silver nanoparticles using an endophytic fungus, curvularialunata and its antimicrobial potential. World J Nano Sci Eng. 1(4):241–247

Ramrakhiani L, Ghosh S (2018) Metallic nanoparticle synthesised by biological route: safer candidate for diverse applications. IET Nanobiotechnol 12:392–404

Rana KL, Kour D, Yadav N, Yadav AN (2020) 10—Endophytic microbes in nanotechnology: current development, and potential biotechnology applications. In: Kumar A, Singh VK (eds) Microbial endophytes. Woodhead Publishing, Sawston

Rani R, Sharma DK, Chaturvedi M, Yadav JP (2017) Antibacterial activity of twenty different endophytic fungi isolated from calotropis procera and time kill assay

Rho H, Hsieh M, Kandel SL, Cantillo J, Doty SL, Kim SH (2018) Do endophytes promote growth of host plants under stress? A meta-analysis on plant stress mitigation by endophytes. Microb Ecol 75:407–418

Rochín-Hernández LS, Rochín-Hernández LJ, Flores-Cotera LB (2022) Endophytes, a potential source of bioactive compounds to curtail the formation-accumulation of advanced glycation end products: a review. Molecules 27:4469

Rodriguez RJ, Henson J, Van Volkenburgh E, Hoy M, Wright L, Beckwith F, Kim Y-O, Redman RS (2008) Stress tolerance in plants via habitat-adapted symbiosis. ISME J 2:404–416

Rodriguez RJ, White JF Jr, Arnold AE, Redman RS (2009) Fungal endophytes: diversity and functional roles. New Phytol 182:314–330

Salleh A, Naomi R, Utami ND, Mohammad AW, Mahmoudi E, Mustafa N, Fauzi MB (2020) The potential of silver nanoparticles for antiviral and antibacterial applications: a mechanism of action. Nanomaterials (Basel) 10:1566

Salunke BK, Sawant SS, Lee S-I, Kim BS (2016) Microorganisms as efficient biosystem for the synthesis of metal nanoparticles: current scenario and future possibilities. World J Microbiol Biotechnol 32:88

Salunkhe RB, Patil S v, Patil CD, Salunke BK (2011) Larvicidal potential of silver nanoparticles synthesized using fungus Cochliobolus lunatus against Aedes aegypti (Linnaeus, 1762) and Anopheles stephensi Liston (Diptera; Culicidae). Parasitol Res 109:823–831

Salvadori MR, Ando RA, Oller Do Nascimento CA, CorrêA B (2014) Intracellular biosynthesis and removal of copper nanoparticles by dead biomass of yeast isolated from the wastewater of a mine in the brazilian amazonia. PLoS One 9:e87968

Sana SS, Kumbhakar D v, Pasha A, Pawar SC, Grace AN, Singh RP, Nguyen VH, Le QV, Peng W (2020) Crotalaria verrucosa leaf extract mediated synthesis of zinc oxide nanoparticles: assessment of antimicrobial and anticancer activity. Molecules 25:4896

Schulz B (2006) Mutualistic interactions with fungal root endophytes. In: Schulz BJE, Boyle CJC, Sieber TN (eds) Microbial root endophytes. Springer, Berlin

Schulz B, Boyle C, Draeger S, Römmert A-K, Krohn K (2002) Endophytic fungi: a source of novel biologically active secondary metabolites* *Paper presented at the British Mycological Society symposium on Fungal Bioactive Compounds, held at the University of Wales Swansea on 22–27 April 2001. Mycol Res 106:996–1004

Seetharaman PK, Chandrasekaran R, Gnanasekar S, Chandrakasan G, Gupta M, Manikandan DB, Sivaperumal S (2018) Antimicrobial and larvicidal activity of eco-friendly silver nanoparticles synthesized from endophytic fungi Phomopsis liquidambaris. Biocatal Agric Biotechnol 16:22–30

Seetharaman PK, Chandrasekaran R, Periakaruppan R, Gnanasekar S, Sivaperumal S, Abd-Elsalam KA, Valis M, Kuca K (2021) Functional attributes of myco-synthesized silver nanoparticles from endophytic fungi: a new implication in biomedical applications. Biology 10:473

Shah A, Lutfullah G, Ahmad K, Khalil AT, Maaza M (2018) Daphne mucronata-mediated phytosynthesis of silver nanoparticles and their novel biological applications, compatibility and toxicity studies. Green Chem Lett Rev 11:318–333

Shankar SS, Ahmad A, Pasricha R, Sastry M (2003) Bioreduction of chloroaurate ions by geranium leaves and its endophytic fungus yields gold nanoparticles of different shapes. J Mater Chem 13:1822–1826
Sharma M, Mallubhotla S (2022) Diversity, antimicrobial activity, and antibiotic susceptibility pattern of endophytic bacteria sourced from Cordia dichotoma L. Front Microbiol 13:879386
Sharma D, Kanchi S, Bisetty K (2019) Biogenic synthesis of nanoparticles: a review. Arab J Chem 12:3576–3600
Sharma G, Agarwal S, Verma K, Bhardwaj R, Mathur V (2023) Therapeutic compounds from medicinal plant endophytes: molecular and metabolic adaptations. J Appl Microbiol 134:lxad074
Singh R, Nalwa HS (2011) Medical applications of nanoparticles in biological imaging, cell labeling, antimicrobial agents, and anticancer nanodrugs. J Biomed Nanotechnol 7:489–503
Singh D, Rathod V, Ninganagouda S, Herimath J, Kulkarni P (2013) Biosynthesis of silver nanoparticle by endophytic fungi Pencillium sp. isolated from Curcuma longa (turmeric) and its antibacterial activity against pathogenic gram negative bacteria. J Pharm Res 7:448–453
Singh D, Rathod V, Ninganagouda S, Hiremath J, Singh AK, Mathew J (2014) Optimization and characterization of silver nanoparticle by endophytic fungi Penicillium sp. Isolated from Curcuma longa (Turmeric) and application studies against MDR E. coli and S. aureus. Bioinorg Chem Appl 2014:408021
Singh T, Jyoti K, Patnaik A, Singh A, Chauhan R, Chandel SS (2017) Biosynthesis, characterization and antibacterial activity of silver nanoparticles using an endophytic fungal supernatant of Raphanus sativus. J Genet Eng Biotechnol 15:31–39
Singh J, Dutta T, Kim K-H, Rawat M, Samddar P, Kumar P (2018) 'Green' synthesis of metals and their oxide nanoparticles: applications for environmental remediation. J Nanobiotechnol 16:84
Singla R, Kumari A, Yadav SK (2019) Impact of nanomaterials on plant physiology and functions. In: Husen A, Iqbal M (eds) Nanomaterials and plant potential. Springer, Cham
Slavin YN, Asnis J, Häfeli UO, Bach H (2017) Metal nanoparticles: understanding the mechanisms behind antibacterial activity. J Nanobiotechnol 15:65
Strobel G, Daisy B (2003) Bioprospecting for microbial endophytes and their natural products. Microbiol Mol Biol Rev 67:491–502
Sumanth B, Lakshmeesha TR, Ansari MA, Alzohairy MA, Udayashankar AC, Shobha B, Niranjana SR, Srinivas C, Almatroudi A (2020) Mycogenic synthesis of extracellular zinc oxide nanoparticles from xylaria acuta and its nanoantibiotic potential. Int J Nanomedicine 15:8519–8536
Sunkar S, Nachiyar CV (2012) Biogenesis of antibacterial silver nanoparticles using the endophytic bacterium Bacillus cereus isolated from Garcinia xanthochymus. Asian Pac J Trop Biomed 2:953–959
Svenningsen NB, Watts-Williams SJ, Joner EJ, Battini F, Efthymiou A, Cruz-Paredes C, Nybroe O, Jakobsen I (2018) Suppression of the activity of arbuscular mycorrhizal fungi by the soil microbiota. ISME J 12:1296–1307
Syed B, Prasad NMN, Satish S (2016) Endogenic mediated synthesis of gold nanoparticles bearing bactericidal activity. J Microsc Ultrastruct 4:162–166
Toju H, Yamamoto S, Tanabe AS, Hayakawa T, Ishii HS (2016) Network modules and hubs in plant-root fungal biomes. J R Soc Interface 13:20151097
Tran QH, Nguyen VQ, Le A-T (2013) Silver nanoparticles: synthesis, properties, toxicology, applications and perspectives. Adv Nat Sci Nanosci Nanotechnol 4:033001
Uddandarao P, Balakrishnan RM (2017) Thermal and optical characterization of biologically synthesized ZnS nanoparticles synthesized from an endophytic fungus Aspergillus flavus: a colorimetric probe in metal detection. Spectrochim Acta A Mol Biomol Spectrosc 175:200–207
Uddandarao P, Balakrishnan RM, Ashok A, Swarup S, Sinha P (2019) Bioinspired ZnS:Gd nanoparticles synthesized from an endophytic fungi Aspergillus flavus for fluorescence-based metal detection. Biomimetics (Basel) 4:11
Verma VC, Kharwar RN, Gange AC (2010) Biosynthesis of antimicrobial silver nanoparticles by the endophytic fungus Aspergillus clavatus. Nanomedicine (Lond) 5:33–40

Verma VC, Anand S, Ulrichs C, Singh SK (2013) Biogenic gold nanotriangles from Saccharomonospora sp., an endophytic actinomycetes of Azadirachta indica A. Juss. Int Nano Lett 3:21

White JF, Belanger F, Sullivan R, Lewis E, Moy M, Meyer W, Bacon CW (2004) Developmental interactions between clavicipitaleans and their host plants. In: Varma A, Abbott L, Werner D, Hampp R (eds) Plant surface microbiology. Springer, Berlin

Xie J, Wu Y-Y, Zhang T-Y, Zhang M-Y, Peng F, Lin B, Zhang Y-X (2018) New antimicrobial compounds produced by endophytic Penicillium janthinellum isolated from Panax notoginseng as potential inhibitors of FtsZ. Fitoterapia 131:35–43

Yang Y, Waterhouse GIN, Chen Y, Sun-Waterhouse D, LI, D. (2022) Microbial-enabled green biosynthesis of nanomaterials: current status and future prospects. Biotechnol Adv 55:107914

Ying S, Guan Z, Ofoegbu PC, Clubb P, Rico C, He F, Hong J (2022) Green synthesis of nanoparticles: current developments and limitations. Environ Technol Innov 26:102336

Zhang T, Yao Y-F (2015) Endophytic fungal communities associated with vascular plants in the high arctic zone are highly diverse and host-plant specific. PLoS One 10:e0130051

Zhang X, Yan S, Tyagi RD, Surampalli RY (2011) Synthesis of nanoparticles by microorganisms and their application in enhancing microbiological reaction rates. Chemosphere 82:489–494

Zhao X, Zhou L, Riaz Rajoka MS, Yan L, Jiang C, Shao D, Zhu J, Shi J, Huang Q, Yang H, Jin M (2018) Fungal silver nanoparticles: synthesis, application and challenges. Crit Rev Biotechnol 38:817–835

The manufacturer's authorised representative in the EU is Springer Nature Customer Service Centre GmbH, Europaplatz 3, 69115 Heidelberg, Germany. If you have any concerns regarding our products, please contact ProductSafety@springernature.com

Printed and bound by CPI Group (UK) Ltd, Croydon, CR0 4YY
01/10/2026
02225914-0014